COMPREHENSIVE CHEMICAL KINETICS

COMPREHENSIVE

Section 1. THE PRACTICE AND THEORY OF KINETICS

Volume 1	The Practice of Kinetics
Volume 2	The Theory of Kinetics
Volume 3	The Formation and Decay of Excited Species

Section 2. HOMOGENEOUS DECOMPOSITION AND ISOMERISATION REACTIONS

Volume 4	Decomposition of Inorganic and Organometallic Compounds
Volume 5	Decomposition and Isomerisation of Organic Compounds

Section 3. INORGANIC REACTIONS

Volume 6	Reactions of Non-metallic Inorganic Compounds
Volume 7	Reactions of Metallic Salts and Complexes, and Organometallic Compounds

Section 4. ORGANIC REACTIONS (6 volumes)

Volume 8	Proton Transfer
Volume 9	Addition and Elimination Reactions of Aliphatic Compounds
Volume 10	Ester Formation and Hydrolysis and Related Reactions
Volume 12	Electrophilic Substitution at a Saturated Carbon Atom
Volume 13	Reactions of Aromatic Compounds

Section 5. POLYMERISATION REACTIONS (3 volumes)

Volume 14	Degradation of Polymers
Volume 14A	Free-radical Polymerisation
Volume 15	Non-radical Polymerisation

Section 6. OXIDATION AND COMBUSTION REACTIONS (2 volumes)

Volume 17	Gas-phase Combustion

Section 7. SELECTED ELEMENTARY REACTIONS (1 volume)

Volume 18	Selected Elementary Reactions

Additional Sections

HETEROGENEOUS REACTIONS

KINETICS AND TECHNOLOGICAL PROCESSES

CHEMICAL KINETICS

EDITED BY

C. H. BAMFORD

M.A., Ph.D., Sc.D. (Cantab.), F.R.I.C., F.R.S.
Campbell-Brown Professor of Industrial Chemistry,
University of Liverpool

AND

C. F. H. TIPPER

Ph.D. (Bristol), D.Sc. (Edinburgh)
Senior Lecturer in Physical Chemistry,
University of Liverpool

VOLUME 17

GAS-PHASE COMBUSTION

ELSEVIER SCIENTIFIC PUBLISHING COMPANY
AMSTERDAM – OXFORD – NEW YORK
1977

ELSEVIER SCIENTIFIC PUBLISHING COMPANY
335 Jan van Galenstraat
P.O. Box 211, Amsterdam, The Netherlands

Distributors for the United States and Canada:

ELSEVIER NORTH-HOLLAND INC.
52, Vanderbilt Avenue
New York, N.Y. 10017

Library of Congress Cataloging in Publication Data

Bamford, C H
Gas-phase combustion.

(Their Comprehensive chemical kinetics ; v. 17)
Bibliography: p.
Includes index.
1. Combustion. I. Tipper, Charles Frank Howlett, joint author. II. Title.
QD501.B242 vol. 17 [QD516] 541'.39'08 [541'.361]
ISBN 0-444-41513-0 76-28370

ISBN: 0-444-41631-5 (Series)
ISBN: 0-444-41513-0 (Vol. 17)

with 177 illustrations and 95 tables

Printed in The Netherlands

COMPREHENSIVE CHEMICAL KINETICS

Contributors to Volume 17

J. A. BARNARD	Department of Chemical Engineering, University College London, London, England
D. J. DIXON	Department of Inorganic, Physical and Industrial Science, The University, Liverpool, England
G. DIXON-LEWIS	Department of Fuel and Combustion Science, The University, Leeds, England
R. T. POLLARD	Dunlop Ltd., Dunlop House, Ryder Street, London, England
G. SKIRROW	Department of Inorganic, Physical and Industrial Chemistry, The University, Liverpool, England
D. J. WILLIAMS	Division of Process Technology, Minerals Research Laboratories, CSIRO, P.O. Box 136, North Ryde, N.S.W. 2113, Australia

Preface

Section 6 deals with the autocatalytic reactions of inorganic and organic compounds with molecular oxygen in the liquid phase, and the highly exothermic processes in the gas phase, collectively known as combustion, which may involve oxygen, other oxidants or decomposition flames and are so important technologically. Catalysis, retardation and inhibition are covered. The kinetic parameters of the elementary steps involved are given, when available, and the reliability of the data discussed.

Volume 17 covers gas-phase combustion, which includes probably the most complex processes investigated by chemists. Chapter 1, about half the book, deals with the oxidation of hydrogen and carbon monoxide, with extensive consideration of all the individual reactions occurring. In Chapter 2, the combustion of hydrocarbons is discussed, with emphasis on the general mechanisms which have been suggested to account for the numerous products of partial oxidation. In Chapter 3, the oxidation of aldehydes, which are important intermediates in combustion of other compounds, is considered, and in Chapter 4, the oxidation of alcohols, ketones, oxirans, ethers, esters, peroxides, amines and halocarbons.

C. H. Bamford
C. F. H. Tipper

Liverpool
March, 1977

Contents

Chapter 1

The Oxidation of Hydrogen and Carbon Monoxide

G. DIXON-LEWIS and D. J. WILLIAMS

1. Introduction

Apart from their own intrinsic interest as two of the simplest combustion reactions, it has long been known that the oxidations of both hydrogen and carbon monoxide may contribute appreciably to the later stages of hydrocarbon oxidation. It is natural therefore that these two simpler oxidations should have been studied very extensively. As a result of such investigations the general pattern of behaviour is now fairly clear in both cases, and in the case of hydrogen the reaction mechanism is understood in some detail. For carbon monoxide, however, although the general behaviour is essentially similar to that for hydrogen, the detailed mechanism is much less clear. The complicating feature here is the large effect of even minute traces of water vapour on the reaction, an effect which has made precise experimentation difficult. The overall pattern is such that it will be convenient to discuss first the oxidation of hydrogen, and then to follow this by discussion of the oxidation of carbon monoxide and hydrogen—carbon monoxide mixtures.

2. General features of the reaction between hydrogen and oxygen

The gas phase reaction between hydrogen and oxygen can be made to take place at temperatures around 500—600 °C if it is induced thermally, and at ordinary temperatures when it is brought about by photochemical means. The general features of the thermally induced reaction have been described by Hinshelwood and Williamson [1], Semenov [2], Jost [3], Lewis and von Elbe [4], Minkoff and Tipper [5] and others. It is one of the best examples of a reaction depending on branching chains and showing the phenomenon of explosion limits. In the region of 550 °C in a silica reaction vessel three such limits are observable — the first at a pressure of a few torr, the second at about 100 torr, and the third at several hundred torr [6—8]. Below the first limit the reaction rate is almost negligibly small; between the first and second limits there is explosion; above the second limit the explosion is replaced by slow reaction, the rate of which increases as the pressure rises; and finally, at the third limit, there is explosion again. The third limit, or boundary of the high pressure explosion region, moves towards lower pressures as the temperature increases. Its occurrence might be expected on purely thermal grounds, as being due to the rate of heat liberation by the reaction becoming greater than the rate of heat loss, so that self-acceleration of the reaction occurs. However, more detailed investigation, which will be discussed later, suggests that it is not a purely thermal type of explosion.

Of greater interest is the occurrence of the explosion region at low pressures [9—12] where, just outside the explosion limit, the rate of

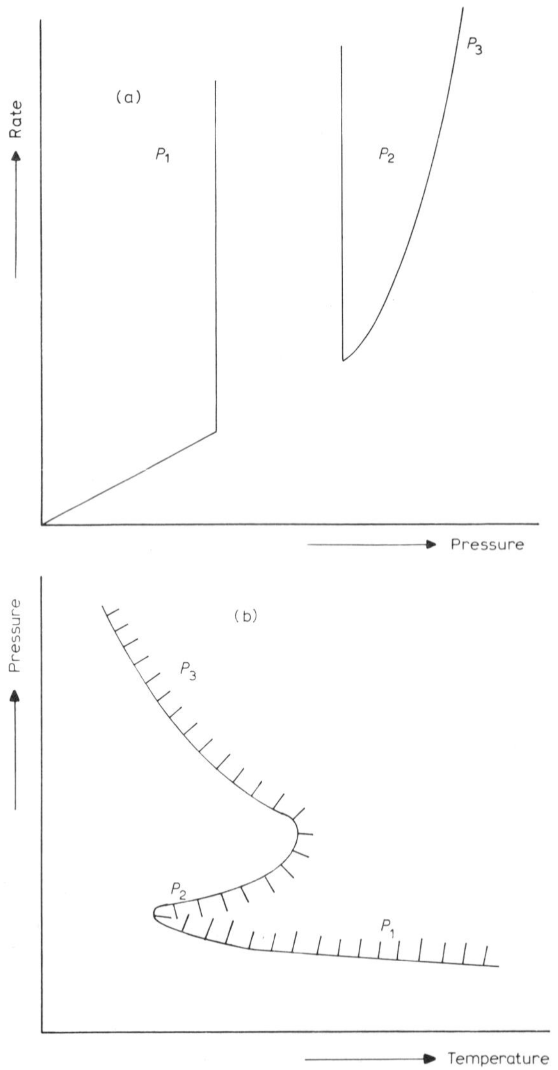

Fig. 1. Explosion relations for a hydrogen—oxygen mixture of fixed composition (diagrammatic).

reaction is quite small. Such a region is characteristic only of a few reactions, all of which seem to proceed by chain mechanisms. The upper boundary of the region, or second limit, moves towards higher pressures on increasing the temperature, and eventually joins up with the third limit. The first limit does not vary much with temperature. As the temperature is reduced the first and second limits move closer together and eventually coalesce giving a low pressure explosion peninsula with its tip in the region of 450 °C. The complete explosion relations for a given hydrogen—oxygen mixture are therefore as shown diagrammatically in Fig. 1. In Fig. 1b, mixtures at pressures and temperatures to the right of the limit line will explode almost immediately; whilst those to the left will undergo a slow reaction. The first, second and third limits are in order of increasing pressure.

The kinetic investigation of the hydrogen–oxygen reaction has involved measurement of the effects of alterations in parameters such as temperature, pressure, mixture composition, and vessel size and surface on the slow reaction rate and on the positions of the explosion limits. The effect of variation of the gas composition on the explosion limits at constant temperature may be shown by plotting the partial pressures of hydrogen at the limit against the partial pressures of oxygen. For certain types of reaction vessel surface such a plot is shown schematically in Fig. 2, where the shaded area represents the explosion region.

With the possible exception of those on the third limit, all the kinetic investigations of the above type have been carried out under essentially isothermal conditions, often with rather low concentrations of chain centres. With such low concentrations of chain centres it is usually necessary only to consider reactions of free radicals with stable molecules,

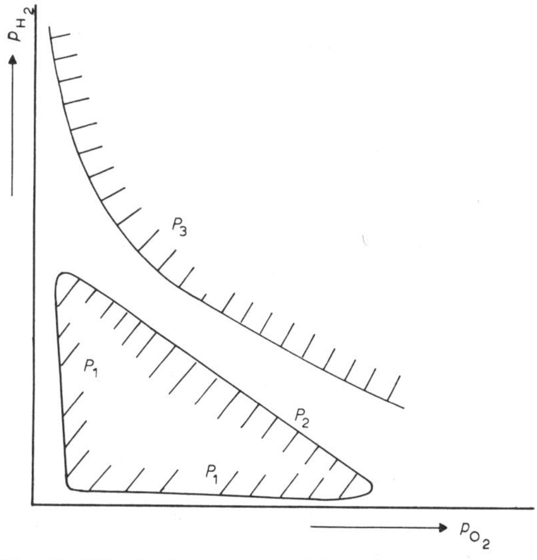

Fig. 2. Effect of gas composition on explosion limits.

References pp. 234—248

even when defining critical conditions leading to limit explosions: such conditions apply particularly when the reaction is carried out in vessels having moderate or high surface chain breaking efficiencies. In these circumstances radical—radical reactions are usually unimportant. Once explosion has set in, however, the concentrations of chain carriers will rapidly increase, and reactions between radicals will become much more probable. This is the case when the oxidation is studied either in flames or shocked gases. In both of these situations the development of the reaction is examined under conditions lying largely within the explosion region.

An intermediate field of study, between the earlier investigations and the flame and shock tube work, is that of the slow reaction and explosion limits in vessels of very low chain breaking efficiency. Here, in contrast with vessels of higher efficiency, radical—radical reactions do begin to be important even in determining limit conditions and slow reaction rates. Into this class fall a number of extensive investigations of the reaction in boric acid coated vessels, which will be discussed. These investigations provide a valuable link between the other two areas of study.

3. Explosion limits and the slow gaseous reaction

3.1 BASIC BACKGROUND

A reaction system as complicated as hydrogen—oxygen turns out to be capable of being explained at a number of different levels of complexity, depending on the range of experimental conditions the explanation is intended to cover. Historically, there may in the late 1940's have been some who, after the then recent publications from the schools of Hinshelwood, Semenov, and Lewis and von Elbe, had felt that little more of major importance was to be gained by further study of the reaction. In this they would, of course, have been wrong. Nevertheless, for discussion of the reaction mechanism, that particular period forms an important turning point, from which it is still profitable to look back at some of what had already been achieved. In the discussion which immediately follows, only the basic type of explosion limit behaviour indicated in Fig. 2 will be considered.

3.2 FIRST EXPLOSION LIMIT

First explosion limits have been investigated by one of two methods. The usual procedure has been to expand a known pressure of mixture from a storage container into the evacuated reaction vessel, the expansion ratio having been previously determined. By a repeated process of trial and error the pressure is found at which admission can just occur without the production of a feeble flash [13]. Very careful observation is necessary since the light emission is very weak, and it may be necessary to work in a darkened room. An alternative method [11] is to follow manometrically the reaction of gases which were initially at a pressure slightly greater than the limit. Under these conditions it is stated that combustion proceeds until the residual pressure of hydrogen and oxygen has dropped to the lower limit, after which subsequent reaction is negligible. The results obtained by the two methods agree satisfactorily, perhaps rather surprisingly in view of the water formed in the second method.

The existence of the lower limit was first demonstrated by Sagulin [9, 10] in 1928, and the limit was further studied by Semenov and co-workers [11]. One of its main characteristics is the large influence of vessel surface and vessel size on the limit pressure. With regard to surface, Frost and Alyea [14] found that rinsing their vessels with potassium chloride raised the limit pressure to several times higher than previously, while Kowalsky [15] was able to produce a correspondingly large decrease in his limits, to a few tenths of a torr, by subjecting his vessel to an electric discharge. Semenova [16] obtained reproducible results by using fresh Pyrex vessels, made from tubing of different width by drawing or blowing so that the surface was freshly formed, and then conditioned or aged by carrying out a set number of explosions. For temperatures above about 440 °C, the limits obtained by Semenova in a vessel of 10.2 mm diameter, and 15 cm long, were again at a fraction of a torr pressure. Her limits were some ten times lower than those found earlier by Hinshelwood and Moelwyn-Hughes [13], who used silica vessels, and were able to obtain moderately reproducible results with these by filling them with oxygen for a few minutes before each experiment. The techniques of both Semenova and Hinshelwood and Moelwyn-Hughes were later used also by Biron and Nalbandjan [17], who confirmed the general trends. The results of Hinshelwood and Moelwyn-Hughes [13] on the effects of mixture composition and vessel size may be summarized as follows.

(a) The effect of mixture composition in a vessel of 18 mm diameter at 550 °C is given in Table 1. The last column shows that the product $p_{H_2} \times p_{O_2}$ is almost constant at the limit.

(b) The effect of vessel diameter on the lower limit of the stoichiometric mixture at 550 °C is given in Table 2. Here the product $p_1 d$, where

TABLE 1
Effect of gas composition on the first explosion limit [13]

p_{H_2} (torr)	p_{O_2} (torr)	$p_{H_2} \times p_{O_2}$
3.12	0.78	2.43
2.30	1.15	2.64
1.65	1.65	2.72
1.07	2.14	2.29
0.86	3.44	2.96

TABLE 2
Effect of vessel diameter on first limits for $2H_2 + O_2$ at 550 °C [13]

d(cm)	p_1(torr)	$p_1 d$
5.4	1.06	5.7
4.9	0.91	4.5
3.2	2.85	9.1
1.8	3.64	6.5

p_1 is the limit pressure and d is the vessel diameter, is approximately constant. Semenova [16] found similar behaviour at a number of temperatures, although her limit pressures themselves were much lower.

These observations all point clearly to the occurrence of a branched chain reaction with chain breaking at the vessel surface. The expression for the rate of a chain reaction is shown by the formal treatments of Semenov and Hinshelwood to have the general form

$$\frac{F(c)}{f_s + f_g + A(1-\alpha)} \qquad (1)$$

where $F(c)$ is determined by the rate of chain initiation, f_s and f_g represent the rates of destruction of chains at the surface and in the gas, respectively, α is the number of centres produced from one initial centre in the branching process, and A is a constant or a function of the concentrations according to circumstances. If there is branching, $(1-\alpha)$ is negative, and the reaction passes into explosion when the negative term exactly balances $f_s + f_g$. At very low pressures f_g will be unimportant and, since f_s decreases with increasing pressure due to increasing resistance to diffusion, a lower limit of explosion becomes possible.

Developing the argument further by means of the treatment of Semenov [18] and Hinshelwood and Williamson [1], let us suppose that in the hydrogen—oxygen system near the lower limit we have two kinds of active particles, X_O and X_H, and let X_O react with hydrogen and X_H with oxygen thus

$$X_O + H_2 = X_H + \cdots$$

$$X_H + O_2 = X_O + \cdots$$

until the chain is broken at the wall of the vessel. In the course of a chain, let there be n_1 collisions between X_O and H_2, n_2 between X_O and O_2, n_3 between X_H and H_2, and n_4 between X_H and O_2, where $n_1 + n_2 + n_3 + n_4 = n$. Ignoring the starting of fresh chains, X_H only appears when X_O

reacts, and vice versa. Hence $n_1 = n_4$. Since

$$\frac{n_1 + n_2}{n_1} = \frac{p_{H_2} + p_{O_2}}{p_{H_2}} \quad \text{and} \quad \frac{n_3 + n_4}{n_4} = \frac{p_{H_2} + p_{O_2}}{p_{O_2}}$$

Then

$$\frac{n}{n_1} = \frac{n}{n_4} = \frac{(p_{H_2} + p_{O_2})^2}{p_{H_2} p_{O_2}}$$

$$n_1 = n_4 = \frac{p_{H_2} p_{O_2}}{(p_{H_2} + p_{O_2})^2} n$$

From the Einstein—Smoluchowsky relation, if d is the diameter of the tube, then

$$n = 1.5\beta^2 d^2 / \lambda^2$$

where λ is the mean free path, and β is a factor which may be used to allow approximately for incomplete destruction of the chains at the wall. Taking as an approximation the mean free paths of the various kinds of particle to be equal, λ is inversely proportional to the total pressure. Hence

$$\lambda = \lambda_0 / (p_{H_2} + p_{O_2})$$

$$n = 1.5\beta^2 d^2 (p_{H_2} + p_{O_2})^2 / \lambda_0^2$$

$$n_1 = n_4 = \frac{1.5\beta^2 d^2}{\lambda_0^2} p_{H_2} p_{O_2}$$

The theory now requires that at some collisions of X_O with H_2 or X_H with O_2 the chain may branch. Let the probability of branching at a given collision be ν. Then the average number of times that a chain branches before it is destroyed at the wall is νn_1, and the condition for explosion is that νn_1 should exceed unity. The limit condition is therefore of the form

$$p_{H_2} p_{O_2} \beta^2 d^2 = \text{const.} \tag{2}$$

Thus for constant vessel diameter and surface it would be expected that the product $p_{H_2} \times p_{O_2}$ is constant, while at constant composition the result should be that $p_1 \times d$ is constant. Both these relationships are observed to good approximation considering the sensitivity of the limits to vessel surface. A further interesting feature of the lower limit phenomenon, which has not yet been discussed, is the effect of inert gases. In the presence of inert gases the collision numbers in the above calculation are altered, and the necessary modification of the calculation leads to the limit condition

$$p_{H_2} p_{O_2} \beta^2 d^2 \left(1 + \frac{D' p_{\text{inert}}}{p_{H_2} + p_{O_2}}\right) = \text{const.} \tag{3}$$

References pp. 234—248

TABLE 3
Effect of inert gases on first limits for $2H_2 + O_2$ in an 18 mm diameter vessel at 550 °C [13]

Nitrogen		Carbon dioxide		Helium		Argon	
p_{N_2} (torr)	$p_{2H_2+O_2}$ (torr)	p_{CO_2} (torr)	$p_{2H_2+O_2}$ (torr)	p_{He} (torr)	$p_{2H_2+O_2}$ (torr)	p_{Ar} (torr)	$p_{2H_2+O_2}$ (torr)
0	3.20	0	4.50	0	3.33	0	4.25
2.12	2.12	6.24	3.12	5.45	2.73	2.90	2.90
3.10	1.55	11.4	2.79	8.70	2.18	4.41	2.21
4.94	1.65					6.72	1.68
5.25	1.05					6.51	1.09
6.15	0.88						

where D' is the reciprocal of the diffusion coefficient of the chain carriers through the inert gas. Inert gases should thus lower the partial pressures of hydrogen and oxygen at the limit, since they hinder diffusion to the wall. Table 3 shows the results of Hinshelwood and Moelwyn-Hughes for the effects of inert gases on the limit for a stoichiometric mixture at 550 °C in an 18 mm diameter bulb. Reasonable straight lines are obtained on plotting $1/(p_{H_2} \times p_{O_2})$ against $p_{inert}/(p_{H_2} + p_{O_2})$, as would be predicted from eqn. (3). With the exception of carbon dioxide, the slopes are approximately proportional to the reciprocals of the diffusion coefficients.

Interesting additional inert gas effects have been observed by Biron and Nalbandjan [17]. Using silica vessels they confirmed the above behaviour; but in experiments under conditions analogous to those used by Semenova, already discussed, they found the critical explosion pressures to be unaffected by argon addition. Their results are given in Table 4. The conditions used by Semenova correspond with extremely low chain breaking efficiency at the vessel wall, and under these conditions the rates

TABLE 4
Effect of argon on first limits in a vessel of low chain breaking efficiency [17]

T = 485 °C		T = 515 °C	
p_{Ar} (torr)	$p_{1.5H_2+O_2}$ (torr)	p_{Ar} (torr)	$p_{1.5H_2+O_2}$ (torr)
—	0.23	—	0.15
0.057	0.228	0.04	0.15
0.12	0.24	0.07	0.14
0.18	0.23	0.105	0.145
0.235	0.235	0.15	0.15

TABLE 5
Effect of temperature on the first limit for $2H_2 + O_2$ in an 18 mm diameter vessel [13]

Temperature (°C)	650	604	550	500	480
p_1 (torr)	2.99	2.74	2.75	3.62	6.20

of diffusion to the wall have comparatively little influence on the rate of chain termination, as discussed later.

Finally, the effect of temperature on the first limit has been measured. Above 550 °C, Hinshelwood and Moelwyn-Hughes [13] find the limit to be practically independent of temperature (Table 5), but below 550 °C the limit increased slowly to about 8 torr at 450 °C, where it joined the second limit. Comparable data in the lower temperature range were found by Kopp et al. [11], namely ~ 9 torr at 440 °C, 4.5 torr at 500 °C and 3 torr at 550 °C. They derive from this an activation energy $E' = -14$ kcal . mole^{-1}, but it is uncertain how much importance should be attached to this figure. Jost [3] points out that if the general theory of the limit already given is correct, so that the rate of chain branching will be proportional to p^2 and the rate of chain breaking roughly independent of temperature, then the limit pressure p would be proportional to $\exp(+\frac{1}{2}E/RT)$, where E is the activation energy of the branching step. In this case $E' = -\frac{1}{2}E$.

For the temperature range 390—560 °C, Semenova [16] finds $E' = -11$ kcal . mole^{-1}.

3.3 SECOND EXPLOSION LIMIT

The second limits must be measured by a special technique in order to avoid passing through the explosion region when admitting the gas mixtures to the reaction vessel. One of the reactant gases, preferably the hydrogen, is admitted to a pressure above that at which explosion can occur, and the required quantity of the other gas is then introduced rapidly from a gas pipette. After allowing a short time for mixing, the gases are withdrawn from the vessel at a predetermined rate such that the explosion pressure can be accurately observed.

One problem in the determination of second limits is that water, a product of the slow reaction, is also a powerful inhibitor of the explosion. In order to reduce errors due to water formation, much of the earlier work on this limit was carried out with potassium chloride coated vessels. With these, and in vessels coated with certain other salts, the limit is much less sensitive to withdrawal rate than it is with a clean Pyrex or a boric acid coated vessel, for example. Pease [19] first noted that potassium chloride coating produces a marked suppression of the slow reaction rate. More recent work by Baldwin et al. [20, 21], which will be discussed later, suggests that the suppression of the limit at low withdrawal rates in clean and boric acid coated vessels is more specifically connected with the detailed mechanism of water formation in the immediate vicinity of the limit. Baldwin et al. controlled their withdrawal rates by insertion of one of a number of calibrated capillary tubes into the withdrawal line. Baldwin and Precious [22] found no suppression with potassium chloride coated vessels. However, some lowering of the limit with such vessels has been observed by Lewis and von Elbe [23] and Egerton and Warren [24] at very low withdrawal rates. The present section will be concerned primarily with results using potassium chloride coated vessels.

TABLE 6
Second limits for $2H_2 + O_2$ in potassium chloride coated vessels

Diameter (cm)	Temperature (°C)	p_2 (torr)	Ref.
2	500	ca. 30	14
2.2	500	31	25
3.9	480	22	23
3.5	500	46	25
7.3	460	21	23
18	460	26	26
7.3	460	18	25
	480	33	23
	500	47	25
	530	86	23
	540	85	25
	570	187	23
	580	ca. 180	25
	590	ca. 250	25
5.5	586	ca. 220	27
1.8	530	68	4
10	530	88	4

The second limits are quite reproducible over long periods using the same apparatus, and the limits from independent investigations also agree well. This is shown by Table 6, which quotes limits for stoichiometric mixtures in potassium chloride coated vessels. According to Willbourn and Hinshelwood [27] the use of a number of similar coatings (KCl, KI, CsCl, CsI) does not alter the limit much. The results of Lewis and von Elbe [23] and Warren [25] in Table 6 also show that the limit is virtually independent of vessel diameter, provided the latter is greater than 4—5 cm. Clearly the limits are not determined primarily by competition between gas phase and wall effects.

There have been a number of measurements of the effects of mixture composition and temperature on the hydrogen—oxygen second limits in potassium chloride coated vessels (e.g. refs. 28, 14, 23, 25, 30). Typical of the results are the explosion regions shown in Figs. 3 and 4. They all

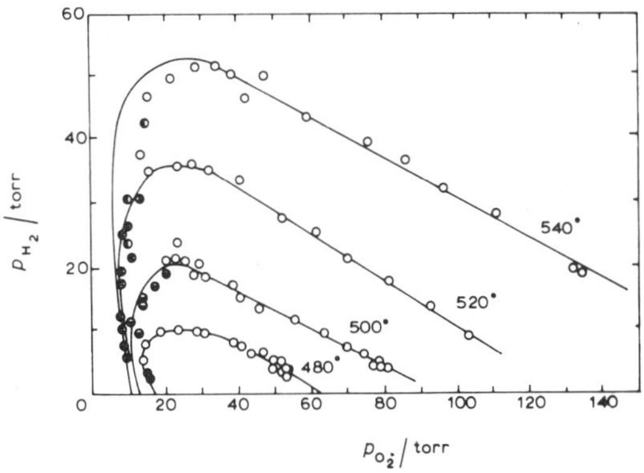

Fig. 3. Low pressure explosion limits of hydrogen—oxygen mixtures (after Frost and Alyea [14]). ○, Withdrawal method; ●, admixture method, 480 °C; ◒, admixture method, 520 °C; ◐, admixture method, 540 °C. KCl coated vessel, 2 cm diameter. (By courtesy of J. Am. Chem. Soc.)

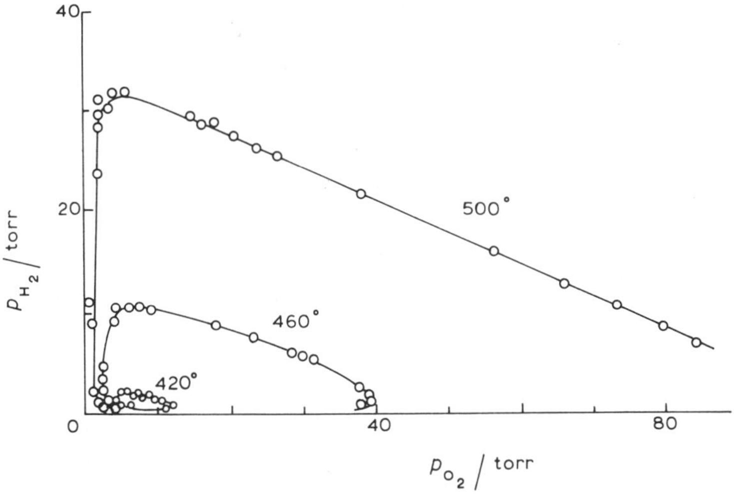

Fig. 4. Low pressure explosion limits in KCl coated vessel, 7.3 cm diameter (after Warren [25]). (By courtesy of The Royal Society.)

confirm the original finding of Grant and Hinshelwood [28] that at fixed temperature the limit condition in the presence of inert gas M is

$$p_{H_2} + k_{O_2} p_{O_2} + k_M p_M = K \tag{4}$$

where k_{O_2}, k_M and K are constants. The constant K increases with temperature according to an Arrhenius law with an apparent activation energy somewhere between 18 and 25 kcal . mole^{-1} depending on the investigator. The more recent investigations tend to favour the lower values.

The general form of eqn. (4), as well as the lack of dependence of the limit on vessel surface and diameter, is readily understood if the gas phase deactivation term in the denominator of eqn. (1) is dominant. Let it be assumed that the reaction of a chain carrier X with O_2 can give rise to different products according to whether it reacts in a bimolecular or termolecular collision thus

$$X + O_2 \xrightarrow{k_1} \alpha Y \rightarrow \text{branching}$$

$$X + O_2 + M \xrightarrow{k_{2,M}} XO_2 + M \rightarrow \text{chain termination}$$

In the second of these reactions M could represent any third molecule which is capable of removing energy from the transition state complex $XO_2^{\ddagger}$ to prevent its re-dissociation.

Clearly the rate of the chain terminating process will increase more rapidly with pressure than the rate of chain branching, so that an upper limit to the explosion region will occur. Equating the two rates at the limit

$$k_1(\alpha - 1)[X][O_2] = \sum_{M} k_{2,M}[X][O_2][M]$$

where M may be any molecule present in the system (H_2, O_2 or inert gas), and $k_{2,M}$ is the corresponding rate coefficient. We may thus write

$$k_1(\alpha - 1)[X][O_2] = k_{2,H_2}[H_2] + k_{2,O_2}[O_2] + \sum_{M \neq H_2,O_2} k_{2,M}[M]$$

and, dividing throughout by k_{2,H_2}, we have

$$[H_2] + k_{O_2}[O_2] + \sum_{M \neq H_2,O_2} k_M[M] = \frac{k_1(\alpha - 1)}{k_{2,H_2}} = \text{const.} \qquad (5)$$

where $k_{O_2} = k_{2,O_2}/k_{2,H_2}$ and $k_M = k_{2,M}/k_{2,H_2}$. Measured values, from second limits results of various workers, for a number of third body coefficients k_M are given in Table 7.

Table 7 shows that the coefficients are almost independent of temperature in the limited temperature range. Although perhaps more apparent than real (see Sect. 6.5 for HO_2 results in extended temperature range), the approximate constancy might have been expected since the activation requirements of the XO_2 forming reaction are unlikely to depend appreciably on the third body, which merely stabilizes by energy removal. The first line of Table 7 gives the ratios of collision numbers Z_M/Z_{H_2}, calculated from gas kinetic theory, for the respective molecules with HO_2. In the case of oxygen and nitrogen these agree well with the measured k_{O_2} and k_{N_2}, suggesting equal efficiencies of H_2, O_2 and N_2 in the deactivation. For the monatomic gases helium and argon, however, the measured coefficients are lower than the calculated ratios of collision

TABLE 7
Third body coefficients for various gases

	O_2	N_2	He	Ar	CO_2	N_2O	H_2O	Surface	Temperature (°C)	Diameter (cm)	Ref.
Z_M/Z_{H_2}	0.4	0.45	0.57	0.39	0.43	0.42	0.38				
	0.4	0.39					11.0	SiO_2	550	5.5	27
	0.33	0.35			0.90		8.25	KCl	580	5.5	27
	0.36						8.1	KCl	553	5.5	27
							10.4	KCl	480	7.4	23
	0.35	0.43	0.36	0.20	1.47		14.3	KCl	530	7.4	23
							14.6	KCl	570	7.4	23
	0.38							KCl	500	7.4	24
							5.5	Hard glass	439—465		29
	0.37							KCl	378	3 l vessel	26
	0.44							KCl	464	3 l vessel	26
		0.42						KCl	443—492	3 l vessel	26
							6.7	KCl	453	3 l vessel	26
	0.38	0.45						KCl	510—570	4.0	30
							5.0	KCl			31
							6.0	Pyrex			22
							6.4 ± 0.7	KCl or B_2O_3	460—540		32
							6.5 ± 1.6				33
			0.32					B_2O_3	500	5.2	289
						1.5 ± 0.05		KCl	500	5.2	289
Best values near 500 °C	0.35	0.44	0.32	0.20	1.47	1.5	6.4 ± 1				

frequencies; while for carbon dioxide the measured coefficient is significantly higher. The high value for water accounts for the self-inhibition effects which may occur in measurements of the second limit. Here the data of Willbourn and Hinshelwood [27] (k_{H_2O} = 8—11) and Lewis and von Elbe [23] ($k_{H_2O} \simeq 14$) appear to be high. The lower results of other workers cover a wide range of both water concentration and composition of the hydrogen—oxygen mixture, and all give values between 5 and 8.

3.4 THIRD EXPLOSION LIMIT

At the first and second limits the observed explosions are of the isothermal branched chain type, and just outside the limits the rate of the slow reaction is extremely small. At the third limit, however, there is no such sharp transition from slow reaction to explosion, and the reaction rate is high even below the limit. For this reason there is some doubt about whether the observed limit explosions are essentially branched chain or thermal phenomena. Coupled with this difficulty is an experimental one. The limits must be measured by making up the required compositions *in the reaction vessel* in such a way that at no time does the mixture pass through the low pressure explosion region (cf. also Section 3.3). The high pressure explosion, if it occurs, is then preceded by an induction period, which becomes shorter at higher pressures over a range of about 50 torr. Ideally, it might be imagined that the shorter the induction period, the more likely is the explosion to be free from thermal effects. On the other hand, a finite time is needed for the smooth admission and mixing of the final constituent in the reaction vessel in such a way that pronounced adiabatic heating does not occur during the admission. For this reason Oldenberg and Sommers [34] measured the induction periods Δt corresponding to a number of limit pressures at constant temperature and extrapolated back to $\Delta t = 0$; while Willbourn and Hinshelwood [27] made a compromise by adopting $\Delta t \simeq 5$ sec as their criterion. Heiple and Lewis [35], on the other hand, measured the lowest pressures at which the explosion would occur at all. From later measurements of the interrelationship of limit pressure and induction period (measured from the start of the oxygen admission), Lewis and von Elbe [23] concluded that the approach to $\Delta t = 0$ was asymptotic, and did not allow definition of a limit at zero induction period. Because of the disturbance caused near the limit by the thermal effect and by the effect of rapid water formation, they discarded third limit data for the investigation of rate coefficients in the overall reaction mechanism.

In all the work on the third limit it has been found that salt coating of the reaction vessel is essential in order to obtain reproducible results. Normally a thick coating of potassium chloride has been used for this purpose, but by altering the thickness of the coating Heiple and Lewis [35] were able to alter the surface efficiencies for chain breaking and

TABLE 8
Dependence of third explosion limit on diameter and temperature with KCl coated vessels [35]

Temperature (°C)	p_3 (torr) at diameters (cm)					p_{small}/p_{large}	d_{large}/d_{small}	$(d_{large}/d_{small})^2$
	4.0	5.8	7.4	8.4	9.9			
531			620		429	1.5	1.4	1.8
540		699			350	2.0	1.7	2.9
550		560			240	2.3	1.7	2.9
560	710			275		2.6	2.1	4.4

thereby to influence their limit pressures. For both thickly and thinly coated vessels it was found that the limit was inversely proportional to a power of the vessel diameter between the first and second, as shown in Table 8. Table 8 also shows the strong dependence of the limit on temperature, and this becomes even more pronounced when the short induction period criterion is also applied. Thus, for a $3H_2 + O_2$ mixture in a 5.5 cm diameter vessel with thick KCl coating, Willbourn and Hinshelwood [27] find that for $\Delta t \simeq 5$ sec, raising the temperature from 580 to 591 °C reduces the limit from 650 to 390 torr.

With both thickly and thinly coated vessels the limit pressures increase with hydrogen content, from about 60 % hydrogen upwards. Thus, Willbourn and Hinshelwood [27] find limits at about 400 and 630 torr for mixtures containing 60 and 80 % hydrogen, respectively, at 586 °C. Addition of both nitrogen and water vapour lowered the total limit pressure, while addition of carbon dioxide hindered the explosion. In the case of water a saturation effect appeared to be obtained: additions of both 5 and 9 torr of water vapour produced much the same effect on the limit pressure. For vessels with thin KCl coatings, addition of helium rapidly raised the limit for a stoichiometric mixture [35], while additions of argon or nitrogen produced smaller effects.

3.5 THE SLOW REACTION BETWEEN THE SECOND AND THIRD LIMITS

The overall rate of the slow reaction may conveniently be measured by following the pressure change accompanying the water formation. Again, of course, the mixtures to be studied must be made up in the reaction vessel in such a way that a low pressure explosion cannot occur. The rates depend markedly on the nature of the vessel surface, the rate in a silica vessel being reduced, for example, 50—2000 times on thickly coating the surface with potassium chloride (Pease [19]). The procedure of salt coating is also reported by Lewis and von Elbe [23] to give much more reproducible rates, and it has been employed by many investigators.

The pattern of kinetic behaviour is influenced by salt coating. Thus in uncoated vessels of porcelain or silica the reaction commences immediately, but the reaction rate increases before consumption of reactants causes it to decrease again. In potassium chloride coated vessels, on the other hand, there is no autocatalysis; the rate apparently reaches its maximum value at the very start and remains constant for some time. There are also differences in the pressure dependences of the reaction rate in the two types of vessel, and in the reaction products which can be isolated.

3.5.1 Uncoated quartz or glass vessels

At 540 °C with a silica bulb of about 250 cm^3 capacity, the dependence of the rate of reaction on the pressures of hydrogen, oxygen, and inert gases has been reported by Hinshelwood and Williamson [1]. Not far above the second limit, with pressures of 50 torr oxygen and 100 torr hydrogen, the rate is very small. The reaction is, however, of a high overall order, which increases with the temperature and may attain a value of about four. The concentration of hydrogen influences the rate more than that of oxygen. With a porcelain vessel at 569 °C the rate was found to be approximately proportional to the cube of the hydrogen concentration, and to a power of the oxygen concentration greater than one. With a silica vessel at 549 °C the rate varied as rather more than the square of the hydrogen concentration, and about as the first power of the oxygen pressure.

The autocatalytic nature of the reaction, described by Hinshelwood and Williamson [1], is in sharp contrast with the effect of water on the surface reaction at lower temperatures, which is "poisoned" by steam, and also with the inhibiting effect of water vapour on the second limit explosions. The autocatalysis has been studied in some detail by Chirkov [36], who used a reaction vessel of Durobax glass with diameter 5 cm and volume 200–250 cm^3. For hydrogen:oxygen ratios of about 2:1 at 550 torr initial pressure and 524 °C, he found the reaction rate w (torr sec^{-1}) to be given in terms of the initial pressure p and the amount of gases reacted x by

$$w = kx(p - x)^2.$$

When the concentration of hydrogen was low, the same expression was found, but with p and x now representing partial pressures of hydrogen, i.e.

$$w = k'[H_2O][H_2]^2$$

A further pair of experiments by Chirkov elegantly illustrates the part played by the water vapour. In one of these he measured the reaction rate for a $2H_2 + O_2$ mixture at 493 °C and 597 torr initial pressure; while, in

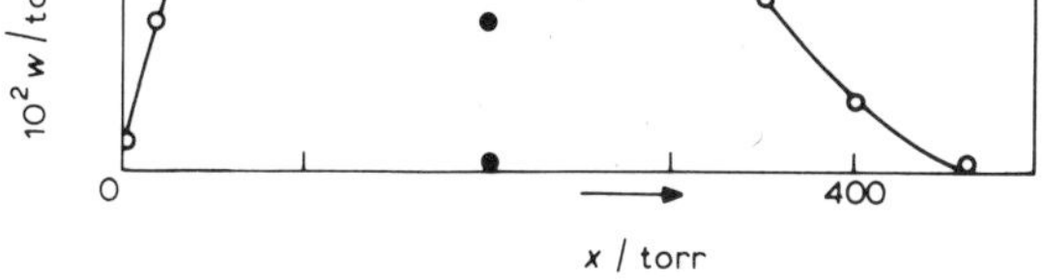

Fig. 5. Rate of reaction of $2H_2 + O_2$ when water vapour is added initially (after Chirkov [36]). ○, Without added water; ●, with added water. (x = amount of gas reacted; (2/3)x = water formed.) (By courtesy of Acta Phys. Chim. U.R.S.S.).

the second, the initial mixture contained 402 torr of hydrogen and oxygen together with 130 torr water vapour. Figure 5 shows that after a very short induction period in the second case compared with the first, the two mixtures react at almost the same rate.

The presence of inert gases also markedly accelerates the reaction and, if water is included among these, the order of effectiveness, according to Hinshelwood and Williamson [1], is He : N_2 : Ar : H_2O = 1 : 3 : 4 : 5, with nitrogen exerting approximately the same effect as an equal addition of oxygen. Hinshelwood and Williamson believed that the accelerating action of steam falls naturally into place as an inert gas effect in this way. However, Jost [3] then finds it difficult to explain the results of Chirkov described above. Further, experiments by Lewis and von Elbe with an uncoated quartz vessel of 3.9 cm diameter at 520 °C [23] suggest that water may have a much stronger influence than above; and it is possible that, in addition to its role as an inert gas, the presence of water may in some way affect the surface conditions. The important experiments of Prettre [37] showing the effect of previously adsorbed hydrogen or water on the reaction rate might lend support to this view. However, the results of both Prettre, and Lewis and von Elbe, were too erratic to permit accurate evaluation. An alternative explanation in terms of a build-up in the concentration of a relatively stable reaction intermediate represents another strong possibility which will be discussed in Sect. 4.2.

The nature of the inert gas effect is shown by studying the effect of vessel diameter on the rate. By comparing rates with two identical silica vessels, one unpacked and the other packed with 17 lengths of 1 cm diameter silica tube, Hinshelwood and Williamson [1] showed the effect to be large. However, a quantitative relation between vessel diameter and reaction rate was not easy to establish because of variation in the nature of the surface. Rates of reaction of $2H_2 + O_2$ mixtures at 560 °C in silica bulbs of varying diameter but identical lengths, were measured for initial pressures of 300 and 600 torr. The differences $v_{600} - v_{300}$ between the rates were taken as quantities which measured the speed of the homogeneous reaction, and were thought to be less influenced by the walls than the individual measured rates. Table 9 shows a reasonable constancy of the quotient $(v_{600} - v_{300})/d^2$, i.e. a velocity proportional to the square of the vessel diameter. Such a relationship, the effect of inert gases, and

TABLE 9
Effect of vessel diameter on rate [1]

Diameter (cm)	1.7	3.2	5.6	7.7
Rate, 600 torr	0.85	3.49	9.35	33.8
Rate, 300 torr	0.18	0.50	0.94	3.45
$(v_{600} - v_{300})/d^2$	0.00232	0.00292	0.00268	0.00512
$(v_{600} - v_{300})/d$	0.0394	0.0935	0.150	0.394

the other observed features clearly indicate a situation with chain breaking at the vessel surface.

The effect of temperature on the reaction rate in clean quartz or Pyrex vessels is large, but attempts at precise quantitative measurement have yielded discordant results. Thus, using a quartz vessel which had been aged by a number of preliminary runs, Oldenberg and Sommers [34], in two separate and internally self-consistent sets of experiments, obtained activation energies of 130 and 170 kcal . mole^{-1} in the temperature range 520—560 °C. With two Pyrex vessels of different shapes, activation energies of 82 and 95 kcal . mole^{-1} were found.

3.5.2 Potassium chloride coated vessels

The rate of the slow reaction is very markedly reduced by coating quartz or Pyrex reaction vessels with potassium chloride, and at the same time the initial increase in rate found with clean quartz vessels is no longer observed. The reaction rate also becomes more reproducible. Similar effects are produced, according to Lewis and von Elbe [23], by coating with $BaCl_2$, $K_2B_4O_7$, $K_2B_4O_7$ + KOH, or Na_2WO_4. According to Willbourn and Hinshelwood [27], the chlorides of the alkali metals all have similar effects with the exception of CsCl which reduces the rate of

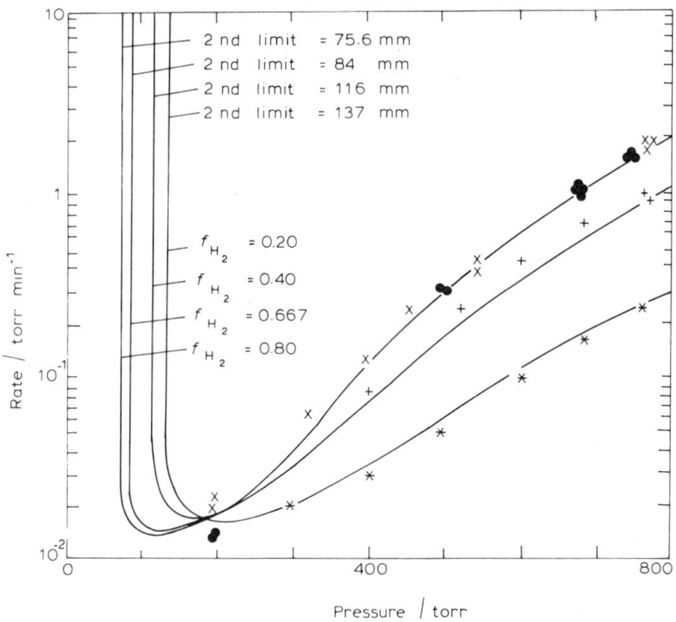

Fig. 6. Initial reaction rates for various mixtures of hydrogen and oxygen at 530 °C in KCl coated vessel, 7.4 cm diameter (after Lewis and von Elbe [23]). ——, Calculated curves; experimental rates: •, f_{H_2} = 0.80; +, f_{H_2} = 0.40; ×, f_{H_2} = 0.667; *, f_{H_2} = 0.20. f refers to mole fraction. (By courtesy of J. Chem. Phys.)

reaction of 300 torr H_2 + 150 torr O_2 at 550 °C to about half that observed with KCl coated vessels. Cullis and Hinshelwood [38] find larger reductions in rate for CsCl surfaces compared with KCl surfaces, a reaction rate of, for example, about 17 torr . sec^{-1} being obtained for 300 torr hydrogen + 100 torr oxygen at 573 °C with KCl, and at 596 °C with CsCl. The apparent activation energy with the CsCl coated vessel lay between 88 and 95 kcal . mole^{-1} depending on the partial pressures of the two gases. With KCl coated vessels the apparent activation energies are somewhat higher, lying in the range 100—135 kcal . mole^{-1} depending on the conditions [34, 37—39]. Lewis and von Elbe [23] find some curvature in their plots of log (rate) against $1/T$, with a gradual increase in slope between 510 and 570 °C. Rates calculated from their reaction mechanism also exhibit the curvature, and they suggest that the failure of some other workers to observe it may have been due to inhibition of the reaction near the third limit by accumulation of water vapour during the manipulation period. This explanation would seem to need confirmation.

Measurements by Lewis and von Elbe [23] of initial reaction rates at 530 °C in a KCl coated Pyrex reaction vessel of diameter 7.4 cm are shown in Fig. 6. For constant total pressure, the rate varies little in hydrogen-rich mixtures but diminishes when the oxygen content increases. The reaction seems more sensitive to the total pressure than it is to mixture composition. The rate diminishes with pressure until the neighbourhood of the second explosion limit is reached. At the limit itself the rate becomes infinite, and very near the limit, within a few torr, Lewis

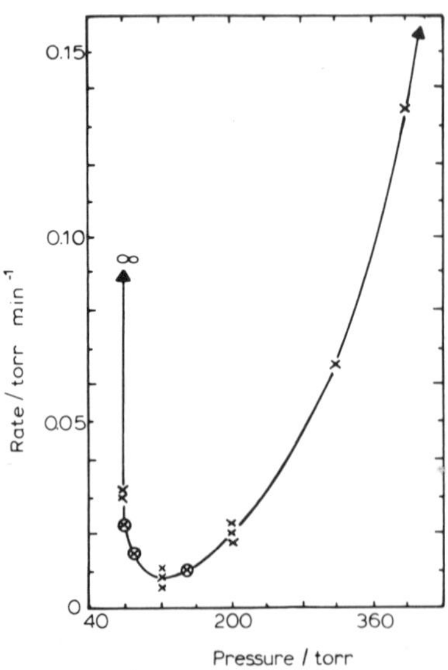

Fig. 7. Initial reaction rates of mixtures of $2H_2$ + O_2 at 530 °C in KCl coated Pyrex vessel, 7.4 cm diameter (after Lewis and von Elbe [23]). ⊗ = duplicated observations. (By courtesy of J. Chem. Phys.)

and von Elbe [23] have been able to detect very small increases in rate with KCl coated vessels (Fig. 7).

The results of Willbourn and Hinshelwood [27] on the influence of inert gases with a 5.5 cm diameter KCl coated vessel at 570 °C also seem to support the view that the rate determining parameter is the total pressure rather than the composition. For equal pressures of H_2, N_2, O_2 and CO_2 added to reacting mixtures, the accelerating effect appears from their curves to be much the same, with H_2O some 2—2½ times as efficient. The relative effects of nitrogen, oxygen and steam are reasonably consistent with those observed by Gibson and Hinshelwood [40] with uncoated porcelain vessels, and already quoted. Prettre [37], using a 6 cm diameter KCl coated vessel at 550 °C, finds similar results to Gibson and Hinshelwood [40] for the relative accelerations produced by nitrogen and argon, namely $N_2 : Ar \simeq 1.3$.

As is the case with uncoated vessels, there is a marked influence of vessel diameter on the rate. The results of Lewis and von Elbe [23] indicate that under many conditions the rate is proportional to a power of the diameter somewhat greater than two, and sometimes even higher than three.

3.5.3 *Silver vessels*

Silver vessels behave quite differently from either uncoated quartz or Pyrex, or salt coated vessels. Here the slow reaction is almost completely suppressed [41]. Added gases have little influence, and the introduction of a silica rod into the vessel fails to initiate any observable chain reaction. The explosion limits are also displaced to higher temperatures.

3.5.4 *Products of reaction*

In addition to the reaction rate, information about the products of the reaction which can be isolated is also important when considering the reaction mechanism. Observations on the reaction products, and in particular on the presence of hydrogen peroxide in these, have been made by Pease [19], Holt and Oldenberg [42], Anzilotti et al. [43], and Linnett and Tootal [44]. Pease passed hydrogen—oxygen mixtures at atmospheric pressure through a Pyrex reaction tube at 520—540 °C, and estimated hydrogen peroxide chemically by bubbling the gases straight from the tube into potassium permanganate solution at room temperature. In vessels pre-rinsed with hot nitric acid and distilled water the reaction velocities were high, and the products contained up to 20 % hydrogen peroxide. Residence times in the furnace of between 6 and 20 sec were used. The longer residence times greatly increased the amounts of water formed, but did not affect the peroxide, thus suggesting a steady state concentration of the latter. A change in H_2/O_2 ratio from 1/3 to 3 doubled the hydrogen peroxide and trebled the water yields. Increase in temperature raised the yields of both products; while at reduced pressure no peroxide was detectable. In very rich mixtures (95 % hydrogen) the yield of peroxide seemed to be favoured by small conversions.

In contrast with these results, when the tube was rinsed with potassium chloride solution the rate of water formation was greatly reduced, and no hydrogen peroxide could be detected at all. It should be noted here that KCl is a good catalyst for the destruction of peroxidic substances, as also is silver.

Holt and Oldenberg [42] used ultra-violet absorption to detect and estimate the hydrogen peroxide actually in the reacting gases at 530—540 °C. For residence times between 3 and 7 sec they found the percentage of H_2O_2 present to increase with residence time. Anzilotti et al. [43] found a similar increase in H_2O_2 with residence time, in contrast with the earlier result of Pease. It is possible that the residence times used by Holt and Oldenberg were too short for a stationary concentration to be achieved.

The experiments of Linnett and Tootal [44] were carried out at 565 °C and a pressure of 230 torr, just above the second limit, with a Pyrex vessel washed with nitric acid and distilled water. The products flowed out of the reaction vessel from the centre of the top via a straight tube of 2 mm diameter, and water and hydrogen peroxide were frozen out in two consecutive traps cooled in liquid air. The contents of the traps were analyzed later. The amounts of water and hydrogen peroxide, expressed as partial pressures in the gases leaving the reaction vessel, are shown in Figs. 8 and 9 as a function of the residence time. The results appear to fall into

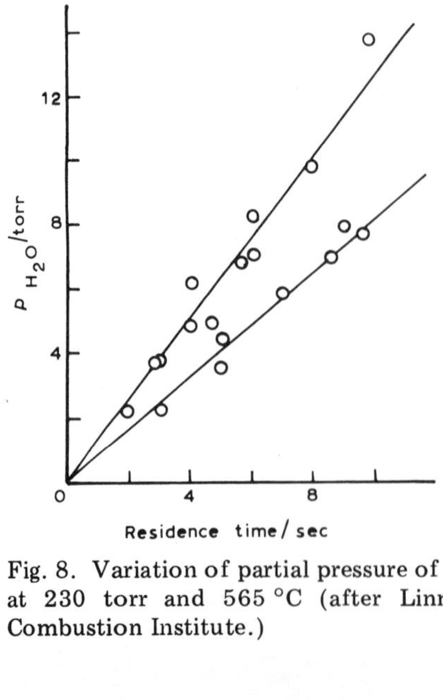

Fig. 8. Variation of partial pressure of water formed with residence time for $2H_2 + O_2$ at 230 torr and 565 °C (after Linnett and Tootal [44]). (By courtesy of The Combustion Institute.)

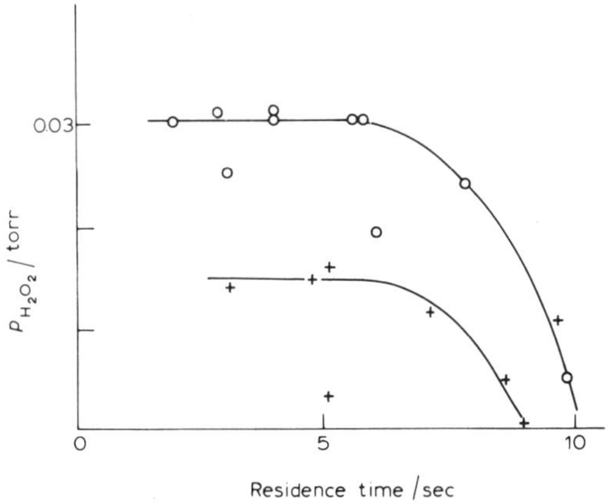

Fig. 9. Variation of partial pressure of hydrogen peroxide formed with residence time for $2H_2 + O_2$ at 230 torr and 565 °C (after Linnett and Tootal [44]). (By courtesy of The Combustion Institute.)

two sets, randomly intermixed, and presumably depending on the state of the vessel wall. In each set the amount of water formed was proportional to the contact time, and a more or less steady state concentration of H_2O_2 was confirmed for residence times up to 5 or 6 sec. The fall-off at longer residence times may have been due to catalytic decomposition in the exit tubes between the furnace and traps, but nevertheless this would need to be extremely rapid. Under the conditions used by Linnett and Tootal, near the second limit, both the water and hydrogen peroxide decrease as the H_2/O_2 ratio is increased from 1.5 to 6.0.

3.6 BASIC MECHANISM OF REACTION

$$OH + H_2 = H_2O + H \qquad (i)$$

$$H + O_2 = OH + O \qquad (ii)$$

$$O + H_2 = OH + H \qquad (iii)$$

$$H + O_2 + M = HO_2 + M \qquad (iv)$$

$$HO_2 \xrightarrow{\text{surface}} \text{destruction} \qquad (v)$$

$$HO_2 \xrightarrow{\text{surface}} \tfrac{1}{2}H_2O_2 + \tfrac{1}{2}O_2 \qquad (va)$$

$$HO_2 \xrightarrow{\text{surface}} \tfrac{1}{2}H_2O \qquad (vb)$$

$$HO_2 + H_2O_2 = H_2O + O_2 + OH \qquad (vi)$$

$$H_2O_2 + M' = OH + OH + M' \qquad (vii)$$

$$HO_2 + H_2 = H_2O_2 + H \qquad (xi)$$

$$HO_2 + H_2 = H_2O + OH \qquad (xia)$$

Formal mechanisms to explain the first and second limits have already been given, and it is generally agreed that the chain branching cycle which controls the low pressure explosions consists of reactions (i)—(iii). Below the first limit these compete unsuccessfully against chain breaking produced by diffusion of the radicals H, OH and O to the vessel wall. The first limit itself occurs when the pressure has risen high enough for the rate of gas phase chain branching just to balance the rate of surface termination.

The gas phase association reaction which competes with the branching cycle at the second limit might be either $O + H_2 + M = H_2O + M$ or $H + O_2 + M = HO_2 + M$. The first of these gives the inert molecule H_2O, and finally ends any possibility of a continuing radical chain. In terms of re-arrangement of existing bonds, it is also not quite such a simple step as that forming the hydroperoxyl radical (cf. Hinshelwood [7]). The HO_2 radical also may give rise to a number of further reactions in favourable circumstances, including the possible formation of hydrogen peroxide. The important association reaction controlling the second limit is therefore assumed to be reaction (iv), with the proviso in the simplest approach that, at this limit in salt coated vessels, all the HO_2 formed diffuses to the wall and is destroyed (reaction (v), with reactions (va) and (vb) representing alternative possibilities). Simple destruction by reaction (v) leads to the limit condition

$$[H_2] + k_{O_2}[O_2] + k_M[M] = 2k_2/k_{4,H_2} \tag{6}$$

so that the apparent activation energy of between 18 and 25 kcal . mole^{-1} mentioned in Sect. 3.3 refers to the difference $(E_2 - E_4)$. The rate of reaction (iv) would not be expected to vary much with temperature, i.e. $E_4 \simeq 0$. Thus, the activation energy E_2 itself will lie approximately within the above range. The whole of the range is permissible energetically, its lower end being slightly greater than the enthalpy change, $\Delta H_2 \simeq +16$ kcal . mole^{-1}.

The gas phase reactions which may be undergone by HO_2 depend on the conditions. It is these elementary steps which are involved in the slow reaction and at the third limit. Salt coated vessels may be assumed to be moderately or highly efficient for the destruction of hydroperoxyl at the surface. In such vessels the radical concentration is likely to be low and reactions between radicals are unlikely to be important. The reactions proposed for the HO_2 under these conditions were (xi) [1, 23], or (xia) [7], the first of which leads to the formation of hydrogen peroxide and has since been shown to be the faster of the two [45]. Assuming that the hydrogen peroxide is decomposed without formation of further chain centres, reactions (iv) and (xi) then form a chain propagating cycle which continues until either a H or HO_2 is destroyed at the vessel surface.

In terms of the explosion limit conditions, the occurrence of reaction

(xi) also implies that the rate of "gas phase" termination is no longer given by the rate of reaction (iv), but becomes only a fraction of this. If this fraction becomes sufficiently small at high temperatures and pressures, it is clearly possible to re-enter the explosion region at an isothermal limit. The observed third limits would correspond with this condition if thermal effects were completely absent.

The ultimate objective of the kinetic studies is to devise a detailed reaction mechanism and obtain rate coefficients for the elementary steps. The approach is to set up the kinetic expression for the rate of formation of the final product (water), to use the stationary state method to eliminate radical concentrations, and then to make quantitative comparison with experiment. In the present mechanism five intermediate species are involved, namely H, O, OH, HO_2 and H_2O_2, and it is necessary to make some simplifying assumptions depending on the problem under discussion. The usual procedure has been to consider the first limit and the low pressure explosion region separately from the slow reaction region and third limit. In the higher pressure region diffusion to the wall of all the chain carriers except HO_2 and H_2O_2 may be neglected, since the first limit where diffusion of H, OH and O is important occurs at very much lower pressures. Conversely, the gas phase reactions of HO_2 (and to a first approximation its formation also) may be neglected at the first limit. Clearly the second limit lies in a region of overlap, where contributions from the important processes on both the high and low pressure sides may need consideration, particularly at the ends of the composition range and at low temperatures where the limit pressure becomes lower. The two regions will be considered separately, but before doing this quantitatively it is necessary to investigate the form of the expressions for the rates of destruction of chain centres at vessel surfaces. A precise knowledge of the form of these expressions is of the utmost importance: the form varies depending on the destructive efficiency of the surface.

3.6.1 Destruction of chain centres at surfaces

It is convenient to express the surface rate of chain breaking as equal to a surface destruction rate coefficient k_s multiplied by the concentration of chain carriers, n. Assuming only first order reactions of the centres in the gas phase also, the rate of development of centre concentration is given by

$$\begin{aligned} \mathrm{d}n/\mathrm{d}t &= n_0 + (f - g - k_s)n \\ &= n_0 + \phi n \end{aligned} \tag{7}$$

where n_0 is the initiation rate per unit volume, f and g are the gas phase branching and termination coefficients, and ϕ is known as the net branching factor.

When the vessel surface is extremely inefficient for the destruction of chain centres, relatively rapid diffusion ensures that the concentration is uniform throughout the vessel. Hence, for small ϵ, simple kinetic theory gives [46]

$$k_{\mathrm{s,ineff}} = \tfrac{1}{4}\epsilon\bar{c}\,(s/v) \tag{8}$$

where $\bar{c}$ is the average velocity of the centres, and s/v is the surface/volume ratio. In this case k_s is independent of pressure, but is directly proportional to ϵ. The stationary concentration of centres in the slow reaction is given by

$$n_s = n_0/(k_s + g - f) \tag{9}$$

For large ϵ, on the other hand, the concentration of centres near the surface will be effectively zero, and concentration gradients will exist within the vessel. The rate of chain breaking will be determined by the rate of diffusion to the wall. The simple formulation above is not strictly correct in these circumstances, and a diffusion equation must be employed. However, by assuming that the concentration of centres at the surface is zero for very efficient surfaces, Bursian and Sorokin [47] have shown that for spherical vessels of radius r an equivalent value of k_s may be used, given by

$$k_{\mathrm{s,eff}} = \pi^2 D/r^2 \tag{10}$$

where D is the diffusion coefficient of the chain centre in the gas. Thus $k_{\mathrm{s,eff}}$ is inversely proportional to pressure and independent of ϵ. The lack of dependence on ϵ explains the comparative reproducibility of results with such vessels.

Between these two extremes the treatment is more complex, and it is difficult to obtain explicit expressions for k_s. Semenov [48] has shown that it is still possible to use eqn. (7) for spherical and cylindrical vessels with volume initiation, even though the situation is diffusion controlled, provided (i) n is replaced by $\bar{n}$, the average concentration of chain centres, and (ii) n_0 is multiplied by a small numerical coefficient ψ which lies between 0.6 and unity, and depends on the vessel shape and surface, i.e.

$$\bar{n}_s = \psi n_0/(k_s + g - f) \tag{9a}$$

The problem of intermediate efficiencies has been discussed by Baldwin [49] in terms of the treatments of Semenov [48] and Lewis and von Elbe [50]. For *cylindrical vessels*, according to the Semenov treatment

$$k_{\mathrm{s,cyl.}} \equiv u_1^2 D/r^2 \tag{11}$$

where u_1 is the smallest root of the equation

$$\begin{aligned}\frac{J_0(u)}{J_1(u)} &= \left[\frac{4\lambda(1-\epsilon/2)}{3\epsilon r}\right]u \\ &= (B/p)u\end{aligned} \tag{12}$$

in which

$$B = \frac{4\lambda_0(1 - \epsilon/2)}{3\epsilon r} \tag{13}$$

λ, λ_0 are the mean free paths at pressure p and at unit pressure, respectively, and $J_0(u)$, $J_1(u)$ are Bessel functions of the first kind of zero and first order, respectively. The value of ψ for cylindrical vessels is

$$\psi_{cyl.} = \frac{4}{u_1^2[1 + u_1^2(B/p)^2]} \tag{14}$$

For *spherical vessels* the corresponding expressions are

$$k_{s,spher} \equiv \theta_1^2 D/r^2 \tag{15}$$

where θ_1 is the first root of the equation

$$1 - \theta \cot\theta = p/B \tag{16}$$

and

$$\psi_{spher} = \frac{6}{[1 - B/p + \theta_1^2(B/p)^2]\theta_1^2} \tag{17}$$

Values of the u_1^2, θ_1^2 and ψ for various values of B/p are given for both cylindrical and spherical vessels in Table 10. If B/p is small the surface destruction is effectively determined by diffusion; while for large B/p the

TABLE 10

Solutions of eqs. (12) and (16) for cylindrical and spherical vessels, respectively [49]

B/p	u_1^2	ψ_{cyl}	θ_1^2	ψ_{spher}
0	5.784	0.692	9.876	0.608
0.05	5.235	0.754	8.910	0.693
0.1	4.750	0.804	8.037	0.760
0.2	3.960	0.872	6.615	0.852
0.3	3.364	0.913	5.522	0.908
0.4	2.910	0.938	4.718	0.939
0.5	2.558	0.954	4.105	0.958
0.6	2.279	0.964	3.637	0.965
0.8	1.866	0.977	2.943	0.978
1.0	1.577	0.984	2.467	0.986
1.2	1.364	0.989	2.123	0.990
1.5	1.134	0.993	1.753	0.993
2.0	0.885	0.996	1.358	0.996
3.0	0.614	0.999	0.935	0.999
5.0	0.380	1.000	0.576	1.000
10.0	0.195	1.000	0.294	1.000
20.0	0.0988	1.000	0.149	1.000
30.0	0.0661	1.000		
50.0			0.060	1.000

TABLE 11
Chain breaking efficiency and diffusion control in 1 cm radius vessel [49]

Pressure (torr)	$\epsilon_{\text{diffusion control}}$	$\epsilon_{\text{diffusion independent}}$
1	$>10^{-1}$	$<10^{-2}$
10	$>10^{-2}$	$<10^{-3}$
100	$>10^{-3}$	$<10^{-4}$
1000	$>10^{-4}$	$<10^{-5}$

destruction is independent of diffusion. It is found that a value of $B/p = 0.1$ only reduces the surface destruction constant k_s about 20 % below the value for complete diffusion control ($B/p = 0$). Using this as a criterion and taking $\lambda_0 \approx 5 \times 10^{-3}$ cm (e.g. for OH at room temperature), then, for a vessel of radius 1 cm, Baldwin calculates values of ϵ above which the destruction will be diffusion controlled at a series of pressures. These are given in the second column of Table 11. Similarly, it was found that the relation between u_1^2 or θ_1^2 and p/B is linear within 20 % for $B/p > 1.0$, i.e. in this range the surface destruction is effectively independent of diffusion. The corresponding values of ϵ at the various pressures are given in column 3 of the Table. For values of ϵ intermediate between the two columns the surface must be regarded as having intermediate efficiency. It is clear from Table 11 that the total pressure is of considerable importance in some circumstances in deciding whether the surface destruction is diffusion controlled, and hence also sensitive to the presence of inert gases. For example, a surface of efficiency around 10^{-3} may give termination independent of diffusion at first limit pressures below 10 torr, but with almost complete diffusion control during slow reaction studies at atmospheric pressure.

The kinetic results will now be examined in the light of these findings.

3.6.2 *Second and third limits (higher pressures)*

To test the application of the theory of branching chains to the third limit, Willbourn and Hinshelwood [27] assumed a reaction mechanism consisting of an initiating step of rate f_1 producing either H or OH, and followed by reactions (i), (ii), (iii), (iv), (vb) and (xia). This leads, by the usual methods, to the following expression for the rate of formation of water in the steady state, viz.

$$d[H_2O]/dt = \frac{f_1 \left\{ \dfrac{1.5k_{5b} + 2k_{11a}[H_2]}{k_{5b} + k_{11a}[H_2]} \right\}}{1 - (2k_2/\Sigma k_4[M]) - \dfrac{k_{11a}[H_2]}{k_{5b} + k_{11a}[H_2]}} \tag{18}$$

and it can be shown that small variations in the mechanism, such as the substitution of (v) for (va), or of (xi) and (vi) for (xia) affect only the numerator of this expression. Explosion limits are obtained when the denominator of eqn. (18) is zero, i.e.

$$\frac{2k_2}{\Sigma k_4[\mathrm{M}]} = \frac{k_{5\mathrm{b}}}{k_{5\mathrm{b}} + k_{11\mathrm{a}}[\mathrm{H_2}]} \tag{19}$$

where

$$\Sigma k_4[\mathrm{M}] = k_{4,\mathrm{H_2}}\{[\mathrm{H_2}] + k_{\mathrm{O_2}}[\mathrm{O_2}] + k_{\mathrm{M}}[\mathrm{M}]\}$$

the constants $k_{\mathrm{O_2}}$ and k_{M} being obtained from experiments on the second limit (cf. Sect. 3.3).

k_{5b} is a rate coefficient for the destruction of HO_2 at the vessel wall, and it was assumed by Willbourn and Hinshelwood that their KCl surface was an efficient surface for HO_2 destruction so that $k_{5\mathrm{b}}$ is diffusion controlled. For a trace component i diffusing through a mixture of gases, the diffusion coefficient D_{i} may be written

$$\frac{1}{D_{\mathrm{i}}} = \sum_{\mathrm{j} \neq \mathrm{i}} \frac{p_{\mathrm{j}}}{\mathscr{D}_{\mathrm{ij}}} \tag{20}$$

where p_{j} denotes pressure and $\mathscr{D}_{\mathrm{ij}}$ denotes a binary diffusion coefficient at unit pressure. We may thus write

$$k_{5\mathrm{b}} \equiv K_{5\mathrm{b}} D_{\mathrm{i}}$$

$$= K_{5\mathrm{b}} \Big/ \left\{ \frac{[\mathrm{H_2}]}{\mathscr{D}_{\mathrm{HO_2-H_2}}} + \frac{[\mathrm{O_2}]}{\mathscr{D}_{\mathrm{HO_2-O_2}}} + \frac{[\mathrm{M}]}{\mathscr{D}_{\mathrm{HO_2-M}}} \right\} \tag{21}$$

Since diffusion coefficients are required relative to hydrogen only, we may now use $D'_{\mathrm{M}} = \mathscr{D}_{\mathrm{HO_2-H_2}}/\mathscr{D}_{\mathrm{HO_2-M}}$ (and similarly for O_2), giving, on substitution of (21) into (19), the explosion condition

$$\frac{[\mathrm{H_2}] + D'_{\mathrm{O_2}}[\mathrm{O_2}] + D'_{\mathrm{M}}[\mathrm{M}]}{1 + k_{\mathrm{O_2}}\rho_{\mathrm{O_2}} + k_{\mathrm{M}}\rho_{\mathrm{M}} - 2k_2/(k_{4,\mathrm{H_2}}[\mathrm{H_2}])} = \frac{k_{4,\mathrm{H_2}} K_{5\mathrm{b}} \mathscr{D}_{\mathrm{HO_2-H_2}}}{2k_2 k_{11\mathrm{a}}} \tag{22}$$

where $\rho_{\mathrm{O_2}} = [\mathrm{O_2}]/[\mathrm{H_2}]$ and $\rho_{\mathrm{M}} = [\mathrm{M}]/[\mathrm{H_2}]$. For a given temperature this reduces to

$$\frac{[\mathrm{H_2}] + D'_{\mathrm{O_2}}[\mathrm{O_2}] + D'_{\mathrm{M}}[\mathrm{M}]}{1 + k_{\mathrm{O_2}}\rho_{\mathrm{O_2}} + k_{\mathrm{M}}\rho_{\mathrm{M}} - \mathrm{K}_{24}/[\mathrm{H_2}]} = C \tag{23}$$

or

$$[\mathrm{H_2}] + k_{\mathrm{O_2}}[\mathrm{O_2}] + k_{\mathrm{M}}[\mathrm{M}] = K_{24} + \frac{1}{C}[\mathrm{H_2}]\{[\mathrm{H_2}] + D'_{\mathrm{O_2}}[\mathrm{O_2}] + D'_{\mathrm{M}}[\mathrm{M}]\} \tag{23a}$$

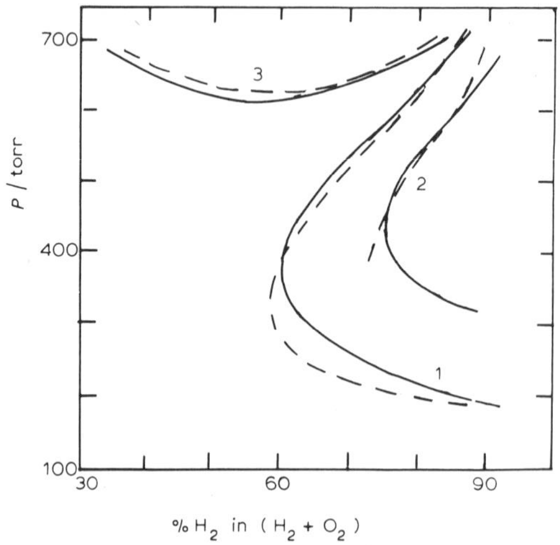

Fig. 10. Influence of hydrogen—oxygen ratio on third limit without additive (curve 1), with 100 mm nitrogen (curve 2) and 100 mm carbon dioxide (curve 3) at 586 °C in KCl coated vessel, 5.5 cm diameter (after Willbourn and Hinshelwood [27]). (By courtesy of The Royal Society.)

where $K_{24} = 2k_2/k_{4,H_2}$ and C is a vessel constant. For a given composition, eqn. (23) is a quadratic in $[H_2]$, the solutions corresponding to the second and third limits.

The influence of hydrogen–oxygen ratio and of additions of 100 mm of nitrogen and carbon dioxide, respectively, on the third limit at 586 °C in a KCl coated vessel of 5.5 cm diameter are shown as the broken lines in Fig. 10. The solid lines show limits calculated from eqn. (23) with $K_{24} = 144$, $C = 1100$ and the values of D'_M and k_M given in Table 12 (results for water vapour are also included in the Table). Values of k_M derived independently from second limit experiments (Table 7) are also given in column 5 of the Table, while columns 3 and 6 give values of D'_M and k_M calculated from kinetic theory. Although the influence of water vapour does present difficulties as discussed below, the general level of agreement, both between theory and experiment and with the independently verifiable second limit mechanism, provides strong evidence for the branched chain mechanism of the third limit in KCl coated vessels. Similar views are held by Lewis and von Elbe [4], Warren [25] and Voevodsky [51].

TABLE 12
Values of coefficients in third limit calculations [27]

M	$D'_{M,obs.}$	$D'_{M,calc.}$	$k_{M,P_3obs.}$	$k_{M,P_2obs.}$	$k_{M,calc.}$
O_2	3.0	3.8	0.38	0.35	0.4
N_2	2.2	3.9	0.25	0.44	0.45
CO_2	2.5	5.2	0.90	1.47	0.43
H_2O	4.0	4.1	8.25	6.4	0.38

Warren's results [25] are also consistent with those of Willbourn and Hinshelwood [27] and Cullis and Hinshelwood [38] regarding the values of the vessel constant C.

In commenting upon the effect of inert gases on the limit, it is interesting to observe that the high value of k_{CO_2} leads to a curve which has quite a different shape from that for nitrogen, and that this effect is also observed experimentally. Water vapour also has a high value of k_{H_2O}, and would be expected to behave similarly. The curves for water, however, resemble those for nitrogen, and it is clear that some other effects are coming into play. Willbourn and Hinshelwood [27] suggested that the water vapour reduced the destructive efficiency of the KCl surface towards HO_2, thus altering the vessel constant C; and in this way they were able to explain their observations by means of a gradual variation of C with concentration of added water. Without some very special absorption relationship, however, it is difficult to reconcile this with the diffusion control of k_5. Voevodsky [51] prefers to believe that the change in effective vessel constant C is due to participation of the gas phase process $HO_2 + H_2O = H_2O_2 + OH$, though because of its high endothermicity ($\Delta H \simeq +30$ kcal . mole^{-1}) there must be some doubt about this also.

With regard to the second limit, the regeneration term

$$\frac{1}{C}[H_2]\{[H_2] + D'_{O_2}[O_2] + D'_M[M]\}$$

in eqn. (23a) will modify the limit pressure from that given by the simple eqn. (6). According to eqn. (10) the constant C in eqn. (23) will vary inversely as the square of the vessel diameter. Hence a small diameter effect on the second limit might be expected in some circumstances. Cullis and Hinshelwood [38] give the temperature dependence of C as $C \propto T^{1.5} \exp(7500/T)$, and if this is used in conjunction with their vessel constant $C = 1053$ at 580 °C for a 6.9 cm diameter vessel, it turns out that, even at 530 °C, the second limit pressure for the mixture $2H_2 + O_2$ is reduced by some 5 or 6 torr when the vessel diameter is reduced from 10 to 2 cm. This will be used in later discussion.

3.6.3 *Slow reaction (higher pressures)*

Following their analysis of the third limit, Willbourn and Hinshelwood [27] went on to study the slow reaction between the second and third limits, and to derive from these studies the form of the function f_1 determining the rate of initiation in eqn. (18). Equation (18) is too complex to be applied directly to the results on the rates. However, all the unknown functions except f_1 can be derived from experiments on the

References pp. 234—248

second and third limits, leaving f_1 as the only unknown. If the reaction rate is expressed as

$$d[H_2O]/dt = f_1 R^* \tag{24}$$

then a similar derivation to that of the preceding section gives

$$R^* = \frac{\left\{\dfrac{0.75C}{[H_2] + D'_{O_2}[O_2] + D'_M[M]} + \dfrac{[H_2]}{K_{24}}\right\}\{[H_2] + k_{O_2}[O_2] + k_M[M]\}}{C\left\{\dfrac{[H_2] + k_{O_2}[O_2] + k_M[M] - K_{24}}{[H_2] + D'_{O_2}[O_2] + D'_M[M]}\right\} - [H_2]} \tag{25}$$

and this can be calculated directly. Thus the variation of the rate of reaction ($f_1 R^*$) with the pressures of hydrogen, oxygen and inert gases can be calculated for different possible forms of the function f_1, and the result compared with the observed variation. It was found that initiating steps like $H_2 + O_2 \rightarrow 2OH$ gave too great a dependence of the rate on the oxygen concentration, and that the best fit seemed to be obtained by writing

$$f_1 = k_0[H_2]\{Z_{H_2}[H_2] + Z_{O_2}[O_2] + Z_M[M]\} \tag{26}$$

where Z_{H_2}, Z_{O_2} and Z_M are the relative collision numbers for hydrogen molecules with hydrogen, oxygen and inert gas M, respectively. This expression was based on the idea that chains are started by dissociation of hydrogen in bimolecular collisions. Using values of Z_M calculated from kinetic theory, the agreement between calculated and observed rates was good for hydrogen—oxygen mixtures alone, and when nitrogen was the inert gas. For carbon dioxide as the inert, the rate equation did not predict a quite large enough increase of rate with hydrogen pressure, while with water as additive it was necessary to use a variable vessel constant as was done in the treatment of the third limit.

From measurements of the influence of temperature on the reaction rates and analysis by the method described above, Cullis and Hinshelwood [38] give mean values of the activation energy of the initiation step, $E_0 = 110$ kcal . mole^{-1} for KCl vessels and $E_0 = 92$ kcal . mole^{-1} for CsCl coated vessels. This would be consistent with initiation by $H_2 + M = H + H + M$, which requires 104 kcal . mole^{-1}. Ashmore and Dainton [39] reached similar conclusions. However, Lewis and von Elbe [52] point out that, since the activation energy of the hydrogen dissociation is at least equal to its heat of dissociation (104 kcal . mole^{-1}), the maximum possible rate of dissociation is only about 10^{-13} of the observed initiation rate. This rules out hydrogen dissociation as the initiating step. Lewis and von Elbe cite this as support for the initiation step which they themselves had proposed earlier [23], namely the dissociation of hydrogen peroxide by reaction (vii). This requires a dissociation energy of only ca. 50 kcal . mole^{-1}, and the objection regarding the dissociation rate is removed. The measurements of Pease [19] had originally suggested the formation of a steady state concentration of H_2O_2, and this has since been confirmed by the measurements of Linnett and Tootal [44], already discussed. Lewis and von Elbe assumed in their mechanism that the steady state concentration of H_2O_2 was built up, after an induction period, by the pair of reactions (xi) and (vi). These, together with the remainder of the reactions (i)—(iv), (va), (vii) and a surface destruction reaction for H_2O_2, made up a complete self-contained mechanism by means of which they were able to calculate explosion limits and reaction rates in excellent agreement with their experiments. The major reactions forming and destroying the peroxide, (xi) and (vi), ensure a direct proportionality between the peroxide and hydrogen concentrations, so that the form of the initiation function f_1 of Willbourn and Hinshelwood may be explained. Unfortunately, more recent measurements of the rate of the slow reaction in boric acid coated vessels, to be discussed later, provide strong evidence that reaction (vi) cannot contribute appreciably to the mechanism.

3.6.4 First and second limits (lower pressures)

The two types of behaviour discussed earlier with respect to surface destruction efficiency are well illustrated at the first limit by the results under the conditions of Hinshelwood and Moelwyn-Hughes [13] (high ϵ) on the one hand, and of Semenova [16] (low ϵ) on the other. The contrasting effects of inert gases in the two types of vessel are shown in Tables 3 and 4. At the first limit the surface destruction refers to removal of H, OH and O.

The situation at the first limit with KCl coated vessels is of some interest. The marked reduction in the rate of oxidation of hydrogen and hydrocarbons when the vessel surface is coated with KCl led to the tacit assumption that KCl is an efficient surface for the destruction not only of HO_2 and H_2O_2 as above, but also for the other chain centres present in the systems. However, the measurements of Warren [25, 53] on first limits in vessels coated with a number of salts showed that KCl is considerably less efficient than a number of other surfaces, e.g. KOH, NaOH, graphite, $MnCl_2$, Al_2O_3, PbO, in this context, and therefore that KCl is not an extremely efficient surface here. Baldwin [54] has also examined the first limit at 500—550 °C with an aged KCl coated vessel. The hydrogen and oxygen in the mixture were varied independently by starting with a standard mixture containing 0.28, 0.14 and 0.58 mole fractions of hydrogen (x), oxygen (y) and nitrogen, respectively, and then interchanging either the hydrogen or the oxygen with the nitrogen while keeping the other reactant constant. The limits for a 51 mm diameter

TABLE 13
Observed and calculated first limits with 51 mm diameter KCl coated vessel at 500 °C [54]

x	y	$P_{1,\text{obs.}}$(torr)	$P_{1,\text{calc.}}$ [a]	
			A	B
0.28	0.72	3.49	4.50	3.48
	0.56	3.99	4.98	3.96
	0.28	6.02	6.73	5.96
	0.14	9.78	9.39	9.76
	0.10	12.58	11.22	12.69
	0.07	16.92	13.85	17.17
0.56	0.14	10.23	10.25	10.26
0.28		9.78	9.39	9.76
0.14		10.03	9.01	9.99
0.10		10.32	9.27	10.27
0.07		10.80	9.50	10.75
0.28	0.56	3.99	4.98	3.96
0.14		4.41	5.17	4.43
0.10		4.75	5.38	4.77
0.07		5.12	5.66	5.18

[a] A, calculated on the assumption that H atoms are efficiently destroyed at the KCl surface.
B, calculated assuming only moderate efficiency for destruction of H atoms, with B of eqn. (13) = 7.8.

vessel at 500 °C are given in Table 13. The limit pressures increase at both low oxygen and low hydrogen mole fractions, implying at least a combination of the chain branching reactions (i)—(iii), the surface destruction of H atoms which react with oxygen, and the surface destruction of either OH or O, both of which react with hydrogen in the branching cycle. According to the best data at present available [55], k_1 is some twenty times larger than k_3 at 520 °C, and surface destruction of O atoms will be assumed to be most important here. Linnett and Marsden [56] have shown KCl surfaces to be quite efficient for the destruction of oxygen atoms at temperatures above 400 °C.

Taking reactions (i)—(v), together with the surface destruction of H and O atoms, the stationary state treatment gives the complete first and second limit explosion boundary as

$$[H_2] + k_{O_2}[O_2] + k_M[M] + \frac{k_H}{[O_2]} + \frac{2k_2k_0}{k_3[H_2] + k_0} = 2k_2/k_{4,H_2} \qquad (27)$$

where k_H and k_O are the surface destruction coefficients which depend on pressure, vessel diameter and other conditions in the manner already outlined. Alternatively, if p_2^0 is the second limit in the absence of surface termination, then p_1 is related to it by

$$\frac{k_H}{2k_2[O_2]} + \frac{k_0}{k_3[H_2] + k_0} = 1 - \frac{p_1}{p_2^0} \tag{28}$$

where $[O_2]$ and $[H_2]$ are partial pressures at the first limit. Developing an equation very similar to (28) by using the treatments discussed earlier for surfaces of intermediate chain breaking efficiency for H atoms, together with high efficiency for O atoms, Baldwin was able to calculate first limits for a wide range of mixture compositions, temperatures, and vessel diameters. Table 13 compares the calculated and measured limits for a range of mixture compositions at one temperature and vessel size. The results (r.m.s. deviation 0.7 %) show the KCl surface to be only of intermediate efficiency for the destruction of H atoms at the first limit pressures. Similar precise agreement was found under all the conditions studied, and the value found for the parameter B of eqn. (13) implies an efficiency ϵ in the region of 10^{-3}.

If the diffusion coefficients of H and O atoms in the gas mixtures can be estimated satisfactorily, then the results of the above treatment can be used also to derive values of the rate coefficients k_2 and k_3. Using estimated hard sphere values ($D_0 = \lambda_0 \bar{c}/3$ where λ_0 is the mean free path at unit pressure and $\bar{c}$ is the mean molecular velocity), Baldwin obtained, at 520 °C

$$k_2 = 2.7 \times 10^6 \text{ l.mole}^{-1}\text{. sec}^{-1}$$

$$k_3 = 2.0 \times 10^7 \text{ l.mole}^{-1}\text{. sec}^{-1}$$

However, theoretical estimates by Weissman and Mason [57] suggest that departures from the hard sphere model would raise the diffusion coefficient of H atoms by some 70 % above the hard sphere value, and this would entail a revision of Baldwin's k_2 to 4.6×10^6 l . mole^{-1} . sec^{-1}. A correction of similar order to k_3 is also indicated.

Kurzius and Boudart [58] have adopted the same approach as above specifically for the determination of k_2 by measuring first limits for mixtures of composition $2H_2 + O_2$ and $9H_2 + O_2$ in a 10.2 cm diameter vessel coated with magnesium oxide. Their analysis considered only reactions (i)—(iii) and the surface destruction of H atoms, for which they assumed the surface to have unit efficiency (the results for the two compositions studied seemed to support this). On this basis, however, their results do not appear to be completely consistent with those of Baldwin. Thus, at 520 °C the parameters derived by Baldwin for KCl coated vessels would predict a limit for $2H_2 + O_2$ at ca. 5.7 torr in a 10.2 cm diameter vessel of unit efficiency. The observed limit was 4.56 torr. This difference is reflected in a higher value for k_2, namely 6.3×10^6 at 520 °C compared with Baldwin's figure of 4.6×10^6 l . mole^{-1} . sec^{-1} already quoted. Over the temperature range 800—1000 K, Kurzius and

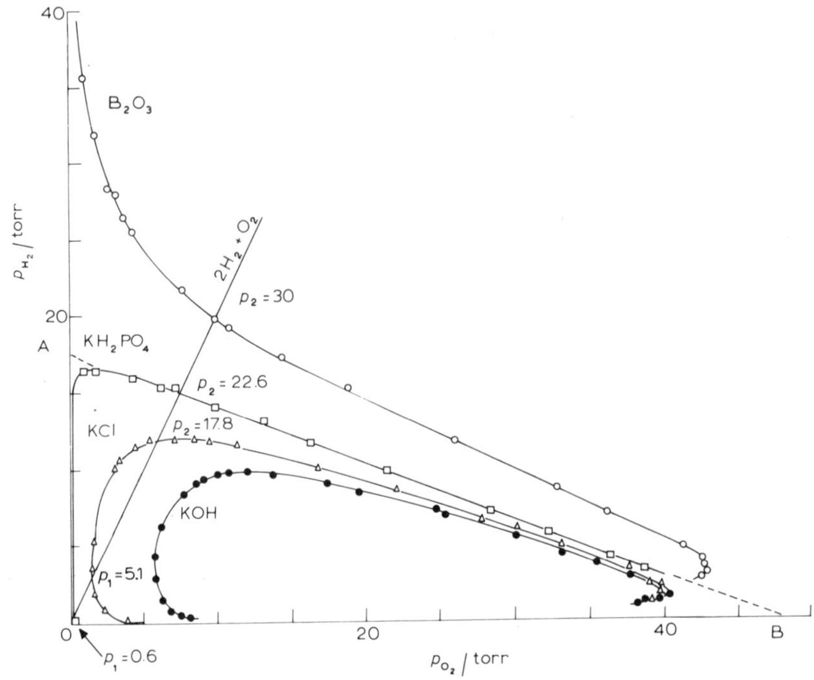

Fig. 11. Effect of variation of coating material on low pressure explosion limits. 7.3 cm diameter vessel at 460 °C (after Warren [25]). AB is the "fundamental" limit line calculated from the B_2O_3 curve. (By courtesy of The Royal Society.)

Boudart give $k_2 = (1.7 \pm 0.4) \times 10^{11} \exp -(8100 \pm 200)/T$ l . mole^{-1}. sec^{-1}, and for the corresponding reaction with D atoms, $k_{2D} = (8.9 \pm 0.22) \times 10^{10} \exp -(7450 \pm 200)/T$. The ratio $k_2/k_{2D} = 1.93 \exp (-650/T)$.

Returning once more to consideration of the second limit, Fig. 11 shows limits measured by Warren [25] at 460 °C in a 7.3 cm diameter vessel coated with various materials. Postponing discussion of the boric acid surface until later, there is clearly some dependence of the limit on the surface coating over the whole composition range, but this becomes particularly prominent for the oxygen deficient mixtures. The limits can again be discussed in terms of eqn. (28), and again KCl shows itself to be a surface of only intermediate efficiency for the destruction of H atoms. The "fundamental" limit line corresponding to eqn. (6) may be calculated from the boric acid curve as discussed later, and is given by the line AB.

In addition to the small dependence of the second limit on vessel surface, Table 6 also shows a small dependence of the limit on the vessel size. Thus, Lewis and von Elbe [4] quote a variation in the second limit for $2H_2 + O_2$ from 88 torr with a 10 cm diameter spherical KCl coated vessel at 530 °C to 68 torr for a 1.8 cm diameter vessel. They invoke a special mechanism for this, involving the postulated reaction $H + O_2 +$

TABLE 14
Recommended values of the ratio $2k_2/k_{4,H_2}$ [72]

Temperature (°C)	470	480	500	520	530
$2k_2/k_{4,H_2}$ (torr)	21.8	26.0	37.0	52.0	61.0

$H_2O_2 = H_2O + O_2 + OH$ competing with reaction (vi) in the small vessels. However, it seems [54] that most, if not all, the difference can be accounted for in other ways. First, as discussed in Sect. 3.6.2 the different degrees of modification of the "fundamental" limit equation by the regeneration term will cause a 5 or 6 torr change in the limit. Secondly, the magnitudes of the surface destruction terms in eqn. (28) may be estimated by extrapolation from the first limit conditions using the treatment already outlined for surfaces of moderate efficiency. The difference between these terms in the two vessels accounts for another 10–12 torr change in the limit, so that the combined effects are essentially capable of explaining the observed change without invoking an important additional reaction step. The best "fundamental" limit lines corresponding to eqn. (6) lead to the values of $2k_2/k_{4,H_2}$ given in Table 14.

3.6.5 Development of the reaction with time within the explosion region

Measurements of the induction periods associated with the reaction near the first limit provide another method for deriving information regarding the rate coefficient k_2. The time development of the reaction in this region was first studied by Kowalsky [15] by means of pressure recordings with a sensitive membrane manometer. The deviation of the membrane was recorded on a moving film by means of a light beam and small reflecting mirror attached to the membrane. The whole development of the reaction within the explosion region lasts less than a second, and in an improved version of the apparatus the optical beam, before reaching the mirror of the membrane, was reflected from another mirror fastened to the end of a vibrating tuning fork of period 1/370 sec. The records thus allowed time resolution of the order of a few milliseconds. Later improvements to the apparatus by Nalbandjan [29] were designed to minimize the time of admission and attainment of uniform pressure in the reactor, and to eliminate any initial shock of the gas on the membrane.

Because of difficulties associated with heat dissipation at the high velocity of the reaction, it was impossible to work at pressures much above the first limit. With the vessel used by Kowalsky [15] this limit for $2H_2 + O_2$ at 480 and 535 °C occurred at 4 and 2.2 torr, respectively. The initial pressures used in the kinetic studies were 4 torr and a little above. The initial stages of the reaction were found to be represented fairly well by $\Delta p = C\,e^{\phi t}$. The net branching factor ϕ increased with temperature and with increasing pressure provided the second limit was not approached too closely. The presence of water added beforehand was found not to affect the course of the reaction, so that the auto-acceleration of the reaction with time could not be ascribed to water catalysis.

The experimental data of Kowalsky on $2H_2 + O_2$ mixtures have been analyzed by Semenov [59] on the basis of reactions (i)—(v), together with surface destruction of H atoms. Because of the lower activation energies of reactions (i) and (iii) compared with reaction (ii), the concentrations of OH and O were assumed to be small compared with H. The variation of H atom concentration could thus be deduced by the method of partial stationary state concentrations [60], giving the net branching factor ϕ at pressure p as

$$\phi = 2k_2 yp \left\{1 - \frac{p}{P_1 + P_2} - \frac{P_1 P_2}{(P_1 + P_2)p}\right\} \tag{29}$$

where y is the mole fraction of oxygen present. It was found that at a fixed temperature the values of k_2 were constant within a few per cent. The average values found were 5.0×10^6 at 485 °C and 6.8×10^6 at 520 °C, both in $\text{l} \cdot \text{mole}^{-1} \cdot \text{sec}^{-1}$. Numerical calculations also showed [59] that, if t is the time corresponding with a fixed pressure drop Δp, then the pressure-dependent constant C is small enough that the product ϕt should be almost constant at a given temperature, regardless of the initial pressure p (provided this is not too close to the first or second limit). This relationship was found to be obeyed by Kowalsky's results. It was later used by Nalbandjan [29] to derive relative values of ϕ from measurements of the induction period τ_i. These too were found to be consistent with eqn. (29) having k_2 constant at a fixed temperature. In terms of the integrated form of eqn. (7)

$$n = n_0(e^{\phi t} - 1)/\phi \tag{30}$$

or, for $e^{\phi t} \gg 1$

$$\log n = \log (n_0/\phi) + \phi t \tag{30a}$$

the small value of C is associated with a comparatively very low rate of primary chain initiation n_0, the effect of which dies out very early in the measured induction time.

Further kinetic measurements of the type made by Kowalsky were carried out by Semenov and co-workers [61] using a vessel washed with hydrofluoric acid and coated with potassium tetraborate. The first limits in this vessel ranged from 0.16 to 0.07 torr between 460 and 600 °C. It was thus possible to penetrate much further into the explosion region than peviously, while at the same time keeping the pressure and reaction velocity low and so avoiding the heat dissipation problem. Initial pressures ranging from 0.3 to 1.2 torr were used. The results, as did those of

TABLE 15
Values of k_2 [61]

Temperature (°C)	460	480	502	522	540	560	580	600
k_2(10^6 l.mole^{-1}.sec^{-1})	1.88	2.65	3.14	4.03	5.30	6.03	7.95	9.76

Kowalsky, agree well with the predictions of the theory. Table 15 gives the values found for the rate coefficient k_2. Near 520 °C the agreement with our upward revision of Baldwin's result is good.

4. Second explosion limits and the slow reaction in vessels having very low surface destruction efficiencies for hydroperoxyl and hydrogen peroxide

The salt coated vessels employed in most of the investigations discussed in the preceding section all have intermediate or high efficiencies for the destruction of chain carriers diffusing to their surfaces. As a result the concentrations of chain carriers during the slow reaction and under immediate pre-explosion conditions are very small. In the present section the characteristics of the reaction in vessels of very low chain breaking efficiency will be considered. Detailed studies of the reaction under these conditions commenced essentially with the discovery of Egerton and Warren [24] in 1951 of the behaviour of the second explosion limit in boric acid coated vessels, and continued with a series of investigations by Baldwin and co-workers of the limits themselves, the slow reaction, and the induction periods in the early stages of the latter.

4.1 SECOND LIMITS IN BORIC ACID COATED VESSELS

The behaviour of the second limit partial pressure plots in boric acid coated vessels has already been shown in Fig. 11. Limit curves at a number of temperatures are also shown in Fig. 12. Clearly the linear relation $p_{H_2} + k_{O_2}p_{O_2} = K$ no longer applies. Instead, as the oxygen concentration decreases the mixture will still explode even though the $[O_2]/[H_2]$ ratio becomes very small: indeed, the limit pressures *increase* with decreasing $[O_2]/[H_2]$ ratio in this region. Egerton and Warren [24] found the limits to be described very closely by an expression of the type

$$p_{H_2} + k_{O_2}p_{O_2} = K + bp_{O_2}^{-1/2} \quad (31)$$

The values found for the constants at a number of temperatures are given in Table 16, and lead to the Arrhenius relations $b = 4.7 \times 10^{11} \exp(-17{,}300/T)$ and $K = 6.3 \times 10^7 \exp(-11{,}000/T)$. The values of k_{O_2} agree well with those for KCl coated vessels, and the value of K at

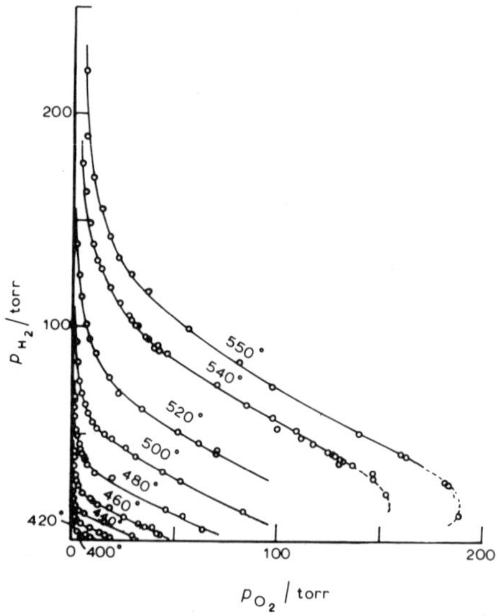

Fig. 12. Second explosion limits of $H_2 + O_2$ in boric acid coated Pyrex vessel, 7.4 cm diameter (after Egerton and Warren [24]). (By courtesy of The Royal Society.)

TABLE 16
Constants in eqn. (31) [24]

Temperature (°C)	k_{O_2}	b	K
400	0.36	1.67	4.15
460	0.37	20.7	15.8
500	0.38	71.5	37.2
540	0.40	230.0	71.6

500 °C (37.2) agrees closely with the value of $K = 37$ found for KCl coated vessels also.

Figure 13 shows the effect of vessel diameter on the limits at two temperatures. Almost identical limits were found with vessels of diameter 2.3, 3.5 and 7.4 cm. This is in contrast with the small but definite dependences observed in KCl coated vessels (e.g. see Table 6).

Second limits in boric acid coated vessels at 500 °C have also been measured by Baldwin et al. [62]. By using the techniques already described starting with a standard mixture containing 0.28, 0.14 and 0.58 mole fractions of hydrogen (x), oxygen (y) and nitrogen, respectively, they were able, by interchanging with nitrogen, to vary the mole fraction of one reactant while keeping the other constant. It was found that the increase in limit pressure at low oxygen mole fractions continued even down to $y = 0.0044$, where p_{O_2} is only 0.7 torr at the limit. Ageing of the

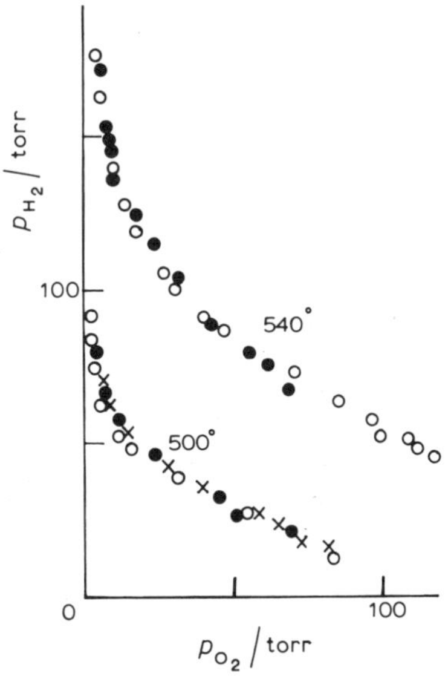

Fig. 13. Effect of diameter of B_2O_3 coated vessel on second limits (after Egerton and Warren [24]). ○, 7.4 cm diameter; X, 3.5 cm diameter; ●, 2.3 cm diameter. (By courtesy of The Royal Society.)

boric acid coating made little difference to the results at 500 °C, the limits, (a) being slightly higher in a freshly coated vessel than in an aged vessel and also, (b) increasing slightly with *decrease* in vessel diameter from 51 to 15 mm.

Similar general behaviour to that described for boric acid coated vessels has also been reported by Dixon-Lewis et al. [63], who used silica vessels washed with hydrofluoric acid and distilled water, and has been found also with H_3PO_4 coated, HNO_3 washed and HF washed Pyrex vessels [53, 62].

As briefly mentioned earlier, there is a marked influence of withdrawal rate on the limits in boric acid coated vessels. Egerton and Warren [24] found with very weak mixtures that, if the evacuation was carried out very quickly, the limits were low and approximately the same as for KCl vessels, while too slow evacuation caused a sluggish ignition which was difficult to observe exactly. The effect of low withdrawal rates on the limits has been studied by Baldwin et al. [21] using both clean Pyrex and boric acid coated vessels at 500 °C, and is shown for the clean Pyrex vessel in Fig. 14. As the withdrawal rate is reduced at high mole fractions of oxygen the explosion is suddenly and completely suppressed at a critical withdrawal rate. At lower mole fractions of oxygen the limit is gradually depressed at the lower rates; and if having passed the normal limit the rate of withdrawal is suddenly increased, explosion could still be made to occur at a lower pressure. Subsidiary experiments in which the limits were approached by heating mixtures initially in the slow reaction region above

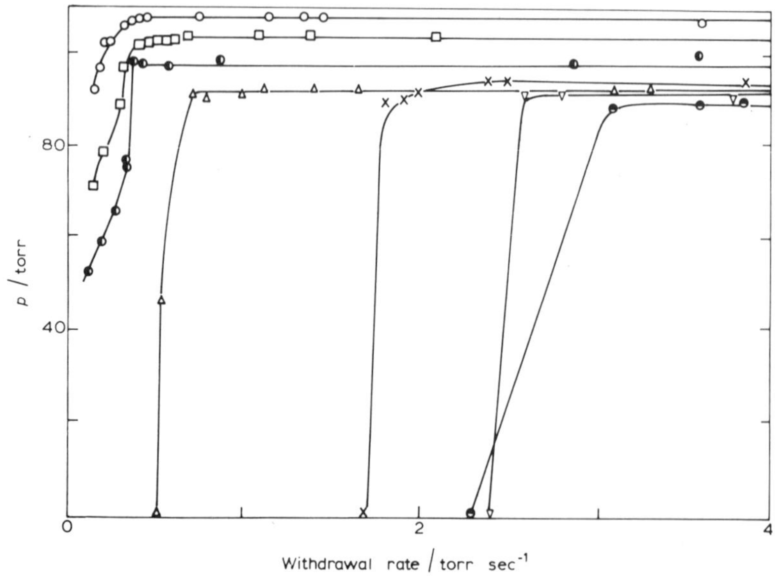

Fig. 14. Effect of withdrawal rate on second limit. 35 mm diameter Pyrex vessel at 500 °C (after Baldwin et al. [21]). $x = 0.28$: ○, $y = 0.72$; □, $y = 0.56$; ◐, $y = 0.42$; △, $y = 0.28$; ×, $y = 0.14$; ▽, $y = 0.10$; ◓, $y = 0.07$. (By courtesy of The Faraday Society.)

the boundary left no doubt that the inhibition was due to water formed as the explosion boundary is approached.

For the purpose of analysis in terms of reaction mechanism, Baldwin et al. [21] defined a critical withdrawal rate in all cases as that giving a limit depressed by at least 5 torr. In both clean Pyrex and boric acid coated vessels the critical rates were found to be proportional to the oxygen mole fraction over a wide range, and almost independent of the hydrogen mole fraction over the range $x = 0.3$—0.8, but increasing significantly at lower x. Only with boric acid coated vessels, however, were the results reproducible for different vessels of the same geometry. Using rigidly standardized manipulation procedures in order to avoid problems due to water formation in the slow reaction, it was possible to measure the effect of vessel diameter on the critical withdrawal rate with both aged and freshly coated vessels. Using the aged vessels little systematic effect was observed: for freshly coated vessels there is a small effect as shown in Figs. 15 and 16.

Below 500 °C the behaviour is even more complex, and Baldwin et al. [20, 21] have investigated the effects of several variations in withdrawal procedure on the limits in a 36 mm diameter vessel. The three factors which could be varied were:

(a) The mixing time prior to withdrawal. Short mixing times of 15 sec were normally used, and tests showed the limits at 500 °C to be independent of mixing time in the range 15—60 sec both with freshly

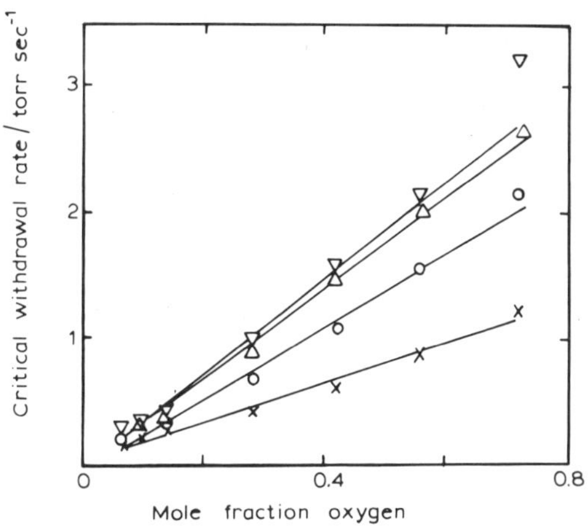

Fig. 15. Variation of critical withdrawal rate with oxygen mole fraction and vessel diameter. Fresh boric acid coated vessels at 500 °C (after Baldwin et al. [21]). $x = 0.28$: ×, 51 mm diam; ○, 36 mm; △, 24 mm; ▽, 15 mm. (By courtesy of The Faraday Society.)

coated vessels, and with aged vessels for all except a few mixtures where small decreases of 1—3 torr were observed. At lower temperatures much longer mixing times could be used.

(b) The complete withdrawal could be carried out smoothly from the initial pressure of 500 torr down to the limit, using calibrated capillary tubes to produce selected reproducible withdrawal rates.

(c) The mixtures could be withdrawn rapidly to within 10 torr (or 100 torr in other experiments) of the limit, followed after a controlled interruption by continued withdrawal at a controlled rate as in (b).

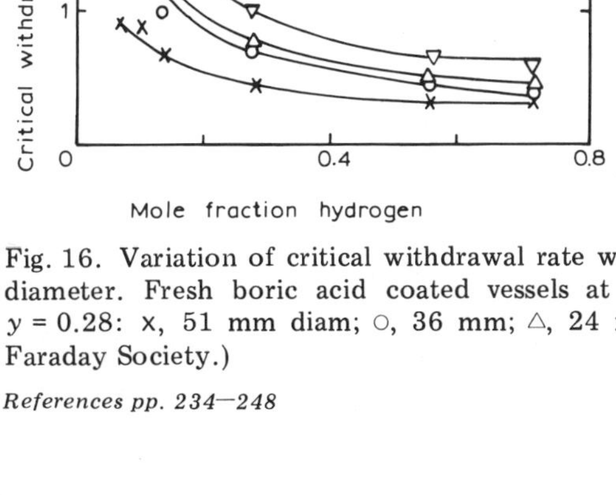

Fig. 16. Variation of critical withdrawal rate with hydrogen mole fraction and vessel diameter. Fresh boric acid coated vessels at 500 °C (after Baldwin et al. [21]). $y = 0.28$: ×, 51 mm diam; ○, 36 mm; △, 24 mm; ▽, 15 mm. (By courtesy of The Faraday Society.)

The results in both aged and freshly coated vessels may be summarized as follows:

(i) At 500 °C the maximum limit is obtained using short mixing times and rapid withdrawal rates. Interruption of the evacuation causes a decrease in the limit for all mixtures, the effect becoming increasingly marked as the mole fraction (y) of oxygen is increased over the range 0.025—0.72. As previously found, use of slow withdrawal rate causes either a decrease in the limit or, at higher y, complete suppression of explosion.

(ii) At 480 °C, the limit rises if the evacuation is interrupted for a short period and falls again as the time of interruption is further increased (Fig. 17). The optimum interruption time increases somewhat as y decreases, varying from about 30 sec for $y = 0.72$ to 2—3 min for $y = 0.025$. With mixtures of low y, the maximum limit can be obtained either by using moderate withdrawal rates and interrupting the evacuation for the optimum period, or by using slow withdrawal rates without interruption. With fast withdrawal rates, even using the optimum interruption time, the limit is significantly below the maximum. With high mole fractions of O_2, use of very slow withdrawal rates may cause complete suppression of the explosion. The pattern is similar to that indicated by Egerton and Warren [24].

(iii) Similar behaviour to (ii) is found at both 460 and 440 °C, except that as the temperature is decreased, the necessary withdrawal rates decrease also, and the interruption times increase significantly. At a given

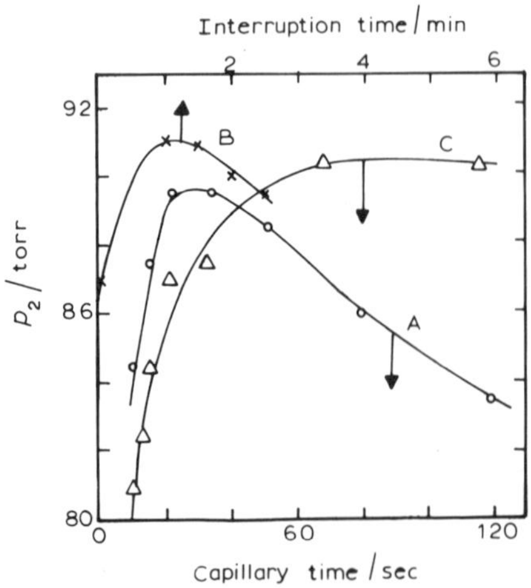

Fig. 17. Variation of second limit with withdrawal rate and interruption period. 36 mm diameter aged boric acid coated vessel at 480 °C. $x = 0.28$, $y = 0.025$ (after Baldwin and Doran [20]). A, Effect of withdrawal rate with interruption period of $2\frac{1}{2}$ min; B, effect of varying interruption period using optimum capillary from A; C, effect of withdrawal rate with no interruption. (By courtesy of the Faraday Society.)

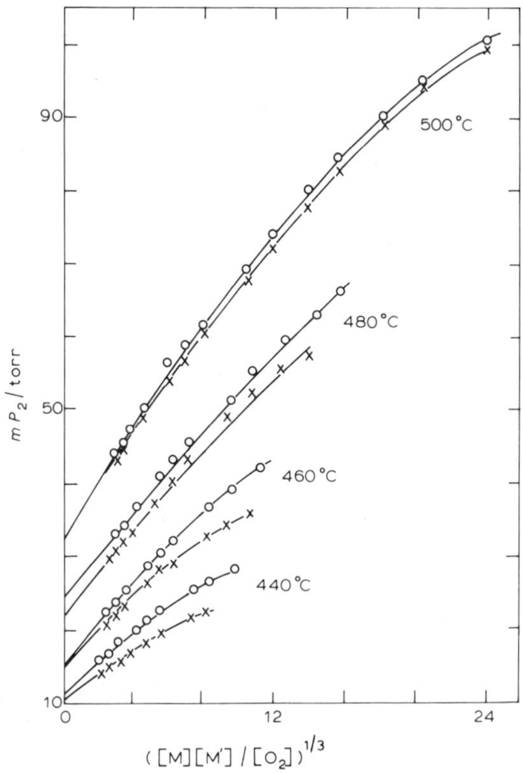

Fig. 18. Maximum second limits for fresh and aged boric acid coated vessels, 36 mm diameter. (after Baldwin and Doran [20]). x = 0.28 (constant), y variable. ○, fresh coating; ×, aged coating. (By courtesy of The Faraday Society.) For explanation of m, [M] and [M′], see eqn. (38) (p. 51).

temperature, the withdrawal rates involved become slower and the necessary interruption times become longer as y decreases. Thus, at 440 °C with $y = 0.025$, the maximum limit could only be obtained by combining the slowest withdrawal rate with an interruption period of 15 min.

Figure 18 shows the maximum limits obtained for both aged and fresh boric acid coated vessels over the temperature range 440—500 °C. The limits are always higher in the freshly coated vessel, and Fig. 18 shows that the discrepancy increases as the temperature decreases.

4.2 SLOW REACTION IN BORIC ACID COATED VESSELS

The slow reaction in aged boric acid coated vessels has been extensively studied by Baldwin and Mayor [45]. While studying the effect of withdrawal rate on the second limit, Baldwin and Mayor observed that in a freshly coated vessel at 500 °C and 500 torr pressure, the rate of reaction is quite small at first (ca. 2 torr in 8 min). As the vessel is used, this rate increases quite slowly over 10—14 days to a value of 6—8 torr in 8 min. Quite suddenly the rate then accelerates over a period of about one day to around 60 torr in 8 min. The rate of the new reaction is very reproducible, the maximum rate (obtained after an induction period) for a standard mixture varying only by ± 5 % over a period of one month. Different vessels of the same diameter also give similar reproducibility, which is also unaffected by leaving the vessel out of use in an evacuated condition, or filled with H_2, O_2, or water vapour, either at room temperature or at 500 °C.

The reaction is autocatalytic, resembling in this respect the situation already encountered with uncoated quartz or glass vessels. However, in contrast with the results of Lewis and von Elbe [23] for a quartz vessel, Baldwin and Mayor [45] found little or no effect of addition of up to 22 torr added water on either the induction period or the maximum rate, irrespective of whether the water was added a short time before, or together with, the reactants themselves. They concluded that the autocatalytic effect cannot be due to poisoning (by absorption of water vapour produced in the reaction) of the ability of the surface to destroy chain centres, as had previously been suggested.

The reaction in aged boric acid vessels shows no significant effect of vessel diameter, either on the maximum rate or the induction period, over the range 15, 24, 36 and 51 mm. For a constant total pressure of 500 torr, Fig. 19 shows the variation of the maximum rate R with (a) oxygen mole fraction y over the range 0.07—0.72, the H_2 mole fraction x being

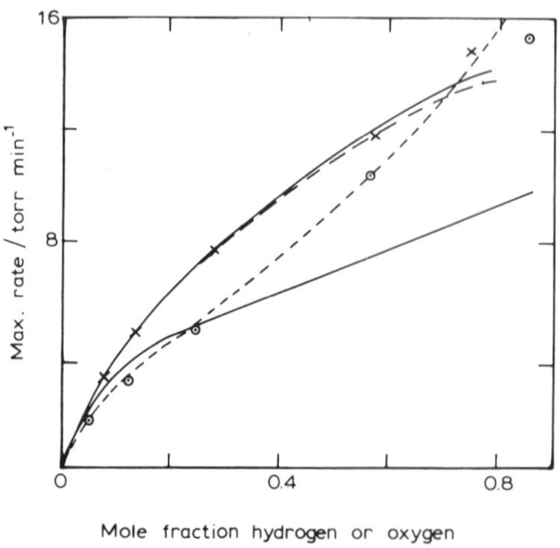

Fig. 19. Variation of maximum rate with mixture composition at a total pressure of 500 torr (after Baldwin and Mayor [45]). ×, x = 0.28, y variable; ○, y = 0.14, x variable; —, calculated excluding reaction (xi); - - -, calculated including reaction (xi). (By courtesy of The Faraday Society.)

constant at 0.28, and (b) hydrogen mole fraction x over the range 0.07—0.86, y being constant at 0.14. The variation of R with x shows a complex effect of H_2. Similar curves for this effect were found at $y = 0.56$; and at a total pressure of 250 torr.

The variation of R with total pressure for the standard mixture at 500 °C is given approximately by $R \propto P^{1.8}$. Addition of nitrogen causes an increase in rate, but the increase is substantially less than with salt coated vessels, e.g. 200 % addition of N_2 only increases the rate by about 50 %. The effect is least marked at low x. Over the temperature range 470—540 °C, the log R versus $1/T$ plot is closely linear, and gives an activation energy of 55.8 ± 0.7 kcal . mole^{-1}. All these properties contrast sharply with the behaviour in porcelain and salt coated vessels described earlier, with which, for example, the activation energy is 100 kcal . mole^{-1} or greater.

The induction period preceding the reaction (defined as the time to maximum rate) is little affected by oxygen mole fraction, total pressure, or inert gas. However, it decreases appreciably with increasing hydrogen mole fraction, and more markedly with increasing temperature. The log τ versus $1/T$ plot gives an activation energy of 59—73 kcal . mole^{-1} depending on the criterion adopted to define the induction period.

Attention has already been drawn to the presence of hydrogen peroxide in the products from the oxidation in Pyrex tubes and its absence for KCl coated tubes. The build up of hydrogen peroxide concentration during the slow reaction in boric acid coated vessels has been investigated by Baldwin et al. [45, 64], and is shown for one set of conditions in Fig. 20. The hydrogen peroxide concentration reaches a maximum at the same time as the reaction rate.

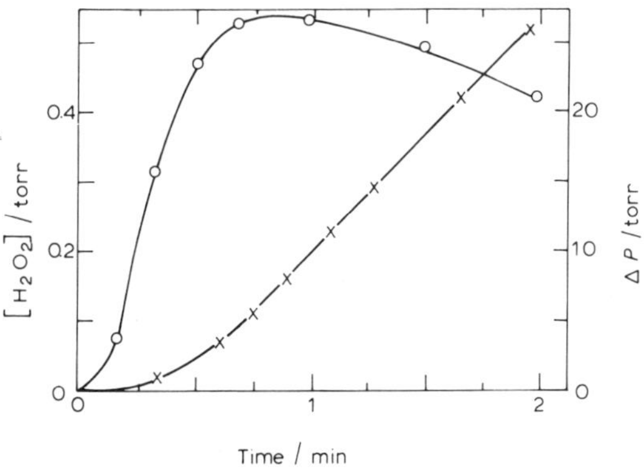

Fig. 20. Variation of pressure change and H_2O_2 concentration with time. 51 mm diam. aged boric acid coated vessel at 500 °C. p_{H_2} = 430 torr, p_{O_2} = 70 torr (after Baldwin et al. [64]). ○, H_2O_2 concentration; ×, pressure change. (By courtesy of The Faraday Society.)

4.3 FURTHER DEVELOPMENT OF THE REACTION MECHANISM

4.3.1 Slow reaction

On the basis of the reaction mechanism already developed in Sect. 3.6, the simplest explanation for the behaviour in boric acid vessels would be to assume a decrease in the surface destruction efficiency of HO_2 as the surface ages, with a consequent increase in the probability of reaction (xi) or (xia). However, Baldwin and Mayor [45] give several reasons why this should not be the sole explanation, among them (i) the contrast between the kinetic characteristics of the reactions in salt coated and aged boric acid vessels, and (ii) the similarity of the second limits in both fresh and aged vessels, which would not be expected if reaction (xi) were more prominent in one than in the other (in fact the slight change that does occur involves a small decrease in the limit as the vessel ages, and is in the wrong direction). The autocatalytic nature of the reaction indicates the formation of a relatively stable reaction intermediate, and this is almost certainly H_2O_2, as evidenced by Fig. 20. A number of further points then arise.

First, since competition between reactions (xi) and (v) or (vb) is excluded by the second limit behaviour referred to in Sect. 4.2 immediately above, the peroxide must be formed by mutual interaction of two HO_2 radicals either at the surface by reaction (va) or in the gas phase by reaction (x) below. Second, the autocatalytic nature of the reaction can only be attributed to the dissociation of H_2O_2 by reaction (vii) or the alternative (viia). Competition between the dissociation and a surface destruction of H_2O_2 would introduce a diameter dependence of the rate which is contrary to the results. A second function of the ageing of the surface therefore must be to eliminate surface destruction of the peroxide. Third, if H_2O_2 always dissociates by (vii) or (viia), the formation of HO_2 by reaction (iv) always leads to a chain propagating cycle

$$H \xrightarrow{\text{(iv)}} HO_2 \xrightarrow{\text{(va) or (x)}} \tfrac{1}{2}H_2O_2 \xrightarrow{\text{(vii)}} OH \xrightarrow{\text{(i)}} H$$

Superimposed on this will be chain branching due to reactions (ii) and (iii), and unless some form of chain termination is introduced the reaction will always be explosive. The termination must be gas phase in order to account for the absence of a diameter effect, and since the observed overall activation energy of 57 kcal . mole^{-1} is close to the expected activation energy of reaction (vii), the terminating reactions probably compete with (vii) for H_2O_2. Possible reactions are (vi), (xiii), (xiv) and (xv).

$$H_2O_2 = 2OH \qquad \text{(viia)}$$

$$HO_2 + HO_2 = H_2O_2 + O_2 \qquad \text{(x)}$$

$$O + H_2O_2 = H_2O + O_2 \qquad \text{(xiii)}$$

$$H + H_2O_2 \quad = H_2O + OH \qquad \text{(xiv)}$$

$$H + H_2O_2 \quad = HO_2 + H_2 \qquad \text{(xiva)}$$

$$OH + H_2O_2 = H_2O + HO_2 \qquad \text{(xv)}$$

The next stage of the treatment involves the derivation of expressions for the maximum reaction rate R for comparison with experiment. Simple analytical expressions can only be obtained if the termination reactions are considered singly in conjunction with reactions (i)—(iv), (va) or (x), and (vii). Using reaction (vi) as the terminating step, the rate expression is quite inconsistent with the experimental observations. This leads to the important conclusion, both for salt coated and boric acid coated vessels, that reaction (vi) is absent from the mechanism. The results of the treatment are consistent with chain termination by a mixture of reactions (xiv) and (xv), and it is possible to predict the effect of oxygen mole fraction on the rate completely in this way (full curve in Fig. 19). However, the effect of hydrogen mole fraction, again shown by the full curve, is not so well predicted, and neither are the effects of inert gas addition or total pressure. The predictions can be brought well into line with observation by inclusion of the regeneration reaction (xi) into the mechanism, as indicated by the dotted curves in Fig. 19. Using the reaction (xiva), the ratios of rate coefficients used to derive the dotted curves (in torr min units) were $k_1k_{14a}/k_4k_{15} = 0.14$, $k_1k_2k_7/k_4k_{15} = 390$, and $k_{11}/k_5 = 10^{-4}$.

Following on this analysis, two further points now become apparent. First, a comparison of the full and dotted curves in Fig. 19 shows that when $x = 0.86$ the effect of reaction (xi) almost doubles the rate. A marked diameter effect on the rate should thus be observed if (va) is the other reaction producing H_2O_2. No trace of this is observed experimentally. There is therefore strong evidence that the formation of H_2O_2 from HO_2 occurs by the gas phase reaction (x) rather than at the vessel surface. Reaction (va) is therefore excluded as a major step. Secondly, the effect of inert gas on the rate provides strong evidence for reaction (vii) rather than (viia). Since all the reactions are gas phase, the influence of inert gas cannot be in preventing diffusion to the surface. Further, since an increase in the concentration of inert gas would reduce the rate of the branching reaction (ii) relative to the propagating step (iv), the acceleration cannot be interpreted in terms of an effect on reaction (iv). The only alternative is an increase in the rate of dissociation of H_2O_2 by the bimolecular reaction (vii), and this is borne out by the quantitative treatment. The major steps responsible for the slow reaction thus appear to be (i)—(iv), (vii), (x), (xi), (xiv) and (xv).

4.3.2 Second limits

The boric acid type of second limit behaviour was shown by Egerton and Warren [24] to be obtained by introducing a quadratic branching step into the mechanism. They proposed reaction (viii), viz.

$$H + HO_2 = OH + OH \tag{viii}$$

However, under quadratic branching conditions the steady state concentration of chain centres is given by

$$dn/dt = n_0 + \phi n + Fn^2 = 0 \tag{32}$$

and this can only occur when $\phi^2 \geqslant 4n_0F$. The explosion condition

$$\phi^2 = 4n_0F \tag{33}$$

thus depends on the initiation rate n_0. The initiating mechanism suggested by Egerton and Warren [24], consisting of reactions (va), (vi) and (viia), leads to an equation of the same form as the observed explosion condition, but it encounters difficulty when the values of K, eqn. (31), for boric acid and KCl coated vessels are compared. Experimentally these values are almost the same at the same temperature, but the mechanism predicts that the boric acid value should be 3/2 times the other. While it is possible to overcome this difficulty, others have now arisen inasmuch as the slow reaction studies have virtually excluded reactions (vi) and (viia). The bimolecular nature of the dissociation of H_2O_2 is supported by other more recent work [65—67]. An alternative mechanism suggested by Dixon-Lewis et al. [63] involved the occurrence of reaction (xi) on the surface, and used reaction (vii) rather than (viia). Although it satisfied the criterion of predicting the same values of K in both B_2O_3 and KCl coated vessels, the difficulty regarding the inclusion of reaction (vi) still remained.

An alternative approach to the second limit mechanism in boric acid coated vessels [62] is to proceed from the slow reaction mechanism developed in the preceding section, reactions (i)—(iv), (vii), (x), (xiva) and (xv). Adding reaction (viii) to these, and omitting the minor termination reaction (xv) at the low values of y, the stationary HO_2 concentration is given by

$$k_8 n^3 - \phi n^2 - an + ab = 0 \tag{34}$$

$$\left.\begin{aligned} \phi &= k_4[O_2][M] - k_2[O_2] \\ a &= k_2k_7k_8[O_2][M']/k_{10}k_{14a} \\ b &= (k_4[O_2][M] + k_2[O_2])/k_8 \end{aligned}\right\} \tag{35}$$

The limiting condition for real solutions to this cubic equation is

$$81k_8^2a^2b^2(1 - \phi/9k_8b)^2 = 12ab\phi^3(1 + 3ak_8/\phi^2)(1 + a/3b\phi) \tag{36}$$

If the terms in brackets can be approximated to unity, this condition becomes

$$\phi^3 = 27k_8^2ab/4 \tag{37}$$

Or

$$[M] = k_2/k_4 + \left(\frac{27k_2k_7k_8^2[M']([M] + k_2/k_4)}{4k_4^2k_{10}k_{14a}[O_2]}\right)^{1/3} \qquad (38)$$

where

$$[M] \equiv [H_2] + k_{O_2}[O_2] + k_{N_2}[N_2] \equiv mP_2$$

This expression fits the experimental results at least as well as eqn. (31) derived from the Egerton-Warren approach. However, the plots of [M] versus $\{[M']([M] + k_2/k_4)/[O_2]\}^{1/3}$ give intercepts close to the value expected (from KCl second limits) for $2k_2/k_4$, whereas eqn. (38) predicts an intercept of k_2/k_4. This difficulty can be overcome by replacing reaction (xiva) by (xiv), when

$$\left.\begin{aligned} \phi &= k_4[O_2][M] - 2k_2[O_2] \\ a &= k_2k_7k_8[O_2][M']/k_{10}k_{14} \\ b &= k_4[O_2][M]/k_8 \end{aligned}\right\} \qquad (39)$$

and the explosion condition becomes

$$[M] = 2k_2/k_4 + B([M][M']/[O_2])^{1/3} \qquad (40)$$

where

$$B = (27k_2k_7k_8^2/4k_{10}k_{14}k_4^2)^{1/3} \qquad (41)$$

A plot of [M] against $([M][M']/[O_2])^{1/3}$ should now give a straight line. On testing their results in this way, Baldwin et al. [62] found that eqn. (40) did not give an entirely satisfactory interpretation consistent with the precision of the results. It was found, however, that an almost precise interpretation could be obtained by re-introducing reaction (xv) into the mechanism, and at the same time making a more rigorous approximation in proceeding from eqn. (37).

Before going on to consider the small differences between fresh and aged boric acid surfaces at the second limit, it is worthwhile to pause at this stage to examine the compatibility of the slow reaction and second limit mechanisms as so far developed. Essentially, three changes have been introduced in considering the second limit behaviour:

(i) Reaction (xiva) $H + H_2O_2 = HO_2 + H_2$ is replaced by reaction (xiv) $H + H_2O_2 = H_2O + OH$. A re-examination of the slow reaction rates by Baldwin and Mayor [45] showed that this substitution did not much affect the prediction of the effect of mixture composition on the rates, but gave an improved prediction of inert gas effects. The slow reaction studies thus provide some support for (xiv), and there is convincing overall evidence that (xiva) is either absent or much less frequent than (xiv). This conclusion is supported by studies of the hydrogen sensitized decomposition of hydrogen peroxide [68—70], from which a ratio $k_{14}/k_{14a} \simeq 8$ is deduced.

(ii) Reaction (xi) $HO_2 + H_2 = H_2O_2 + H$ (or (xia) $HO_2 + H_2 = H_2O + OH$) plays an important part in the slow reaction at 500 torr, but does not contribute to the second limit. Again, a detailed analysis by Baldwin and Mayor [45] shows that this situation is possible, but only if reaction (xi) is used, and not (xia). This distinction is discussed again in more quantitative terms later.

(iii) Reaction (viii) $H + HO_2 = OH + OH$ is essential for the interpretation of the second limit, but does not contribute appreciably to the slow reaction at 500 torr. This situation can be justified in qualitative terms if H_2O_2 is formed from HO_2 via the gas phase reaction (x), since this process will be favoured relative to (viii) as the HO_2 concentration increases at the higher pressures. Quantitatively, Baldwin and Mayor [45] have been able to show that at 500 torr and 500 °C, reaction (viii) cannot increase the rate by more than a few per cent, and the conclusion is supported by the detailed numerical studies to be discussed later.

The role of hydrogen peroxide at the second limit is of some interest because of the inclusion of the initiation rate in the limit condition with quadratic branching (cf. eqn. (38)). Thus, the rise in limit with initial increase in manipulation time, shown in Fig. 17, is most marked at low values of the oxygen mole fraction y, where the quadratic branching effect is most important. The increase is almost certainly associated with the build-up of H_2O_2. In support of this, the increase in optimum interruption time as the temperature falls (about 1 min at 480 °C, 4 min at 460 °C, and 15 min at 440 °C) corresponds with an activation energy of 70 kcal . mole^{-1}, a value similar to those of 57 kcal . mole^{-1} for the slow reaction, 59—73 kcal . mole^{-1} for the induction period preceding the slow reaction, and 46—50 kcal . mole^{-1} for the dissociation of H_2O_2 [67].

Since water is much more efficient than either H_2, N_2 or O_2 as a third body in reaction (iv) (see Table 7), the simplest interpretation of the suppression of the limits for manipulation times greater than the optimum is that it is associated with water formation by the slow reaction as the limit is approached. This interpretation is supported by the fairly successful calculation [21] of critical withdrawal rates at 500 °C using rate coefficients derived from the slow reaction studies at 500 torr. At 500 °C this depression is the only effect observable, and there is never any rise in limit above that at fast withdrawal rates. At this temperature therefore, the quadratic branching is fully developed.

Even with aged vessels at 440 °C, and using fast withdrawal rates, Baldwin and Doran [20] found some quadratic branching effects to be still present, and these became more marked in freshly coated vessels. It seems therefore that some H_2O_2 is present in both cases. However, at 440 °C an interruption period of 15 min is required to give the maximum limits with fast withdrawal rates, and even these limits are some 2 torr lower than can be obtained using slow withdrawal rates with the same interruption period. It is probable therefore that the limits with fast withdrawal at 440 °C are limits in the absence of H_2O_2 formed by the gas phase mechanism. For quadratic branching to occur, however, some initiating process must be present, and it appears that surface initiation must be assumed. Since the quadratic branching on fresh surfaces is significantly higher than on aged surfaces at 440 °C, the surface initiation must be greater in fresh vessels than in aged vessels. Further, the surface initiation is likely to have a lower activation energy than the homogeneous dissociation of H_2O_2, so that its importance will decrease at higher temperatures. In aged vessels at 500 °C it has become insignificant, giving limits independent of vessel diameter. With fresh surfaces some small but significant effect remains, giving slightly higher limits than in aged vessels and a small increase in limit with decrease in vessel diameter, as is observed.

To conclude the discussion of the role of H_2O_2 at the second limit, it is interesting to note that Forst and Giguère [71] find that H_2O_2 inhibits the limit at 447 °C in clean Pyrex vessels. At first sight this appears to contradict the conclusions already reached, particularly since there is no obvious terminating step which added H_2O_2 introduces into the mechanism. However, using reactions (i)—(iv), (vii), (viii), (x), (xiv) and (xv) and writing stationary concentrations for H, OH, O and HO_2 at an arbitrary concentration of H_2O_2, the concentration of HO_2 radicals is given [20] by

$$an^3 - bn^2 - cn + d = 0 \tag{42}$$

where

$$n = k_8[HO_2]/k_4[O_2]$$

$$a = 1 - R_{OH}$$

$$b = M - M_0 + R_{OH}\left(M + \frac{R_H M_0}{2}\right)$$

$$c = AM'R_H(1 - R_{OH})/[O_2]$$

$$d = AM'R_H\left\{M + R_{OH}\left(M + \frac{R_H M_0}{2}\right)\right\}\Big/[O_2]$$

and

$$R_H = k_{14}[H_2O_2]/k_2[O_2]$$

$$R_{OH} = k_{15}[H_2O_2]/k_1[H_2]$$

$$M_0 = 2k_2/k_4$$

$$A = k_2k_7k_8^2/k_4^2k_{10}k_{14}$$

If cn can be neglected, the explosion condition becomes (cf. eqns. (34)–(37))

$$M = M_0 - R_{OH}\ M + \frac{R_H M_0}{2} + \left[\frac{27(1-R_{OH})^2 AM'R_H \left\{ M + R_{OH}\left(M + \frac{R_H M_0}{2} \right) \right\}}{4[O_2]} \right]^{1/3} \quad (43)$$

Since R_H and R_{OH} are proportional to $[H_2O_2]$, the negative term is proportional to $[H_2O_2]^{1-2}$, while the positive term is proportional to $[H_2O_2]^{1/3-1}$. Thus at sufficiently low concentrations of H_2O_2 the limit will be raised, passing through a maximum and then decreasing at higher concentrations of H_2O_2. Using rate coefficients at 440 °C which are consistent with those quoted in the following section, namely k_{14}/k_2 = 430, k_{15}/k_1 = 5.5, k_2/k_4 = 6 and A = 0.864 in torr units (H_2 = 1), calculated quantitative effects of H_2O_2 are shown in Fig. 21. Curves A, B and C include quadratic branching in the mechanism as above, and the steady state peroxide concentrations at the uninhibited limits with fully developed quadratic branching are shown by the vertical arrows. Curve D shows the inhibition in the absence of quadratic branching. As expected, the effects of reaction (viii) are particularly prominent at low oxygen

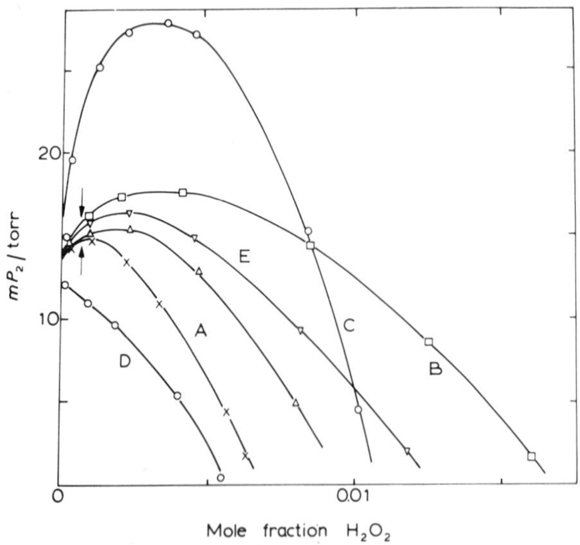

Fig. 21. Calculated effect of H_2O_2 on second limit at 440 °C (after Baldwin and Doran [20]). A, $x = 0.10$, $y = 0.44$; B, $x = 0.40$, $y = 0.44$; C, $x = 0.40$, $y = 0.10$; D, $x = 0.10$, $y = 0.44$ no quadratic branching; E, $x = 0.40$, $y = 0.44$, $k_{14a}/k_{14} = 0.1$; F, $x = 0.40$, $y = 0.44$, $k_{14a}/k_{14} = 0.2$. (By courtesy of The Faraday Society.)

mole fraction. However, they are not inappreciable for curves A and B either.

4.3.3 Quantitative treatment of limits, rates and induction periods

Recapitulating for convenience, the complete mechanism developed to account for the kinetic features of the $H_2 + O_2$ reaction in boric acid coated vessels is

$$OH + H_2 = H_2O + H \tag{i}$$
$$H + O_2 = OH + O \tag{ii}$$
$$O + H_2 = OH + H \tag{iii}$$
$$H + O_2 + M = HO_2 + M \tag{iv}$$
$$H_2O_2 + M' = OH + OH + M' \tag{vii}$$
$$H + HO_2 = OH + OH \tag{viii}$$
$$HO_2 + HO_2 = H_2O_2 + O_2 \tag{x}$$
$$HO_2 + H_2 = H_2O_2 + H \tag{xi}$$
$$O + H_2O_2 = H_2O + O_2 \tag{xiii}$$
$$H + H_2O_2 = H_2O + OH \tag{xiv}$$
$$H + H_2O_2 = HO_2 + H_2 \tag{xiva}$$
$$OH + H_2O_2 = H_2O + HO_2 \tag{xv}$$

The reaction rate is effectively controlled by the rate of dissociation of H_2O_2, and the induction period is determined by the rate of build-up of this species. Since H_2O_2 is the least reactive chain centre the partial stationary state procedure of Semenov [60] may be used, in which a differential equation is set up for the H_2O_2 concentration, and stationary state equations for the other species. Thus

H atoms

$$\begin{aligned} k_1[OH][H_2] + k_3[O][H_2] + k_{11}[HO_2][H_2] &= k_4[H][O_2][M] \\ &+ k_2[H][O_2] + k_8[H][HO_2] + k_{14}[H][H_2O_2] \\ &+ k_{14a}[H][H_2O_2] \end{aligned} \tag{44}$$

O atoms

$$k_2[H][O_2] = k_3[O][H_2] + k_{13}[O][H_2O_2] \tag{45}$$

OH radicals

$$\begin{aligned} k_2[H][O_2] + k_3[O][H_2] + 2k_7[H_2O_2][M'] + 2k_8[H][HO_2] \\ + k_{14}[H][H_2O_2] = k_1[OH][H_2] + k_{15}[OH][H_2O_2] \end{aligned} \tag{46}$$

HO_2 radicals

$$k_4[H][O_2][M] + k_{14a}[H][H_2O_2] + k_{15}[OH][H_2O_2]$$
$$= 2k_{10}[HO_2]^2 + k_8[H][HO_2] + k_{11}[HO_2][H_2] \tag{47}$$

H_2O_2

$$d[H_2O_2]/dt = \theta + k_{10}[HO_2]^2 + k_{11}[HO_2][H_2]$$
$$-k_7[H_2O_2][M'] - k_{14}[H][H_2O_2] - k_{14a}[H][H_2O_2]$$
$$-k_{15}[OH][H_2O_2] - k_{13}[O][H_2O_2] \tag{48}$$

where θ is the rate of primary initiation, assumed to produce H_2O_2. If required, the rate of formation of water is given by

$$d[H_2O]/dt = k_1[OH][H_2] + k_{13}[O][H_2O_2] + k_{14}[H][H_2O_2]$$
$$+ k_{15}[OH][H_2O_2] \tag{49}$$

The solution of eqns. (44)—(48) is not straightforward [72]. After a preliminary reduction by linear algebra, the problem resolves itself into a numerical one of calculating values of $d[H_2O_2]/dt$ corresponding to given $[H_2O_2]$ and mixture composition; and this in turn involves solution of a cubic equation in $G = k_{10}^{1/2}[HO_2]$. The solution shows the following parameters to be behaviour determining

$$R_1 = \theta \qquad R_2 = k_7 \qquad R_3 = k_2/k_4$$
$$R_4 = k_{14}/k_2 \qquad R_5 = k_{15}/k_1 \qquad R_6 = k_{13}/k_3$$
$$R_7 = k_{11}/k_{10}^{1/2} \qquad R_8 = k_8/k_2k_{10}^{1/2} \qquad R_9 = k_{14a}/k_2$$

Clearly, the complexity of the system of eqns. (44)—(48) is such that although the earlier mathematical analyses of Baldwin et al. [45, 62] were able to provide strong evidence for the reaction mechanism, the quantitative application nevertheless suffered some limitations. These limitations have been largely removed by a later computer treatment, which optimized the set of ratios R_1—R_9 so as to give the best simultaneous prediction of the induction periods, the maximum reaction rates and the second limits over a wide range of conditions at a given temperature. The sensitivity of the three measurable quantities to the various ratios was first investigated. With R_2—R_9 set close to their final values at 500 °C, the effect of varying each in turn is shown in Table 17. P_2, τ and rate of reaction was found to be sensitive to R_2 and R_3; while in addition the induction periods were sensitive to R_7 and to a lesser extent R_4, the second limits to R_8, and the slow reaction rates to R_4, R_7 and, at low $[H_2]/[O_2]$ ratios, R_5. The primary initiation rate R_1 may affect the induction period calculation chiefly: its optimum value is around 10^{-4} mole . l^{-1} . s^{-1} at 500 °C, but the sensitivity is not high.

The optimization process was made more realistic by two types of independent measurement which accurately fix R_2, R_3, R_9/R_4, and to

TABLE 17
Sensitivity of second limit, induction period and reaction rate to parameters R_2 to R_9 [72]

Effect of 10 % increase in	Second limit		Induction period	Reaction rate		
	A	B	C	D	E	F
$R_2 = k_7$	+1.1	−4.1	−6.4	+8.2	+11.8	+10.0
$R_3 = k_2/k_4$	+10.0	−27.8	−5.3	+0.0[a]	−2.6	+12.4
$R_4 = k_{14}/k_2$	−0.5	+2.3	−2.3	−5.7	—	—
$R_5 = k_{15}/k_1$	−0.1	+0.3	−0.3	−1.0	—	—
$R_7 = k_{11}/k_{10}^{1/2}$	+0.2	−1.1	−5.7	+3.7	+6.8	−1.1
$R_8 = k_8/k_2k_{10}^{1/2}$	+2.5	—	−0.1	+0.4	+0.2	+0.2
$R_9 = k_{14a}/k_2$	−0.5	+1.7	−0.3	−1.0	−2.0	0.0

[a] There is no increase in rate for the standard mixture, but for most mixtures there is a small increase in rate, usually 1—2 %.
A, % increase in limit for standard mixture (x = 0.28, y = 0.14).
B, % increase in optimum value of $k_8/k_2k_{10}^{1/2}$.
C, % increase in induction period for standard mixture.
D, % increase in reaction rate for standard mixture.
E, % increase in optimum value of k_{14}/k_2.
F, % increase in optimum value of k_{15}/k_1.

some extent R_5. First, the parameter R_3 was obtained from measurements (previously discussed) of the second limit in KCl- and other salt-coated vessels, correction being made if necessary for the occurrence of reaction (xi) and for surface termination of H atoms (cf. Sect. 3.6.4 and Table 14). Secondly, the parameter R_2 at the temperatures of interest may be accurately determined from independent studies of the decomposition of H_2O_2 in the presence of N_2 and H_2 over the temperature range 440—560 °C [67]. Here the sequence of reactions (vii), (i) and (xiv) gives rise to a chain decomposition of H_2O_2, the initiation rate being that of reaction (vii) and leading to a value for R_2. At high H_2 concentrations the chain length is determined by a competition between reactions (xiv) and (xiva), with the latter reaction terminating the chain. From the chain length under these conditions, R_9/R_4 = 0.143 ± 0.015 at 440—500 °C [68, 69]. Similarly at low H_2 concentrations the chain terminating step (xv) may clearly compete with reaction (i). Assuming no formation of O atoms by reaction (xvi)

$$OH + OH = O + H_2O \qquad \text{(xvi)}$$

and subsequent termination by (xiii) under these conditions, the ratio $R_5 = k_{15}/k_1$ = 5.0 ± 1.0 was found [68, 69], again with no significant temperature variation between 440 and 500 °C.

These independent measurements considerably reduce the number of adjustable parameters in the main optimization process. For the scheme given so far, optimization of the second limits gives R_8 as a single adjustable parameter, while the induction periods give R_7, and the slow reaction rates give R_4 and R_5 similarly. For the temperature range 460—530 °C, the R values so obtained are given in Table 18, while comparisons of observed and calculated induction periods (defined as the time to half maximum rate), maximum reaction rates and second limits are shown in Tables 19, 20 and 21, respectively. The fact that such good agreement is obtained at 470—500 °C over such a wide composition range confirms the validity of the treatments. Outside this temperature range a number of experimental difficulties combine to make the treatment less satisfactory, so that many, and at 460 °C all, of the parameters used in the computation of the induction periods were estimated by extrapolation. The larger r.m.s. deviations at the ends of the temperature range may be due at 460 °C to surface destruction of H_2O_2, since k_7 decreases by a factor of 5 between 500 and 460 °C. At the higher temperatures (520 and 530 °C) and at the highest reaction rates, self-heating effects at the maximum rate may give too long an apparent induction period. Allowance for self heating effects at 500 °C, together with allowances for the pressure change accompanying H_2O_2 formation, lead to the ratios given in column B of Table 18 [73].

Three further points are worthy of mention.

(i) The parameter $R_6 = k_{13}/k_3$ was arbitrarily set equal to zero in the original computations [72], and led to the combined effect of reactions (xiii) and (xv) being included in R_5. Some evidence for this came from the independent value of $R_5 = 5.0 \pm 1.0$, quoted above, from the sensitized H_2O_2 decomposition studies. The rate coefficient k_{13} has, however, recently been estimated by Albers et al. [74] to be 2.8×10^{10} $\exp(-3{,}200/T)$, leading to $k_{13}/k_3 = 12.0$ at 500 °C.

(ii) There is considerable evidence from flame studies that reaction (viii) is not the only reaction which may occur between H and HO_2. Of the alternative possibilities

$$H + HO_2 = H_2 + O_2 \tag{xx}$$

and

$$H + HO_2 = O + H_2O \tag{viiia}$$

reaction (viiia) is kinetically equivalent to (viii) in the present context. Reaction (xx), on the other hand, is a recombination step. Recent work [73] has shown that the interpretation of the second limit is improved by including reaction (xx), with consequent revision also of k_8.

For $k_{13} = 0$, the optimum values of the ratios involving k_8 at 500 °C were $k_{20}/k_8 = 0.14$ and $k_8/k_2 k_{10}^{1/2} = 0.498$. For $k_{13}/k_3 = 12.0$, a complete optimization at 500 °C leads to $k_{20}/k_8 = 0.17$, together with the values of R_4, R_5, R_7, R_8 and R_9 given in column C of Table 18.

TABLE 18

Optimized ratios of rate coefficients (l.mole.sec units; M = H_2 in reactions (iv) and (vii)) [72, 73]

Temp (°C)	460	470	480	500			520	530
				A	B[a]	C[a]		
$R_1 = \theta$	6.6×10^{-8}	6.5×10^{-8}	8.9×10^{-8}	1.2×10^{-7}	(1.2×10^{-7})	(1.2×10^{-7})	2.4×10^{-7}	2.4×10^{-7}
$R_2 = k_7$	7.2	11.2	17.1	38.6	(38.6)	(38.6)	83.5	121.0
$R_3 = k_2/k_4$	1.97×10^{-4}	2.35×10^{-4}	2.77×10^{-4}	3.84×10^{-4}	(3.84×10^{-4})	(3.84×10^{-4})	5.26×10^{-4}	6.09×10^{-4}
$R_4 = k_{14}/k_2$	330	306	281	270	(249)	(236)	230	221
$R_5 = k_{15}/k_1$	6.2	6.0	5.7	4.7	(5.1)	(3.7)	6.2	5.2
$R_6 = k_{13}/k_3$	0	0	0	0	(0)	(12.0)	0	0
$R_7 = k_{11}/k_{10}^{1/2}$	1.38×10^{-2}	1.78×10^{-2}	2.13×10^{-2}	3.37×10^{-2}	(3.03×10^{-2})	(3.03×10^{-2})	5.02×10^{-2}	6.09×10^{-2}
$R_8 = (k_8 + k_{8a})/k_2 k_{10}^{1/2}$	0.797	0.720	0.593	0.367	(0.498)	(0.572)	0.279	0.208
$R_9 = k_{14a}/k_2$	52	49	46	43	(39)	(39)	37	35
$R_{10} = k_{20}/(k_8 + k_{8a})$	0	0	0	0	(0.14)	(0.17)	0	0

[a] Columns B and C incorporate further refinements of treatment compared with ref. 72 (see text).

TABLE 19
Observed and calculated induction periods (sec) at 460—530 °C [72]

H_2 (torr)	O_2 (torr)	Temperature (°C) 460		470		480		500		520		530	
		Obs.	Calc.	Obs.	Calc.	Obs.	Calc.	Obs.	Calc.	Obs.	Calc.	Obs.	Calc.
140	360	582	448	288	275	181	176	72.5	69.6	—	—	—	—
	280	565	432	260	266	175	171	69.5	67.2	—	—	—	—
	140	452	388	227	239	151	153	60.5	59.3	27.2	22.8	18.2	14.4
	70	397	334	196	205	125	131	50.0	50.2	22.9	19.6	15.2	12.4
	35	276	276	157	169	104	108	38.0	41.3	19.7	16.3	13.2	10.4
430	70	195	152	91	93	58.5	60	23.5	23.6	10.6	9.8	7.4	6.4
280		250	211	131	129	82	83	32.0	32.2	15.2	13.1	10.2	8.4
70		628	487	310	301	179	190	72.0	72.5	31.5	27.0	21.2	16.8
35		815	650	410	407	239	256	93.0	96.6	42.4	34.3	26.9	21.2
35	140	978	697	450	437	288	274	101.5	104.1	43.1	36.7	29.1	22.6
35	280	1000	737	461	461	294	289	109.5	109.6	47.6	38.1	30.7	23.4
70		696	586	380	362	244	229	90.0	88.4	39.2	32.2	26.6	19.9
220		480	340	216	210	148	135	57.5	53.9	—	—	—	—
% r.m.s. deviation		27.9		3.8		4.8		3.8		16.1		21.8	

TABLE 20
Observed and calculated rates (torr min^{-1}) [72]

H_2 (torr)	O_2 (torr)	Temperature (°C) 500 R(obs.)	500 R(calc.)	480 R(obs.)	480 R(calc.)	470 R(obs.)	470 R(calc.)
140	360	17.0	15.8	5.35	4.88	3.26	2.96
	280	15.2	14.0	4.94	4.58	2.99	2.80
	140	9.72	9.44	3.62	3.52	2.29	2.21
	70	6.07	5.99	2.43	2.42	1.56	1.55
	35	3.77	3.70	1.57	1.54	1.00	1.00
430	70	15.7	16.3	7.01	7.51	4.77	5.09
280		10.3	10.5	4.53	4.64	3.06	3.08
70		3.90	3.97	1.41	1.42	0.84	0.87
35		2.74	2.81	0.83	0.87	0.48	0.51
35	140	3.82	4.07	1.06	1.09	0.59	0.63
35	280	5.13	5.35	1.24	1.25	0.68	0.71
70		8.86	8.64	2.26	2.35	1.43	1.38
220		21.9	20.0	7.80	7.44	5.14	4.69
% r.m.s. deviation		4.8		4.7		5.3	

TABLE 21
Observed and calculated second limits (torr) [72]

Mole fractions		500 °C		480 °C	
H_2	O_2	*P*(obs.)	*P*(calc.)	*P*(obs.)	*P*(calc.)
0.28	0.72	82.0	80.7	57.0	56.8
	0.56	83.0	81.4	57.0	57.4
	0.42	84.5	82.8	58.0	58.6
	0.28	86.5	85.7	59.5	60.8
	0.14	93.5	92.6	64.5	66.0
	0.10	97.5	97.0	71.5	69.2
	0.07	104.0	102.3	74.0	73.2
	0.035	116.5	115.6	84.0	82.9
	0.025	123.0	123.7	89.5	88.9
	0.0175	132.0	133.8	95.0	96.3
	0.0125	140.5	145.1	—	—
% r.m.s. deviation		1.6		1.6	

(iii) The computer treatment has also led to a reconsideration of the distinction between reactions (xi) and (xia), already briefly mentioned in Sect. 4.3.2. It was found that either of these reactions provides an almost equally acceptable interpretation of the induction periods and maximum rates. However, the earlier mathematical treatment due to Baldwin and Mayor [45] showed that the inclusion of reaction (xi) should lead to a marked variation of H_2O_2 concentration with changing initial H_2 concentration, whereas little variation would be expected with reaction (xia). Careful measurement of the H_2O_2 yields from a number of compositions at 500 °C [64] led to the results given in Table 22. This provides decisive evidence in favour of reaction (xi) as the controlling step in the $H_2 + O_2$ reaction. However, since the value of k_{11a} required for the interpretation of the induction periods is almost 10 times that for k_{11}, it is not possible to exclude entirely the possibility that from the point of view of HO_2 consumption reactions (xi) and (xia) are of equal importance.

To conclude this section, the treatment outlined gives a remarkably good account of the experimental observations over a wide range of H_2 + N_2 + O_2 compositions at 470—500 °C, and to a slightly lesser extent at somewhat higher and lower temperatures. The agreement between 470 and 500 °C (r.m.s. deviation < 5 %) is such as to generate considerable confidence in the validity of the treatment and in the rate coefficient ratios given in Table 18.

TABLE 22

Observed and calculated concentrations of hydrogen peroxide at 500 °C in 51 mm diameter aged boric acid coated vessel [64] (All concentrations in torr)

H_2	O_2	N_2	Calc.		Obs. (H_2O_2)
			$(H_2O_2)_{11}$	$(H_2O_2)_{11a}$	
140	360	0	0.797	0.534	0.722
140	280	80	0.709	0.471	0.636
140	140	220	0.480	0.304	0.454
140	70	290	0.297	0.176	0.271
140	35	325	0.174	0.094	0.163
430	70	0	0.533	0.177	0.511
280	70	150	0.417	0.179	0.378
70	70	360	0.222	0.163	0.197
35	70	395	0.166	0.139	0.145
220	280	0	0.922	0.517	0.851
70	280	150	0.477	0.374	0.439
35	280	185	0.308	0.267	0.271

5. Studies of the reaction in shock tubes and flames

The kinetic investigations of the hydrogen—oxygen reaction so far described have mostly involved gases reacting more or less homogeneously in static systems. These have been studies of the positions of the explosion limits, and time-resolved studies in the slow reaction region. Inside the explosion region the reaction times are by definition much shorter, and the Russian induction period measurements at pressures just above the first limit, again in a more or less homogeneous static system (p. 37), represent the only early attempts at studying the reaction in this area. More extended investigations at higher temperatures in the explosion region (to the right of the junction of the second and third limits in Fig. 1b) have had to await the development of techniques for the study of such fast reactions. A major objective here must be either (i) to contrive a precise time origin in relation to the total reaction time at the high temperature, i.e. extremely rapid heating, or (ii) to follow the history of the reaction during the heating period. The first approach is used in shock tube studies, and the second is realized in studies of flame systems. Given the reaction mechanism already developed, studies using these techniques have been most fruitful in providing further information about the elementary processes.

5.1 BACKGROUND OF SHOCK TUBE STUDIES

The techniques involved in the use of shock waves for the study of chemical reactions have been described by Bradley [75], by Gaydon and Hurle [76], and by Greene and Toennies [77]; and their application to the hydrogen—oxygen system has recently been reviewed by Schott and Getzinger [78]. Here the initial heating occurs in times much less than microseconds, and the ensuing reaction is studied in the flowing shocked gases as they pass an observation station. Measurements of the shock velocity serve to relate the immediate post-shock temperature and pressure with the pre-shock conditions, and to relate particle time in the shocked gas with measured laboratory time. To avoid complication due to the thermochemical effects of the reaction itself, the reactants are normally heavily diluted with an inert gas such as argon. Thus the reaction is again studied under essentially (though not always precisely) isothermal conditions. In this connection Mirels [79], and later Belles and Brabbs [80], have drawn attention to the effects of boundary layer growth in the flowing gases behind incident shocks. The development of the boundary layer progressively reduces the effective flow velocity behind the shock front, and so causes progessive increases in gas temperature, density and residence time compared with the uniform flow situation. Most of the earlier derivations (pre-1970) of reaction rate coefficients from shock tube results assumed uniform flow, and did not include corrections for these effects. Such corrections may be considerable [80], particularly for processes with high activation energies, leading to high apparent values for the reaction rates.

In hydrogen—oxygen mixtures the development of the reaction with time has very often been followed spectroscopically using absorption by the hydroxyl radical [78, 81], and less frequently by studying OH in emission [82]. Other quantitative spectroscopic techniques have used absorption by H or O atoms [83, 84] and IR emission from water vapour [85—87], or have measured emission intensities on addition of small amounts of indicators such as carbon monoxide [80, 88—92]. Interferometry [93—96] and schlieren techniques [97—99] have also been used to follow the reaction, but high dilution with inert gas diminishes the sensitivity of these methods. Chemical reaction times which can most conveniently be studied by shock tube methods are of the order of 10^{-5}–10^{-3} sec, following the much more rapid passage of the shock front and subsequent thermal relaxation of the shocked gases. Here it should be noted that vibrational relaxation times may not be absolutely negligible in the context of the early part of the reaction, particularly for oxygen [100]. However, Belles and Lauver [101] and Asaba et al. [81] considered in some detail the effect of slow O_2 relaxation on H_2—O_2 ignition, and concluded that it could only be small.

The reaction following the passage of a shock front through a mixture containing hydrogen and oxygen shows an initial induction period during which there is an exponential growth of both intermediate and product concentrations. The reaction rate and the intermediate concentrations continue to rise until they are limited by consumption of reactants. Following this, a gradual decay of intermediate concentrations, e.g. OH, towards their final equilibrium values may be observed. The last two of these phases may be observed also in more detail in flame systems: they will be discussed in Sect. 5.4. During the early stages of the reaction the conditions in shock tubes are much less complicated than later, and in recent years studies of the initial acceleration of the rates in shocked gases have provided much valuable information on the rates of elementary processes at high temperatures.

5.2 EXPONENTIAL ACCELERATION RATES AND INDUCTION PERIODS

The acceleration following the appearance of any detectable reaction in the shocked gases is so rapid that the precise definition of the induction time is not too important. Measured induction times are of the order of a few to a few hundred microseconds. Reflected shock studies of the ignition in the hydrogen—oxygen system at pressures around five atmospheres and temperatures extending upwards from about 850 K show two distinct types of behaviour. Above 1100 K, with conditions similar to those used earlier by Schott and Kinsey [102], Miyama and Takeyama [103] observed an induction period τ_1, at the end of which there was a single increase in OH absorption simultaneously with a pressure rise. The

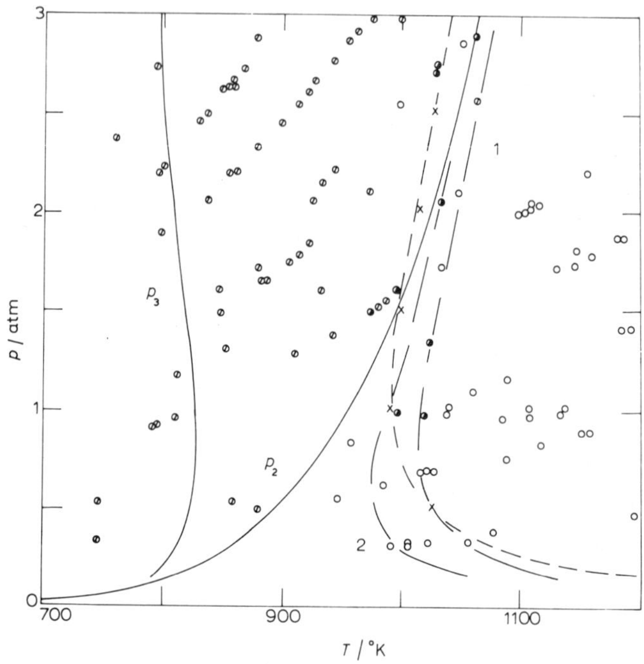

Fig. 22. Explosion limits in $H_2 + O_2$ (after Voevodsky and Soloukhin [98], and Meyer and Oppenheim [107]). ○, "Sharp" ignition; ⊘, "mild" ignition; ◑, intermediate cases. Solid lines: P_2 = extended second limit; P_3 = third limit. Broken lines give calculated pressures for $2H_2 + O_2$ [107]: - - - -, $\tau = 100$ μsec; — —, curve 1, $(\partial\tau/\partial T)_p$ = 1 μsec. K^{-1}; — —, curve 2, $(\partial\tau/\partial T)_p$ = 2 μsec. K^{-1}. (By courtesy of The Combustion Institute.)

earlier observation of Schott and Kinsey [102] of the constancy of the product $\tau_1[O_2]$ at constant temperature was confirmed. Below 1100 K, however, the first appearance of OH absorption after an induction period τ_1 was not accompanied by a pressure rise. The latter only occurred after a longer induction period τ_2, at the end of which there was a second increase in OH absorption also. There was no correlation between τ_2 and oxygen concentration: instead the product $\tau_2[H_2]$ was found to be constant. Other authors [97, 104, 105] have found similar evidence for a change in the mechanism of ignition, while schlieren observations by Saytzev and Soloukhin [106], Voevodsky and Soloukhin [98, 99], and Meyer and Oppenheim [107] showed a change from a single source, "sharp" ignition to a multiple source, "mild" ignition as the temperature was reduced. Meyer and Oppenheim found that in the "mild" ignitions there was at first practically no pressure rise, and the latter only became apparent after a relatively much longer period of time (of the order of 100 μsec compared with much shorter induction times for "strong" ignition). The regions of the p—T diagram in which the two types of ignition occur are shown in Fig. 22. The transition temperatures lie close to, but always on the high temperature side of, the extrapolation of the second limit line. Meyer and Oppenheim [107] have related the transition limit with a critical value of the gradient $(\partial\tau/\partial T)_p$ of the induction period with temperature at constant pressure. This gradient increases markedly as the second limit is approached from the high temperature side, and the transition to "mild" ignition is regarded as due to interaction in these circumstances between the chemistry and the gas dynamics of the shock process. In chemical terms though, the mechanistic changes implied by the extrapolated second limit line are the important ones.

During sufficiently early stages of the induction period the radical concentrations are low, the consumption of reactants is very small, and only those elementary steps which are first order in the radical concentration need to be considered. Further, the ignition times in the shocked gases are so short that diffusion processes and wall reactions cannot make themselves felt. Following a transient situation in which primary initiation by reactions such as $H_2 + O_2 = 2OH$ must be important, the processes controlling the major part of the ignition in the high temperature, low pressure region, to the right of the extended second limit line, are principally reactions (i), (ii) and (iii) (p. 55). In this region then, the chain branching can be studied in a relatively uncomplicated environment [102]. In the lower temperature, higher pressure region, to the left of the extension, some additional process must be considered. Reaction (iv) will have become more important in the higher pressure range, and, because of the imposed restriction that the new reaction must be first order in radical concentration, the additional process is generally considered to be reaction (xi). In the light of the discussion in Sect. 4 and the demonstration in Table 17 of the sensitivity of the induction periods in B_2O_3 coated vessels to the parameter $k_{11}/k_{10}^{1/2}$, this restriction may be too severe for an accurate treatment of the measured higher pressure, shock-initiated induction times.

Approximate analytical solutions of the full set of differential equations for the kinetics of radical growth by reactions (i)—(iv) and (xi), valid also at high temperature, have been given by Brokaw [108]. The solution is more difficult than that encountered for closed vessel studies at lower temperatures, since increasing the temperature causes the rate coefficient k_2 to increase more rapidly than k_1 or k_3: indeed, above about 1500 K, k_2 becomes greater than k_3. Under these conditions the OH concentration, and particularly the O atom concentration, in the quasi-steady state may become large enough to invalidate the normal application of the partial stationary state approach. The solution without this treatment gives an exponential radical growth $c_i = A_i \exp(\phi t)$, (with i = H, OH, O or HO_2) and seeks the net branching factor ϕ as the single positive root of the determinantal equation

$$\begin{vmatrix} -(k_1[\mathrm{H_2}]+\phi) & k_2[\mathrm{O_2}] & k_3[\mathrm{H_2}] & 0 \\ k_1[\mathrm{H_2}] & -(k_2[\mathrm{O_2}]+k_4[\mathrm{O_2}][\mathrm{M}]+\phi) & k_3[\mathrm{H_2}] & k_{11}[\mathrm{H_2}] \\ 0 & k_2[\mathrm{O_2}] & -(k_3[\mathrm{H_2}]+\phi) & 0 \\ 0 & k_4[\mathrm{O_2}][\mathrm{M}] & 0 & -(k_{11}[\mathrm{H_2}]+\phi) \end{vmatrix} = 0 \qquad (50)$$

i.e.

$$\begin{aligned} &\phi^4 + \{(k_1+k_3+k_{11})[\mathrm{H_2}] + (k_2+k_4[\mathrm{M}])[\mathrm{O_2}]\}\,\phi^3 \\ &+ \{k_1k_3[\mathrm{H_2}]^2 + (k_1+k_3)k_4[\mathrm{H_2}][\mathrm{O_2}][\mathrm{M}] \\ &+ (k_1[\mathrm{H_2}] + k_2[\mathrm{O_2}] + k_3[\mathrm{H_2}])k_{11}[\mathrm{H_2}]\}\,\phi^2 \\ &+ k_1k_3[\mathrm{H_2}]^2\{(k_4[\mathrm{M}] - 2k_2)[\mathrm{O_2}] + k_{11}[\mathrm{H_2}]\}\,\phi \\ &- 2k_1k_2k_3k_{11}[\mathrm{H_2}]^3[\mathrm{O_2}] = 0 \end{aligned} \qquad (50a)$$

In eqn. (50) the coefficient k_{11} is very small compared with k_1, k_2, k_3 or $k_4[\mathrm{M}]$, and can therefore be neglected in the sums containing it. There are then three possible solutions as follows.

(a) $k_4[M] > 2k_2$

This regime corresponds with the "mild" ignitions to the left of the extended second limit line in Fig. 22. Here the ignition lags are long and the positive ϕ is very small. We may neglect the terms higher than first power in ϕ giving

$$\phi \simeq \frac{2k_2k_{11}[\mathrm{H_2}]}{(k_4[\mathrm{M}] - 2k_2)} \qquad (51)$$

the approximation becoming less exact near the extended second limit line where the denominator is zero. Equation (51) provides a basis for the correlation $\tau_2[\mathrm{H_2}]$ = constant, observed for constant temperature by Miyama and Takeyama [103].

(b) $k_4[M] = 2k_2$

To a first approximation this condition marks the boundary between "mild" and "strong" ignitions. Here the coefficient of ϕ in eqn. (50a) is zero (neglecting the term in $k_{11}[\mathrm{H_2}]$). Neglecting the terms in ϕ^3 and ϕ^4 also

$$\phi \simeq \left\{ \frac{2k_1 k_2 k_3 k_{11} [H_2]^2 [O_2]}{k_1 k_3 [H_2] + (k_1 + k_3) k_4 [O_2][M]} \right\}^{1/2} \tag{52}$$

(c) $2k_2 > k_4 [M]$

This corresponds to the important region of "strong" ignitions with short delays, the measurement of which has provided much data on the chain branching process. In this region all terms involving k_{11} may be neglected, leading to the cubic equation

$$\begin{aligned} &\phi^3 + \{(k_1 + k_3)[H_2] + (k_2 + k_4[M])[O_2]\}\phi^2 \\ &\quad + \{k_1 k_3 [H_2]^2 + (k_1 + k_3) k_4 [H_2][O_2][M]\}\phi \\ &\quad - k_1 k_3 [H_2]^2 \{2k_2 - k_4[M]\}[O_2] = 0 \end{aligned} \tag{53}$$

This is the same equation as deduced by Kondratiev [109] and others [78, 81] starting from reactions (i)—(iv) alone. At sufficiently low densities or high temperatures k_1, k_3 and $2k_2 \gg k_4[M]$, and eqn. (53) becomes

$$\begin{aligned} &\phi^3 + \{(k_1 + k_3)[H_2] + k_2[O_2]\}\phi^2 + k_1 k_3 [H_2]^2 \phi \\ &\quad - 2k_1 k_2 k_3 [H_2]^2 [O_2] = 0 \end{aligned} \tag{53a}$$

Above 1000 K the measured ϕ are of the same order as $k[x]$ when $[x]$ constitutes about 0.1 % of the overall molar density. For $[H_2] \gg [O_2]$ we then have from (53) and (53a)

$$\phi = (2k_2 - k_4[M])[O_2] \tag{54}$$

$$\simeq 2k_2 [O_2] \tag{54a}$$

With such very hydrogen-rich mixtures the partial stationary state treatment becomes valid for [OH] and [O], and eqn. (54) is identical with eqn. (29) if surface termination of H and O atoms are omitted from the latter by putting $P_1 = 0$. Equation (54a) is the basis of the ignition delay correlation $\tau_1 [O_2]$ = constant at constant temperature used by Schott and Kinsey [102].

For $[H_2] \ll [O_2]$, eqns. (53) and (53a) give

$$\phi = \{k_1 k_3 (2k_2 - k_4[M])/(k_2 + k_4[M])\}^{1/2}[H_2] \tag{55}$$

$$\simeq (2k_1 k_3)^{1/2} [H_2] \tag{55a}$$

Thus, measurements of the exponential growth constant in very lean mixtures give information about the product $k_1 k_3$.

Lastly, information about the sum $(k_1 + k_3)$ may be obtained from measurements using intermediate compositions. Schott [91] determined values of k_2, $k_1 k_3$ and $(k_1 + k_3)$ from direct measurements of ϕ using time-resolved studies at a number of compositions, and then attempted to derive values for the individual rate coefficients. However, because of the form of the coupling between k_1 and k_3, the sensitivity of the measurements to their sum was not high enough to give a satisfactory result. Some alternative procedures to give k_2 and k_3 involve using independent estimates of $k_4[M]$ and/or k_1 [88, 89, 110, 111], while yet another approach [92] has used additions of CO to the H_2—O_2—Ar mixture, thus allowing the reaction

$$OH + CO = CO_2 + H \tag{xxiii}$$

to occur in parallel with reaction (i), but allowing no parallel for reaction (iii). The effect in eqn. (53) is to replace the terms $k_1[H_2]$ by $(k_1[H_2] + k_{23}[CO])$. At small $[H_2]$ we then have

$$\phi = \left\{\frac{(k_1[H_2] + k_{23}[CO])k_3[H_2](2k_2 - k_4[M])[O_2]}{k_{23}[CO] + (k_2 + k_4[M])[O_2]}\right\}^{1/2} \tag{56}$$

$$\simeq \left\{\frac{2k_2k_3[H_2][O_2](k_1[H_2] + k_{23}[CO])}{k_{23}[CO] + k_2[O_2]}\right\}^{1/2} \tag{56a}$$

If in addition $k_{23}[CO] \gg k_2[O_2]$, then

$$\phi \simeq (2k_2k_3[H_2][O_2])^{1/2} \tag{57}$$

thus allowing an independent determination of k_3; whereas if $k_2[O_2] \gg k_{23}[CO]$ then k_1 and k_{23} can be found, since

$$\phi \simeq (2k_1k_3)^{1/2}[H_2] \text{ for } k_1[H_2] \gg k_{23}[CO] \tag{55a}$$

$$\simeq (2k_3k_{23}[H_2][CO])^{1/2} \text{ for } k_{23}[CO] \gg k_1[H_2] \tag{58}$$

Equations (55)—(58) have been used by Brabbs et al. [92] to assist in the selection of four mixtures suitable for examination in order to determine the four primary rate coefficients. For the mixtures selected, Table 23 shows the sensitivities of the growth constants to each of the five reaction rates, calculated from the modified eqn. (53). Table 24 gives a selection of the final results. The rate coefficients themselves were obtained by means of an iterative procedure based on eqn. (53), and using initial independent estimates of k_1, k_3, k_4 and k_{23} in order to derive the first value of k_2. Boundary layer effects in the shock tube were allowed for in the initial determination of the growth constants. The apparent k_2 determined without these corrections were some 20—60 % larger than the values given in Table 24, with an apparent activation energy of only 11.9 instead of 16.3 kcal . mole^{-1}.

An alternative, and experimentally less demanding approach to the time-resolved studies for the determination of the growth constants is the measurement of overall induction times τ_i for the appearance of a fixed, detectable signal from a reaction intermediate or final product. Referring to Sect. 3.6.5 and eqn. (30a), the method in its simplest form depends on constancy of the product $\phi\,\tau_i$ at a fixed temperature — a condition which in turn requires a small ratio n_0/ϕ and an early disappearance of the perturbing effect of the primary initiation transient on the exponential

TABLE 23
Mixture compositions and growth constant sensitivities [92]

	Mixture number			
	1	2	3	5
Reaction	$OH + H_2 \rightarrow H_2O + H$	$H + O_2 \rightarrow OH + O$	$O + H_2 \rightarrow OH + H$	$OH + CO \rightarrow CO_2 + H$
Composition (%)				
H_2	0.21	5	0.1046	0.1035
CO	0.11	6	10.0	6.01
O_2	10.0	0.5	0.503	10.0
CO_2	5.0	—	4.99	5.0
Ar to 100 %				
Sensitivities				
$\partial \ln \phi / \partial \ln (k_1 [H_2])$	0.34	0.01	0.00	0.07
$\partial \ln \phi / \partial \ln (k_2 [O_2])$	0.33	1.00	0.64	0.21
$\partial \ln \phi / \partial \ln (k_3 [H_2])$	0.48	0.06	0.39	0.49
$\partial \ln \phi / \partial \ln (k_4 [O_2][M])$	−0.17	−0.07	−0.06	−0.06
$\partial \ln \phi / \partial \ln (k_5 [CO])$	0.02	0.00	0.04	0.29

growth. Figure 23 shows a typical semi-logarithmic plot of the growth of the measured signal with time. Clearly, for the simplest application of the induction time method the log of the measured signal at the end of the induction period must be large compared with the intercept of the straight line portion on the vertical axis. However, the final signal must also remain small enough for the mathematical treatment to retain its validity.

Schott and Kinsey in 1958 [102] were the first to use induction time measurements in shocked H_2—O_2—Ar mixtures in order to derive kinetic

TABLE 24
Experimental results and rate coefficients for hydrogen—oxygen ignitions [92]
(a) Reaction $H + O_2 \rightarrow OH + O$ (Mixture 2 of Table 23)

T(K)	P(atm)	$\phi(10^4\ sec^{-1})$	$k_2(10^8\ l.mole^{-1}.sec^{-1})$
1166	1.248	1.16	1.14
1176	1.614	1.37	1.12
1180	1.444	1.26	1.12
1216	1.488	1.67	1.42
1239	1.141	1.39	1.48
1246	1.335	2.03	1.87
1286	1.197	2.50	2.58
1292	1.203	2.44	2.50
1310	1.412	2.66	2.38
1344	1.255	2.71	2.75
1369	1.275	3.19	3.23
1393	1.517	3.60	3.13
1409	1.538	4.07	3.53

TABLE 24—*continued*
(b) Reaction $O + H_2 \rightarrow OH + H$ (mixture 3 of Table 23)

T(K)	P(atm)	$\phi(10^4\ sec^{-1})$	$k_3(10^8\ l.mole^{-1}.sec^{-1})$
1172	1.383	0.548	4.71
1212	1.435	0.696	4.99
1250	1.271	0.755	5.33
1255	1.480	0.929	6.32
1266	1.492	1.05	7.64
1272	1.509	1.01	6.40
1297	1.106	0.910	7.97
1315	1.349	1.08	6.88
1327	1.366	1.09	6.32
1335	1.586	1.28	6.59
1353	1.277	1.21	8.08
1360	1.406	1.23	6.50
1422	1.234	1.32	7.45
1436	1.258	1.43	8.20
1454	1.273	1.44	7.47
1498	1.310	1.68	8.52
1504	1.312	2.17	16.03
1543	1.088	1.77	12.63
1575	1.125	2.12	16.72
1612	1.146	2.21	15.64

TABLE 24—*continued*
(c) Reaction $OH + H_2 \rightarrow H_2O + H$ (mixture 1 of Table 23)

T(K)	P(atm)	$\phi(10^4\ sec^{-1})$	$k_1(10^9\ l.mole^{-1}.sec^{-1})$
1083	1.405	0.828	2.48
1115	1.461	1.36	4.87
1117	1.458	1.41	5.36
1130	1.290	1.22	1.63
1152	1.420	1.36	1.22
1170	1.345	2.08	4.14
1180	1.558	2.20	3.11
1186	1.373	1.66	1.29
1195	1.183	1.62	1.36
1242	1.451	2.33	1.54
1280	1.500	3.10	2.25
1284	1.286	2.99	2.71
1285	1.285	3.10	3.01
1344	1.254	3.66	3.29
1353	1.380	3.40	2.07
1370	1.391	4.19	3.23
1422	1.341	4.55	3.41
1444	1.239	3.98	2.64
1454	1.133	4.34	3.93
1472	1.265	4.96	4.04

(*c*) *Reaction* $OH + H_2 \rightarrow H_2O + H$ (mixture 1 of Table 23)—*continued*

T(K)	P(atm)	$\phi(10^4\ sec^{-1})$	$k_1(10^9\ l.mole^{-1}.sec^{-1})$
1511	1.056	4.50	4.25
1533	1.074	4.70	4.31
1554	1.097	4.57	3.62
1573	1.108	5.65	5.75
1596	1.130	5.36	4.56
1596	1.127	5.91	5.86

information about the reaction. Their induction times were taken as the time between the passage of the shock front and the appearance of a detectable OH signal in absorption (estimated to correspond with $X_{OH} \simeq 10^{-8}$). Their kinetic analysis was simpler than that just discussed in that it employed the partial stationary state approach with reaction (ii) rate-controlling, as had previously been done at lower temperatures [59—61]. The approach leads to eqns. (54) at all H_2/O_2 rates, and hence to the results

$$\tau_i[O_2] = \text{constant}/2k_2 \tag{59}$$

and

$$\log(\tau_i[O_2]) = A + B/T \tag{60}$$

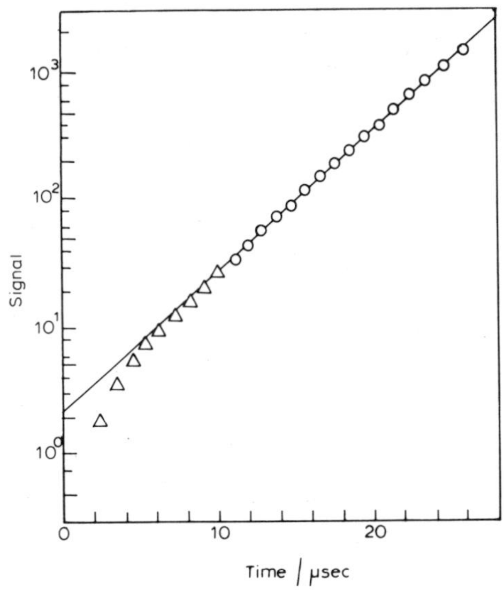

Fig. 23. Semi-logarithmic plot of growth of radical concentration with time (after Schott [91]). Mixture composition: 0.25 % H_2, 0.76 % O_2, 2.03 % CO, 96.96 % Ar. Reflected shock temperature 2168 K. Pre-shock pressure 100 torr, ○, data from zig-zag oscillogram recording CO + O emission; △, data from high sensitivity, single trace oscillogram. (By courtesy of The Combustion Institute.)

where A and B are constants. Over the range of compositions $0.5 \leq [H_2]/[O_2] \leq 5$ studied by Schott and Kinsey, this relationship was found to be approximately obeyed in the temperature range 1100—2600 K and at pressures below two atm. Studies over wider composition ranges, however, [81, 108] showed the inadequacy of the partial stationary state treatment, and led to the development of the more complete set of eqns. (50)—(55) for the growth constants. Similar analytical solutions, with assumed primary initiation steps also included in the mechanism, have been used by Gardiner and co-workers [81, 82, 112—115] as the basis of a multi-parameter fit to induction time observations over a wide range of conditions. Ripley and Gardiner [112] found the direct dissociation of molecular hydrogen and oxygen to be too slow to act as the primary initiation step, for which they proposed exchange initiation by some such reaction as

$$\text{(o)}\ H_2 + O_2 \rightarrow H + HO_2 \quad \text{or } OH + OH$$

Their optimum agreement with experiment between 1400 and 2500 K was found using the rate coefficients (l . mole . sec units)

$$k_0 = 2.5 \times 10^9 \exp(-19{,}500/T)$$

$$k_1 = 4 \times 10^{10} \exp(-2850/T)$$

$$k_2 = 8 \times 10^{11} \exp(-8800/T)$$

$$k_3 = 1.2 \times 10^{10} \exp(-4600/T)$$

Figure 24 shows some of their calculated OH profiles during the first 75 μsec of the induction period, and illustrates clearly the effect of the two assumed primary initiation steps.

Qualitative reference has already been made to the existence of the two types of ignition behaviour in the hydrogen—oxygen system (Fig. 22), and an approximate analytical treatment of the region on the high pressure, low temperature side of the extended second limit line led to eqn. (51) for the growth constant. In this region, however, quantitative treatments either by way of analytical solution or by numerical integration of the rate equations have not been successful in predicting the temperature dependence of the induction times [99, 116]. Using values of rate coefficients derived from other sources, the theory employing reactions (o)—(iv) and (xi) predicts much too rapid a transition from short to longer induction times on reducing the temperature so as to cross the extended second limit line. The difficulty can be overcome by allowing a freer fit of all the rate coefficients [98, 105, 116], but there are then large discrepancies with other types of experiment. The reason for the discrepancies is thought to lie in certain features of the gas dynamic effects associated with reflected shock waves [116—118].

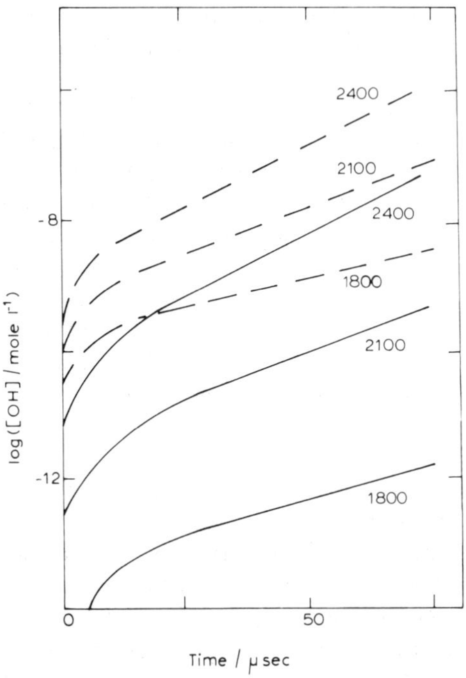

Fig. 24. OH concentration profiles, showing the effect of the exchange initiation reaction on the growth of OH during the first 75 μsec of the induction period (after Ripley and Gardiner [112]). Profiles calculated for 1:1:98, $H_2:O_2$: Ar mixture at 1800, 2100 and 2400 K. Pre-shock pressure 10 torr. - - -, including exchange initiation; —, excluding exchange initiation. (By courtesy of J. Chem. Phys.)

5.3 BACKGROUND TO FLAME STUDIES

A flame may be defined as a localized reaction zone which is able to propagate itself sub-sonically through the material supporting it. Most flames are concerned with exothermic reactions of this type, in which typically reactants at near ambient temperatures are converted more or less adiabatically to combustion products at 1000 K or above. Detailed kinetic studies have principally been confined to premixed flames, in which a well-defined reactant mixture at a known initial temperature is converted into combustion products in full chemical equilibrium at the final flame temperature. Assuming adiabatic combustion, the final conditions may be calculated thermodynamically.

The linear burning velocity S_u is defined as the normal velocity of approach of the unburnt gas towards the flame front. Alternatively, the mass burning velocity M is the mass rate of consumption of reactant mixture per unit area of flame surface. By continuity, M is constant through a one-dimensional flame, and is given by

$$M = \rho S = \rho_u S_u = \text{const.} \tag{61}$$

Here ρ and S are the density and corresponding normal linear flow velocity at any point in the flame, and the subscript u refers to the unburnt gas.

If now the unburnt gas flow velocity in the y direction is S_u, then the flame front will be in the x, z plane, and the gas properties will depend only on the distance y through the flame. For measurement of these gas properties, the flame reaction zone must be thick enough to give adequate spatial resolution along the y-direction. This is achieved by studying either slow-burning flames at atmospheric pressure, or alternatively flames at sub-atmospheric pressures. Experimental techniques for studying flame profiles are described by Fenimore [119], by Fristrom and Westenberg [120], and by Dixon-Lewis and Williams [121]. The profile measurements which may be carried out are more varied than in the shock tube situation, since the flame may be stabilized on a burner to give a stationary flowing reaction system, with the reaction zone itself fixed in the laboratory system of co-ordinates. Extraction of samples from the flame over extended periods thus becomes possible.

The flow velocities in flame systems are such that transport processes (diffusion and thermal conduction) make appreciable contributions to the overall flows, and must be considered in the analysis of the measured profiles. Indeed, these processes are responsible for the propagation of the flame into the fresh gas supporting it, and the exponential growth zone of the shock tube experiments is replaced by an initial stage of the reaction where active centres are supplied by diffusion from "more reacted" mixture slightly further downstream. The measured profiles are related to the kinetic reaction rates by means of the continuity equations governing the one-dimensional flowing system. Let W_i represent the concentration ($g \cdot cm^{-3}$) of any quantity i at distance y and time t, and let F_i represent the overall flux of the quantity ($g \cdot cm^{-2} \cdot sec^{-1}$). Then continuity considerations require that the sum of the first distance derivative of the flux term and the first time derivative of the concentration term be equal to the mass chemical rate of formation q_i of the quantity, i.e.

$$\partial F_i/\partial y + \partial W_i/\partial t = q_i \qquad (62)$$

An equation of type (62) exists for each species present in the gas, and for the energy of the mixture. The time derivatives vanish in the stationary flame equations.

Now let w_i be the weight fraction of species i in the element of gas mixture considered. Then the molar concentrations c_i, from which the reaction rates are calculated, are given by

$$c_i = \rho w_i/m_i \qquad (63)$$

where m_i is the molecular weight.

For a species, the flux F_i consists of two parts, (i) a convective term Mw_i representing the mean mass flow of i, and (ii) a diffusion term j_i.

Thus,

$$\begin{aligned} \mathbf{F}_i &= \mathrm{M}w_i + \mathrm{j}_i \\ &= \mathrm{M}G_i \end{aligned} \tag{64}$$

where G_i is the weight fraction of i in the overall mass rate of flow [122].

In considering the energy, the appropriate total flux is of the form $\{\mathrm{M}\,\Sigma_i\,(G_iH_i) - \lambda\,\partial T/\partial y\}$, where H_i is the enthalpy per gram of species i and λ is the thermal conductivity of the mixture. The chemical rates of production of heat are given by the terms $M\,\Sigma\,(\partial G_i/\partial y)H_i$ in the first distance derivative of this expression, so that for an adiabatic stationary system the conservation of energy is given simply by the equation

$$\mathrm{d}/\mathrm{d}y\{\mathrm{M}\,\Sigma\,(G_iH_i) - \lambda\,\mathrm{d}T/\mathrm{d}y\} = 0 \tag{65}$$

The exact form of the expressions for the diffusional fluxes j_i depends on the degree of sophistication used in representing the transport phenomena. A precise approach, including also the calculation of the thermal conductivity of gas mixtures, and based on the Chapman-Enskog kinetic theory, has been described by Dixon-Lewis [122]. However, simpler approaches involving the form $\mathrm{j}_i = -\rho D_i \mathrm{d}w_i/\mathrm{d}y$ may also give satisfactory representation in many cases [119—121, 123].

The interpretation of measured flame profiles by means of the continuity equations may be approached in one of two ways. The direct experimental approach involves the use of the measured profiles to calculate overall fluxes, reaction rates, and hence rate coefficients. Its successful application depends on the ability to measure the relevant profiles, including concentrations of intermediate products. This is not always possible. In addition, the overall fluxes in the early part of the reaction zone may involve large diffusion contributions, and these depend in turn on the slopes of the measured profiles. Thus accuracy may suffer. The lining up on the distance axis of profiles measured by different methods is also a problem, and, in quantitative terms, factor-of-two accuracy is probably about the best that may normally be expected from this approach at the position of maximum rate. Nevertheless, examination of the concentration dependence of reaction rates in flames may still provide useful preliminary information about the nature of the controlling elementary processes [119—121]. Some problems associated with flame profile measurements and their interpretation have been discussed by Dixon-Lewis and Isles [124]. Radical recombination rates in the immediate post-combustion zones of flames are capable of measurement with somewhat higher precision than above.

The second approach to the interpretation of flame profiles is to assume a reaction mechanism and data, solve the conservation equations to obtain the flame properties, and then compare these with experiment.

Even in cases where the first method has been successfully applied, this can provide a stringent test for the accuracy of the derived data. A number of alternative methods for the numerical solution of the systems of flame equations associated with complex reaction mechanisms are now available [123, 125–130].

5.4 MAIN REACTION ZONE AND RECOMBINATION REGION IN HYDROGEN–OXYGEN IGNITION

Superficially, the passage of reacting gases through a flame zone is exactly analogous to the post-induction phase of shock tube ignition, and both will be considered together. In both cases high concentrations of radical intermediates develop, and both the reaction rate and these concentrations continue to rise until they are limited by consumption of reactants. Following this, a gradual decay of the intermediate concentrations occurs towards their final equilibrium values. The major difference between the reaction kinetics in these phases of the ignition and the kinetics considered in previous sections of this chapter is that now the radical concentrations are high enough for it to be necessary to consider radical–radical elementary processes as major contributors to the overall scheme. It is convenient to consider a number of aspects of the ignition in order of increasing kinetic complexity.

5.4.1 Radical recombination in fuel-rich systems. Partial equilibration concepts

A number of flame-photometric methods have been developed by Sugden and co-workers [131–134] to measure hydrogen atom concentrations in the burnt gas from hydrogen–oxygen flames. When small quantities of a sodium (or similar) salt are added to a flame, and if the flame temperature is high enough, thermal sodium D-line emission occurs. At low concentrations this emission is proportional to the concentration of the metal atoms. However, if lithium salts are added to the flame, some hydroxide is formed [135] by the process

$$Li + H_2O \rightleftharpoons LiOH + H$$

This reaction is sufficiently rapid for the maintenance of equilibrium. Thus the total amount of lithium added to the flame, $[Li]_0$, is equal to $[Li] + [LiOH]$, and if the amount of free lithium $[Li]$ at a position in the flame is measured spectroscopically the concentration of lithium hydroxide can be deduced. Since the water vapour concentration in the burnt gas is known, it is then possible to deduce the concentrations of H atoms from the equilibrium expression. In sufficiently hot flames the concentration of free lithium may be estimated, after calibration of the system using an equilibrium burnt gas where [H] is known, by measurement of the intensity of the thermal emission [131]. More recent developments of the method using atomic absorption spectroscopy to measure the lithium concentrations [136] have extended its range of application to cooler flames also.

In flames with lower final flame temperatures where the thermal emission from added metal atoms is less, a chemiluminescent effect [134] may occur. Here, there is a rapid rise of intensity in the reaction zone followed by a steady decay towards the thermal level. The chemiluminescence is due to excitation of the metal (in this case sodium) by the reactions

$$H + H + Na = H_2 + Na^*$$

$$H + OH + Na = H_2O + Na^*$$

The intensity I of the emission can be shown to be

$$I = C_1 [H]^2 + C_2 [H][OH] \tag{66}$$

where C_1 and C_2 are constants involving instrument, quenching, and rate coefficient factors. From this intensity the relative concentrations of H atoms in the burnt gas can be deduced.

In the burnt gas recombination region of fast, fuel-rich hydrogen—nitrogen—oxygen flames the observed intensities of chemiluminescence for sodium and other metal additives were found [134] to obey the relation

$$\frac{1}{I^{1/2}} - \frac{1}{I_0^{1/2}} = kt = \frac{ky}{S_b} \tag{67}$$

where k is a constant, S_b is the linear burnt gas velocity, and I_o is the intensity at the time or distance origin. From (66), the corresponding kinetic relation, if reaction (i) of the H_2—O_2 scheme is effectively equilibrated so that $[OH] \propto [H]$, is

$$\frac{1}{[H]} - \frac{1}{[H]_0} = k't \tag{68}$$

Kaskan [137], using UV absorption by OH as the diagnostic method, found a similar relation to (68) for [OH] in flames, while Schott and Bird [138] found the relation to be applicable also to the decay of OH following shock tube ignition in rich mixtures.

In fast flames and shock tube flows such as are considered here, the concentration gradients in the recombination region are such that diffusion effects can be neglected. The recombination can also be considered as taking place in the presence of effectively constant concentrations of the bulk species H_2, H_2O and N_2 or Ar. As was first pointed out by Sugden and co-workers [133] the radical concentrations do not behave independently during the approach to full equilibrium. The observed relationships are consistent with a recombination region in which H, OH and O (and to a lesser extent also the minor constituent O_2 in rich flames) are equilibrated amongst themselves by means of the rapid forward and reverse reactions (i), (ii) and (iii) of the main $H_2 + O_2$ scheme, even though the concentrations of all the radicals are above their concentrations at full equilibrium. This is therefore a *partial equilibrium* situation. The decay of the *pool* of radicals towards full equilibrium occurs by the slower forward recombination steps (xvii)—(xix), viz.

$$OH + H_2 \rightleftharpoons H_2O + H \quad \text{(i)}$$

$$H + O_2 \rightleftharpoons OH + O \quad \text{(ii)}$$

$$O + H_2 \rightleftharpoons OH + H \quad \text{(iii)}$$

$$H + H + M \rightleftharpoons H_2 + M \quad \text{(xvii)}$$

$$H + OH + M \rightleftharpoons H_2O + M \quad \text{(xviii)}$$

$$H + O + M \rightleftharpoons OH + M \quad \text{(xix)}$$

though because of the low O atom concentrations in rich systems, the last of these will not be too important. The partial equilibrium situation arises because of the high rates of the bimolecular steps (i), (ii) and (iii) compared with the termolecular recombination reactions, and the realization of this led to a major simplification in the treatment of the recombination region in flames and shock tubes. The radical pool concept will be discussed further in Sects. 5.4.3 and 4. The constant k' in eqn. (68) is a pseudo-second order recombination rate coefficient. Its value will change with the nature of the bulk constituents which provide the major part of the "chaperon" molecules M, and with the relative amounts of H, OH and O in the recombining mixture.

Even neglecting the rather small contribution of reaction (xix) to the recombination in rich H_2—O_2 systems, the breakdown of the constants k' into their constituent third order rate coefficients is a matter of some difficulty. Three chaperon molecules, H_2, H_2O and the inert diluent, are involved in each of the reactions (xvii) and (xviii); and for some of these it is difficult in flames to vary sufficiently the burnt gas compositions in which the recombination occurs. Further, because of the equilibration of reaction (i), it is impossible to distinguish reaction (xvii) with $M = H_2O$ from (xviii) with $M = H_2$. The following discussion gives a likely overall picture based on results at present available, though detailed confirmation is necessary in some areas.

The recombination rate coefficients k_{17} and k_{18} have been considered in some detail by Baulch et al. [55]. Shock tubes have certain advantages for the study of these at high temperatures, since the attainment of the high temperature is independent of the heat liberated by the reaction. A number of shock tube investigations have been made of the dissociation of

TABLE 25
Third order recombination rate coefficients from shock tube dissociation studies of $H + H + M = H_2 + M$ ($l^2 . mole^{-2} . sec^{-1}$)

M = H	H_2	Ar	Temp. (K)	Ref.
$5 \times 10^{13}\ T^{-1.0}$	$7.5 \times 10^{12}\ T^{-1.0}$		2950—5330	140
		$7.5 \times 10^{11}\ T^{-1.0}$	3430—4600	140
$1 \times 10^{13}\ T^{-1.0}$	$3.0 \times 10^{12}\ T^{-1.0}$	$1.5 \times 10^{12}\ T^{-1.0}$	2800—5000	141
	$2.6 \times 10^{12}\ T^{-1.0}$	$6.4 \times 10^{11}\ T^{-1.0}$	2500—5400	142
1.02×10^{10}			2750	142
1.2×10^{10}			3140	142
9.1×10^{9}			3500	142
5.1×10^{9}			4200	142
2.5×10^{9}			4840	142
		$2 \times 10^{12}\ T^{-1.0}$	3800—5300	144
	1.2×10^{9}		2925	143
	5.4×10^{8}		3540	143
	1.3×10^{9}		3630	143
	6.3×10^{8}		3850	143
	7.3×10^{8}		4500	143
	4.9×10^{8}		5920	143
6.1×10^{10}			2695	143
4.6×10^{10}			3000	143
2.6×10^{10}			3355	143
5.1×10^{9}			3740	143
5.4×10^{9}			4020	143
1.0×10^{9}			4660	143
3.2×10^{8}			5585	143
$2 \times 10^{13}\ T^{-1.0}$	$2.5 \times 10^{12}\ T^{-1.0}$	$1 \times 10^{12}\ T^{-1.0}$	2900—4700	145
	1.75×10^{15} exp $(-4.5 \times 10^{-4} T)$	6.1×10^{15} exp $(-6.3 \times 10^{-4} T)$	2500—7000	146

hydrogen and the recombination of H atoms with both H_2 and Ar as the chaperon molecules (e.g. refs. 140—146). A selection of results is given in Table 25. The measurements of k_{17,H_2} by Rink [141], Sutton [142], Hurle [143] and Jacobs et al. [145] are seen to be in reasonable accord.

At lower temperatures the recombination of H atoms has been measured mostly in fast flow systems, with initial dissociation of molecular hydrogen either thermally on a hot wire, or by means of a microwave discharge. In substantial agreement with previous work by Larkin and Thrush [147], Ham et al. [148], and Walkauskas and Kaufman [149] have recently found $k_{17,H_2} = (3.0 \pm 0.2) \times 10^9$ at room temperature. Their results were very reproducible over an extended period, and surface recombination at the wall of the flow tube contributed only a few per cent to the observed decay. Combining the room temperature result with data at lower temperatures (down to 70 K) gave approximately a $T^{-0.6}$ temperature dependence, leading to the expression $k_{17,H_2} = 9.2 \times 10^{10} T^{-0.6}$. Although this temperature dependence is lower than that found with the shock tube experiments considered above, the low

TABLE 26
Rate coefficients at low temperatures for $H + H + M = H_2 + M$ [149]
(Parameters A and B refer to the expression: $k = AT^{-B}$)

M	k_M at 295 K (10^9 ($l^2.mole^{-2}.sec^{-1}$))	k_M/k_{H_2} at 295 K	A	B
H_2	3.0	1	9.2×10^{10}	0.6
He	2.6	0.87	2.54×10^{10}	0.4
Ar	3.4	1.14	3.26×10^{11}	0.8
			(1.0×10^{12})	(1.0)
N_2	3.4	1.13	5.65×10^{12}	1.3
CH_4	5.7	1.89	5.35×10^{12}	1.2
CO_2	6.1	2.02	5.49×10^{14}	2.0
SF_6	7.2	2.41	2.07×10^{14}	1.8

temperature expression nevertheless extrapolates satisfactorily to the region of the shock tube results.

Walkauskas and Kaufman [149] have also measured the recombination rate coefficients at low temperatures for a number of chaperon molecules other than molecular hydrogen. The temperature dependence of the coefficients was found not to be the same for all the chaperons. Table 26 gives the rate coefficients and the efficiencies of the chaperons relative to molecular hydrogen, both at room temperature, together with the coefficient A and exponent B to be used in order to calculate the rate coefficients themselves from the expression $k = AT^{-B}$. For argon, the use of the unbracketed parameters A and B at shock tube temperatures leads to somewhat high values of $k_{17,Ar}$ compared with the shock tube expressions of Table 25. The expression of Jacobs et al. [145] which uses the bracketed parameters A and B in Table 26, fits both the shock tube and room temperature results.

An interesting feature of the shock tube results on the recombination (Table 25) is the high chaperon efficiency of H atoms at temperatures around 3000 K. This efficiency is found to fall off rapidly above 3000 K,

TABLE 27
Hydrogen atom recombination following shock ignition of rich H_2—O_2—diluent mixtures

$M = N_2$	Ar	H_2O	Temp. (K)	Ref.
	1×10^9		ca. 2100	150
	$(2 \pm 1) \times 10^8$		1700	138
6×10^8	7.5×10^8	$\leqslant 10^{10}$	1700	151
$(3.8 \pm 0.5) \times 10^8$	$(3.8 \pm 0.5) \times 10^8$	$(2.3 \pm 0.3) \times 10^9$	1220—2370	152
	1.0×10^9		1300—1700	153

with approximately a T^{-7} temperature dependence according to Hurle [143]. It has been found also to be small ($k_{17,H} < 2.5 \times 10^9$) at room temperature.

Recombination following the ignition of hydrogen—oxygen mixtures behind shock waves has been studied extensively by Schott [150], Schott and Bird [138], White and Moore [94], Getzinger and Blair [151], Gay and Pratt [152] and Mallard and Owen [153]. Here of course the observed effects are more complex than in the dissociation—recombination experiments. White and Moore [94] studied mixtures very rich in hydrogen, and assumed the recombination to be entirely due to reaction (xvii). There is some doubt about their definition of k_{17}, since their numerical values all seem to be about a factor of two higher than those found by others for similar mixtures. If their results are divided by two, they find $k_{17,mean} = (1.8 \pm 0.2) \times 10^9$ at 1600—2100 K in a mixture of $7H_2 + O_2$; while for a mixture containing excess argon ($8H_2 + O_2 + 91Ar$) they found a mean $k_{17} = (7 \pm 1) \times 10^8$ on the same basis. For argon, the expression of Jacobs et al. [145] leads to $k_{17,Ar} = 5 \times 10^8$ at 2000 K, in reasonable agreement with White and Moore's result also.

Other shock tube recombination results following ignition in rich mixtures are given in Table 27. The most complete approach is probably that of Getzinger and Blair [151], who studied both rich and lean mixtures. The extension of the recombination studies to lean mixtures will be considered in Sect. 5.4.3. In the rich mixtures, Getzinger and Blair's mean value of $k_{17,Ar} = 7.5 \times 10^8$ at 1700 K is about 25 % higher than would be predicted by the bracketed parameters for argon in Table 26; while their value of $k_{17,N_2} = 6 \times 10^8$ at 1600 K is some 50 % higher than would be predicted by the nitrogen parameters in the Table. Although both discrepancies are within the uncertainty of the measurements, it is also possible that in the case of argon the expression of Jacobs et al. [145] underestimates the rate coefficient in the intermediate temperature range, and that an exponent B varying from about 0.8 at room temperature to slightly greater than one at 3000—4000 K would give a more precise fit. In both cases more data are needed, particularly in the 1000—2000 K temperature range.

Turning now to rich flames, recent analyses by Halstead and Jenkins [154] of a number of recombination results with hydrogen—diluent—oxygen flames, some containing added steam as diluent, have used $k_{17,H_2} = 7 \times 10^8$, from shock tube work [143, 146], as an input parameter. They found $k_{17,N_2} = (1.9 \pm 0.7) \times 10^9$, $k_{17,Ar} = (1.8 \pm 0.7) \times 10^9$, and $(k_{17,H_2O} + k_{18,H_2}/K_1) = (3.6 \pm 0.4) \times 10^9$, at 1900 K. Here K_1 is the equilibrium constant of reaction (i). The mean values for nitrogen and argon are two to three times those indicated by the above discussion of the shock tube and fast flow work; and taking account of the details of the analysis carried out by Halstead and Jenkins, it seems likely that the discrepancy is associated with an erroneous value of $(k_{17,H_2O} + k_{18,H_2}/$

K_1), which originated from the experiments with water vapour as diluent. These experiments are particularly difficult to interpret, since the addition of water vapour affects, at one and the same time, both the "chaperon" composition of the mixture and the relative contributions of reactions (xvii) and (xviii) to the recombination.

Both the shock tube results of Getzinger and Blair [151] and the flame results of Halstead and Jenkins [154] suggest $k_{17,N_2} = k_{17,Ar}$ at around 2000 K. The fast flow results of Walkauskas and Kaufman [149] give $k_{17,N_2} = k_{17,Ar}$ at room temperature. A good approximation for both gases between 300 and 2000 K is therefore likely to be given by the expression of Jacobs et al., $k_{17,M=Ar,N_2} = 1.0 \times 10^{12} T^{-1.0}$. A numerical re-examination of the H_2—N_2—O_2 flame results of Halstead and Jenkins at 1900 K has been made by Dixon-Lewis and Greenberg [155] on the assumptions (i) that $k_{17,N_2} = 1.0 \times 10^{12} T^{-1.0}$, (ii) that $k_{17,H_2} = 9.2 \times 10^{10} T^{-0.6}$, and (iii) that $k_{18,H_2O} = 5k_{18,N_2}$ [55].

For flames containing a large excess of hydrogen, reaction (xviii) is of little importance. Its importance increases as the composition approaches stoichiometric from the rich side; and for a range of compositions for which its contribution to the recombination varied from approximately 25—50 %, the optimum values of $(k_{17,H_2O} + k_{18,H_2}/K_1)$ and k_{18,N_2} at 1900 K were found to be 5.2×10^9 and 4.9×10^9, respectively. The further assumption that $k_{18,H_2} = k_{18,N_2}$ then led to $k_{17,H_2O} = 4.8 \times 10^9$ also, i.e. $k_{17,H_2O}/k_{17,H_2} = 4.8$. This result, the recombination results* of Kaskan [137] on a very rich flame at a lower temperature (1200—1320 K), and recombination results for rich flames at around 1050 K [156] (discussed below) are all quite consistent with the above expressions for k_{17,H_2} and k_{17,N_2}, together with $k_{17,H_2O} = 6 \times 10^{13} T^{-1.25}$. At 300 K this last expression gives $k_{17,H_2O} = 4.8 \times 10^{10}$, in good agreement with the value of $(4.5 \pm 1.0) \times 10^{10}$ reported by Eberius et al. [157].

5.4.2 *Main reaction zone in fuel-rich systems*

The burning velocity, and the temperature and composition profiles in a low temperature, fuel-rich hydrogen—nitrogen—oxygen flame at atmospheric pressure having an unburnt gas composition $X_{H_2,u} = 0.1883$, $X_{N_2,u} = 0.7657$ and $X_{O_2,u} = 0.0460$, with $T_u = 336$ K, were measured by Dixon-Lewis et al. [156]; while the burning velocities of a number of flames having compositions not too far from this were also examined by Dixon-Lewis and co-workers [158, 159]. In a number of these flames the main reaction zone extended from approximately 600—1060 K, and the predominantly recombination zone from about 1060—1080 K. The maxi-

* With some correction of calibration for changes in *f*-number of OH (see Sect. 5.4.3).

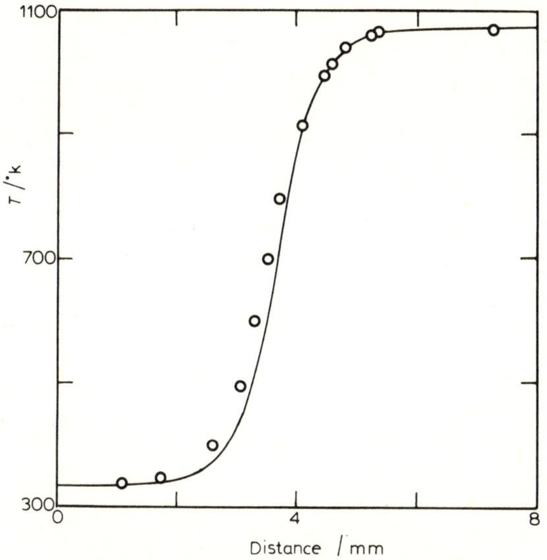

Fig. 25. Computed and measured temperature profiles for "standard" flame having initial conditions: $X_{H_2,u}$ = 0.1883, $X_{N_2,u}$ = 0.7657, $X_{O_2,u}$ = 0.0460, T_u = 336 K. ○, observations of Dixon-Lewis et al. [156]; line computed using set 2 of rate coefficients in Table 30.

mum radical concentration will be seen to occur at 1030—1040 K (Figs. 25 and 26). Following the approach mentioned earlier in which detailed flame properties were computed corresponding with assumed reaction mechanisms and rate coefficients, the principal reactions determining the

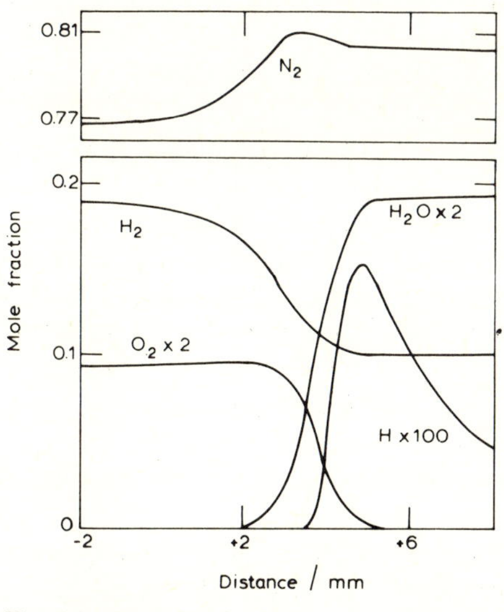

Fig. 26. Computed mole fraction profiles. Conditions as in Fig. 25.

flame properties were shown [158—160] to be the forward reactions

$$OH + H_2 \rightleftharpoons H_2O + H \quad \text{(i)}$$

$$H + O_2 \rightleftharpoons OH + O \quad \text{(ii)}$$

$$O + H_2 \rightleftharpoons OH + H \quad \text{(iii)}$$

$$H + O_2 + M \rightleftharpoons HO_2 + M \quad \text{(iv)}$$

$$H + HO_2 \rightleftharpoons OH + OH \quad \text{(viii)}$$

$$H + HO_2 \rightleftharpoons O + H_2O \quad \text{(viiia)}$$

$$H + H + M \rightleftharpoons H_2 + M \quad \text{(xvii)}$$

$$H + OH + M \rightleftharpoons H_2O + M \quad \text{(xviii)}$$

$$H + O + M \rightleftharpoons OH + M \quad \text{(xix)}$$

$$H + HO_2 \rightleftharpoons H_2 + O_2 \quad \text{(xx)}$$

$$OH + HO_2 \rightleftharpoons H_2O + O_2 \quad \text{(xxi)}$$

$$O + HO_2 \rightleftharpoons OH + O_2 \quad \text{(xxii)}$$

together with the reverse reactions (—i) and (—iii). The mechanism was also consistent with burning velocity and structure measurements by Dixon-Lewis et al. [161] on a flame of similar composition at a pressure of about 1/8 atmosphere. Reactions such as (xx)—(xxii) are suggested by fast flow studies of the reaction of H atoms with O_2 at room temperature [162—165].

The determination of reliable rate coefficients from individual flame studies is again a matter of considerable difficulty. The direct experimental approach discussed in Sect. 5.3 demands not only the difficult derivation of reaction rates, but also the measurement of absolute concentration profiles for intermediate species like H and OH. Even if curves of relative concentrations of these species can be determined and properly aligned with other measurements in the system, the absolute calibrations present considerable problems. Prior to the above measurements of Dixon-Lewis et al. [156, 161], Fenimore and Jones had probed several fuel-rich hydrogen—nitrogen—oxygen flames burning at atmospheric [166] and at reduced pressures [167] on water-cooled burners. They determined rates of disappearance of oxygen at high temperatures, and measured H atom concentrations in the same region by determining the rate of formation of HD from traces of D_2O added to the gases entering the flame. The calibration of the H atom concentration here depends on the value assumed for the rate coefficient k_{-1De} for

$$H + D_2O \rightarrow OD + HD \quad \text{(—iDe)}$$

Fenimore and Jones assigned to this the value $k_{-1De} = 10^{12} \exp(-12{,}750/T)$: then, assuming the disappearance of oxygen to be solely by reaction (ii), they obtained the values for k_2 given in Table 28. So far as

TABLE 28
Mean values of k_2 [166, 167]

T(K)	k_2(10^8 l. mole^{-1} . sec^{-1})
1100	1.5
1285	2.9
1324	3.8
1340	4.4
1420	7.2
1500	10.0

can be estimated, their calibration rate coefficient k_{1De} is high by approximate factors of 1.5, 2 and 2.5 at 1100, 1285 and 1500 K, respectively.

TABLE 29
Rate coefficients from hydrogen—oxygen flames [168]
(a) Reaction $H + O_2 = OH + O$

T(K)	k_2(10^7 l.mole^{-1}.sec^{-1})
770	0.48
795	0.8
840	1.0
875	1.4
905	2.1
935	3.0
960	4.0
980	4.9
995	5.7
1015	6.2
1025	6.3
1040	6.2

(b) Reaction $OH + H_2 = H_2O + H$

T(K)	k_1(10^8 l.mole^{-1}.sec^{-1})
476	0.65
553	1.0
615	1.9
700	3.6
765	5.1
852	4.7
967	7.5
1150	12.0
1190	16.0
1245	14.0
1370	16.0
1495	19.0

More recently, Eberius et al. [168] have sampled rich hydrogen—oxygen flames at 10.6 torr pressure, again stabilized on a water-cooled porous plate. They measured the molecular species mass spectrometrically, OH by UV absorption, and H atoms by sampling into an ESR cavity. Precautions were taken to allow for H atom recombination in the ESR probe, and the measured profile was found to be in good agreement with one computed by solution of the time dependent flame equations. Reaction rates were determined directly from the profiles of the stable species. Decay of oxygen was interpreted in terms of reactions (ii) and (iv); and the formation of water in the early part of the flame in terms of reaction (i). These interpretations led to values of k_1 and k_2 given in Table 29. The calibration of the H atom concentrations to allow for probe losses still leaves room for some doubt.

Following earlier attempts at estimating rate coefficients by a similar direct approach, Dixon-Lewis et al. [169—171] have recently favoured the methods using independent computation of the detailed properties of the adiabatic flame for comparison with experiment. These computations use the reaction rate coefficients as input parameters, fixing those which are supposedly reliably known and adjusting the remainder so as to optimize the agreement with experiments. Clearly, for a scheme of the complexity of that given, recourse must be had to a wide variety of experimental data. Initially, for the very fuel-rich flame of which the detailed structure was measured, it was assumed that OH and O, once formed, reacted immediately by reactions (i) and (iii); while HO_2 was assumed to react immediately by (viii) or (xx). The flame could then be considered as being controlled by four reaction cycles

$$H + O_2(+ 3H_2) \xrightarrow{k_2} 2H_2O + 3H \qquad \text{(iia)}$$

$$H + O_2 + M(+ H + 2H_2) \xrightarrow{ak_4} 2H_2O + 2H + M \qquad \text{(iva)}$$

$$H + O_2 + M(+ H) \xrightarrow{(1-a)k_4} H_2 + O_2 + M \qquad \text{(ivb)}$$

$$H + H + M \xrightarrow{k_{17}} H_2 + M \qquad \text{(xvii)}$$

where $a = k_8/(k_8 + k_{20})$.

Since the ratio $2k_2/k_4$ is reliably known from second explosion limit work, the three kinetic unknowns in the system are now k_2, k_8/k_{20} and k_{17}. Again initially, k_2 was assigned the fixed value 2.05×10^{11} exp $(-8{,}250/T)$. It was found that the best fit of the burning velocity, the relative H atom concentration decay profile in the recombination region (measured by intensity of sodium chemiluminescence), and the temperature and composition profiles were obtained with $k_8/k_{20} = 5 \pm 1$ and $k_{17} = (4.5 \pm 1.5) \times 10^9$, assuming equal efficiencies of all the molecules in the flame as "chaperons" in the recombination. Both rate coefficients were assumed not to vary with temperature. Independent examinations of the effect of changes in the unburnt H_2/N_2 ratio at constant oxygen, and of the mole fraction of oxygen in the unburnt gas at constant H_2/N_2 ratio [158, 159] led to the conclusion that further chain breaking steps involving OH and O should be included in the mechanism. Reactions (xviii), (xix), (xxi) and (xxii) fulfil this function, and a contribution from reaction (viiia) is also not excluded. Detailed assignment of rate coefficients to these elementary steps is clearly beyond the scope of the experimental data so far presented. However, for the range of recombination rate coefficients $k_{17} = (4.5 \pm 1.5) \times 10^9$, and with reasonable values for k_{18} and k_{19}, the dependence of burning velocity on mixture composition led to the result that $(k_8 + k_{8a})/k_{20}$ lies in the range 6.5 ± 1.0, apparently independent of the ratio k_{21}/k_{22}, and again assumed independent of temperature in the flame reaction zone. These values of $(k_8 + k_{8a})/k_{20}$ are considerably higher than that found when the chain breaking reactions of OH and O were neglected. The median value of $k_{17} = 4.5 \times 10^9$ gave $(k_8 + k_{8a})/k_{20} = 6.7$.

Another important feature of this analysis was that for fixed values of k_{17}, and for the imposed condition of satisfactory prediction of measured burning velocities, the H atom concentration profiles in specific flames were not appreciably affected by the particular combination selected from the adjustable parameters concerned with reactions (viii), and (xviii)—(xxii), i.e. the rate coefficients k_{18} and k_{19}, and the ratios k_{8a}/k_8, $(k_8 + k_{8a})/k_{20}$, k_{21}/k_{20} and k_{22}/k_{20}. This implies that, despite somewhat incomplete characterization at this stage, the flame and the computational approach may be used to study the reactions of its radical species with trace additives. Such an analysis with D_2O, D_2 and CO_2 as the trace additives, has been used by Dixon-Lewis [172] to obtain information about the rate coefficients k_{1Da}, k_1 and k_{23},

$$OH + HD \rightleftharpoons HOD + H \qquad \text{(iDa)}$$

$$OH + CO \rightleftharpoons CO_2 + H \qquad \text{(xxiii)}$$

For a fixed k_{17}, these rate coefficients may be determined with an accuracy of ±5—15 % depending on the precision of the experimental data. For $k_{17} = 4.5 \times 10^9$ the values found were $k_{1Da} = (9.6 \pm 0.5) \times 10^8$, $k_1 = (2.7 \pm 0.4) \times 10^9$ and $k_{23} = (2.4 \pm 0.12) \times 10^8$, all at 1050 K.

Several inconsistencies with independent data now arise:

(i) k_1 and k_{23} are both somewhat higher than the average indicated by other investigations at comparable temperatures [172]. Since previous flame computations [160] had shown that lower values of k_{17} lead to higher radical concentrations in the flame, this suggests a lower value of k_{17} than that quoted. However, using the initially fixed expression for k_2, values of k_{17} below 3.0×10^9 began to produce discrepancies between the shapes of the computed and measured relative H atom concentration profiles [160].

(ii) Recent independent measurements of k_{17} confirm $k_{17} < 3 \times 10^9$ (cf. Sect. 5.4.1).

(iii) Following the establishment of both reactions (viii) and (xx) as part of the flame mechanism, a recent re-analysis by Baldwin et al. [73] of the second limits of hydrogen—nitrogen—oxygen mixtures in boric acid coated vessels has given values of $(k_8 + k_{8a})/k_{20} = 7.1$ or 6.0 at 773 K, depending on values assumed for k_{13} (see Sect. 4.3.3). Measurements of the same ratio at room temperature have given values varying between 0.6 [165] and 2 [162] at 293 K. Assigning the same activation energy to both reactions (viii) and (viiia) and taking values of 0.66 and 6.6 at 293 and 773 K, respectively gives a maximum activation energy difference $E_8 - E_{20} \simeq 2.2$ kcal . mole^{-1}. Using this in combination with the mean and the lowest of the high temperature values in turn gives $(k_8 + k_{8a})/k_{20} = 27.5$ (or 25.0) exp $(-1100/T)$.

Acceptance of the independent estimates of k_{17} and $(k_8 + k_{8a})/k_{20}$ forces one to the conclusion that k_2 must be reduced from its earlier, fixed value. It turns out that an excellent fit to the whole range of experimental data may be obtained in this way. Putting $k_{17,\text{ all molecules}} = 1.5 \times 10^9$ exp $(+250/T)$ and retaining the same temperature dependence as before for k_2 led to the Arrhenius expressions $k_2 = 1.44$ (or 1.58) $\times 10^{11}$ exp $(-8{,}250/T)$ corresponding to the two values of $(k_8 + k_{8a})/k_{20}$ given above. These values of k_2 are independent of the absolute values of k_8 and k_{20}—k_{22}. The calculations assumed, for conservation purposes, that the radical pool in the very hydrogen-rich flames consisted entirely of H atoms, and the calculation of the (small) concentrations of OH and O by means of quasi-steady state relations was appended separately for estimation of the chain breaking effects associated with these species.

A more refined method of calculation has recently been developed [173] which includes also all the reverse reactions in the mechanism, as well as reaction (xvi).

$$OH + OH \rightleftharpoons O + H_2O \qquad \text{(xvi)}$$

This method integrally employs the quasi-steady state assumptions to relate the concentrations of H, OH and O in the overall radical pool, and can be applied to either fuel-rich or fuel-lean flames. Concentrations of HO_2 were also calculated using the quasi-steady state condition, but because these were mostly much smaller than the other radical concentrations they were considered in the same manner as OH and O in the simpler method. Both methods lead to similar results for the low temperature, fuel-rich flames considered at present, indicating that the reverse reactions other than (—i) and (—iii) are relatively unimportant over most of these reaction zones. Three internally consistent sets of rate coefficients on which the more refined treatments may be based are given

TABLE 30
Equilibrium constants and rate coefficients used in computation of hydrogen—nitrogen—oxygen flames [155] (Constants are expressed as AT^{B} exp $(-C/T)$ in l.mole.sec units)

Reaction no.	Reaction		Forward rate coefficient			Equilibrium constant		
			A	*B*	*C*	*A*	*B*	*C*
(i)	$OH + H_2$	$\rightleftharpoons H_2O + H$	2.04×10^{10}	0	2550	0.21	0	−7640
(ii)	$H + O_2$	$\rightleftharpoons OH + O$		See below		300.0	−0.372	8565
(iii)	$O + H_2$	$\rightleftharpoons OH + H$	1.8×10^{10}	0	4700	2.27	0	938
(iv)	$H + O_2 + M$	$\rightleftharpoons HO_2 + M$		See below		7.449×10^{-4}	0	−23380
(viii)	$H + HO_2$	$\rightleftharpoons OH + OH$		See below		227.4	−0.372	−19625
(viiia)	$H + HO_2$	$\rightleftharpoons O + H_2O$		See below		21.0	−0.372	−28203
(xvi)	$OH + OH$	$\rightleftharpoons O + H_2O$	5.75×10^{9}	0	390	9.25×10^{-2}	0	−8578
	$H + H + H_2$	$\rightleftharpoons 2H_2$	9.2×10^{10}	−0.6	0			
(xvii)	$H + H + N_2$	$\rightleftharpoons H_2 + N_2$	1.0×10^{12}	−1.0	0	2.4×10^{-4}	0	−52590
	$H + H + O_2$	$\rightleftharpoons H_2 + O_2$	1.0×10^{12}	−1.0	0			
	$H + H + H_2O$	$\rightleftharpoons H_2 + H_2O$	6.0×10^{13}	−1.25	0			
	$H + OH + M$	$\rightleftharpoons H_2O + M$						
(xviii)	$M = H_2, N_2, O_2$		9.77×10^{11}	−0.71	0	5.943×10^{-5}	0	−59910
	$M = H_2O$		4.89×10^{12}	−0.71	0			
	$H + O + M$	$\rightleftharpoons OH + M$						
(xix)	$M = H_2, N_2, O_2$		6.2×10^{10}	−0.6	0	5.65×10^{-4}	0	−51570
	$M = H_2O$		3.1×10^{11}	−0.6	0			
(xx)	$H + HO_2$	$\rightleftharpoons H_2 + O_2$		See below		0.32	0	−29210
(xxi)	$OH + HO_2$	$\rightleftharpoons H_2O + O_2$		See below		7.97×10^{-2}	0	−36530
(xxii)	$O + HO_2$	$\rightleftharpoons OH + O_2$		See below		0.758	0	−28190

TABLE 30—*continued*
Additional forward rate coefficients

Reaction No.		Set 1			Set 2			Set 3		
	$(k_8 + k_{8a})/k_{20}$ =	6.1			12.0 exp (—540/T)			27.5 exp (—1,100/T)		
	σ_H(Å) =	3.5			3.5			2.71		
		A	*B*	*C*	*A*	*B*	*C*	*A*	*B*	*C*
(ii)		1.7×10^{11}	0	8250	1.42×10^{11}	0	8250	1.46×10^{11}	0	8250
(iv)	[a] $M = H_2$	2.6×10^{11}	—0.488	0	1.03×10^{12}	—0.72	0	8.0×10^{11}	—0.675	0
(viii)		8.8×10^{10}	0	0	1.6×10^{11}	0	540	2.72×10^{11}	0	1100
(viiia)		9.6×10^{9}	0	0	1.0×10^{10}	0	540	3.0×10^{10}	0	1100
(xx)		1.6×10^{10}	0	0	1.4×10^{10}	0	0	1.1×10^{10}	0	0
(xxi)		1.2×10^{10}	0	0	8.5×10^{9}	0	0	1.6×10^{10}	0	0
(xxiia)		[b] 2.6×10^{10}	0	0	1.6×10^{10}	0	540	8.2×10^{10}	0	1100
(xxiib)		[b] 4.8×10^{9}	0	0	1.4×10^{9}	0	0	3.3×10^{9}	0	0

[a] Chaperon efficiencies relative to H_2 are 0.35, 0.44 and 6.5 for O_2, N_2 and H_2O, respectively.
[b] $k_{22} = k_{22a} + k_{22b}$.

in Table 30. Equilibrium constants were taken from JANAF Thermochemical Tables [174], with those for reactions (i)—(iii) represented by the expressions due to del Greco and Kaufman [175].

Although the more refined calculation involves absolute values of k_8 and k_{20}—k_{22}, their precise values are not critical in the present context. The estimation of the absolute values will be discussed in Sect. 5.4.3. The important features in the present context are still the ratios $(k_8 + k_{8a})/k_{20}$, k_{21}/k_{20}, k_{22}/k_{20} and k_{8a}/k_8. The three sets of coefficients in Table 30 correspond with $(k_8 + k_{8a})/k_{20} = 6.1$, $12.0 \exp(-540/T)$ and $27.5 \exp(-1{,}100/T)$, respectively, with k_{21}/k_{20} and k_{8a}/k_8 assumed independent of temperature. In sets 1 and 3, k_{22}/k_{20} was put equal to $0.3(k_8/k_{20} + 1)$. This expression arbitrarily relates k_{22} with k_8 and k_{20} by means of a ratio of collision numbers. The major factor in estimating the remaining independent ratios is the variation of the burning velocity with initial $[H_2]/[N_2]$ ratio [158, 159]. Using the above expression for k_{22}/k_{20}, it became virtually impossible not to predict too large a change in burning velocity, however small the values chosen for k_{8a}/k_8 and k_{21}/k_{20}. The burning velocity and flame property variation were best reproduced using the smaller ratio $k_{22}/k_{20} = 0.1(k_8/k_{20} + 1)$, as in set 2, and this led to a lower k_2 than the mean of the values in sets 1 and 3. Although an unambiguous choice of a combination of k_{21}/k_{20}, k_{22}/k_{20} and k_{8a}/k_8 is not possible on the basis of the limited data considered here so far, the value of k_2 giving the optimum fit for a given $(k_8 + k_{8a})/k_{20}$ will not be much affected by the precise combination selected.

To obtain the expression for k_{4,H_2}, values at 773 K were related with k_2 by way of the second explosion limit result that $2k_2/k_{4,H_2} = 37.0$

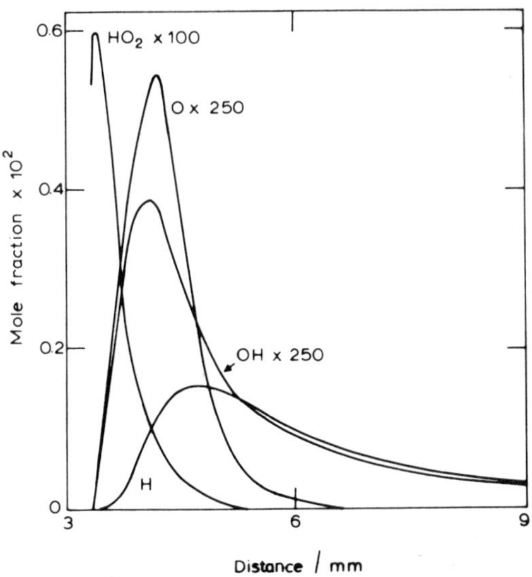

Fig. 27. Computed mole fractions of free radicals. Conditions as in Fig. 25.

torr [25, 72]. A $k = AT^B$ temperature dependence was then deduced by combining the results with $k_{4,H_2} = 1.7 \times 10^{10}$ at 298 K [176]. Because of paucity of information, "chaperon" efficiencies in reactions (iv), (xviii) and (xix) were assumed to remain constant throughout the temperature range of interest. This assumption is, however, at variance with the more detailed information now available on H atom recombination [149], and indeed with some of that which is becoming available for reaction (iv) (cf. Table 41).

Figures 25—27 show the temperature and composition profiles calculated for the "standard" flame by the refined treatment using set 2 of the rate coefficients of Table 30. Figure 25 also includes for comparison a number of points representing the observed temperature profile. Agreement is excellent. The composition profiles for the stable species in the flame were measured by means of a mass spectrometric probe, using the unburnt gas ratios of each species concentration to that of nitrogen as calibration standards. Realistic comparison is then in terms of these ratios, and is shown in Fig. 28. The relative intensities of sodium chemiluminescence in the recombination region of the low temperature flames are proportional to the square of the H atom concentrations. A comparison between theory and experiment on this basis, with intensities normalized with respect to the maximum H atom concentration and the

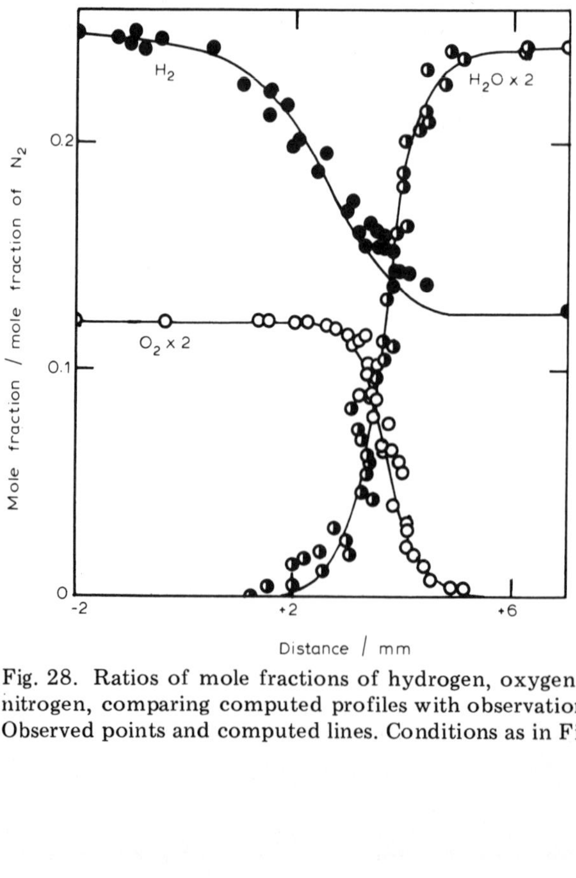

Fig. 28. Ratios of mole fractions of hydrogen, oxygen and steam to mole fraction of nitrogen, comparing computed profiles with observations of Dixon-Lewis et al. [156]. Observed points and computed lines. Conditions as in Fig. 25.

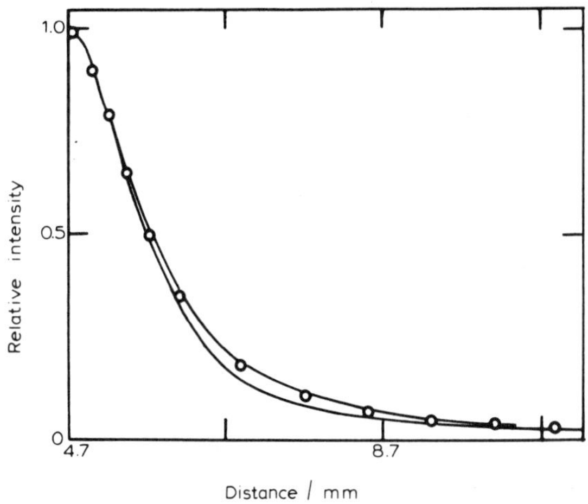

Fig. 29. Comparison of computed relative chemiluminescent intensities with profiles observed by Dixon-Lewis et al. [156]. Conditions as in Fig. 25. ○, Computed profile; lines indicate approximate error limits on observations.

peak measured intensity, is shown in Fig. 29. The curvature of the calculated line depends not only on the recombination rate coefficients (chiefly k_{17}), but also on the diffusion coefficient of H atoms in the flame system. Representing the intermolecular potentials by the Lennard-Jones 12:6 model, and with ϵ_H/k = 37.0 K [177], optimum agreement was found with $\sigma_H \simeq 3.5$ Å, and this value was used in the overall calculation. It is, however, some 25—30 % higher than the molecular diameter recommended by Svehla [177], thence giving H atom diffusion coefficients some 25 % lower in the $H_2 + N_2 + H_2O$ mixture. The kinetic

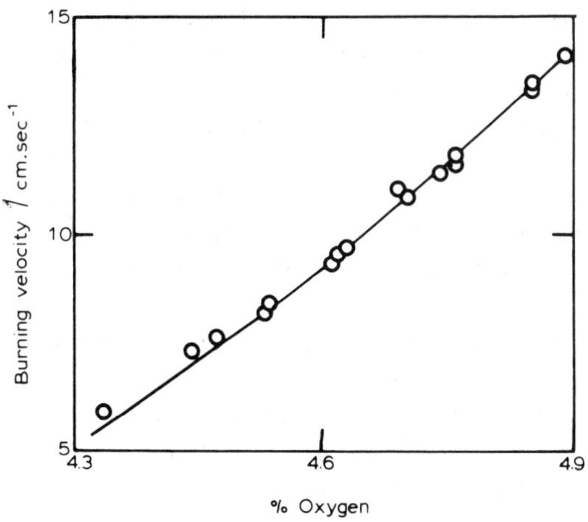

Fig. 30. Burning velocities of hydrogen + nitrogen + oxygen flames having $X_{H_2,u}/X_{N_2,u}$ = 0.246 and T_u = 336 K (after Dixon-Lewis et al. [158]). (By courtesy of The Royal Society).

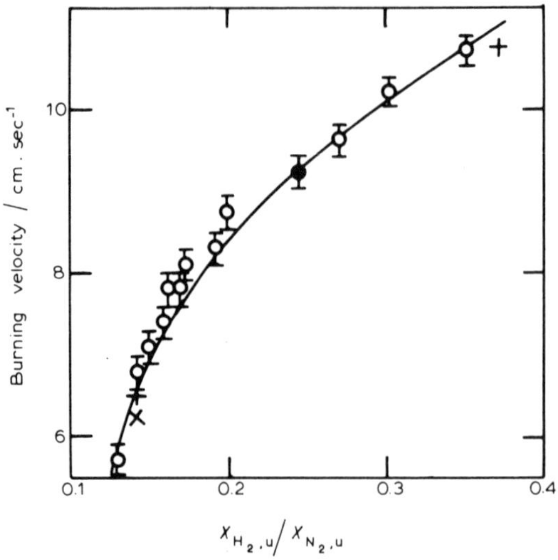

Fig. 31. Burning velocities of hydrogen + nitrogen + oxygen flames having $X_{O_2,u}$ = 0.0460 and T_u = 336 K, showing dependence on the initial mole fraction ratio $X_{H_2,u}/X_{N_2,u}$ (after Dixon-Lewis et al. [158]). Line represents values calculated by Dixon-Lewis et al.; ● and x, additional calculations using sets 1 and 3 of rate coefficients in Table 30; ● and +, additional calculations using set 2 of Table 30 (+ at each end). (By courtesy of The Royal Society.)

rate coefficients and the overall flame properties other than the H atom profile are not much affected by the substitution. In the case of H atoms, the lower diffusion coefficient (higher σ_H) gave a higher $X_{H,max}$ and a larger curvature to the profile in the recombination region.

For the range of rich, slow burning flames considered, Fig. 30 and 31 show the effect of composition on burning velocity. In Fig. 30 the ratio $X_{H_2,u}/X_{N_2,u}$ was kept constant and the initial oxygen concentration was varied. In Fig. 31 $X_{H_2,u}/X_{N_2,u}$ was varied at constant oxygen. The complete lines in both figures were calculated using an early set of rate coefficients, with $k_{17} = 4.5 \times 10^9$. The major composition effects are observed in Fig. 31, in which all the flames have nearly the same final temperature. Using the sets of rate coefficients given in Table 30, the calculated burning velocities at the middle and ends of this line are as indicated in the legend.

Turning now to the radical concentrations, a further important feature of the more recent theorectical results with lower recombination rate coefficients is that, although the H atom concentration profile retains the same shape as before, the absolute concentrations are now higher. The peak mole fraction rises from $X_{H,max} = 1.07 \times 10^{-3}$ (for $k_{17} = 4.5 \times 10^9$ and $\sigma_H = 2.25$ Å) to 1.56×10^{-3} (for k_{17} as in Table 30 and $\sigma_H = 3.5$ Å). This results in a corresponding reduction in the rate coefficients k_{1Da}, k_1 and k_{23}, mentioned earlier, to $k_{1Da} = (6.6 \pm 0.4) \times 10^8$, $k_1 = (1.85 \pm 0.3)$

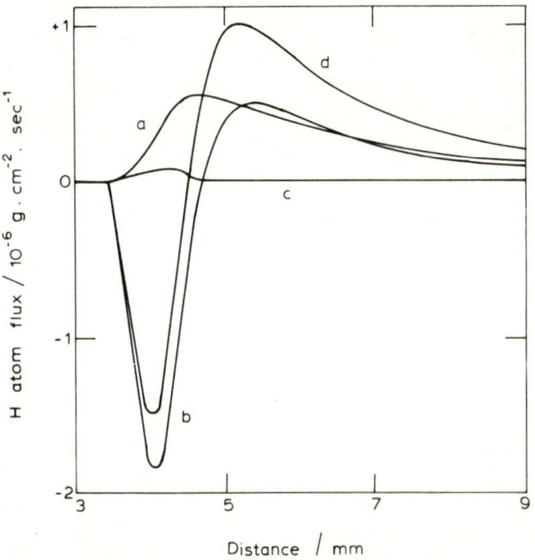

Fig. 32. Computed fluxes of hydrogen atoms in flame of Fig. 25. (a) Convective flux, Mw_H (see eqn. (64)); (b) ordinary diffusional flux, j_H^c; (c) thermal diffusional flux, j_H^T; (d) overall flux, MG_H.

x 10^9 and k_{23} = (1.65 ± 0.1) x 10^8 at 1050 K. A further 30 % reduction of k_{17} below Table 30 was also investigated. It necessitated a still further reduction of k_2 to around 1.2 x 10^{14} exp (—8,250/T) in order to fit the flame properties, and an increase in $X_{H,max}$ to 1.77 x 10^{-3}. Optimization of the agreement of k_1 and k_{23} with independent estimates [178] thus further supports the values of k_{17} in Sect. 5.4.1.

Finally, the way in which the dominant reactions change as the gases pass through the flame front is worthy of special note. Figure 32 shows the hydrogen atom fluxes in the "standard" flame, with positive values denoting fluxes from left to right, or from cold to hot in the actual flame. The gradient of curve d at any position defines the rate of formation of H atoms at that position in the flame. At low temperatures this gradient is negative and the molecular oxygen concentration is high: cycles (iva) and (ivb) (and the similar cycles using reactions (xxi) and (xxii)) are dominant in this region of the flame. Between about 900 and 1050 K the slope is positive, and here the chain branching cycle is competing successfully with the termination steps. Above 1050 K virtually no oxygen is left and the gradient again becomes negative. This is a region where recombination is principally due to reaction (xvii), with assistance also from (xviii). Additional features are the apparently minor roles of the H_2O_2-forming reactions (x) and (xi) in the rich flame mechanism. This has been discussed by Dixon-Lewis [160]. For the most probable rate coefficients, the concentrations of HO_2, H and H_2 are such that reaction (xi) never becomes important, while reaction (x) may occur appreciably only in a very small region at the start of the reaction zone (see Fig. 27).

5.4.3 Radical recombination in near-stoichiometric and fuel-lean systems

The decay of the hydroxyl radical concentration in the burnt gas of a number of lean hydrogen—air flames supported on a water-cooled porous plate burner was measured by Kaskan [179] using UV absorption. Flame temperatures lay between 1300 and 1650 K. Assuming equilibration of reactions (i), (ii) and (iii) according to the partial equilibrium hypothesis, the observed decay was too fast to be accounted for by reactions (xvii) to (xix). Fenimore and Jones [180] have probed a number of lean hydrogen flames at reduced pressures on a similar porous plate burner, measuring H atom concentrations by studying the rate of reaction of traces of added nitrous oxide by

$$H + N_2O \rightarrow OH + N_2$$

They found the heat release rate to be proportional to the product $[H][O_2][H_2O]$, and the dependence of H on pressure and mass flow to be also consistent with the removal of H by reaction (iv). Similar conclusions about the recombination were reached by Getzinger and Schott [181] from shock tube experiments, in which OH concentrations were measured and used to calculate total radical concentrations by means of the partial equilibrium assumption.

Quantitative studies of the recombination following shock induced ignition of lean hydrogen—oxygen mixtures have been used, notably by Getzinger et al. [85, 151, 181] to give rate coefficients for reaction (iv). The calibration of the OH absorption requires great care. Since the ignitions are carried out in the presence of a large excess of inert diluent, the results depend mostly on reaction (iv) with M = diluent. In the interpretation it was assumed that the HO_2 formed is rapidly removed in essentially irreversible bimolecular reactions that do not change the number of moles in the system [181]. Mean results are given in Table 31, relating principally to the temperature range 1300—1900 K. Within the narrow range from 1300—1600 K the temperature dependence is within the uncertainty of the results.

The hypothesis that the HO_2 formed in reaction (iv) is rapidly removed (thus preventing its re-dissociation) has recently been examined for flame systems by Dixon-Lewis et al. [182]. On the assumption of equilibration of the fast, bimolecular, electron spin conserving reactions (i), (ii) and (iii), it is possible to compute concentration profiles for all the chemical species in the recombination region of a wide variety of flame systems. The calculation requires knowledge of the rate coefficients k_4, k_8, k_{8a} and k_{17}—k_{22}, which control the rate of electron spin removal (recombination). The rate of recombination via HO_2 is calculated as the difference

TABLE 31
Third order rate coefficients for $H + O_2 + M = HO_2 + M$ from shock ignition of lean H_2—O_2—diluent mixtures

M = N_2	Ar	H_2O	O_2	Temp. (K)	Ref.
	$(1.42 \pm 0.24) \times 10^9$			1500	181
	$3.2 \times 10^{12}\ T^{-1.0}$			1500—2200	96
2.1×10^9	2.2×10^9	5.4×10^{10}	$\leqslant 4.3 \times 10^9$	1300—1900	151
	$(3.0 \pm 1.3) \times 10^9$			1400—1900	85

between the forward and reverse rates of reaction (iv), with the small concentration of HO_2 in the systems given by the quasi-steady state equation

$$[HO_2] = \frac{\{k_4[H][O_2][M] + k_{-20}[H_2][O_2] + k_{-21}[H_2O][O_2] + k_{-22}[OH][O_2] + k_{-8}[OH]^2 + k_{-8a}[O][H_2O]\}}{\{k_{-4}[M] + (k_8 + k_{8a} + k_{20})[H] + k_{21}[OH] + k_{22}[O]\}} \quad (69)$$

Comparison of the computed profiles with experiment may in principle be used to establish values of some of the unknown rate coefficients. The radical pool in this computation includes molecular oxygen as a bi-radical. The validity of the partial equilibrium assumptions will be discussed in Sect. 5.4.4.

Using the calculation technique described, with k_4, k_{17} and k_{18} taking the values in Table 30, a survey of a number of published flame recombination investigations in both rich and lean systems leads to the assessment, shown in Table 32, of the relative importance of the net contributions of the three primary recombination steps at approximately the centre of each range of measurement. Clearly, results with sufficiently fuel-rich flames should be capable of providing reliable values of k_{17}, while in lean flames recombination is principally by way of HO_2 formation. On the other hand, k_{18} is always more difficult to measure reliably, since reaction (xviii) is not the exclusive recombination step in any system. The recombination in lean flames depends also on the fate of the intermediate HO_2. This in turn depends on the rate coefficients k_{-4}, k_8, and k_{20}—k_{22}.

The reactions of H atoms with HO_2 have been discussed by Dixon-Lewis and co-workers [159]. The numerical side of their argument is modified slightly here to accommodate new information obtained from a recent re-interpretation by Baldwin et al. [73] of their second limits in boric acid coated vessels (Sect. 4.3.3 and Table 18). This gives $(k_8 + k_{8a})^2/k_2^2 k_{10} = 0.325$ at 773 K to correspond with $(k_8 + k_{8a})/k_{20} = 12.0 \exp(-540/T)$. Two values of k_{10} are available at room temperature: (i) $(1.8 \pm 0.2) \times 10^9$ due to Foner and Hudson [184], and (ii) $(2.2 \pm 0.3) \times 10^9$ due to Paukert and Johnston [185]. Assuming $k_{10} = 2 \times 10^9$

TABLE 32
Relative importance of primary association reactions in flame recombination regions

Flame	p(atm)	$(H_2/O_2)_u$	$(N_2/O_2)_u$	V_b[a] (cm.sec^{-1})	Approx. temp. range of study	Approx. % primary recombination by (xvii) H + H + M	(xviii) H + OH + M	(iv) H + O_2 + M	Ref.
A	1.0	4.16	4.59	133	1680—1825	74	26	V. small	154
B	1.0	4.19	3.865	157	1750—1840	74	26	V. small	154
C	1.0	3.30	5.48	118	1825—1850	60	40	0.2	154
D	1.0	2.70	6.09	107	1740—1840	45	53.5	1.5	154
E	1.0	5.22	4.97	88	1580—1660	87	13	V. small	154
F	1.0	4.44	5.77	76	1540—1650	83	17	V. small	154
G	1.0	2.93	7.37	65	1540—1640	65	35	0.5	154
H	1.0	2.05	3.76	27.5	1655—1680	7	67	26	137
I	0.5	3.48	3.76	51.0	1190—1320	92	8	0.2	137
J	0.5	2.38	3.76	35.3	1420—1540	52	44	4.0	137
K	1.0	1.00	3.76	33.4	1520—1530	V. small	0.7	99.3	179
L	1.0	1.60	3.76	16.8	1500—1530	V. small	2.0	98.0	179
M	0.45	1.00	3.76	18.8	1370—1410	V. small	0.6	99.4	179
N	0.45	1.60	3.76	11.9	1320—1435	V. small	1.6	98.4	179
O	1.0	1.67	4.00	168	1925—2150	1.0	33	65	183
P	1.0	1.54	3.61	168	1950—2160	0.6	29	70	183
Q	1.0	1.43	3.29	168	1950—2160	0.5	25	74	183
R	1.0	1.33	3.00	168	1960—2160	0.4	22	77	183

[a] V_b gives linear burnt gas velocity corrected to standard conditions of 298 K/p atm.

independent of temperature, and using $k_2 = 3.3 \times 10^6$ then leads to $(k_8 + k_{8a}) = 8.5 \times 10^{10}$ and $k_{20} = 1.42 \times 10^{10}$ at 773 K. Assuming k_{20} to be also independent of temperature, we obtain the Arrhenius form $(k_8 + k_{8a}) = 1.7 \times 10^{11} \exp(-540/T)$, and at 293 K, $(k_8 + k_{8a}) = 2.47 \times 10^{10}$. These figures give the sum $(k_8 + k_{8a} + k_{20}) = 3.9 \times 10^{10}$ at 293 K. Albers [186] finds $(k_8 + k_{8a} + k_{20}) \leqslant 2 \times 10^{10}$ at 293 K, and his results thus suggest somewhat lower values than the above for the three rate coefficients at room temperature. A small additional activation energy (ca. 650 cal. mole^{-1}) for both reactions (viii) and (xx) would permit satisfaction of Alber's criterion as well as the conditions at 773 K. Alternatively, set 3 of the rate coefficients in Table 30 already gives a sum at room temperature which satisfies Alber's criterion completely. However, both alternatives also require a rather high pre-exponential factor A_8.

Reactions (xxi) and (xxii) may a priori be expected to become more important in lean flames, and eventually to overtake reactions (viii) and (xx). The radical concentrations in lean flames are probably such that reaction (xxi) dominates. However, because both reactions (xxi) and (xxii) increase in importance together, their separation is again difficult.

The key to the situation lies in considering flames K, L, M and N of Table 32. In each of the pairs K and M, and L and N, the initial gas compositions are the same, and the OH concentrations in the recombination regions studied also cover the same range. The difference between the members of each pair is that the flames K and L burn at one atm pressure, while flames M and N burn at 0.45 atm. This pressure difference alters the balance of competition in the denominator of eqn. (69) between re-dissociation of HO_2 and its further reaction with H, OH and O. Using approximate values for all the rate coefficients concerned, it turns out that in the 0.45 atmosphere flames all the primary HO_2 formed in reaction (iv) effectively undergoes full recombination. Hence the measured [OH] profiles here depend virtually entirely on the value of k_4, and may be used for its determination. Having thus determined k_4, the measurements in the flame at one atmosphere may then be used to investigate k_{21} and k_{22}.

An initial difficulty with this approach was that Kaskan's recombination results for both lean and rich flames [137, 179] were obtained using OH absorption measurements, and the absolute calibration caused problems due to some uncertainty about the absorption coefficients. However, Professor Kaskan (private communication) has kindly provided the information that the f_{00}-value of OH used in his original publication was Oldenberg and Rieke's original value [139] of 12.3×10^{-4}, and he has also provided estimates of factors by which his published concentrations must be multiplied to allow for Doppler broadening of the emission lines from the source lamp (1.1) and pressure broadening of the absorption lines in the flame (1.36 for flames at 1 atm and 1500 K; 1.18

for flames at 0.5 atm and 1500 K). Now the recombination in the fuel-rich flames I and J of Table 32 is mostly controlled by reactions (xvii) and (xviii), whose rate coefficients have already been discussed in Sects. 5.4.1 and 2. By adjusting the calibration of the measured [OH] for these flames so that the gradients of the profiles of $[OH]^{-1}$ match the corresponding computed profiles, we can then estimate a calibration factor for the OH concentrations in all the flames. This is, of course, performed after the appropriate line broadening corrections have been applied, and is essentially a kinetic determination of the *f*-value. This calibration will be discussed in more detail elsewhere. After some optimization by iteration between flames H, I, J, K, L, M and N of Table 32, it leads to an f_{00}-value for OH of 9.5×10^{-4}. Adopting this calibration, and assuming the "chaperon" efficiencies given in Table 30 for reaction (iv),* Kaskan's recombination results in flames M and N at 0.45 atm are consistent with $k_{4,H_2} = 1.03 \times 10^{12}\ T^{-0.72}$ (Table 30), leading to $k_{4,H_2} = 5.6 \times 10^9$ at 1400 K.

Finally in connection with the OH calibration, it should be noted that the *f*-value derived here is identical with a recent re-determination by Rouse and Engleman [189] using the same method as Oldenberg and Rieke, as well as with Oldenberg and Rieke's original value when the latter is corrected for changes in the thermochemistry of OH and for vibration—rotation interaction. It is about 6 % higher than the mean of a number of recent determinations from the radiative lifetime. It is, however, some 33 % higher than the value found by Golden et al. [190], who generated OH from H + NO_2 in a discharge-flow system. If the higher value is correct, this will in turn have repercussions on some of the other determinations of rate coefficients to be discussed in Sect. 6.

To continue the present kinetic discussion, if k_4, k_8, k_{8a}, k_{17}—k_{20}, and k_{22} are given values as in Table 30, then the lines in Fig. 33 show recombination results for flame L of Table 32, computed using a number of assumed values of k_{21}. The points show the measurements of Kaskan, recalibrated as above for atmospheric pressure. Comparison of theory and experiment yields $k_{21} = (8 \pm 4) \times 10^9$ at about 1530 K. In constructing Table 30, k_{21} was assigned the value 8.5×10^9, and was assumed to be independent of temperature. The rate coefficients k_8 and k_{20}—k_{22} in the Table were obtained by iteration between the lean flame recombination results and the rich, lower temperature flame structure results discussed in Sect. 5.4.2.

Again using the rate coefficients from Table 30, Table 33 shows the

* It has recently become apparent (cf. Sect. 6.5, Table 41 and Fig. 41) that the chaperon efficiency of nitrogen (relative to $H_2 = 1$) varies with temperature, and that $k_{4,N_2}/k_{4,H_2}$ may be in the region of 0.28 at 1500 K. However, because of the very high chaperon efficiency of water vapour, the change of $k_{4,N_2}/k_{4,H_2}$ from 0.44 to 0.28 only affects the average $k_{4,M}/k_{4,H_2}$ by about 5 % in the burnt gas of these flames.

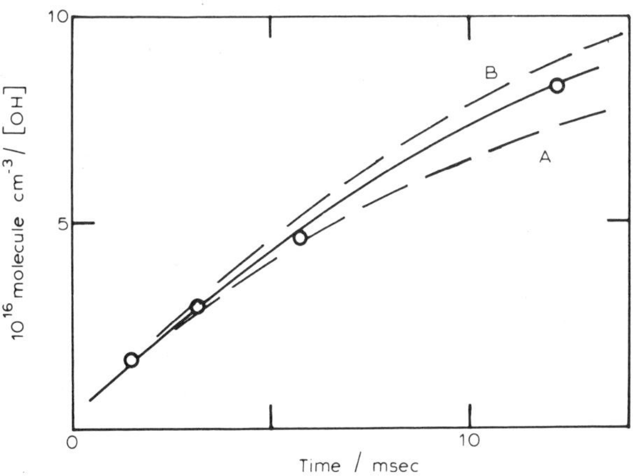

Fig. 33. Recombination in lean hydrogen + nitrogen + oxygen flames. Comparison of measured points of Kaskan [179] for flame L of Table 32, re-calibrated as described in text, with computed lines. Solid line, rate coefficients as in set 2 of Table 30; broken lines, as set 2 of Table 30, but with k_{21} = 4 x 10^{12} (curve A) and 1.2 x 10^{13} (curve B).

fate of the HO_2 formed in flames K to R of Table 32. It transpires that in all these flames a good half or more of the HO_2 emerging from the forward reaction (iv) undergoes eventual complete recombination. Regarding the determination of k_{21}, it also turns out that intermediate temperature flames like K and L offer the best opportunity in terms of competition between reactions (xx), (xxi) and (xxii). At the higher temperatures used by Friswell and Sutton [183], the competition of the re-dissociation of HO_2 with reaction (xxi) should be more favourable for the determination of k_{21}. However, the combination of the temperature dependence of reaction (viii) with the higher concentrations of H relative to OH which occur in these flames, causes reaction (viii) to dominate the

TABLE 33
Fate of hydroperoxyl in lean flames of Table 32, at approximate mid-points of range of investigation

Flame	Approximate % HO_2 reacting			
	By redissociation	With H	With OH	With O
K	46	1.6	49	3.2
L	48	3.5	46	2.1
M	10.4	14	65	10.7
N	14.2	22	57	6.4
O	36	46	14	4
P	46	34	16	4.5
Q	42	35	17	5.5
R	46	30	18	6

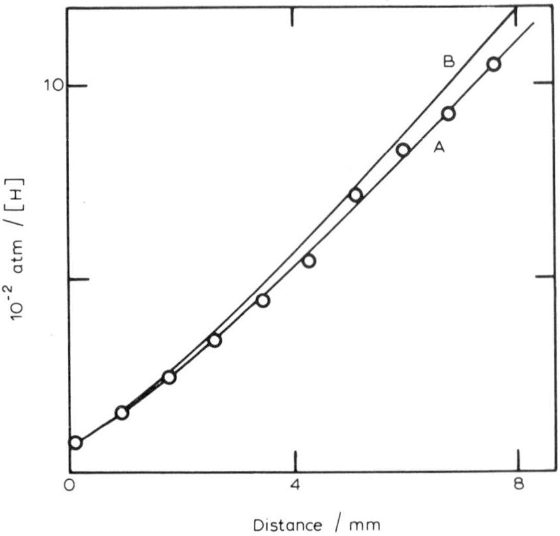

Fig. 34. Recombination in lean hydrogen + nitrogen + oxygen flames. Comparison of measured points of Friswell and Sutton [183] for flame O of Table 32 with lines computed using rate coefficients as in set 2 of Table 30. Temperature ranges: line A, 1833—2152 K; line B, 1797—2129 K.

recombination. Because of this, and because the re-dissociation reaction (—iv) is still not large enough to dominate the fate of the HO_2 at their temperatures, the analysis of their results given by Friswell and Sutton is incorrect.

An additional matter of importance in the analysis of high temperature recombination results (> 2000 K) is the degree of dissociation into atoms and radicals at full equilibrium. To illustrate this, and to draw attention to the necessity for very precise temperature measurement in such investigations (ideally, measurement of the temperature profile in the recombination region, in order to eliminate errors due to heat losses), Fig. 34 shows recombination profiles for flame O of Table 32. The lines 1 and 2 show profiles calculated, again using the rate coefficients of Table 30, but on the assumption that recombination occurs over temperature ranges differing by only about 20 K. Friswell and Sutton, whose results are shown by the points in Fig. 34, quote a single temperature of 2130 K, measured by the method of sodium D-line reversal. Bearing in mind the accuracy of this method of temperature measurement above 2000 K, their recombination results are reasonably in accord with the rate coefficients of Table 30.

Lastly, the parameters given in Table 30 for reaction (xviii), when taken in conjunction with the other parameters in the Table, are consistent with both the flame structure and flame recombination data [155]. However, as already discussed, k_{18} is the least directly accessible of the

TABLE 34
Third order rate coefficients for H + OH + M = H_2O + M from shock ignition of near-stoichiometric H_2—O_2—diluent mixtures

M =	N_2	Ar	H_2O	H_2	Temp. (K)	Ref.
		$10^{10}-10^{11}$			1000—2600	150
		$\leqslant(6 \pm 4) \times 10^9$			1400—2000	138
		$(5.4 \pm 2.7) \times 10^9$			1307—1846	187
	8.6×10^9	3.3×10^9	6.6×10^{10}	$<1.6 \times 10^{11}$	1630—1750	151
		$\leqslant 1.5 \times 10^{11}\ T^{-0.5}$			1930—2165	96
	$(1.1 \pm 0.3) \times 10^{10}$	$(2.7 \pm 0.7) \times 10^9$	$(5.0 \pm 1.3) \times 10^{10}$		1220—2370	152

three more important primary recombination rate coefficients. Similar remarks apply to the evaluation of k_{18} from shock tube results. Results from this source are given in Table 34. Those for argon, nitrogen and steam are in moderate agreement amongst themselves when account is taken of error limits.

5.4.4 Partial equilibrium and quasi-steady state hypotheses in the flame and shock tube kinetics

The kinetic analyses of the recombination region in both the flame and the shock-induced ignition is very much simplified, and indeed only became practicable initially, with the use of the partial equilibrium (p.e.) assumptions already described in Sect. 5.4.1). By considering the growth of a radical pool consisting of H and O atoms, hydroxyl radicals, and molecular oxygen as one moves backwards through the flame from the hot end it is possible, as already indicated, to calculate profiles of temperature and all the species concentrations in the system. The p.e. assumptions on reactions (i), (ii) and (iii) are in this case used to divide the radical pool into its separate components at each step of the integration, while the overall size of the pool is determined by its (backward) growth consequent upon the recombination steps. The complete p.e. approach can thus only be used to examine the recombination region. On the other hand, by using an alternative construction of the radical pool it is also presumably possible to introduce kinetic control of one or more of reactions (i)—(iii), if desired, while keeping the remaining steps in a balanced condition.

At the other end of the spectrum of possible approximations lies the detailed kinetic consideration of the growth and decay of each radical species, without approximation. Although this was possible analytically in the treatment of the early stages of ignition of shocked hydrogen—oxygen mixtures, numerical attempts to deal with the later stages of ignition may encounter mathematical difficulties due to "stiffness" in the differential equations. The straightforward integration of the stationary flame equations also becomes impractical due to the occurrence of more than one unknown boundary condition at the start of a working integration [173].

An extremely useful intermediate approach, which is capable of handling the whole flame reaction zone, is that employing the quasi-steady state (q.s.s.) assumption, referred to in Sect. 5.4.2. In this case a radical pool consisting only of H, OH and O is considered. The growth of the overall pool is now effectively determined by reaction (ii), and its decay by the recombination steps. Its sub-division into the separate components is carried out in rich flames by way of the q.s.s. assumptions on OH and O. In more precise terms, the overall mass flux of free radicals is expressed as a mass flux of H atoms by defining the composite mass flux fractions (cf. eqn. (64))

$$G_{rad} = G_H + \tfrac{1}{8}G_O + \tfrac{1}{17}G_{OH}$$

$$G^*_{H_2} = G_{H_2} - \tfrac{1}{4}G_O - \tfrac{2}{17}G_{OH}$$

$$G^*_{H_2O} = G_{H_2O} + \tfrac{9}{8}G_O + \tfrac{18}{17}G_{OH}$$

Then, using a procedure similar to that described in Sect. 5.4.2, the growth and decay of the composite fluxes are controlled by reaction cycles like (iia), (iva), (ivb), (xvii), etc. The gradients of the mass fluxes of OH, O and HO_2 in the stationary, one-dimensional flame are given by the equations

$$\partial \mathbf{F}_{OH}/\partial y = q_{OH} \qquad (62a)$$

$$\partial \mathbf{F}_{O}/\partial y = q_{O} \qquad (62b)$$

$$\partial \mathbf{F}_{HO_2}/\partial y = q_{HO_2} \qquad (62c)$$

where q represents the overall mass rate of formation of the species, and for constant values of these mass flux gradients the following conditions hold, viz.

$$\sum_i \partial q_{OH}/\partial X_i \cdot \partial X_i/\partial y + \partial q_{OH}/\partial T\, \partial T/\partial y = 0$$

$$\sum_i \partial q_{O}/\partial X_i \cdot \partial X_i/\partial y + \partial q_{O}/\partial T\, \partial T/\partial y = 0 \qquad (70)$$

$$\sum_i \partial q_{HO_2}/\partial X_i \cdot \partial X_i/\partial y + \partial q_{HO_2}/\partial T\, \partial T/\partial y = 0$$

The q.s.s. condition is then inserted at the working hot boundary, which represents a perturbation of full equilibrium, by introducing $q_{OH} = q_O = q_{HO_2} = 0$ there, together with trial values for the unknown q_{O_2} at the boundary. This last quantity provides the single boundary condition which must be guessed. For each q_{O_2}, the remaining conditions governing the hot boundary composition are provided by the various atom conservation equations and the conservation of energy. The validity of the q.s.s. assumption depends on the net rates of formation q_{OH}, q_O and q_{HO_2} remaining always a small difference between large rates of formation and removal by the elementary reaction processes. The application of the overall procedure to flame computation, and its adaptation for fuel-lean flames, is described by Dixon-Lewis et al. [173].

The range of validity of the partial equilibrium assumptions in specific flames may now be examined by comparison of the H, OH, O and O_2 profiles computed on this assumption with those computed by means of the q.s.s. condition. The p.e. assumption gives profiles which continue to rise indefinitely on integration backwards from the hot boundary of the flame. It can also be shown that the q.s.s. overall radical profile, represented by ($X_H + 2X_O + X_{OH}$) approaches the similar p.e. profile (i.e.

$X_H + 2X_O + X_{OH}$ again) from underneath as the gases move from the cold to the hot side of the flame, and that the q.s.s. molecular oxygen profile approaches the corresponding p.e. profile from above. For given recombination kinetics therefore, the p.e. profile gives a maximum possible rate of rise of the overall radical concentration on moving backwards from the hot boundary. However, the distribution of the pool between H, O and OH may be such that, for example, the comparatively small oxygen atom concentration appreciably overshoots its p.e. value in rich flames. Attention has been drawn by Dixon-Lewis [123] to the departure of the [H]/[OH] ratio from its p.e. value in a fuel-rich hydrogen—nitrogen—oxygen flame, while Hamilton and Schott [188] have also shown the possibility of oxygen atom overshoots in hydrogen—oxygen shock tube kinetics, particularly in rich mixtures.

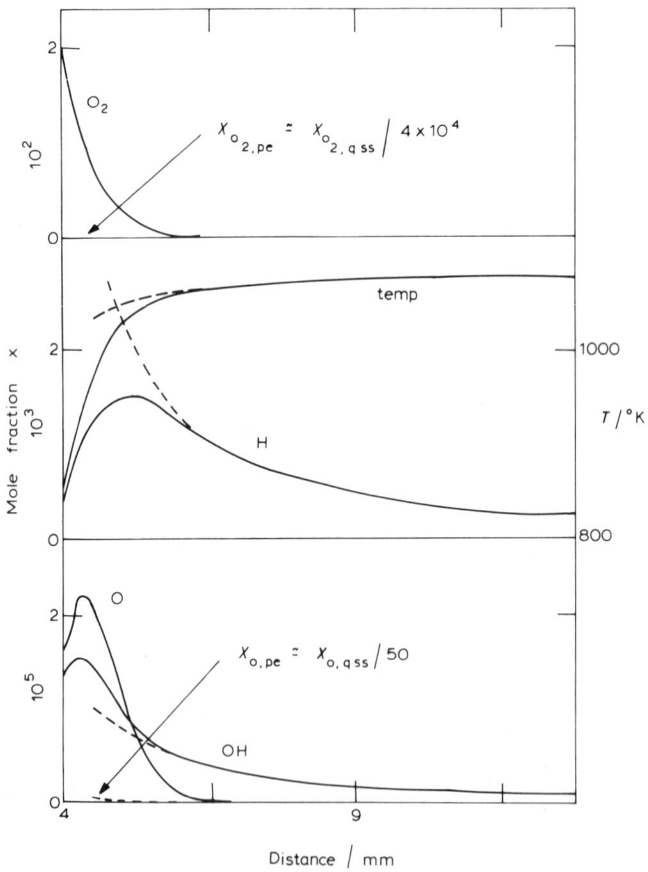

Fig. 35. Computed quasi-steady state and partial equilibrium profiles for "standard" flame. Conditions as in Fig. 25. Solid lines, q.s.s. profiles; broken lines, p.e. profiles (only marked when distinguishable from q.s.s.).

For the fuel-rich flame already illustrated in Figs. 25—29, Fig. 35 compares the radical profiles, the molecular oxygen profiles and the temperature profiles calculated by the p.e. approach with those obtained from the full flame calculation based on the q.s.s. condition. For both atomic and molecular oxygen the concentrations in the reaction zone are clearly very far from those given by the p.e. calculation. The q.s.s. calculation leads to an O atom "spike" with concentrations up to 50 or 60 times the p.e. value. At a distance of 9.5 mm in Fig. 35 the q.s.s. O atom concentration is still some 25 % above the p.e. value; while even at much greater distances (20.0 mm) the low q.s.s. molecular oxygen mole fraction of about 1.5×10^{-9} is still some twenty times above that at partial equilibrium. For OH, the p.e. assumption has higher validity than for O or O_2, with the q.s.s. condition still, however, producing some overshoot above the p.e. case. Limited experience to date suggests that for lean and stoichiometric flames, where the concentrations of OH and O are relatively much higher, the overshoot phenomena occur to a much smaller extent, if at all. The departures from the p.e. profiles are probably similar to that for H atoms in Fig. 35.

From the view-point of determination of recombination rate coefficients using measurements of H atom concentrations for example, the overshoot phenomena mentioned do not invalidate the p.e. approach, since the concentrations of the overshooting species are too low to contribute to the overall radical concentrations in the recombination region. It is more likely that the conditions in many actual flames are such that the p.e. assumption will predict slightly too rapid a recombination rate from a given set of rate coefficients. In some circumstances, however, O atom overshoot may influence the accuracy of prediction of rates of O atom reactions in flames using the p.e. assumptions. This may need careful consideration, for example, before attempting to calculate nitric oxide formation by the Zeldovich mechanism.

6. Rate coefficients of elementary processes

Because of the complexity and subtleties of the complete system of some twenty steps now established as constituting the hydrogen—oxygen reaction mechanism, studies of the type already discussed, which have been instrumental in establishing the mechanism, are not always most suitable for the determination of the rate coefficients of the elementary steps. In addition, most of these studies belong by their very nature to a fairly limited temperature regime, or more particularly to a limited range of reciprocal temperature. Fortunately the overall studies have been supplemented by direct studies of many of the elementary processes, or at least of much simpler-systems, over more extended temperature ranges. Many of the more direct studies have taken place at or near room temperature, and for long there was little information available between about 300 and 700 K — representing quite a large range of reciprocal temperature. For some of the more important reactions in the H_2/O_2 system, however, this gap also has now been filled.

Studies of the overall reaction or the elementary processes in this lower temperature range demand some means of perturbing the purely thermal system. Such perturbation may take the form of (a) a continuous perturbation in a static system, leading to a steady reaction rate, as for example in the early studies of the mercury-photosensitized reaction described by Hinshelwood and Williamson [1], (b) a continuous perturbation such as may be produced by an electric, r.f., or microwave discharge in a fast flowing system, followed by measurement of the chemical change as a function of distance along the tube, or (c) a short-lived perturbation of a static system from an external source, e.g. flash photolysis, pulse radiolysis, followed by direct observation of the chemical relaxation of the system as a function of time. In the case of the discharge methods, the radio frequency or microwave techniques are to be preferred to the electrical discharge, since they cannot produce contamination from electrodes. Methods which have been used for the direct observation of the transient species involved include optical absorption spectroscopy, isothermal calorimetric probe techniques (e.g. refs. 147, 191, 192), electron spin resonance spectroscopy (e.g. ref. 193), molecular beam sampling into the ionizing region of a mass spectometer (e.g. ref. 194) and techniques using indicators in order for example to induce measurable chemiluminescent emission which is proportional to the concentration of the transient (e.g. ref. 195).

As techniques for the study of very fast reactions have become more readily available, there has during the last decade been a corresponding vast increase in the amount of data on fast elementary reaction steps. The data relating to the hydrogen—oxygen system has recently been thoroughly collated by Baulch et al. [55], and it is proposed here only to add new information in selected areas, and in relation to the more recent flame and other work described in Sect. 5. As further more precise measurements have become available for certain elementary steps over continuous and large temperature ranges, it has become clear from the experimental side that representation of the rate coefficients over the entire temperature range is not always simple. For entropy reasons, plots of $\log_{10} k$ vs. reciprocal temperature may on occasion exhibit curvature corresponding with an apparent activation energy change of some few kcal . $mole^{-1}$ between say 500 and 2000 K [196]. If precise rate coefficients are required, it is therefore necessary to use a more complex expression than $k = A \exp(-E/RT)$ for representation, or alternatively to use a two parameter fit to the Arrhenius expression (or some alternative) over more limited temperature ranges.

Some of the high temperature rate data for this reaction has already been given in Tables 24 and 29. Further absolute measurements, including those at lower temperatures, are summarized in Table 35, and the whole is plotted in standard Arrhenius form in Fig. 36. The solid line in the figure corresponds with the simple Arrhenius expression $k_1 = 2.2 \times 10^{10} \exp(-2{,}590/T)$ recommended by Baulch et al. [55] for the temperature range 300—2500 K. Although it fits the data moderately over this temperature range, considerable deviation from the data of Smith and Zellner [197] occurs at lower temperatures.

Also relevant to the discussion are data on the ratio k_1/k_{23}, where (xxiii) is the reaction $OH + CO = CO_2 + H$. These are summarized in Table 36 and plotted as $\log_{10}(k_1/k_{23})$ versus $10^3/T$ in Fig. 37. Taking log (k_{OH+H_2}/k_{OH+CO}) to be represented by the straight line in Fig. 37 gives $k_1/k_{23} = 77.5 \exp(-2{,}210/T)$, and combining with $k_{23} = 1.5 \times 10^4 T^{1.3} \exp(+385/T)$ [215] then leads to

$$k_1 = 1.17 \times 10^6 T^{1.3} \exp(-1{,}825/T) \qquad (71)$$

Figure 38 shows the absolute rate coefficients plotted as $\log_{10}(k_1 T^{-1.3})$ versus $10^3/T$, with the solid line corresponding with the expression (71). Rejecting the measurements of Avramenko and Lorentso [198] and Schott [150] from the evaluation a priori, the results of Browne et al. [203], Eberius et al. [168], and Westenberg and de Haas [208] all lie systematically below this predicted line. However, the above expression for k_{23} recommended by Baulch and Drysdale [215] also lies above the measurements of Westenberg and de Haas [208] on that reaction, and for the single case where the measurements of the latter authors on reactions (i) and (xxiii) have been carried out at nearly the same temperature, their ratio k_1/k_{23} is close to the evaluation of Fig. 37. All their results may therefore be systematically low.

Turning now to the results of Browne et al. [203] and Eberius et al. [168], both of these are from flame studies, and they are the only results quoted which require precise measurements of absolute concentrations of OH. Since these measurements have been made by UV absorption, uncertainties about the oscillator strength (*f* number) of OH may therefore affect both sets of results. Further investigation is therefore still needed. The question of calibration of UV absorption measurements to give absolute concentrations of OH has already been considered in Sect. 5.4.3. At present, suggested error limits on the values of k_1 calculated from eqn. (71) are ±20 % at 250 K, increasing to ±50 % above 1000 K and up to 2500 K.

TABLE 35
Absolute measurements of k_1

k_1 ($l.mole^{-1}.sec^{-1}$)	Temp. (K)	Method and comments	Ref.
$4.2 \times 10^9\ T^{1/2} \exp(-5{,}000/T)$	378—489	D.F. OH by discharge through water vapour, with H_2 added downstream. [OH] measured by UV absorption. Source of OH at fault (199). Results invalid.	198
$3 \times 10^{11} \exp(-3{,}020/T)$	1000—2600	Shock tube. $H_2/O_2/Ar$ mixtures. [OH] by UV absorption (absolute concentration required). Interpretation by comparing maximum [OH] with that calculated on basis of assumed reaction mechanism.	150
$(4.3 \pm 1.0) \times 10^6$	310	D.F. Discharge in H_2/Ar or H_2/He. OH from H + NO_2, and measured by UV absorption. 1—5 torr pressure. Large excess H_2 makes reaction effectively first order in OH. Hence absolute concentrations of OH only necessary for estimate of small second order contribution to OH decay from OH + OH → O + H_2O.	175
5.0×10^6	300	D.F. H_2 discharge at pressures <1 torr. H_2 saturated with	200
6.4×10^6	300	water vapour. Decay of H atoms at wall followed by ESR. Interpretation by mathematical model. Uncertainty probably large.	
9.6×10^8	900	Low pressure flame. [H], [O], [OH] measured by ESR.	201
12.6×10^8	943	Poor spatial resolution.	
15.7×10^8	993	[H_2O] profile determined by freezing technique. Rate coefficients involve determination of d[H_2O]/dt	
20.5×10^8	1052	from latter (see Sect. 5.3)	
$(3.9 \pm 0.2) \times 10^6$	300	D.F. Discharge in H_2/Ar at ca. 1 torr. OH from H + NO_2, and measured by ESR. Bulk of H_2 added downstream of	202

TABLE 35—*continued*

k_1 (l . mole^{-1} . sec^{-1})	Temp. (K)	Method and comments	Ref.
		NO_2 inlet port. Large excess H_2 in measurement region makes reaction effectively first order in OH.	
$4 \times 10^{10} \exp(-2{,}850/T)$	1400—2500	Shock tube. (see Sect. 5.2). Induction period by UV absorption signal from OH.	112
$(4.0 \pm 0.2) \times 10^6$	301	F.P. $H_2O/H_2/Ar$ mixtures at pressures ⩽200 torr. [OH] by kinetic spectroscopy (relative concentrations only needed). Correction for $OH + OH \rightarrow O + H_2O$ required only at lowest [H_2].	205
$1.5 \times 10^{10} \exp(-2{,}500/T)$	1000—1700	Flames in rich C_2H_2/O_2 mixtures with added H_2. Profiles by sampling and mass spectroscopy. [OH] by UV absorption. Requires absolute [OH], and uses oscillator strengths from refs. 190 and 204.	203
$(4.66 \pm 0.32) \times 10^6$	295	F.P. 1 % $H_2O/H_2/He$ mixtures at 5.35×10^{-6} mole.cm^{-3} gas concentration (corresponds with 100 torr at 300 K). [OH] by kinetic spectroscopy. (only precise relative concentrations needed since reaction effectively first order in OH). ±10 % corrections for 2nd order decay by $OH + OH \rightarrow O + H_2O$ and for tailing of photolysis flash.	205
$(5.69 \pm 0.19) \times 10^6$	295		
$(4.98 \pm 0.14) \times 10^6$	296		
$(5.71 \pm 0.29) \times 10^6$	296		
$(3.86 \pm 0.21) \times 10^6$	300		
$(4.15 \pm 0.35) \times 10^6$	300		
$(4.24 \pm 0.19) \times 10^6$	300		
$(4.71 \pm 0.18) \times 10^6$	305		
$(8.57 \pm 1.42) \times 10^6$	332		
$(1.40 \pm 0.03) \times 10^7$	358		
$(3.40 \pm 0.09) \times 10^7$	420		
$(7.88 \pm 0.64) \times 10^7$	498		
$(6.39 \pm 0.43) \times 10^7$	495		
$(6.88 \pm 0.20) \times 10^7$	495		
$(6.92 \pm 0.10) \times 10^7$	495		
$(6.39 \pm 0.24) \times 10^7$	495		
5.0×10^6	304	Stirred flow reactor. H_2 discharge at ca. 1 torr. OH from $H + NO_2$. Excess H_2 added. H_2O followed by mass spectrometry.	206
3.6×10^7	403		
6.5×10^7	504		
$(1.85 \pm 0.3) \times 10^9$	1050	Flame study. $H_2/N_2/O_2$ at atmospheric pressure (see Sect. 5.4.2).	

TABLE 35—*continued*

k_1(l. mole^{-1}. sec^{-1})	Temp. (K)	Method and comments	Ref.
4.3×10^6 (±15 %)	298	F.P. OH by pulsed vacuum UV photolysis of H_2O, and measured by resonance fluorescence. Relative values only needed (effective 1st order decay of OH in presence of large excess H_2).	207
4.6×10^6 10.6×10^6 19.6×10^6 8.2×10^7 2.2×10^8 4.0×10^8	298 352 403 518 628 745	D.F. Discharge in H_2/He mixtures at 1—3 torr. OH form H + NO_2, and measured by ESR. Bulk of H_2 added downstream of discharge. Negligible contribution from $OH + OH \rightarrow O + H_2O$. Small contribution from 1st order decay of OH at wall.	
$1.84 \times 10^{11} \exp(-5{,}350/T)$	1200—1800	Shock tube. Relative [OH] in rich mixtures (10 H_2/O_2/89 Ar) by UV absorption. Measures [OH] overshoot using internal calibration, and derives $k_2/k_1 = 0.77 \exp(-2{,}900/T)$ by method of ref. 188. Allows for boundary layer effects. Combined with $k_2 = 1.42 \times 10^{11} \exp(-8{,}250/T)$ from eqn. (72) to give expression quoted in column 1. Overshoot not large. Conditions chosen so that magnitude not much affected by recombination reactions.	111
$1.1 \times 10^{10} \exp(-2{,}310/T)$	210—460	F.P. of mixtures containing 0.1—0.5 torr H_2O or 10 torr N_2O in H_2 [OH] by time-resolved resonance absorption. Relative concentrations only needed in presence of excess H_2. Corrections at lower temperatures for contributions from $OH + OH + M = H_2O_2 + M$ and $OH + H + M = H_2O + M$	197

TABLE 35—*continued*

k_1 (l. mole^{-1} . sec^{-1})	Temp. (K)	Method and comments	Ref.
(4.2 ± 0.4) x 10^6 (1.1 ± 0.1) x 10^7 (3.1 ± 0.3) x 10^7	298 348 425	Flow system. H_2/Ar at total pressure 15—20 torr containing 0.01—0.03 torr H_2O. OH by repetitive pulsed vacuum UV photolysis of H_2O at $\lambda \geqslant 1050$ Å. [OH] by resonance fluorescence. Relative [OH] only needed.	475

D.F., discharge-flow; F.P. flash photolysis.

TABLE 36
Values of k_{OH+H_2}/k_{OH+CO}

k_1/k_{23}	Temp. (K)	Method and comments	Ref.
20	1217—1345	Rich $H_2/O_2/CO_2$ flame at atm pressure. Measured ratio $k_{H+D_2O}/k_{H+CO_2} = 0.33 \exp(+3{,}900/T)$. If assumed $k_{H+H_2O} = k_{H+D_2O}$, then gives k_1/k_{23} = 12.5. Value in column 1 allows for isotope effects on equilibrium constants and rate coefficients according to ref. 172.	209
0.429 0.615 0.875 1.21	473 523 573 623	Photolysis of H_2O/CO mixtures at 185 nm in presence and absence of H_2. CO_2 analyzed by gas chromatography.	210
20	1773	Flame study. No details available. Attributed to Wagner.	170
0.048 0.056 0.09 ± 0.02 0.40 ± 0.02 0.6 ± 0.1	300 305 333 420 495	From absolute values of individual rate coefficients determined at these temperatures.	205
4.3 ± 0.9	713	Decomposition of H_2O sensitized by H_2 and by CO gives k_{15}/k_1 = 5.0 ± 1.0 and k_{15}/k_{23} = 21.3 ± 1.0, respectively, where (xv) is $OH + H_2O_2 = H_2O + HO_2$. [$H_2O_2$] determined by colorimetric method.	211

TABLE 36—*continued*

k_1/k_{23}	Temp. (K)	Method and comments	Ref.
11.3 ± 2.2	1050	Rich $H_2/O_2/N_2$ flame at atm pressure. Study of reaction of H with added traces of D_2O, D_2 and CO_2. Mass spectrometric probe.	172
5	870—1000	Shock tube study of decomposition of H_2O_2 accelerated by H_2 and CO. H_2O_2 monitored by UV absorption.	212
0.81	553	Photolysis of H_2O/CO mixtures at 185 nm in presence and absence of H_2. CO_2 analyzed by gas chromatography.	213
1.5×10^{-3}	210	From absolute values of individual coefficients.	197, 214
0.94	520	From absolute values.	208
46 exp $(-1{,}350/T)$	1500—2000	From absolute rate expressions.	111

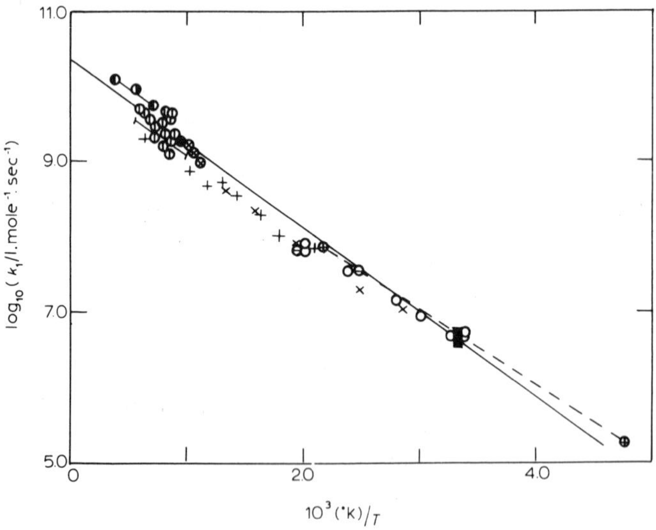

Fig. 36. Arrhenius plot of k_1. ⊕, Smith and Zellner [197]; ■, Kaufman and del Greco [175], Wise et al. [200], Dixon-Lewis et al. [202], Greiner [205], Wong and Belles [206], Stuhl and Niki [207], Westenberg and de Haas [208]; ○, Greiner [205]; ⊖, Wong and Belles [206]; x, Westenberg and de Haas [208]; +, Eberius et al. [168]; ⊗, Balakhnin et al. [201]; ●, see Sect. 5.4.2; ⦶, Brabbs et al. [92]; ◐, Ripley and Gardiner [112]; ◑, Gardiner et al. [111]; ⊢⊣, Browne et al. [203].

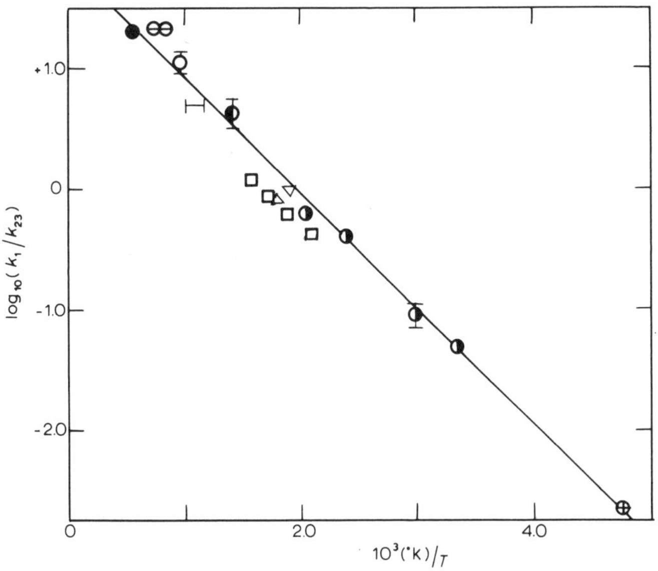

Fig. 37. Arrhenius plot of k_1/k_{23}. ⊕, Smith and Zellner [197, 214]; ◑, Greiner [205]; □, Ung and Back [210]; ◐, Baldwin et al. [211]; ├─┤, Kijewski and Troe [212]; ○, Dixon-Lewis [172]; ⊖, Fenimore and Jones [209]; ●, Wagner (see ref. 170); △, Baulch et al. [213, 215]; ▽, Westenberg and de Haas [208].

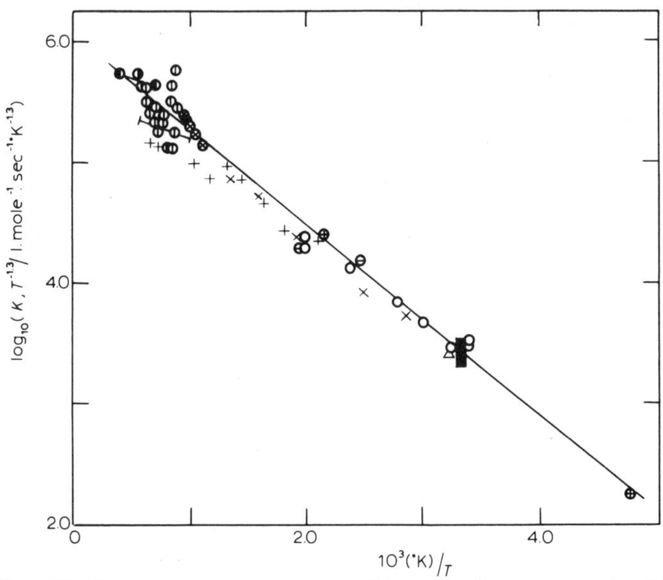

Fig. 38. Temperature dependence of k_1 with three parameter fit. ⊕, Smith and Zellner [197]; ■, del Greco and Kaufman [199], Wise et al. [200], Dixon-Lewis et al. [202], Greiner [205], Wong and Belles [206], Stuhl and Niki [207], Westenberg and de Haas [208]; △, Kaufman and del Greco [175]; ○, Greiner [205]; ⊖, Wong and Belles [206]; ×, Westenberg and de Haas [208]; +, Eberius et al. [168]; ⊗, Balakhnin et al. [201]; ●, see Sect. 5.4.2; ⦶, Brabbs et al. [92]; ◐, Ripley and Gardiner [112]; ◑, Gardiner et al. [111]; ├─┤, Browne et al. [203].

6.2 REACTION (ii) $H + O_2 \rightarrow OH + O$

Values of k_2 from a variety of sources have already been presented in Sect. 3.6 and 5.2, and in Tables 15, 24, 28 and 29. Some additional data from low pressure flame studies is given in Table 37. There is also a considerable body of data from shock tube investigations, reported by Baulch et al. [55], for which it is not clear that corrections were applied for boundary layer effects in the shock tube (see Sect. 5.1). This data has been omitted from consideration here — although much of it is in reasonable accord with the expression recommended below for k_2.

Figure 39 shows the range of data which is regarded as the most reliable basis for evaluation. The upward pointing arrow leading from the single point due to Baldwin [54] indicates the probable magnitude of the correction of his original value to allow for deviations from the kinetic theory hard sphere model when calculating the diffusion coefficient of H atoms (see Sect. 3.6.4). The results of Fenimore and Jones [166, 167] also have corrections indicated by downward pointing arrows. These corrections are a consequence of the fact that they used the reaction of H atoms with D_2O in order to calibrate the H atom concentrations in their flame. They used an incorrect rate coefficient k_{H+D_2O} (see Sect. 5.4.2), and substitution of a more valid rate coefficient based on eqn. (71), the

TABLE 37
Values of k_2

k_2 (l . mole^{-1} . sec^{-1})	Temp. (K)	Method and comments	Ref.
1.1×10^7 1.7×10^7 2.6×10^7 4.0×10^7	900 943 993 1052	Low pressure flame in furnace. 1 : 1 H_2/O_2 mixture at 2.86 torr. Average [H] and [O_2] by ESR. [H_2O] by freezing down in calibrated volume. Assumed all oxygen consumed by reaction (ii).	201
3.2×10^6 4.0×10^6 5.0×10^6 6.4×10^6 7.3×10^6 8.5×10^6	825 843 860 878 893 905	Low pressure flame in furnace. H_2/O_2 mixtures with traces of D_2 at 3—6 torr. H_2, D_2 and HD measured by mass spectrometry. [H_2O] by freezing down in calibrated volume. Extent of isotopic exchange related to competition of reaction (ii) with isotopic exchange $H + D_2 \rightarrow HD + D$ to give ratio of rate coefficients. Values for k_2 are based on $k_{H+D_2} = 2 \times 10^{13} \exp(-5{,}500/T)$ (ref. 217).	216

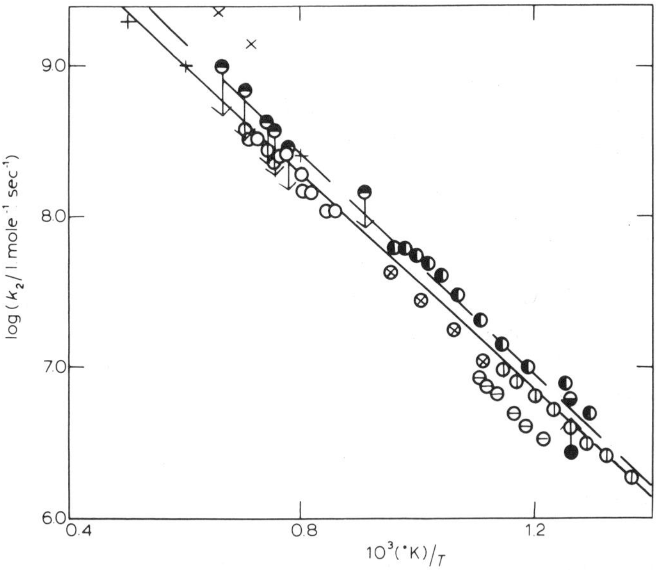

Fig. 39. Arrhenius plot of k_2. ◒, Semenov [59]; ⍖, Baldwin [54]; ⊖, Buneva et al. [216]; ⦶, Karmilova et al. [61]; ◐, Eberius et al. [168]; ⊗, Balakhnin et al. [201]; ⍒, Fenimore and Jones [166, 167]; ○, Brabbs et al. [92]; ×, Ripley and Gardiner [112]; +, Schott [219]. Broken line, recommendation of Baulch et al. [55]; solid line, eqn. (72).

isotopic rate coefficient ratio $2k_{1\,\mathrm{D\,a}}/k_1$ from ref. 172, and the equilibrium constant $K_{1\,\mathrm{D\,a}}$ for

$$\mathrm{OH + HD \rightleftharpoons HOD + H} \qquad \text{(i Da)}$$

leads to corrections of the magnitudes shown. These changes, as well as the flame computation and experimental results of Dixon-Lewis et al. (Sect. 5.4.2 and Table 30) strongly suggest that the expression recommended by Baulch et al. for k_2 predicts rate coefficients at 700—1500 K which are some 25 % too high. The expression recommended here is

$$k_2 = 1.42 \times 10^{11} \exp(-8{,}250/T) \qquad (72)$$

with suggested error limits of ±20 % between 700 and 1500 K. Above about 1500 K, it is estimated that the application of the boundary layer correction to the (unquoted) shock tube results of Hirsch and Ryason [218], Myerson and Watt [83], and Browne et al. [96] would bring them into approximate agreement with this expression. However, there remain two discordant results for this high temperature region. First, the expression predicts rate coefficients which are only about one third of those deduced by Ripley and Gardiner [112] (Sect. 5.2). Such a discrepancy is certainly larger than could be accounted for purely by their neglect of boundary layer effects. Secondly, the results of Gutman, Schott et al. [88—91] using reflected shocks and "end-on", spatially integrated O + CO emission measurement, are skew to the recommended line, giving a smaller temperature dependence between 1250 and 2500 K. Their results lead to the expression [219] $k_2 = 1.22 \times 10^{14}\ T^{-0.907}$ exp (—8,315/T), and at around 1250 K are about 20—25 % higher than those of Brabbs et al. [92]. At 700—800 K the expression quoted by Schott [219] also predicts values of k_2 which are double those predicted by eq. (72) — a situation completely inconsistent with the flame results of Dixon-Lewis et al. (cf. Section 5.4.2 and Table 30). The expression due to Schott [219] is therefore not valid in the lower temperature range.

Extrapolating eqn. (72) towards lower temperatures gives $k_2 = 0.163$ at 300 K, and dividing this by the equilibrium constant $K_2 = 1.0 \times 10^{-11}$ leads to $k_{-2} = 1.63 \times 10^{10}$ at this temperature. Direct measurements on the reverse reaction by discharge-flow techniques [175, 220—223] have given $k_{-2} = (2.3 \pm 1.0) \times 10^{10}$ at 300 K.

6.3 REACTION (iii) $O + H_2 \rightarrow OH + H$

A number of investigations of this reaction in discharge-flow systems have been carried out, and the data have again recently been reviewed by Baulch et al. [55]. Results from several investigations are summarized in Table 38, and plotted in Fig. 40. The latter also includes the shock tube results of Brabbs et al. [92], given in Table 24. Following Baulch et al. [55], the ordinate in Fig. 40 is $\log_{10}(kT^{-1})$. Shock tube results which are uncorrected for boundary layer effects are again not included. With such corrections, the results of Browne et al. [96], Wakefield [115], Dean and Kistiakowsky [230] and Jachimowski and Houghton [110] would come into good agreement with the expression recommended by Baulch et al. [55], viz.

$$k_3 = 1.8 \times 10^7\ T \exp(-4{,}480/T) \tag{73}$$

for the temperature range 400—2000 K. Suggested error limits are ±30 % over the whole range. There is at present no obvious reason for amending this evaluation. It is also worthy of note that the product of expressions (71) and (73) gives

$$k_1 k_3 = 2.1 \times 10^{13} T^{2.3} \exp(-6{,}305/T) \tag{74}$$

For the temperature range 1200—2500 K this expression converted to simple Arrhenius form gives an apparent average activation energy E_p of 20.4 kcal . mole^{-1}, with a value of $k_1 k_3 = 9.5 \times 10^{18}$ l^2 . mole^{-2} . sec^{-2} at 1600 K. Experimentally, Schott [91] finds $E_p = 20.0$ kcal . mole^{-1} and $k_1 k_3 = 15.0 \times 10^{18}$ l^2 . mole^{-2} . sec^{-2} at 1600 K from reflected shock measurements in the above temperature range. In view of the uncertainties in the reflected shock technique, this agreement with the prediction of eqn. (74) is very good.

TABLE 38
Measurements of k_3

k_3(l . mole^{-1} . sec^{-1})	Temp. (K)	Method and comments	Ref.
$(1.5 \pm 0.2) \times 10^5$ $(3.2 \pm 0.4) \times 10^5$ $(9.0 \pm 1.0) \times 10^5$ $(2.6 \pm 0.2) \times 10^6$ $(6.0 \pm 0.6) \times 10^6$ $(2.2 \pm 0.2) \times 10^7$	409 441 494 554 623 733	N_2 discharge flow system. N atoms titrated with NO to produce O atoms. Relative [O] by air afterglow intensity. Observed rate coefficients divided by 2 to allow for fast consecutive reaction (—ii) $O + OH \rightarrow H + O_2$.	195
1.2×10^5 1.0×10^5 2.4×10^5 5.1×10^5 2.0×10^6 1.7×10^6 7.8×10^6 8.9×10^6	397 400 428 450 506 508 596 600	N_2 discharge-stirred flow reactor. O atoms from N + NO. H_2 added. [O] by mass spectrometry. Observed rate coefficients corrected for consecutive reactions. Claimed accuracy ±50 %.	224
$(1.8 \pm 0.6) \times 10^8$	993	Thermal flow system. H_2/O_2 mixtures with $[H_2]/[O_2]$ = 1.0—1.5. Average [OH], [O], [H] and $[O_2]$ by ESR. H_2O trapped downstream. Method of $[H_2]$ determination not clear. Steady state analysis applied to complex reaction mechanism to give k_{OH+H_2}/k_{O+H_2} = 7.0.	201
$1.3 \times 10^{10} \exp((-4{,}700 \pm 80)/T)$	373—478	N_2 discharge-flow system. O atoms from N + NO. [O] and $[H_2]$ estimated by ESR and mass spectrometry. H_2 added through moveable inlet. Observed rate coefficients divided by 2 to allow for fast consecutive reaction (—ii).	225
$(1.2 \pm 0.1) \times 10^4$	320	N_2 discharge-flow system. O from N + NO. H_2 added downstream. Relative [O] by air afterglow intensity. Observed rate coefficient corrected for fast consecutive reaction (—ii).	226
$(2.4 \pm 0.2) \times 10^5$ $(1.32 \pm 0.04) \times 10^6$ $(6.8 \pm 0.3) \times 10^6$ $(5.9 \pm 0.5) \times 10^7$ $(8.9 \pm 0.6) \times 10^7$	423 514 613 812 910	Discharge-flow. <0.5 % O_2 in Ar or He. H_2 added through moveable inlet. [O] by ESR. Revision of earlier work [227].	165

TABLE 38—*continued*

k_3(l.mole^{-1}.sec^{-1})	Temp. (K)	Method and comments	Ref.
1.8×10^5 7.6×10^5 1.1×10^6 2.6×10^6	408 458 480 520	O_2/Ar discharge-flow system. H_2 added through moveable inlet. [O] by ESR. Observed rate coefficients divided by 2 to account for fast consecutive reaction (—ii).	228
$(5.1 \pm 1.0) \times 10^{11} \exp(-8{,}250/T)$	1400—1900	Shock tube. O atom overshoot in rich H_2—O_2—CO—CO_2—Ar mixtures measured using calibrated O + CO emission. Incident shock with boundary layer corrections. Obtains $k_3/k_2 = (3.6 \pm 0.7)$ over the T range. Combined with eqn. (72) to give expression in column 1.	229

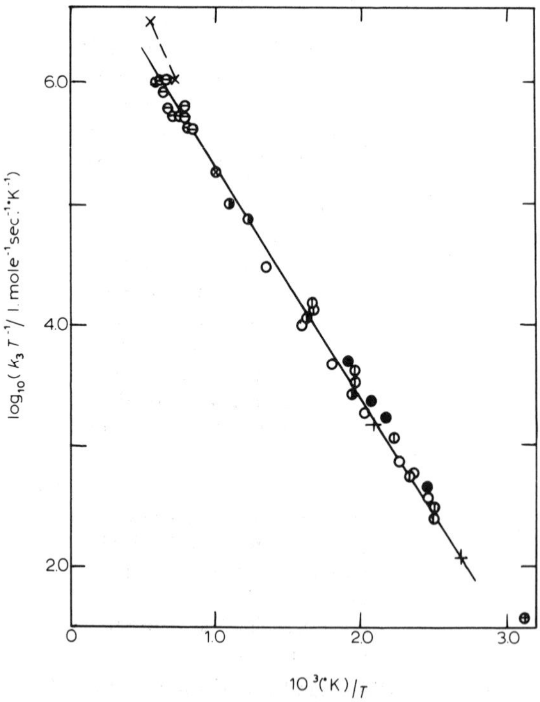

Fig. 40. Temperature dependence of k_3 with three parameter fit. ⊕, Campbell and Thrush [226]; +, Hoyermann et al. [225]; ⦶, Wong and Potter [224]; ○, Clyne and Thrush [195]; ●, Balakhnin et al. [201]; ◑, Westenberg and de Haas [165]; ⊗, Balakhnin et al. [228]; ⊖, Brabbs et al. [92]; x———x, Schott et al. [229].

6.4 REACTION (xvi) $OH + OH \rightarrow O + H_2O$

The available reliable information on the rate coefficient of reaction (xvi) depends almost entirely on fast flow-discharge studies, and, with the exception of one recent shock tube result, direct measurements are confined to near 300 K. Even here there is a factor of two or three disagreement between authors. Results are summarized in Table 39. Uncertainties arise from two major causes. First, the second order gas phase decay of OH is accompanied by a first order heterogeneous decay. Optimization of the separation of the observed decay into its first and second order components is difficult, and this may account for some of the reported discrepancies [222]. Secondly, all the investigations have used the $H + NO_2$ reaction as the source of OH, with the NO_2 added downstream of the discharge. The relevant elementary steps causing the growth and decay of OH are then

$$H + NO_2 \rightarrow OH + NO \qquad \text{(xxiv)}$$

$$OH + OH \rightarrow O + H_2O \qquad \text{(xvi)}$$

$$O + OH \rightarrow O_2 + H \qquad \text{(—ii)}$$

with small contributions also [222, 231—233] from

$$OH \xrightarrow{\text{wall}} \text{removed}$$

and

$$OH + NO_2 \rightarrow HNO_3$$

Neglecting the last two reactions, and assuming both reaction (xxiv) and the mixing of the NO_2 with the main gas stream to be infinitely fast, for $k_{-2} \gg k_{16}$ we find that a small steady state concentration of O atoms quickly builds up and the overall stoichiometry becomes

For excess H

$$3OH \xrightarrow{k_{16}} H_2O + O_2 + H$$

i.e.

$$d[OH]/dt = -3k_{16}[OH]^2 \qquad (75)$$

and for excess NO_2

$$2OH + NO_2 = H_2O + O_2 + NO$$

i.e.

$$d[OH]/dt = -2k_{16}[OH]^2 \qquad (76)$$

Still assuming an infinite mixing rate, Westenberg and de Haas [233] have modelled the effect of including a finite but fast rate for reaction (xxiv), and found that the full excess H atom stoichiometry is only achieved for $[H]_0/[NO_2]_0 \geqslant 3$. Additionally, however, they found that $[OH]_0$ never builds up to $\Delta[NO_2]$ as was originally assumed by del Greco and Kaufman [199] for their calibration of the OH concentration;

TABLE 39
Measurements of k_{16}

k_{16} (l. mole^{-1}. sec^{-1})	Temp.	Method and comments	Ref.
$(7.5 \pm 2) \times 10^8$	310	H_2/He D.F. system. OH by titrating H with NO_2, and measured by UV absorption. Calibrated by extrapolating back to NO_2 inlet where assumed $[OH]_0 = \Delta[NO_2]$. OH decay unaffected by Ar, N_2, O_2, NO and H_2O. Recalibration [221] using *f*-value of OH from ref. 190 gives $k_{16} = 8.5 \times 10^8$. [OH] decay 2nd order.	199
$(1.55 \pm 0.12) \times 10^9$	300	H_2/He and H_2/Ar D.F. system. OH from H + NO_2, and measured by moveable ESR. (calibration against NO). CO also added and final $[O_2]/[CO_2]$ ratios measured by mass spectrometry. Results support higher value of k_{16} than ref. 199. [OH] decay 2nd order.	202
$(1.25 \pm 0.05) \times 10^9$	300	H_2/Ar D.F. system. OH from H + NO_2. [OH] by ESR. Excess NO_2 used and CO added. $[CO_2]$ by mass spectrometry. Found OH removed by reaction (xvi), and also by reaction with NO_2, e.g. $OH + NO_2 \rightarrow HNO_3$. Latter conclusion confirmed by Mulcahy and Smith [232].	231
$(5.1 \pm 1.6) \times 10^8$	~300	H_2/Ar discharge flow system. OH from H + NO_2. NO_2 added at moveable inlet. [OH] by fixed ESR. Boric acid coating on flow tube. [OH] decay resolved into 1st and 2nd order contributions. First order contribution due to OH → wall. Confirmed by Mulcahy and Smith [232].	222

TABLE 39—*continued*

k_{16}(l . mole^{-1} . sec^{-1})	Temp.	Method and comments	Ref.
$(1.4 \pm 0.2) \times 10^9$	300	H_2/Ar D.F. system. OH from H + NO_2. NO_2 added at moveable inlet. [OH] by fixed ESR. Uncoated and B_2O_3 coated flow tube, 3.3 cm diameter. [OH] decay resolved into 1st and 2nd order contributions, and effects of non-infinite rate of OH production on $[OH]_0$ investigated theoretically for different initial H/NO_2 ratios (see text).	233
$6.7 \times 10^6\ T \exp(-1{,}230/T)$ (or $3.2 \times 10^{10} \exp(-2{,}950/T)$)	1500—2000	Shock tube. Relative [OH] in lean mixtures (1 H_2/10 O_2/89 Ar) by UV absorption. Measures [OH] overshoot using internal calibration. Overshoot sensitive principally to k_3/k_{16}, though sensitivity not high. Result quoted in ref. 111 corresponds with k_3/k_{16} = 2.7 exp (−3,250/T). Combined here with k_3 from eqn. (73) to give the unbracketed value quoted in column 1.	111

and Westenberg and de Haas imply that the difference between their own results and del Greco and Kaufman's may be due at least partly to this cause. Although Kaufman [221] later revised the del Greco—Kaufman result upwards slightly by using an absolute calibration based on the f-value of OH due to Golden et al. [190], the latter also depends on the production of OH from H + NO_2, and the difference between the two results for k_{16} is still large. If the arguments put forward in Sect. 5.4.3 regarding the f-value of OH are accepted, the optimum value of the rate coefficient at 300 K would become $k_{16} = 1.2 \times 10^9$ l . mole^{-1} . sec^{-1}.

At higher temperatures, Albers et al. [74] have studied the reverse reaction (−xvi), again in a discharge-flow system. The discharge was passed through O_2/He mixtures, and water vapour was added downstream through a moveable inlet. [O] and [H] were measured by ESR. It was found that $\Delta[H]/\Delta[O] = 0.62 \pm 0.06$, indicating that the only significant reactions were (—xvi) and (—ii). On this basis $d[O]/dt = -3k_{-16}[O][H_2O]$. Values of k_{-16} are given in Table 40, together with values of k_{16} calculated using the equilibrium constant. Again, there is at present insufficient new evidence to justify revision of the expression for k_{16} recommended by Baulch et al. [55], viz.

$$k_{16} = 6.3 \times 10^9 \exp(-550/T) \tag{77}$$

However, favouring $k_{16} = 1.2 \times 10^9$ instead of 1.0×10^9 at 300 K leads to

$$k_{16} = 5.6 \times 10^9 \exp(-460/T) \tag{77a}$$

Suggested error limits on the calculated rate coefficients are ±50 % between 300 and 2000 K.

TABLE 40
Values of k_{-16} [74] and k_{16}

Temp (K)	$k_{-16}/10^5$ (l. mole^{-1}. sec^{-1})	$k_{16}/10^9$ (l. mole^{-1}. sec^{-1})
753	3.44	3.38
773	5.94	4.33
814	6.88	2.84
814	7.28	3.00
814	7.75	3.20
849	15.8	4.19
859	13.1	3.08
935	31.1	3.20
1045	94.5	3.64

6.5 REACTION (iv) $H + O_2 + M \rightarrow HO_2 + M$

Since hydroperoxyl is not a stable molecule, reaction (iv) must be considered together with one or more elementary steps which remove HO_2 from an experimental system.

6.5.1 Room temperature and below

At room temperature and below, in fast flow systems, the additional reactions are those of H atoms with HO_2

$$H + HO_2 \rightarrow OH + OH \tag{viii}$$

$$H + HO_2 \rightarrow O + H_2O \tag{viiia}$$

$$H + HO_2 \rightarrow H_2 + O_2 \tag{xx}$$

Reactions (viii) and (viiia) are in turn followed by further reactions of OH and O. When M = Ar or He, these further reactions are principally

$$OH + OH \rightarrow O + H_2O \qquad \text{(xvi)}$$

$$O + OH \rightarrow O_2 + H \qquad \text{(—ii)}$$

leading to an overall stoichiometry of (viii) or (viiia)

$$H + HO_2 = \tfrac{2}{3}(H_2O + O_2 + H)$$

If, however, molecular hydrogen is present in sufficient quantity, reaction (i)

$$OH + H_2 \rightarrow H_2O + H \qquad \text{(i)}$$

may also occur, giving an overall stoichiometry

$$H + HO_2(+2H_2) = 2H_2O + 2H$$

These subsequent reactions of OH and O are all sufficiently fast that [O] and [OH] never become comparable with [H]. Reactions (xxi) and (xxii) of OH and O with HO_2 do not therefore become important in these room temperature systems where the initial radical is the H atom. Reactions (viii) and (xx) are also sufficiently fast that $[HO_2]$ never becomes large enough for reaction (x) to need consideration.

Clyne and Thrush [162] produced H atoms in a flow tube of 28 mm internal diameter, by means of a microwave discharge either in a stream of pure hydrogen or in streams of argon or helium containing 1 % H_2, with total pressure ca. 2 torr. Oxygen was added downstream of the discharge, and a trace of NO was added just upstream of an observation point. The concentration of H was determined by monitoring the HNO emission intensity. Four different reaction times were obtained by using four different oxygen inlet positions upstream of the observation point. It was found that HO_2, OH and O quickly reached their pseudo-stationary concentrations, so that using the overall stoichiometries above for the various reaction paths of HO_2, one may write

$$\frac{[H_2O]\text{ formed}}{[H]\text{ used}} = \frac{a\{\frac{2}{3}(k_8 + k_{8a})\} + (1-a)\{2(k_8 + k_{8a})\}}{2k_{20} + \frac{4}{3}a(k_8 + k_{8a})} \qquad (78)$$

where

$$a \simeq k_{16}[OH]/(k_1[H_2] + k_{16}[OH])$$

the approximation becoming precise if $k_{8a} = 0$.

Clyne and Thrush indeed found the ratio $[H_2O]$ formed/[H] used to increase in mixtures containing more molecular hydrogen. In their experiments using Ar or He containing 1 % H_2 initially, the amount of molecular hydrogen remaining after the discharge is small. In these cases they found $[H_2O]$ formed/[H] used = 0.29 ± 0.05, and taking $a = 1$ they deduced $k_{20}/(k_8 + k_{8a}) = 0.51 \pm 0.21$. It is worth noting here, however, that using values of k_1 and k_{16} which have since become available, it is probable that $a = 0.9$ would give a better representation of their experimental conditions. For this situation the mean ratio $k_{20}/(k_8 + k_{8a}) \simeq 0.7$. Both these results are in good agreement with the room temperature value of between 0.5 and 1 estimated by Bennett and Blackmore [164], who used molecular hydrogen containing a trace of oxygen as the carrier gas, coupled with absolute measurements of [H] by ESR. Dodonov et al. [163], using probe sampling and mass spectrometry to measure [H], [O], [OH] and $[H_2O]$ in dissociated H_2/He mixtures at ca. 21 torr, also found $k_{20}/(k_8 + k_{8a}) < 1$, but their ratio k_{8a}/k_8 is about 11, whereas other investigators find this ratio to be small (e.g. ref. 159). Baulch et al. [55] consider that the high k_{8a}/k_8 may be due to loss of OH during sampling.

The major disagreement concerning the values of the ratio $k_{20}/(k_8 + k_{8a})$ comes from the work of Westenberg and de Haas [234] who found $k_8 : k_{8a} : k_{20} = 0.27 : 0.11 : 0.62$ as their preferred results. i.e. $k_{20}/(k_8 + k_{8a}) \simeq 1.6$. Their method relied on ESR measurement of [H], [OH] and [O], and their stationary state analysis of the kinetic system in terms of reactions (iv), (viii), (viiia), (xx), (xvi) and (—ii) led to expressions for $d(\ln[H])/dt$, $[OH]_s/[O]_s$ and $[OH]_s[O]_s$ in terms of k_4, the ratios $k_8 : k_{8a} : k_{20}$ and the further rate coefficients k_{-2} and k_{16} (these two need to be known a priori). $[H_2O]$ was not measured, and possible first order wall loss of the intermediate species HO_2, OH and O was not considered. On the rather intuitive grounds that the method is much less direct than that of Clyne and Thrush, with certain of its assumptions more open to question, the author's preference is towards the lower values of the ratio $k_{20}/(k_8 + k_{8a})$.

Returning now to the consideration of k_4, the results of the low pressure discharge-flow experiments ($p \leqslant 2$ or 3 torr [162—165, 234, 235] give linear plots of ln[H] against time, and for M = Ar or He the overall stoichiometry (excluding reaction (i)) leads to

$$-d[H]/dt = 2k_{4,M}[H][O_2][M]\left\{1 - \frac{k_8 + k_{8a}}{3(k_8 + k_{8a} + k_{20})}\right\} \tag{79}$$

Fortunately, the whole range of values of the ratio $k_{20}/(k_8 + k_{8a})$ quoted above only gives a 10 % variation in the factor in brackets at the end of this expression. For $k_{20}/(k_8 + k_{8a}) = 0.51$ [162] we have

$$-\frac{1}{[O_2][M]}\frac{d(\ln[H])}{dt} = 1.56k_{4,M} \tag{80}$$

Values of k_4 near or below room temperature for a number of "chaperon" molecules M are given in Table 41. Added to the values from discharge—flow experiments are a number of results obtained recently from

TABLE 41

Third order rate coefficients (10^9 $l^2 . mole^{-2} . sec^{-1}$) for $H + O_2 + M = HO_2 + M$ at lower temperatures

Temp. (K)	M = Ar	He	H_2	N_2	H_2O	CH_4	Ref.
203		8.3					237
213		7.1					237
220	11.8			31.6			238
225	12.7 ± 1.1						162
226				32.0			237
234		6.1					237
244	14.5 ± 1.1						162
262		6.1					237
293	12.0 ± 3.0						220
293	8.0 ± 0.7	7.6 ± 0.7			189 ± 75		162
293	13.5						147
293		22.0					163
297	5.65	5.47	23.2				235
298			17 ± 4				176
298	5.9 ± 0.7						236
298	5.76 ± 0.80	5.76 ± 0.80		19.6 ± 2.8		90 ± 27	237
298	2.2	2.8	4.4				239
298	7.4 ± 0.8	6.9 ± 0.7	21.7 ± 4.3	20.0 ± 2.5		154 ± 67	238
300	6.8 ± 1.0	6.8 ± 1.0					234
357		4.59					237
434		3.89					237
203—404		$(2.44 \pm 0.37) \exp\{(238 \pm 46)/T\}$					237
220—360	$(2.48 \pm 0.40) \exp\{(345 \pm 64)/T\}$						238

The results of Dorfman et al. [176, 236] and Wong and Davis [238] for argon and hydrogen give $k_{4,Ar}/k_{4,H_2} = 0.35$ at 298 K, and the results of Wong and Davis [238] for nitrogen and hydrogen give $k_{4,N_2}/k_{4,H_2} = 0.92$ at room temperature (cf. Table 7, where second limit measurements at higher temperatures give $k_{4,Ar}/k_{4,H_2} = 0.2$ and $k_{4,N_2}/k_{4,H_2} = 0.44$). Changes in chaperon efficiency with temperature have also been found by Walkauskas and Kaufman [149] for H atom recombination (cf. Table 26).

pulse radiolysis or flash photolysis experiments (Dorfman et al. [176, 236]; Kurylo [237]; Wong and Davis [238]; Ahumada et al. [239]), in which measurement of [H] was by the very sensitive absorption of the Lyman-α line. Concentrations of 10^{-8} to 10^{-9} M of H atoms are readily detectable by this method [176]. In the experiments of Dorfman et al. [176, 236], for example, H atom concentrations of $<4 \times 10^{-8}$ M were used in the presence of 1—5 torr of oxygen and 100—1500 torr of the "chaperon" gas. Under these conditions, the decay was first order in [H] and was very much more rapid than in the absence of oxygen: hence there was no significant contribution from wall effects. The H atom concentrations were also small enough for the further reactions (viii), (viiia) and (xx) of HO_2 to be unable to contribute significantly, at least at "chaperon" pressures above 300 torr (cf. high temperature reaction in salt-coated vessels, Sect. 3.6.2). This spectrophotometric technique thus seems to be remarkably free from side effects and should give correspondingly reliable results. The conclusion is supported by the agreement between the results of Dorfman et al. [176, 236], Kurylo [237] and Wong and Davis [238]. The lower results of Ahumada et al. [239] may be due to their having used a mercury-photosensitized production of H atoms.

6.5.2 *High temperatures*

At high temperatures, direct evaluations of k_4 are available from shock tube work (Table 31) and from the measurements of Kaskan [179] on the decay of [OH] in the burnt gas of suitable low pressure, fuel-lean hydrogen—air flames (cf. Sect. 5.4.3 and Tables 32 and 33). With calibration of the [OH] measurements as described in Sect. 5.4.3, and with the chaperon efficiencies of N_2, O_2 and H_2O taking the values 0.44, 0.35 and 6.5 relative to hydrogen, these last experiments give $k_{4,H_2} = 5.6 \times 10^9$ l^2 . mole^{-2} . sec^{-1} at 1400 K.

Indirect estimates of k_{4,H_2} in the temperature range 733—803 K may be obtained by combining the ratios k_2/k_4 from second limit measurements (Table 18) with values of k_2 from eqn. (72). The available results for M = H_2, N_2, Ar and He are shown in Fig. 41, in which $\log_{10} k$ is plotted against $\log_{10} T$. Those for M = H_2 and M = N_2 are represented reasonably well over the temperature range 250—1500 K by

$$k_{4,H_2} = 1.03 \times 10^{12} T^{-0.72} \tag{81}$$

$$k_{4,N_2} = 3 \times 10^{14} T^{-1.7} \tag{82}$$

while those for M = He obey the relation

$$k_{4,He} = 4.6 \times 10^{11} T^{-0.77} \tag{83}$$

in the smaller temperature range 200—800 K. Turning attention to the extended temperature range in the case of argon, the second limit results near 773 K (assuming $k_{4,Ar}/k_{4,H_2} = 0.2$) do not appear to be consistent

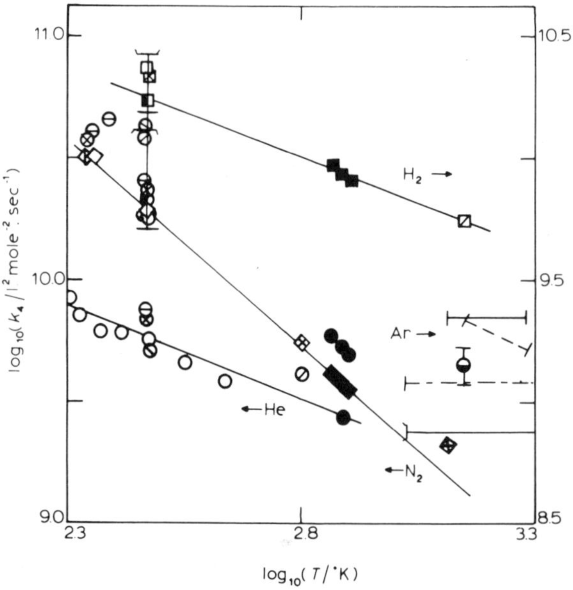

Fig. 41. Temperature dependence of k_4. M = Ar: ⊖, Clyne and Thrush [162]; ⊗, Wong and Davis [238]; ○, Kurylo [237], Moortgat and Allen [235]; ◑, Hikida et al. [236]; ◐, Westenberg and de Haas [234]; ⊘, Clyne [220]; ⦸, Larkin and Thrush [147]; ●, using $k_{4,Ar}/k_{4,H_2}$ = 0.2 from second limits [72]; ⊢⊣, Getzinger and Blair [151]; ⊢--⊣, Browne et al. [96]; ◒, Getzinger and Schott [181]; ⊢-- ⊣, Gay and Pratt [152]. M = H_2: □, Moortgat and Allen [235]; ◧, Bishop and Dorfman [176]; ⊠, Wong and Davis [238]; ■, Baldwin et al. [72]; ⧄, Kaskan (see Sect. 5.4). M = N_2: ◇, Kurylo [237], also Wong and Davis [238] at 298 K; ◇, Wong and Davis [238]; ♦, using $k_{4,N_2}/k_{4,H_2}$ = 0.44 from second limits [72]; ⦶, Getzinger and Blair [151]; ⊗, using $k_{4,N_2}/k_{4,H_2}$ = 0.55 from NO_2-sensitized reaction [33]; ⟩⊢⟨, Gay and Pratt [152]. M = He: ○, Kurylo [237], also Moortgat and Allen [235] at 298 K; ⊖, Clyne and Thrush [162]; ⊗, Westenberg and de Haas [234], Wong and Davis [238]; ⦸, Hack et al. [281]; ⊘, using $k_{4,He}/k_{4,H_2}$ = 0.41 from NO_2—sensitized reaction [33]; ●, using $k_{4,He}/k_{4,H_2}$ = 0.32 from second limits [289]. Vertical lines at room temperature indicate total ranges for H_2 and Ar.

with the shock tube results at around 1500 K. Clearly this is an area where further work is needed.

6.6 FURTHER REACTIONS OF HO_2 WITH H, O, OH AND HO_2 (REACTIONS (viii), (viiia), (x), (xx), (xxi) AND (xxii))

Because of the difficulty of observing the HO_2 radical directly, firm data on all of these reactions is sparse, and in the case of reaction (xxii) non-existent. The disproportionation reaction (x) has been investigated at room temperature (i) by using an r.f. discharge in H_2O_2 at ca. 0.5 torr as the source of HO_2 for a flow system, with measurement of HO_2 by direct molecular beam sampling into a mass spectrometer [184], and (ii) by producing HO_2 by photolysis of H_2O_2 (or O_3 in the presence of H_2O_2)

using a molecular modulation technique, and monitoring HO_2 by UV absorption spectroscopy [185]. Both studies gave results within the range $k_{10} = (2.0 \pm 0.4) \times 10^9$ l . mole^{-1}. sec^{-1} at 300 K. The only other information about k_{10} is due to Troe [327], who estimated k_{10}/k_{15} = 0.2—0.4 at around 1000 K (see Sect. 6.8). The activation energy E_{10} is found to be small, and for the purpose of deriving absolute values of $(k_8 + k_{8a})$ and k_{11} at 773 K from the ratios of Table 18, it was initially assumed that $E_{10} = 0$.

The measurement of the ratio $k_{20}/(k_8 + k_{8a})$ at room temperature has been described in the preceding section, and both this ratio and the absolute rate coefficients k_8, k_{8a} and k_{20} have been discussed at some length in Sect. 4.3.3 and 5.4. Flame considerations indicate that the ratios k_{8a}/k_8 and $k_{22}/(k_8 + k_{20})$ are small (each $\leqslant 0.1$), but no firm data are available. Further, the only "measured" value of k_{21} is that of $(8 \pm 4) \times 10^9$ at about 1530 K, derived in Sect. 5.4 from the lean flame recombination results of Kaskan [179]. Like reaction (x), both of the reactions (xx) and (xxi) are likely to have only small activation energies, and for flame computations it has been assumed that $E_{20} = E_{21} = 0$. The scarcity of data in this particular area reflects the complexity of the systems. Alternative possibilities regarding E_{10}, and their effects on k_8 and k_{20}, will be discussed together with reaction (xi) in Sect. 6.8.

6.7 REACTIONS OF H_2O_2 WITH O, H AND OH (REACTIONS (xiii), (xiv), (xiva) AND (xv))

The reactions of hydrogen peroxide with both O and H atoms have been studied by Albers et al. [74] in the temperature range 300—800 K, using a discharge-flow system. The H_2O_2 was always added downstream of the discharge. Analysis was by ESR and mass spectrometry.

6.7.1 Reaction with O atoms

Oxygen atoms in He or Ar carrier gas at 2—5 torr were produced from N + NO, or by passing molecular oxygen through the discharge. When sufficient excess of H_2O_2 was present the [O] decays, measured with ESR, were first order with respect to O, and increased in proportion to $[H_2O_2]_0$ as long as the decays were limited to 10—20 %. The overall stoichiometry was found to be

$$H_2O_2 + 2.5O = 2O_2 + 0.5H_2O + H$$

and this may be explained by equal probabilities for reactions (xiii) and (xiiia), viz.

$$O + H_2O_2 \rightarrow H_2O + O_2 \qquad \text{(xiii)}$$

$$O + H_2O_2 \rightarrow OH + HO_2 \qquad \text{(xiiia)}$$

if (xiiia) is followed by (xxii) and (—ii). The implication of a relatively frequent occurrence of the mechanistically difficult re-arrangement (xiii) is, however, somewhat surprising. For $[H_2O_2]_0 = 10[O]_0$, it is also difficult to see why reaction (xv)

$$OH + H_2O_2 = H_2O + HO_2, \qquad \text{(xv)}$$

does not contribute, since $k_{15} = 0.1\ k_{-2}$ at just above 600 K. The reaction of some OH with H_2O_2 can produce results kinetically equivalent to reaction (xiii). Having regard to the overall stoichiometry, the decay rate measurements lead to an overall expression

$$k_{13} + k_{13a} = \frac{1}{2.5}\frac{1}{[H_2O_2]_0}\,d(\ln[O])/dt$$

$$= 2.8 \times 10^{10} \exp\{-(3200 \pm 300)/T\} \qquad (84)$$

if only reactions (xiii) and (xiiia) contribute to the removal of H_2O_2.

The reaction of oxygen atoms with H_2O_2 has also been studied more recently by Davis et al. [240], who used flash photolysis of O_3 as the O atom source, and resonance fluorescence for the concentration measurement. Applying the factor of 2.5 to their results, as in eqn. (84), leads to

$$k_{13} + k_{13a} = (6.7 \pm 1.0) \times 10^8 \exp\{-(2{,}125 \pm 261)/T\} \qquad (84a)$$

between 283 and 368 K. At 298 K the expressions (84) and (84a) give $k_{13} + k_{13a} = 6.1 \times 10^5$ and 5.4×10^5, respectively.

6.7.2 *Reaction with H atoms*

The reaction of H_2O_2 with H atoms may proceed by two alternative routes

$$H + H_2O_2 = H_2O + OH \qquad \text{(xiv)}$$

$$H + H_2O_2 = H_2 + HO_2 \qquad \text{(xiva)}$$

and here again considerable uncertainty exists. At 713—773 K, Baldwin et al. [68] found $k_{14a}/k_{14} = 0.143 \pm 0.015$, with no significant temperature dependence (cf. Sect. 4.3.3 and Table 18). Thus, if their mechanism for the H_2-sensitized decomposition of H_2O_2 is correct, reaction (xiv) occurs the more frequently in this temperature range. A similar conclusion was also previously reached by Hoare and Peacock [241] using another method. On the other hand, Albers et al. [74] have used their discharge-flow system to examine the reactions of D atoms with H_2O_2, and find the yield of HOD from the primary reaction at 421 K to be only about 10 % of the yield of HD, i.e. $k_{14aD}/k_{14D} \simeq 10$ at this temperature. These data imply $E_{14} - E_{14a} \simeq 8$ kcal . mole^{-1} between 420 and 750 K.

Although, when H_2O_2 is added in large excess to gas flows containing H or D atoms, the observed H or D atom decays are first order with

References pp. 234—248

respect to [H] or [D], the rates are not those of the primary reactions (xiv), (xiva), or their D atom analogues (xivD) and (xivaD), viz.

$$D + H_2O_2 \rightarrow OH + HOD \qquad \text{(xivD)}$$

$$D + H_2O_2 \rightarrow HD + HO_2 \qquad \text{(xivaD)}$$

These primary steps are followed principally by the rapid reactions (viii) or (viiiD) of H or D atoms with HO_2, and by the reactions (xv) or (xvD) of OH or OD with H_2O_2, viz.

$$D + HO_2 \rightarrow OD + OH \qquad \text{(viiiD)}$$

$$OD + H_2O_2 \rightarrow HOD + HO_2 \qquad \text{(xvD)}$$

A reaction chain is therefore set up. However, by adding oxygen atoms to the system in concentrations comparable with $[H_2O_2]$, Albers et al. [74] were able to arrest the attack of OH or OD on H_2O_2 in favour of the faster reaction (—ii),

$$O + OH \rightarrow O_2 + H \qquad \text{(—ii)}$$

since $k_{-2}/k_{15} \geqslant 10$. In this way the attack on the H_2O_2 is limited to the primary reactions (xiv). Unfortunately, if this is done, then for every H or D atom removed by reactions (xiv), one H atom is produced by (—ii); while for every H or D atom entering reactions (viii), a similar H or D atom is returned. In the case of the $H + H_2O_2$ reaction the decay of H atoms thus becomes virtually zero; while in the case of the $D + H_2O_2$ reaction there is a decay of D atoms due to reactions (xivD). The sum $(k_{14D} + k_{14aD})$ could be measured satisfactorily in the temperature range 294—464 K, and a single measurement of the decay of H_2O_2 in the H +

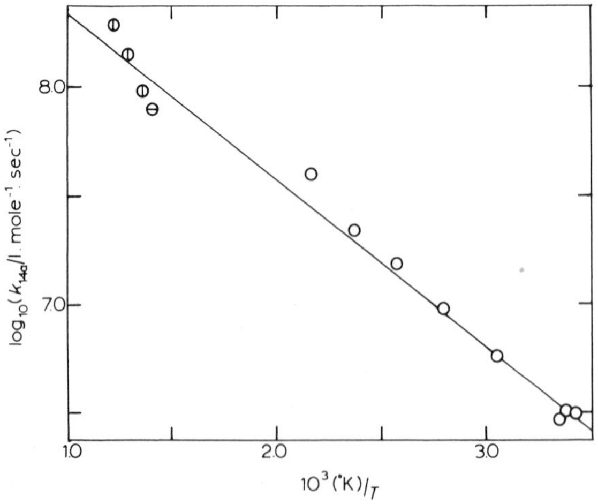

Fig. 42. Arrhenius plot of k_{14a}. ○, Albers et al. [74]; ◐, Baldwin et al. [72, 73]; ⊖, Baldwin et al. [69].

H_2O_2 reaction served to give k_{14a}/k_{14aD} = 0.43 at 375 K. If we assume this ratio to remain constant over the temperature range 294—464 K, a series of values of k_{14a} may be obtained. In Fig. 42 these are combined with values of k_{14a} obtained (i) from the data of Baldwin et al. [72, 73] in Table 18, together with the expression (72) for k_2, and (ii) one value at 713 K from a measurement of the inhibitory effect of oxygen on the hydrogen sensitized decomposition of H_2O_2 [69] (the measurements led to a value of $k_{4,H_2}/k_{14a}$ which may be combined with eqn. (81)). Figure 42 shows a reasonable Arrhenius line, giving

$$k_{14a} = 1.4 \times 10^9 \exp(-1800/T) \qquad (85)$$

This is essentially the same expression as that recommended by Baulch et al. [55].

The apparent contribution of reaction (xiv) itself to the observations of Albers et al. was small and not well defined. The work therefore provides no low temperature data for this reaction. On the other hand, Klemm et al. [522] have studied the overall reaction of H atoms with H_2O_2 at 283—353 K, using flash photolysis and resonance fluorescence, under conditions where only reactions (xiv) prevailed. They found $(k_{14} + k_{14a})$ = $(3.1 \pm 1.2) \times 10^9 \exp\{-(1390 \pm 140)/T\}$ in this temperature range, and, in contrast with Albers et al., suggested by comparison with high temperature results that the dominant reaction below 1000 K is reaction (xiv). At 300 K their mean overall result is approximately an order of magnitude higher than those of Albers et al. Clearly confirmation is needed. At 773 K, Baldwin et al. [73] find k_{14} = 7.8×10^8 $l.mol^{-1}.sec^{-1}$ (column C, Table 18 and expression (72) for k_2).

6.7.3 *Reaction with OH radicals*

The reaction of OH with H_2O_2 (reaction (xv)) is a major step involved in the production of the HO_2 radical when H_2O_2 is subjected to a discharge or to flash photolysis. Because of the speed of reaction (xv), this is a good method of producing HO_2 for further kinetic studies [184, 185]. The reaction rates in the temperature range 300—460 K have been measured by Greiner [242], who used kinetic spectroscopy to follow the OH concentrations following flash photolysis of H_2O_2/Ar mixtures. The decay of OH is interpreted in terms of reactions (xv) and (xvi). Results are given in Table 42.

TABLE 42
Values of k_{15} at low temperatures [242]

Temp. (K)	303	360	396	462	462	462
$k_{15}(10^8 l.mole^{-1}.sec^{-1})$	5.7	7.2	8.1	13.0	14.0	15.0

At higher temperatures around 700—800 K, linked values of k_{15}/k_1 and k_{13}/k_3 may be derived from studies of the slow reaction rates in H_2/O_2 mixtures [72], the inhibition of the second explosion limits in the latter by H_2O_2 [71], or the H_2-sensitized decomposition of H_2O_2 when the hydrogen concentrations are low [67—70]. All these aspects have been discussed in Sects. 4.3.2 and 4.3.3. Following the demonstration by Albers et al. [74] of an appreciable contribution from one or both of reactions (xiii) and (xiiia) — the chain breaking effects of which are not the same — there remains some doubt about separating the effects of reactions (xiii) and (xv) in the above studies. Assuming $k_{13}/k_3 = 0$, all three of the high temperature investigations give $k_{15}/k_1 = 5.0 \pm 1.0$ over the whole range from 700—800 K. However, taking $k_{13}/k_3 = 12.0$ at 773 K (column C, Table 18), the value of k_{15}/k_1 from the slow reaction studies becomes 3.7 at this temperature. Using expression (71) for k_1, the former ratio gives $k_{15} = (3.2 \pm 0.6) \times 10^9$ at 773 K; while the latter gives $k_{15} = 2.3 \times 10^9$.

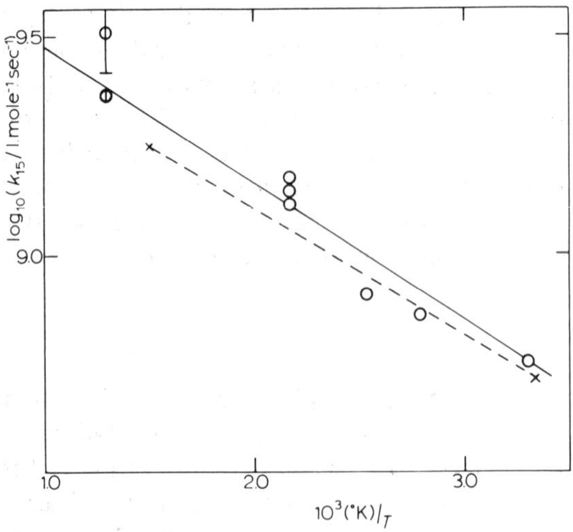

Fig. 43. Arrhenius plot of k_{15}. ○, Greiner [242]; ⦶, Baldwin et al. [72, 73]; x— — —x, Hack et al. [281].

Figure 43 shows an Arrhenius plot for reaction (xv), and leads to

$$k_{15} = 6.1 \times 10^9 \exp(-720/T) \tag{86}$$

This expression gives preference to the lower value of k_{15} at 773 K, and as a result it predicts lower values than the recommendation of Baulch et al. [55] at higher temperatures. Equation (86) agrees closely with the expression $k_{15} = (4.8 \pm 1.0) \times 10^9 \exp\{-(670 \pm 70)/T\}$ recently given by Hack et al. [281].

6.8 REACTION (xi) $HO_2 + H_2 = H_2O_2 + H$, AND FURTHER CONSIDERATION OF k_8, k_{10} AND k_{20}

The direct estimation of k_{11} from the ratios $k_{11}/k_{10}^{1/2}$ of Table 18 involves prior knowledge of k_{10} at the temperature concerned. In estimating k_8 from $k_8/k_2k_{10}^{1/2}$ (Sect. 5.4.3), it was assumed that $E_{10} = 0$ and $k_{10} = 2 \times 10^9$ l . mole^{-1} . sec^{-1} at all temperatures. Following a similar procedure on the ratio $R_7 = k_{11}/k_{10}^{1/2}$ leads to $k_{11} = 1.36 \times 10^3$ l . mole^{-1} . sec^{-1} at 773 K. It should perhaps be re-emphasized here that the values of R_7 in Table 18 were optimized assuming no contribution from the possible alternative step

$$HO_2 + H_2 = H_2O + OH \tag{xia}$$

to the slow reaction rates or induction periods in the H_2/O_2 system (Sect. 4.3.3). Thus the ratios R_7 give maximum values of $k_{11}/k_{10}^{1/2}$ consistent with observation.

Also given in Table 18 are optimized values of $R_9 = k_{14a}/k_2$. At 773 K, $R_9 = 89$; and using $k_2 = 3.3 \times 10^6$ l . mole^{-1} . sec^{-1} gives $k_{14a} = 1.29 \times 10^8$ l . mole^{-1} . sec^{-1}. Since reaction (xiva) is the reverse of reaction (xi), multiplication by the equilibrium constant $K_{11} = 2.6 \times 10^{-5}$ then leads to $k_{11} = 3.39 \times 10^3$ l . mole^{-1} . sec^{-1} at 773 K — 2.5 times greater than the result above. The discrepancy between the two values of k_{11} disappears if we put $k_{10} = 1.25 \times 10^{10}$ at 773 K, leading in combination with the room temperature value to

$$k_{10} = 4 \times 10^{10} \exp(-900/T) \tag{87}$$

Equation (87) confirms that the activation energy E_{10} is small.

If eqn. (85) is multiplied by the expression for the equilibrium constant K_{11} between 300 and 800 K, we find

$$k_{11} = 6 \times 10^8 \exp(-9{,}300/T) \tag{88}$$

Both expressions (85) and (88) rely heavily on the work of Albers et al. [74] for their temperature dependence. The value of $E_{11} = 18.6$ kcal . mole^{-1} is some 3—4 kcal . mole^{-1} lower than would be expected from the values of $k_{11}/k_{10}^{1/2}$ obtained by Baldwin et al. [72] over the smaller temperature range from 743—803 K.

It remains to consider next the values of k_8 and k_{20} which result when k_{10} is represented by eqn. (87). Corresponding with $(k_8 + k_{8a})^2/k_2^2 k_{10} = 0.325$ and $(k_8 + k_{8a})/k_{20} = 6.0$ at 773 K, we have $(k_8 + k_{8a}) = 2.1 \times 10^{11}$ and $k_{20} = 3.5 \times 10^{10}$. The value of $(k_8 + k_{8a})$ is such that E_8 and E_{20} must both be very close to zero. However, putting $E_8 = E_{20} = 0$ gives $(k_8 + k_{8a} + k_{20}) = 2.4 \times 10^{11}$ at 293 K; and this is about twelve times the maximum value of 2×10^{10} found by Albers [186]. Clearly there is still some considerable uncertainty in all the rate coefficients of these reactions of HO_2, and it is not yet possible to formulate a fully consistent and reasonable picture from the rather complex situation.

Finally, the reactions of H_2O_2 and HO_2 must be considered also in the light of the work of Meyer et al. [326], Troe [327] and Kijewski and Troe [212], who examined the decomposition of H_2O_2, highly diluted with argon, in shock waves at temperatures between 950 and 1450 K. The decay of H_2O_2 was followed using UV absorption at 2900 Å. Around 2300 Å, on the other hand, an additional absorption to that of H_2O_2 was observed, corresponding with a species which is formed and then consumed again during the reaction. The additional absorption was attributed to HO_2.

Troe [327] has interpreted the time variation of the absorption in terms of a reaction mechanism consisting of (vii), (xv), (x) and (xxi), similar to the mechanism proposed by Baldwin et al. [67—69] but including also reaction (xxi), viz.

$$H_2O_2 + Ar = OH + OH + Ar \quad \text{(vii)}$$

$$OH + H_2O_2 = H_2O + HO_2 \quad \text{(xv)}$$

$$HO_2 + HO_2 = H_2O_2 + O_2 \quad \text{(x)}$$

$$OH + HO_2 = H_2O + O_2 \quad \text{(xxi)}$$

For several assumed combinations of values of the ratios $C_2 = k_{10}/k_{15}$ and $C_3 = k_{21}/k_{15}$, and for a series of initial conditions defined by the parameter $C_1 = k_{15}[H_2O_2]/k_7[Ar]$, Troe carried out numerical integrations of the series of differential equations controlling the system, in order to determine the maximum HO_2 concentrations and the concentrations of H_2O_2 remaining at the time of the maximum (defined by $\beta' = [HO_2]_{max}/[H_2O_2]_0$ and $\alpha = [H_2O_2]/[H_2O_2]_0$, respectively). For each combination of C_2 and C_3, the values of β' were then normalized to give $\beta = \beta'/\beta'_{C_1=150}$. Plots of the calculated α and β versus C_1 were then prepared. Next, the UV absorption by the H_2O_2 at 2300 Å may be calculated from that at 2900 Å by use of a calibration factor from the H_2O_2 spectrum. The additional absorption at 2300 Å may then be used to give $\gamma' = \epsilon.\beta'_{obs}$, a *relative* value of β'_{obs} which again can be normalized to give $\beta_{obs} = \gamma'/\gamma'_{C_1=150}$. Comparison of the calculated and measured plots of α and β versus C_1 then allow estimates of the correct values of C_2 and C_3 to be made, and following this, a determination of the absorption coefficient of HO_2 at 2300 Å. Such plots are shown in Figs. 44 and 45. As a result of the analysis, Troe [327] estimated $k_{10}/k_{15} = 0.2$—0.4 and $k_{21}/k_{15} = 0.4$—2.4 at 1100 K. At this temperature, eqn. (86) gives $k_{15} = 3.2 \times 10^9$ l . mole^{-1} . sec^{-1}. The work thus confirms that both reactions (x) and (xxi) have at most very small activation energies. The decay of H_2O_2 in the experiments over the temperature range 950—1450 K led to $k_{7,Ar} = 1.6 \times 10^{13} \exp(-21{,}500/T)$.

Following the work with pure argon as diluent, Kijewski and Troe [212] have studied the pyrolysis of H_2O_2 in the presence of hydrogen at

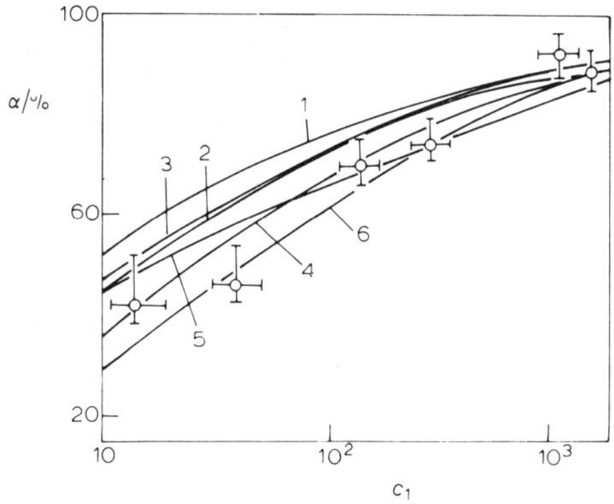

Fig. 44. Decomposition of H_2O_2 in argon at 1100 K (after Troe [327]). Plots of $\alpha = [H_2O_2]/[H_2O_2]_0$ at point of maximum HO_2 concentration against $C_1 = k_{15}[H_2O_2]/k_7[Ar]$. Calculated curves for series of values of $C_2 = k_{10}/k_{15}$ and $C_3 = k_{21}/k_{15}$. Curve 1, $C_2 = 0.8$, $C_3 = 2.4$; curve 2, $C_2 = 0.8$, $C_3 = 0.4$; curve 3, $C_2 = 0.4$, $C_3 = 2.4$; curve 4, $C_2 = 0.4$, $C_3 = 0.4$; curve 5, $C_2 = 0.2$, $C_3 = 2.4$; curve 6, $C_2 = 0.2$, $C_3 = 0.4$. Points represent observations. (By courtesy of Bunsengesellschaft für physikalische Chemie.)

temperatures between 870 and 1000 K, again using UV absorption. The decomposition of the H_2O_2 is accelerated in exactly the same way as that observed by Baldwin et al. [69] at somewhat lower temperatures. At around 930 K, the sensitized rate reached an upper limit of 21 ± 6 times the rate of the decomposition in pure argon. This limiting rate was approached at $[H_2]/[Ar] \approx 0.1$. Using the mechanism proposed by Baldwin et al., consisting of reactions (vii), (i) and (xiv), the chain length at high $[H_2]$, and hence the ratio of the limiting rate to the unsensitized rate, is determined by a competition between reactions (xiv) and (xiva), with the latter reaction terminating the chain (see Sect. 4.3.3). The observed

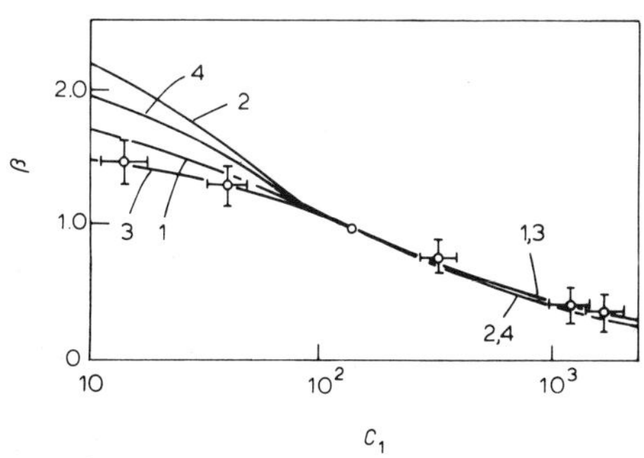

Fig. 45. Decomposition of H_2O_2 in argon at 1100 K (after Troe [327]). Plots of β against C_1. Calculated curves for series of values of C_2 and C_3. Curve 1, $C_2 = 0.4$, $C_3 = 2.4$; curve 2, $C_2 = 0.4$, $C_3 = 0.4$; curve 3, $C_2 = 0.2$, $C_3 = 2.4$; curve 4, $C_2 = 0.2$, $C_3 = 0.4$. Points represent observations, normalized with respect to observation at $C_1 = 150$. (By courtesy of Bunsengesellschaft für physikalische Chemie.)

acceleration leads to a mean $k_{14a}/k_{14} = 0.06$ at 930 K, in good agreement with the value of $k_{14a}/k_{14} = 0.143$ at 713—773 K coupled with $E_{14} - E_{14a} = 8$ kcal . mole^{-1} (cf. Sect. 6.7).

$$H_2O_2 + M = OH + OH + M \quad \text{(vii)}$$

$$OH + H_2 = H_2O + H \quad \text{(i)}$$

$$H + H_2O_2 = H_2O + OH \quad \text{(xiv)}$$

$$H + H_2O_2 = H_2 + HO_2 \quad \text{(xiva)}$$

$$H + HO_2 = OH + OH \quad \text{(viii)}$$

$$H + HO_2 = H_2 + O_2 \quad \text{(xx)}$$

However, against this simple analysis must be set a number of difficulties regarding the measured maximum HO_2 concentrations in the sensitized systems. Kijewski and Troe [212] have carried out a quasi-stationary state analysis of the sensitized decomposition on the basis of reactions (i), (vii), (viii), (xiv), (xiva), (xx) above, and (x), (xv) and (xxi), and obtained the equations

$$\frac{k}{k_0} = \frac{1 + \frac{k_{21}}{k_{15}} X + \frac{k_1}{k_{15}} Z \left\{ \frac{k_{14} + k_{14a} + k_{20} X}{k_{14} + k_{14a} + (k_8 + k_{20}) X} \right\}}{1 + \frac{k_{21}}{k_{15}} X + \frac{k_1}{k_{15}} Z \left\{ \frac{k_{14a} + (k_{20} - k_8) X}{k_{14} + k_{14a} + (k_8 + k_{20}) X} \right\}} \quad (89)$$

$$\frac{k_{10}}{k_7} YX^2 = \frac{1 - \frac{k_{21}}{k_{15}} X + \frac{k_1}{k_{15}} Z \left\{ \frac{k_{14a} - (k_8 + k_{20}) X}{k_{14} + k_{14a} + (k_8 + k_{20}) X} \right\}}{1 + \frac{k_{21}}{k_{15}} X + \frac{k_1}{k_{15}} Z \left\{ \frac{k_{14a} + (k_{20} - k_8) X}{k_{14} + k_{14a} + (k_8 + k_{20}) X} \right\}} \quad (90)$$

where

$$\frac{k}{k_0} = -\frac{1}{2k_7 [M] [H_2O_2]} \frac{d[H_2O_2]}{dt} \quad (91)$$

$$X = [HO_2]/[H_2O_2] \quad (92)$$

$$Y = [H_2O_2]/[M] \quad (93)$$

$$Z = [H_2]/[H_2O_2] \quad (94)$$

At 950 K, $k_{21}/k_{15} \simeq 3$ and $k_1/k_{15} \simeq 0.5$.

With $[H_2] = 0$, $k = k_0$ from eqn. (89). At low temperatures, when X is small, we also have, for $[H_2] = 0$,

$$X = \frac{[HO_2]}{[H_2O_2]} = \left(\frac{k_7 [M]}{k_{10} [H_2O_2]} \right)^{1/2} \quad (95)$$

With increasing $[H_2]$, if the simplification $(k_{21}/k_{15})X \ll 1$ is introduced at $T < 1000$ K, one obtains at first

$$\frac{k}{k_0} \simeq 1 + \frac{k_1}{k_{15}} Z \left\{ \frac{k_{14} + k_8 X}{k_{14} + k_{14a} + (k_8 + k_{20})X} \right\} + O(Z^2) \tag{96}$$

Then, at larger concentrations of H_2, with $Z \gg 1$, k/k_0 becomes independent of Z, as is observed in the experiments.

$$\frac{k}{k_0} \simeq \frac{k_{14} + k_{14a} + k_{20} X}{k_{14a} + (k_{20} - k_8)X} \tag{97}$$

$$= \frac{k_{14} + k_{14a}}{k_{14a}} \text{ at low } X \tag{97a}$$

The expression in curly brackets in eqn. (96) is close to unity, and both eqns. (96) and (97) are in essential agreement with the expressions derived by Baldwin et al. [69], who considered the initial rates of decomposition on the assumption that all HO_2 formation leads to chain termination. Baldwin et al. obtained

$$\frac{k}{k_0} = \frac{1 + \dfrac{k_1}{k_{15}} Z}{1 + \dfrac{k_1}{k_{15}} Z \left\{ \dfrac{k_{14a}}{k_{14} + k_{14a}} \right\}} \tag{98}$$

For the HO_2 yield at $Z \gg 1$ and $T < 1000$ K (small X), eqn. (90) predicts

$$\frac{k_{10}}{k_7} Y X^2 \simeq \frac{k_{14a} - (k_8 + k_{20})X}{k_{14a} - (k_8 - k_{20})X} \tag{99}$$

and if $k_8 \gg k_{20}$, then X is again given by eqn. (95). This would mean that at constant $[H_2O_2]$ and [M], the HO_2 concentration should be approximately the same as in the unsensitized decomposition, in agreement with observation [212]. At 950 K we also have $k_8/k_{20} \simeq 7$ (cf. Sect. 4.3.3 and 5.4.2).

Unfortunately, the argument just presented also depends on the further condition that the numerator in eqn. (99) should remain positive, i.e. $k_{14a}/(k_8 + k_{20}) > X$. Now in the experiments with large amounts of added hydrogen, such that $[H_2]/[H_2O_2]_0 \gg 40$, the observed ratio $[HO_2]/[H_2O_2]$ at $[HO_2]_{max}$ was in the region 0.05—0.1 at temperatures between 900 and 1000 K, and this would seem to require that $k_{14a}/(k_8 + k_{20}) \gg X \simeq 0.08$. However, if we take $R_7/R_8 = k_{11}k_2/(k_8 + k_{8a})$ from Table 18, and then divide by $k_2 K_{11}$, the resulting ratio $k_{14a}/(k_8 + k_{8a})$ at 773 K becomes 3.75×10^{-3} (column A, Table 18) or 2.16×10^{-3} (column C). An increase from the larger figure at 773 K to 8×10^{-2} at 950 K would imply $E_{14a} - E_8 \simeq 25$ kcal . mole^{-1}; and if $E_{14a} = 3.6$ kcal . mole^{-1} (cf. eqn. (85)) this is impossible. Again then, it

TABLE 43
Rate coefficients in hydrogen—oxygen system, expressed as $k = AT^B \exp(-C/T)$ (l. mole sec. units)

Reaction			A	B	C (K)	Temp. range (K)	Error in log k
(i)	$OH + H_2$	$= H_2O + H$	1.17×10^6	1.3	1825	250—2500	±0.1
(ii)	$H + O_2$	$= OH + O$	1.4×10^{11}	0	8250	700—1500	±0.1
(iii)	$O + H_2$	$= OH + H$	1.8×10^7	1.0	4480	400—2000	±0.1
(iv)[a]	$H + O_2 + H_2$	$= HO_2 + H_2$	1.03×10^{12}	—0.72	0	300—1500	±0.1
	$H + O_2 + N_2$	$= HO_2 + N_2$	3×10^{14}	—1.7	0	300—1500	±0.1
	$H + O_2 + He$	$= HO_2 + He$	4.6×10^{11}	—0.77	0	200—800	±0.1
(vii)	$H_2O_2 + N_2$	$= OH + OH + N_2$	1.2×10^{14}	0	22900	700—1500	±0.1[b]
(x)	$HO_2 + HO_2$	$= H_2O_2 + O_2$	$(2.0 \pm 0.4) \times 10^9$			300	—
(xi)[c]	$HO_2 + H_2$	$= H_2O_2 + H$	6×10^8	0	9300	300—800	±0.3
(xiii)	$O + H_2O_2$	$= H_2O + O_2$	No recommendation — see text and eqns. (84)				
(xiiia)	$O + H_2O_2$	$= OH + HO_2$					
(xiv)	$H + H_2O_2$	$= H_2O + OH$	7.8×10^8			773	±0.3
(xiva)	$H + H_2O_2$	$= H_2 + HO_2$	1.4×10^9	0	1800	300—800	±0.3
(xv)	$OH + H_2O_2$	$= H_2O + HO_2$	6.1×10^9	0	720	300—800	±0.1
(xvi)	$OH + OH$	$= O + H_2O$	5.6×10^9	0	460	300—2000	±0.2
(xvii)	$H + H + M$	$= H_2 + M$	See Table 26 for M = H_2, He, Ar, N_2, CH_4, CO_2, SF_6 at 70—300 K				
	$H + H + N_2$, Ar	$= H_2 + N_2$, Ar	1.0×10^{12}	—1.0	0	300—4700	
	$H + H + H_2$	$= H_2 + H_2$	9.2×10^{10}	—0.6	0	300—2500	
	$H + H + H_2O$	$= H_2 + H_2O$	6.0×10^{13}	—1.25	0	300—2500	
(xviii)	$H + OH + N_2$	$= H_2O + N_2$	1.6×10^{16}	—2.0	0	300—2000	
(xix)	$H + O + M$	$= OH + M$	No recommendation.				
(xxi)	$OH + HO_2$	$= H_2O + O_2$	$(8 \pm 4) \times 10^9$			1500	
(xxii)	$O + HO_2$	$= OH + O_2$	$<0.1\ (k_8 + k_{20})$				

[a] At 773 K chaperon efficiencies relative to H_2 are 0.35, 0.44 and 6.5 for O_2, N_2 and H_2O, respectively (see also Tables 7, 31 and 41).
[b] Error in log k increases to ±0.3 at 1500 K.
[c] This expression is obtained by combining the expression for k_{14a} with the equilibrium constant $K_{11} (= k_{11}/k_{14a})$. Combining

further with the ratio $k_{11}/k_{10}^{1/2}$ at 773 K (Table 18), it leads to $k_{10} = 1.25 \times 10^{10}$ at 773 K, and in combination with the tabulated value of k_{10} at room temperature, to eqn. (87) for k_{10}. Again using the ratios of Table 18, we then find $(k_8 + k_{8a}) = 2.1 \times 10^{11}$ and $k_{20} = 3.5 \times 10^{10}$ at 773 K (see text for comment on compatibility with room temperature value of $k_8 + k_{8a} + k_{20}$).

The alternative simple approach is to assume $E_{10} = 0$ and $k_{10} = 2 \times 10^9$ at all temperatures. This is of course incompatible with the values of k_{14a} and $k_{11}/k_{10}^{1/2}$ already quoted. However, it leads to more acceptable values of $(k_8 + k_{8a})$ and k_{20}, and if we accept Clyne and Thrush's mean value of $(k_8 + k_{8a})/k_{20}$ at room temperature we find, for $E_{20} = 0$

		A	*B*	*C*	Temp. range (K)
(viii)	$H + HO_2 = OH + OH$ }	1.7×10^{11}	0	540	300—800
(viiia)	$H + HO_2 = O + H_2O$ }				
(xx)	$H + HO_2 = H_2 + O_2$	1.4×10^{10}	0	0	300—800

The data for reaction (viii) refer to the sum $(k_8 + k_{8a})$. Flame computation suggests $k_{8a} < 0.1\, k_8$. Using the data given, the sum of the rate coefficients at room temperature is approximately twice the maximum value of $(k_8 + k_{8a} + k_{20}) = 2 \times 10^{10}$ found by Albers [186]. The discrepancy is removed if both the calculated coefficients are multiplied by $1.5 \exp(-330/T)$.

appears that the results or interpretation in this area are not entirely consistent.

6.9 RECOMBINATION REACTIONS

The recombination reactions

$$H + H + M = H_2 + M \qquad \text{(xvii)}$$

$$H + OH + M = H_2O + M \qquad \text{(xviii)}$$

have been discussed in Sect. 5.4. To add to that discussion, Zellner et al. [309] have recently found $k_{18,M} = 5.4 \times 10^{10}$, 8.3×10^{10}, 1.7×10^{11} and 3.25×10^{11} $l^2 \cdot mole^{-2} \cdot sec^{-1}$ for M = He, Ar, N_2 and CO_2, respectively, at 300 K, using a discharge flow—resonance fluorescence technique for the measurement of [OH]. In conjunction with the higher temperature results in Sect. 5.4, this gives $k_{18,N_2} = 1.6 \times 10^{16} T^{-2.0}$, i.e. a steeper temperature dependence than that given in Table 30. This is unlikely appreciably to affect the flame calculations of Sect. 5.4.

6.10 RECOMMENDED RATE COEFFICIENTS

The complete set of recommended rate coefficient expressions for the hydrogen–oxygen system is summarized in Table 43. Based on the foregoing discussion, the table includes two alternative, reasonably internally consistent, sets of parameters for the reactions of HO_2 with other species. Although neither is completely consistent with the whole range of experimental observations, these sets do, in combination with Table 30, give an impression of the degree of uncertainty still encountered in this particular area.

7. The reaction between deuterium and oxygen

Compared with the reaction between hydrogen and oxygen, there have been relatively few studies of the deuterium—oxygen system. Early studies by Hinshelwood et al. [243] dealt with the second explosion limits and the slow reaction in silica vessels; while at about the same time Frost and Alyea [14] measured the second limits in a KCl coated Pyrex vessel of 20 mm diameter. More recently Linnett and Selley [244] have determined the relative efficiencies of a number of molecules in reaction (ivD)*

$$D + O_2 + M = DO_2 + M \qquad \text{(ivD)}$$

from studies of the second limit in larger KCl coated vessels (6.0 cm diameter and 8.0 cm long, ellipsoid), and Baldwin et al. [245] have made

* Numbering in the D_2—O_2 reaction will be the same as in H_2—O_2, but with the following letter D.

TABLE 44
Third body coefficients in reaction (iv) (H_2 = 1)

Gas	D_2	O_2	N_2	Ar	CO_2	CF_4	SF_6	H_2O	D_2O	Surface	Temp (°C)	Diameter (mm)	Ref.
$k_{4D,M}/k_{4,H_2}$	0.75	—	0.59	0.2	1.49	1.0	2.41	—	—	KCl	530	60	244
	(0.73)	0.39	0.56	—	—	—	—	—	—	KCl	540	35	245
	(0.73)	—	—	—	—	—	—	7.5	6.4	KCl	540	51	245
$k_{4,M}/k_{4,H_2}$	(0.75)	—	0.43	0.18	1.53	2.29	2.65	—	—	KCl	530	60	244
	(0.73)	0.36	0.45	—	—	—	—	—	—	KCl	540	35	245
	(0.73)	—	—	—	—	—	—	7.1	5.56	KCl	540	51	245
	(0.73)	—	—	—	—	—	—	6.4	4.9	B_2O_3	500	51	245

similar investigations with both KCl coated and aged boric acid coated vessels, in order to find the efficiencies of both H_2O and D_2O in reactions (iv) and (ivD). Table 44 gives efficiencies $k_{4D,M}$ relative to k_{4,H_2}. They are obtained by multiplying the observed efficiencies $k_{4D,M}/k_{4D,D_2}$ by the collision frequency ratios $Z_{DO_2-D_2}/Z_{DO_2-H_2}$ and $Z_{DO_2-H_2}/Z_{HO_2-H_2}$.

A more recent study of the $D_2 + O_2$ reaction by Baldwin et al. [246] has involved measurements of the second limits, and the induction periods and maximum rates of the slow reaction in an aged boric acid coated vessel of 52 mm diameter. Maximum concentrations of D_2O_2 in the slow reaction were also determined. The kinetic parameters of the oxidation process were then determined by a computer optimization treatment similar to that described in Sect. 4.3.3 for the $H_2 + O_2$ reaction. Excluding the primary initiation rate θ which is necessary for the calculation of induction periods, but which needs to be only approximately defined, there are a minimum of seven significant parameters (cf. Table 18).

In order to make their optimization procedure more realistic in the case of the $H_2 + O_2$ reaction, Baldwin et al. [72] used independent measurements of (a) second limits in KCl coated vessels in order to give k_2/k_4 (after correction if necessary for reaction (xi) and for surface termination of H atoms (Sect. 3 and Table 14)), and (b) the homogeneous decomposition of H_2O_2 in the presence of hydrogen. The latter measurements give accurate values for k_7, and k_{14a}/k_{14} or k_{15}/k_1 at high or low hydrogen concentrations respectively, and the combined measurements leave just three adjustable parameters to be determined by the results with the B_2O_3 coated vessels.

Unfortunately in the case of the $D_2 + O_2$ reaction, similar independent measurements cannot easily be made at 500 °C, firstly because of lack of information about the surface termination of D atoms in KCl coated vessels (such information would involve a detailed study also of the first

TABLE 45
Rate coefficient ratios for H_2 and D_2 reactions at 500 °C. (l. mole . sec units) [246]

Ratio	$H_2 + O_2$	$D_2 + O_2$
k_2/k_4	3.84×10^{-4} (M = H_2)	5.50×10^{-4} (M = D_2)
k_7	38.6 (M = H_2)	33.5 (M = D_2)
k_{14}/k_2	265	167
k_{15}/k_1	4.7	5.4
$k_{11}/k_{10}^{1/2}$	0.0337	0.0145
$k_8/k_2k_{10}^{1/2}$	0.367	0.466
k_{13}/k_3 [a]	0	0

[a] Assumed (cf. Table 18).

limit), and, secondly, because of lack of ready availability of pure D_2O_2. Instead, Baldwin et al. [246] estimated $k_{2D}/k_{4D,M=H_2}$ at 500 °C by comparing the second limits of the $D_2 + O_2 + N_2$ and $H_2 + O_2 + N_2$ systems at 540 °C (where surface destruction of D and H atoms is unimportant) to give the ratio $(k_2/k_{4,H_2})/(k_{2D}/k_{4D,D_2})$ at that temperature. They then used Table 44 and the temperature dependence of the ratio k_2/k_{2D} measured by Kurzius and Boudart [58] (cf. Sect. 3.6.4) to give the ratio $(k_2/k_{4,H_2})/(k_{2D}/k_{4D,H_2})$ at 500 °C. In addition, k_{7D} was estimated by comparing the yields of D_2O_2 and H_2O_2 at the maximum rate in the slow reactions of the D_2 and H_2 systems at 500 °C. With this initial input, the optimization procedures applied to the second limits, maximum rates and induction periods led to the values of $k_{8D}/k_{2D}k_{10D}^{1/2}$, $k_{11D}/k_{10D}^{1/2}$, k_{14D}/k_{2D} and k_{15D}/k_{1D} given in Table 45, which also compares the parameters for the hydrogen and deuterium reactions.

Of the remaining studies of the overall reaction between deuterium and oxygen, the measurements of the first explosion limit by Kurzius and Boudart [58], and their deduction of an expression for k_{2D} in the temperature range 800--1000 K, have already been discussed in Sect. 3.6.4. A shock tube study of the development of the ignition, as well as of deuterium dissociation in the presence of argon and the isotopic exchange reactions (xxiv) and (xxv)

$$H + D_2 = HD + D \qquad \text{(xxiv)}$$

$$D + H_2 = HD + H \qquad \text{(xxv)}$$

in the temperature range 1700—3100 K, has recently been reported by Appel and Appleton [247]. They used the Lyman-α line to measure D atom concentrations. In the case of the ignitions, exponential growth constants were measured for a series of mixture compositions and temperatures, and an analysis similar to that described in Sect. 5.2 was made. As previously found by Schott [91] in connection with the $H_2 + O_2$ system, the difficulty of separating k_{1D} from k_{3D} led to uncertainties in these rate coefficients. The results for k_{2D} agreed within ±20 % with the expression $k_{2D} = 8.9 \times 10^{10} \exp(-7{,}450/T)$ due to Kurzius and Boudart [58].

Attention has already been drawn in Sect. 5.4.2 to the possibility of using suitable well-characterized flames for the examination of the reactions of the flame radicals with trace additives. Assuming $k_{-1Da} = 0.5\, k_{-1De}$,

$$OH + HD \rightleftharpoons H + HOD \qquad \text{(iDa)}$$

$$OD + HD \rightleftharpoons H + D_2O \qquad \text{(iDe)}$$

a detailed analysis [172] of the HD profiles measured when traces of heavy water were added to three H_2—N_2—O_2 flames leads directly to

References pp. 234—248

TABLE 46
Rate coefficients of isotopic reactions

Reaction		k	k_H/k_D	A[a]	C (K)[a]	Temp. range (K)	Ref.
(iDa)	$OH + HD = H + HOD$	$(6.6 \pm 0.4) \times 10^8$	2.8			1050	See text
(iDc)	$OH + D_2 = D + HOD$		3.64			473	210
			1.76			573	210
		1.27×10^6	(3.2)[b]			300	248
		1.24×10^6	(3.5)			300	207
		$(1.33 \pm 0.24) \times 10^6$	(3.3)[c]			300	197
				7.6×10^9	2560	210—460	197
(—iDa)	$H + HOD = OH + HD$	3.65×10^6		(5.4×10^{10})	(10805)[d]	1050	
(—iDe)	$H + D_2O = OD + HD$	7.3×10^6		(1.08×10^{11})	(10805)[d]	1050	
(iiD)	$D + O_2 = OD + O$			2.9×10^{10}	7150	823—973	249
				$(8.9 \pm 0.22) \times 10^{10}$	7450 ± 200[e]	800—1000	58
(iiiD)	$O + D_2 = OD + D$	$(3.4 \pm 0.7) \times 10^4$	(4.5)	2.0×10^{10}	5500	416	250
		$(1.06 \pm 0.02) \times 10^5$				454	
		$(1.48 \pm 0.13) \times 10^5$				467	
		$(3.8 \pm 0.2) \times 10^5$				508	
		$(1.74 \pm 0.15) \times 10^6$				604	
		$(1.10 \pm 0.13) \times 10^7$				739	
		$(7.0 \pm 0.5) \times 10^7$				968	

(ivD)	$D + O_2 + Ar = DO_2 + Ar$		0.9 ± 0.2			225	162
			0.9 ± 0.2			244	
			0.7 ± 0.2			293	
	$D + O_2 + He = DO_2 + He$		0.6 ± 0.2			293	
	$D + O_2 + H_2 = DO_2 + H_2$		0.86			773	246
(viiD)	$D_2O_2 + D_2 = OD + OD + D_2$	33.5				773	246
(xivaD)	$D + H_2O_2 = HD + HO_2$			7×10^9	2100 ± 200	294—464	74
(xviD)	$OD + OD = O + D_2O$	$(9.5 \pm 1.0) \times 10^8$	(1.6)			300	251
(xviiD)	$D + D + Ar = D_2 + Ar$	$(2.9 \pm 0.54) \times 10^9$	(1.55 ± 0.43)			300	252
	$D + D + He = D_2 + He$	$(4.0 \pm 0.65) \times 10^9$	(1.47 ± 0.34)			300	252
	$D + D + D_2 = D_2 + D_2$	$(5.5 \pm 0.4) \times 10^9$	(1.22 ± 0.15)			77	148
		$(2.2 \pm 0.1) \times 10^9$	(1.33 ± 0.08)			298	

[a] Parameters A and C refer to two parameter expression $k = A \exp(-C/T)$, l. mole . sec units.
[b] Ratios in parentheses are calculated from individual rate coefficients measured by the same author.
[c] Obtained from $k_1/k_{1De} = 1.44 \exp(+250/T)$ between 210 and 460 K [197].
[d] It is assumed here that $k_{-1De} = 2k_{-1Da}$. Parameters A and C are derived from $k_1 = 2.2 \times 10^{10} \exp(-2{,}590/T)$ [55], $k_1/k_{1Da} = 2.4 \exp(+155/T)$ (see text) and $K_{1Da} = 0.17 \exp(+8{,}060/T)$ [172].
[e] Combined with result for reaction (ii) [58], gives $k_2/k_{2D} = 1.92 \exp(-650/T)$.

k_{-1Da}, and then, by use of the equilibrium constant K_{1Da}, to k_{1Da} = $(6.6 \pm 0.4) \times 10^8$ at 1050 K (cf. Sect. 5.4.2). The addition of a trace of D_2 to one of the three H_2—N_2—O_2 flames then led to an HD profile in which HD formation was controlled by the isotopic exchange reaction (xxiv), and its removal by reaction (iDa) at lower temperatures than those operating in the D_2O experiment. At the temperatures of the D_2O experiment (near the final flame temperature) the ratio [OH]/[H] is the partial equilibrium ratio determined only by K_1; but at the lower temperature the OH concentration is controlled more by the forward rate of reaction (i), so that the oxidation parameter controlling the HD profile in the D_2 experiment is the ratio k_1/k_{1Da}. Assuming A_1/A_{1Da} to be the ratio of collision numbers Z_{OH+H_2}/Z_{OH+HD}, the expression $k_1/k_{1Da} = 2.4 \exp(+155/T)$ was found. This ratio is independent of the precise calibration of the radical profiles in the flame, and the absolute value of k_{1Da} above corresponds with $k_1 = (1.85 \pm 0.3) \times 10^9$ l . mole^{-1} . sec^{-1} at 1050 K.

A number of the elementary processes in the H_2—D_2—O_2 system have also been studied at lower temperatures using the techniques, such as discharge-flow, described in Sect. 6. A selection of results is given in Table 46.

8. Nitrogen oxides and hydrogen oxidation

The free radical nature of the nitric oxide molecule allows it readily to form association products with other free radicals. Both the nitroxyl radical (HNO) and nitrogen dioxide (NO_2) may be regarded as simple association products of this type, and the high reactivities of both these species towards H, OH or O are able to produce several interesting and important effects when one of the two nitrogen oxides is added to, or formed in, the hydrogen—oxygen system.

8.1 CATALYSIS OF RADICAL RECOMBINATION

Chain termination may occur due to catalysis of atom and free radical recombination by the succession of general steps

$$A + NO + M = ANO + M \tag{xxvi}$$

$$B + ANO = AB + NO \tag{xxvii}$$

where A or B may be H, OH or O. The actual termination step is (xxvi), and when A is a general free radical, reactions of this type are responsible for the well-known inhibiting effect of NO in chain reaction systems.

When A and B are both H atoms, the catalyzed recombination occurs by way of the nitroxyl radical, through reactions (xxviii) and (xxix), viz.

$$H + NO + M \rightleftharpoons HNO + M \tag{xxviii}$$

$$H + HNO = H_2 + NO \tag{xxix}$$

At high H atom concentrations, when $k_{29} \gg k_{-28}[M]$, the decay of H atoms is first order in H, and is controlled by the rate of reaction (xxviii). These are the conditions which have been used in studies of the catalyzed recombination in discharge-flow systems [253—256]. Values of k_{28} for H_2, He, Ar, CO_2, N_2O, SF_6 and H_2O are summarized by Baulch et al. [257], who recommend the expression

$$k_{28,H_2} = 5.4 \times 10^9 \exp(+300/T) \qquad (100)$$

in the temperature range 230—700 K, with error limits of ±50 %. The values at room temperature from the discharge-flow work agree well with an independent determination by Hikida et al. [236] using the pulse radiolysis technique with measurement of [H] by Lyman-α absorption. This is a direct measurement of k_{28,H_2} under conditions where there is no subsequent reaction (xxix). Measurements by Ahumada et al. [239], Atkinson and Cvetanović [469], and Moortgat and Allen (private communication to Ahumada et al. [239]) are also in good agreement.

At low H atom concentrations ($k_{29} \ll k_{-28}$M) reaction (xxviii) is equilibrated, the H atom decay becomes second order in the radical concentration, and the overall rate is controlled by reaction (xxix). With the assumption of partial equilibration of reactions (i), (ii) and (iii) of the hydrogen—oxygen system in the recombination region of rich, atmospheric pressure $H_2/N_2/O_2$ flames, and with the addition of reaction (xxx)

$$OH + HNO = H_2O + NO \qquad (xxx)$$

to the catalyzed recombination mechanism, Bulewicz and Sugden [258], and later Halstead and Jenkins [259], interpreted the decay of [H] in such flames to give [259] $k_{29} = (4.8 \pm 1.2) \times 10^9$ and $k_{30} = (3.6 \pm 1.2) \times 10^{10}$ l . mol^{-1} . sec^{-1} at 2000 K. The added nitric oxide showed no marked decomposition in the flames — a result also found by Day et al. [260]. Day et al. found also that the addition of the nitric oxide decreased the burning velocity of their low temperature, fuel-rich $H_2/N_2/O_2$ flames.

Data on reactions (xxvi) when A is O or OH are reviewed by Baulch et al. [257].

8.2 OXIDATION OF HYDROGEN BY NITROGEN DIOXIDE

The direct oxidation of hydrogen by nitrogen dioxide has been studied by Ashmore and Levitt [261], and Rosser and Wise [262]. The reaction proceeds rapidly at 600—700 K without change of pressure, according to the overall stoichiometry

$$H_2 + NO_2 = H_2O + NO$$

The rate is much greater than the rate of decomposition of NO_2 to NO and oxygen. The reaction is inhibited by nitric oxide, and the extent of the inhibition indicates it to be a chain process almost entirely. The oxidation is accelerated by the presence of traces of NOCl, but the accelerated reaction is still inhibited by nitric oxide. At high $[H_2]/[NO_2]$ ratios the unsensitized rate is given [261, 262] by

$$-d[NO_2]/dt = \frac{k_A[NO_2][H_2]^{1.5 \pm 0.15}}{[NO] + k_B[NO_2]} \qquad (101)$$

where k_A is constant and k_B may vary slowly with $[H_2]$.

The results may be explained by means of the free radical straight chain mechanism

$$H_2 + NO_2 = H + HNO_2 \qquad (xxxi)$$

$$H + NO_2 = OH + NO \qquad (xxxii)$$

$$OH + H_2 = H_2O + H \qquad (i)$$

$$OH + NO_2 + M = HNO_3 + M \qquad (xxxiii)$$

$$OH + NO + M = HNO_2 + M \qquad (xxxiv)$$

where, again, reactions (xxxii) and (xxxiv) are of the general types (xxvii) and (xxvi) respectively. Reaction (xxxii) is very rapid (see Sect. 8.3), and reactions (xxxiii) and (xxxiv) proceed at comparable, slower rates [261, 262].

The sensitization by NOCl may be explained by the addition to the above mechanism of reactions (xxxv) and (xxxvi)

$$NOCl + M = NO + Cl + M \qquad (xxxv)$$

$$Cl + H_2 = HCl + H \qquad (xxxvi)$$

The overall reaction also shows an ignition boundary in the temperature range 790—840 K. The ignitions are thermal in nature, and the rate law is quantitatively similar to that found in the slow reaction [261].

8.3 SENSITIZATION OF THE HYDROGEN—OXYGEN SYSTEM

It has long been known [263—265] that traces of nitrogen dioxide can lower the ignition temperatures of certain hydrogen + oxygen mixtures by more than 200 K. Nitrosyl chloride [266, 267], nitrous oxide [4, 268], ammonia [4, 269, 270], cyanogen [4], chloropicrin [271] and nitric oxide [267] are other sensitizers which can have the same effect. For a given pressure and composition of the H_2—O_2 mixture, the ignitions with CCl_3NO_2, NOCl and NO_2 occur between a lower and upper sensitizer pressure limit of ignition. The variation of these sensitizer limits with temperature, total pressure of reactants, addition of inert gases, and the diameter of the reaction vessel, have been well established [263—271]. The solid line in Fig. 46 shows a typical ignition boundary for sensitization by nitrogen dioxide. The sensitizer limits are virtually identical in the $H_2 + O_2 + NO_2$, $H_2 + O_2 + NOCl$ and $H_2 + O_2 + CCl_3NO_2$ systems

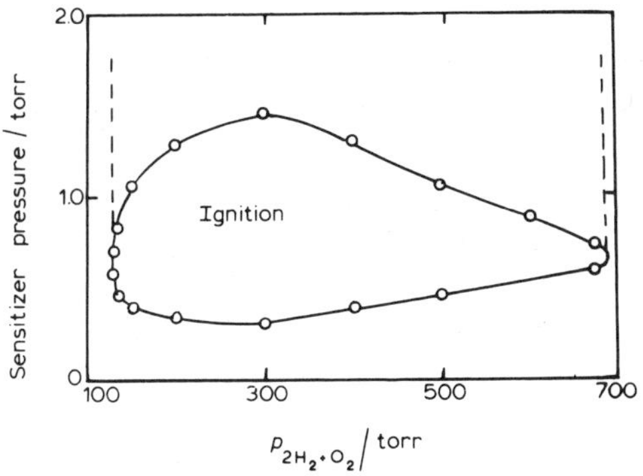

Fig. 46. Boundaries for sensitized ignitions at 364 °C. Solid line, NO_2 sensitization in KCl coated vessel, 7.0 mm diameter [265]; broken lines, NO sensitization (diagrammatic — no upper sensitizer limit when NO is admitted to reaction vessel first, i.e. not premixed with H_2 and O_2).

[272]. With nitric oxide, however, there is no upper sensitizer limit of ignition when the nitric oxide is placed in the reaction vessel and the mixture of $2H_2 + O_2$ run in from a mixing vessel [267, 272]. There is only a lower limit, as illustrated by the broken line in Fig. 46. If, on the other hand, the nitric oxide is premixed with the H_2 and O_2 in the mixing vessel, there is then both a lower and an upper sensitizer limit as with nitrogen dioxide.

Outside the ignition limits there is a slow reaction between the hydrogen and oxygen [273]. The rate of this reaction decreases on moving immediately away from both the lower and upper limits (sensitized reaction region). However, at higher sensitizer pressures above the upper limit, the rate increases again somewhat (catalyzed reaction region).

With CCl_3NO_2, NOCl or NO_2 as sensitizer, there is an induction period of several seconds between admitting the reactants to the reaction vessel and the onset of slow reaction or explosion. This induction period is absent when the reaction is sensitized by non-premixed nitric oxide (cf. above). By means of photometric studies of the NO_2-sensitized system, Ashmore and Levitt [274, 275] showed that the NO_2 was removed during the induction period at a rate which could be predicted by eqn. (101) belonging to the $H_2 + NO_2$ reaction. At the end of the induction period, when p_{NO_2} has fallen from its initial value p_0 to some value p_e which is characteristic of the experimental conditions, the rate of NO_2 removal increases rapidly. If p_0 is within the sensitizer limits for ignition, ignition follows when p_{NO_2} has fallen to some lower value p_i. On the other hand, if p_0 lies above the upper sensitizer limit, the acceleration declines, and p_{NO_2} reaches a stationary value p_s: in this event only a slow pressure decrease corresponding with slow reaction between the hydrogen and oxygen is observed. If $p_s > p_i$, then slow reaction is obtained; whereas $p_i > p_s$ corresponds with conditions inside the ignition region.

By detailed studies of the structure of the transitions from p_e to p_i, and of the accompanying pressure changes, Ashmore and Tyler [276] were able to show that, near the lower sensitizer limits, the ignitions were thermal in nature, while close to the upper limit they were nearly isothermal, branched chain ignitions. This was indicated by the observation that the pressure decrease near the lower limit was preceded by a pressure pulse or increase which could only be reasonably explained by self-heating in the early part of the reaction prior to the ignition. Such pressure increases near the upper limit were less pronounced and occurred much less frequently. In addition, the rates of reaction just outside the upper limit were much smaller than at the lower boundary. The rates near the limits for 100 torr $2H_2 + O_2$ at 360 °C are shown in Fig. 47.

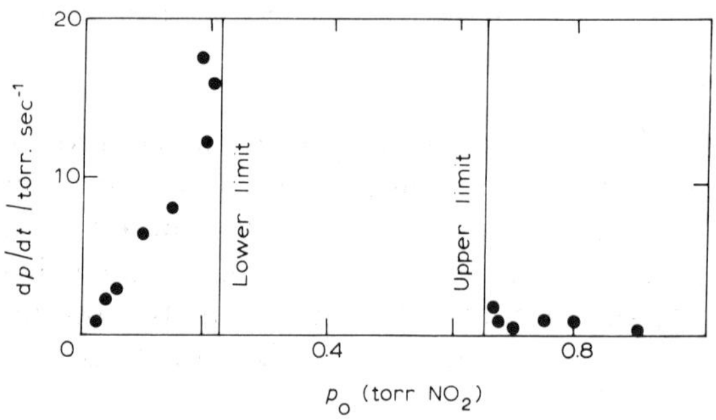

Fig. 47. Rates of pressure change outside the ignition limits (after Ashmore and Tyler [276]). 100 mm $2H_2 + O_2$; 20 mm diameter quartz vessel; temp. = 360 °C. (By courtesy of The Combustion Institute.)

For other conditions constant, Ashmore and Tyler [276] found that the nitrogen dioxide partial pressures p_e at the end of the initial induction period were independent of vessel diameter, and of the thermal conductivity of the gas mixture (varied by substituting He for N_2 as inert gas). This, combined with the rate law for the disappearance of NO_2 during the induction period, led to the conclusion that the reaction during this phase is simply the isothermal reaction between H_2 and NO_2. Further, a detailed examination of the pressure changes showed that the pressure pulse before slow reactions does not begin before p_e. The accelerated removal of NO_2 at p_e is therefore not due to thermal effects; and due to the abrupt nature of the acceleration it was concluded that the latter must be brought about by an increase in chain centre concentration due to net chain branching. The rate of change of concentration of chain centres n in a system with rates of initiation n_0, positive branching fn, linear termination gn, and quadratic termination δn^2, is given by

$$\begin{aligned} \mathrm{d}n/\mathrm{d}t &= n_0 + (f-g)n - \delta n^2 \\ &= n_0 + \phi n - \delta n^2 \end{aligned} \tag{102}$$

TABLE 47
Relative values of p_s for various inert gases at 633 K [33]

M	H_2	O_2	H_2O	N_2	CO_2	He
p_s (relative)	1.0	0.33	6.6	0.55	1.39	0.41

During the initial induction period the net branching factor ϕ is negative, giving a non-branched chain system (cf. Sect. 8.2). However, if ϕ increases during the induction period, because of the changes in concentration of NO_2 and NO, then a sudden increase in chain centre concentration would occur as ϕ passes through zero and becomes positive. This is taken to occur at p_e.

The kinetics of the reactions that control p_s were studied [33] using reactant pressures high enough to lie well outside the ignition region. It was found that p_s was independent of vessel diameter, and independent of the initial pressure, p_0, of nitrogen dioxide (and hence also of p_{NO}, since $p_{NO} = p_0 - p_s$, and usually $p_s \simeq 0.1\ p_0$), provided p_0 was less than about 1 torr. By a series of experiments in which the pressures of oxygen, hydrogen and inert gas (CO_2, N_2, He, H_2O) was varied, it was found that p_s is directly proportional to the oxygen pressure and to the inert gas pressure, i.e. $p_s \propto p_{O_2} p_M$. For the same pressures of different inert gases, the relative values of p_s are given in Table 47. These are virtually identical with the third body coefficients associated with the unsensitized hydrogen—oxygen second limit, and leave little doubt about the participation of reaction (iv) in the establishment of p_s. If NO is reoxidized to NO_2 by the fast reaction (xxxvii), then it is possible [33] to explain the observed kinetics in terms of reactions (xxxii), (iv) and (xxxvii), viz.

$$H + NO_2 = OH + NO \qquad \text{(xxxii)}$$

$$H + O_2 + M = HO_2 + M \quad \text{(slow)} \qquad \text{(iv)}$$

$$HO_2 + NO = OH + NO_2 \quad \text{(fast)} \qquad \text{(xxxvii)}$$

This leads to

$$p_s = \frac{\Sigma k_{4,M}[M]}{k_{32}}[O_2] \qquad (103)$$

in agreement with experiment. Equation (103) also explains why p_s is independent of p_0 for $p_0 < 1$ torr. At higher values of p_0 the reaction

$$2NO + O_2 \rightleftharpoons 2NO_2 \qquad \text{(xxxviii)}$$

becomes increasingly important, and eventually at high enough pressures dominates the free radical reactions [273, 275], causing p_s to increase. However, this is outside the range of the sensitized ignitions, where p_s is independent of p_0.

The complete reaction scheme which Ashmore and Tyler [276] regard as most satisfactory for explaining the sensitization phenomena, including the ignition, is

$$OH + H_2 = H_2O + H \quad \text{(i)}$$

$$H + O_2 = OH + O \quad \text{(ii)}$$

$$O + H_2 = OH + H \quad \text{(iii)}$$

$$H + O_2 + M = HO_2 + M \quad \text{(iv)}$$

$$HO_2 \rightarrow \text{destruction} \quad \text{(v)}$$

$$HO_2 + HO_2 = H_2O_2 + O_2 \quad \text{(x)}$$

$$HO_2 + NO = OH + NO_2 \quad \text{(xxxvii)}$$

$$O + NO_2 = NO + O_2 \quad \text{(xxxix)}$$

$$H + NO_2 = OH + NO \quad \text{(xxxii)}$$

$$OH + NO_2 + M = HNO_3 + M \quad \text{(xxxiii)}$$

$$OH + NO + M = HNO_2 + M \quad \text{(xxxiv)}$$

$$H_2 + NO_2 = H + HNO_2 \quad \text{(xxxi)}$$

During the initial induction period, when ϕ is negative, the major reactions are those of the H_2—NO_2 system. The branching reaction (ii) is outweighed by the fast reaction (xxxix) of O atoms, and by reaction (iv) and its successors. Reaction (xxxix) becomes less important as $[NO_2]$ decreases, and at the same time the increasing concentration of NO favours reaction (xxxvii) of HO_2 at the expense of (v) or (x). Since reactions (xxxiii) and (xxxiv) have similar rates (see Sect. 8.2) the effect of replacement of NO_2 by NO on these termination steps is negligible. ϕ therefore increases during the induction period.

Regarding the chain termination by reactions of HO_2, the lack of dependence of p_e on vessel diameter suggests that reaction (x) rather than a surface reaction (v) is the major contributor. This also accounts for the thermal nature of the ignitions at the lower sensitizer limits, where the NO and NO_2 concentrations are also low. Near the upper sensitizer limit the NO concentration at the end of the induction period will be comparatively high. This favours the removal of HO_2 by reaction (xxxvii) rather than (x), and at the same time favours reaction (xxxiv) rather than (x) as the chain terminating step. With the replacement of quadratic by mainly linear termination, the nature of the ignition changes from purely thermal to nearly isothermal.

A valuable feature of the mechanism is that all the elementary steps are well known in other systems. In particular, there is no necessity to add a new branching step to the hydrogen—oxygen scheme, and indeed, apart from (xxxvii), the radical reactions of NO and NO_2 are all of the general types (xxvi) and (xxvii). The key to the sensitization lies in reaction (xxxvii). At the time the above mechanism was proposed, the occurrence of reaction (xxxvii) was strongly suggested by the detailed study of p_s already described, and independently by kinetic observations on the H_2O_2/NO system [277]. Recent measurements using discharge-flow or flash photolysis techniques confirm that reaction (xxxvii) is fast, with $k_{37} \simeq 10^8$ l. mole^{-1} . sec^{-1} at room temperature [278—281]. Hack et al. [281] give $k_{37} = (2 \pm 1) \times 10^{10} \exp(-1{,}430/T)$ in the temperature range $298 < T < 670$ K; while Glänzer and Troe [328] found $k_{37} = (4.5 \pm 1.0) \times 10^9$ l. mole^{-1} . sec^{-1} at 1350—1700 K in a shock tube study of HNO_3—NO_2 mixtures.

Returning to consideration of the sensitized hydrogen—oxygen system outside the ignition region, the values of the stationary pressures, p_s, led, by eqn. (103), to $k_{4,H_2}/k_{32} = 0.10 \pm 0.01$ at 633 K [33]. In combination with eqn. (81) this leads to $k_{32} = 1.0 \times 10^{11}$ l. mole^{-1}. sec^{-1} at that temperature. At 298 K, Phillips and Schiff [282] find k_{32} $(3.0 \pm 0.03) \times 10^{10}$.

There is little doubt that the action of the other sensitizers mentioned involves the initial production of nitric oxide, followed by a similar series of reactions.

8.4 OTHER REACTIONS WITH NITROGEN OXIDES

Of the successive stages of reduction of nitrogen dioxide

$$NO_2 \rightarrow NO \rightarrow N_2O \rightarrow N_2$$

in combustion systems, the facile conversions are those from NO_2 to NO, and from N_2O to N_2, in which the number of N atoms per molecule does not increase. This is a consequence of the high stability of the NO molecule with respect to N atoms ($\Delta H^\circ_{f,298}$ = +21.6 and +113 kcal . mole^{-1} for NO and N, respectively). In flame systems, the comparative lack of reactivity of NO as a supporter of combustion was shown by the spectroscopic examination by Wolfhard and Parker [283] of the flames of hydrogen with N_2O, NO and NO_2, respectively. The OH radical was observed in all the flames. In the N_2O supported flame, NO was observed in the burnt gases at all mixture strengths, and the Schumann—Runge bands of oxygen were observed in lean flames. These are probably not connected with the main flame reaction (see below), but are due instead to the side reactions

$$O + N_2O = N_2 + O_2 \tag{xl}$$

$$O + N_2O = 2NO \tag{xli}$$

With nitric oxide as oxidant, OH was the only detectable intermediate, and no NH was found in absorption. Further, whatever the mixture strength, only very small quantities of NO were observed in the burnt gas, while for fuel-lean mixtures the appearance of the Schumman—Runge bands in absorption demonstrated the presence of oxygen in this region. The nitric oxide is virtually all decomposed in the reaction zone. From such observations, Adams et al. [284] concluded that the flame proceeds by the decomposition of the nitric oxide, followed by rapid reaction of the resulting oxygen with hydrogen. The conclusion is apparently supported by comparison of the flame temperature with that of the decomposition flame of pure NO. The latter can be obtained on a 3 cm diameter burner by preheating the NO to 1300 K [284], and results in a theoretical final flame temperature of 3020 K. This is about 100° lower than the flame temperature of the stoichiometric H_2 + NO mixture at S.T.P. However, more recent shock tube studies, to be discussed in Sect. 8.4.2, show that pure NO does not decay in shocked gases at temperatures below 3000 K, whereas NO in the presence of H_2 does decay. Flame speeds and theoretical flame temperatures for H_2 + NO mixtures are given by Adams and Stocks [285], and by Magnus et al. [308].

More revealing in relation to the comparative lack of reactivity of nitric oxide are the observations of Wolfhard and Parker on $H_2 + NO_2$ flames. Here nitric oxide is found strongly in absorption in the burnt gas of both rich and lean flames, showing that it does not play a major part in the reaction. This conclusion is supported by measurement of the flame temperature of the stoichiometric mixture for $H_2 + \frac{1}{2}NO_2 = H_2O + \frac{1}{2}N_2$ [286]. Theoretically this should be 2890 K if the stoichiometry is as quoted. The measured flame temperature by line reversal was 1780 K.

The reactions of hydrogen with nitrous and nitric oxides in closed vessels and in shock tubes will now be discussed.

8.4.1 Reaction of hydrogen with nitrous oxide

The thermal oxidation of hydrogen by nitrous oxide in silica vessels at temperatures between 823 and 1023 K was studied by Melville [287, 288]. The reaction was much faster than the decomposition of nitrous oxide, and the products were mainly N_2 and H_2O. The rate was determined by measuring the pressure drop when the water was (continuously) absorbed on P_2O_5. At pressures between 50 and 400 torr [286] the rate was directly proportional to $[N_2O]$ and nearly independent of $[H_2]$, except when $[N_2O]$ was high. The reaction was faster in wider vessels, though the increase was less than proportional to the square of the diameter. The apparent activation energy was 36 kcal . mole^{-1}. Additions of nitrogen or argon had no effect on the rate. At pressures below 50 torr the characteristics of the reaction were similar, except that the rate was then proportional to $[N_2O]^2$, and was markedly retarded by packing the reaction vessel [288]. The apparent activation energy at the lower pressures was 49 kcal . mole^{-1}. Photochemical experiments using mercury photosensitization for the production of H atoms gave a chain length similar to that for the thermal reaction when the rate of initiation for the latter was based on the rate of decomposition of pure N_2O. This suggests a chain process with reaction (xlii) as the initiating step in the thermal case. At higher pressures (above 50 torr), the photochemical rate varied as the square root of the light intensity, showing that the chains end by self-neutralization. Under these conditions the suggested mechanism for the thermal reaction was

$$N_2O + M = N_2 + O + M \tag{xlii}$$

$$O + H_2 = OH + H \tag{iii}$$

$$H + N_2O = N_2 + OH \tag{xliii}$$

$$OH + H_2 = H_2O + H \tag{i}$$

$$H + H + M = H_2 + M \tag{xvii}$$

the slight change in the kinetics at lower pressures being explained by a replacement of the gas phase termination step by termination through surface destruction of H atoms. The terminating steps will be considered further below.

Following an initiating process consisting of reactions (xlii) and (iii), the chain propagating step (xliii) seems the natural one leading to the major observed products. It is exothermic ($\Delta H^\circ_{298} = -61.6$ kcal . mole^{-1}), and so should compete successfully with the alternative endothermic process (xliv)

$$H + N_2O = NH + NO \qquad \Delta H^\circ_{298} \simeq +30 \text{ kcal.mole}^{-1} \tag{xliv}$$

The occurrence of reaction (xliii) must be considered also in the light of the work of Baldwin et al. [289], who added N_2O (5—20 %) to slowly reacting mixtures of hydrogen and oxygen diluted with helium in aged boric acid coated vessels at 773 K. N_2 formation was measured as a function of the pressure change. The value of $d[N_2]/d(\Delta P)$ decreased very slightly with increasing $[H_2]$, was almost proportional to $[N_2O]$, and was inversely proportional to $[O_2]$ and to the total pressure. The results suggest that reaction (xliii) is effectively the only process removing N_2O. In this case the $H_2 + O_2$ mechanism of Sect. 4.3 leads, with certain justifiable simplifying assumptions, to eqn. (104) for the relative rates of formation of N_2 and loss of H_2, viz.

$$-d[N_2]/d[H_2] = k_{43}[N_2O]/\Sigma k_4[O_2][M] \tag{104}$$

Since $\Delta[H_2] = 2\Delta P$, eqn. (104) is consistent with observation. Using $k_2/k_{4,H_2} = 3.84 \times 10^{-4}$ at 773 K (cf. Table 18), the experiments lead to $k_{43}/k_2 = 0.64 \pm 0.07$ and $k_{43} = 2.1 \times 10^6$ l . mole^{-1} . sec^{-1} at 773 K (cf. Table 43 for k_2).

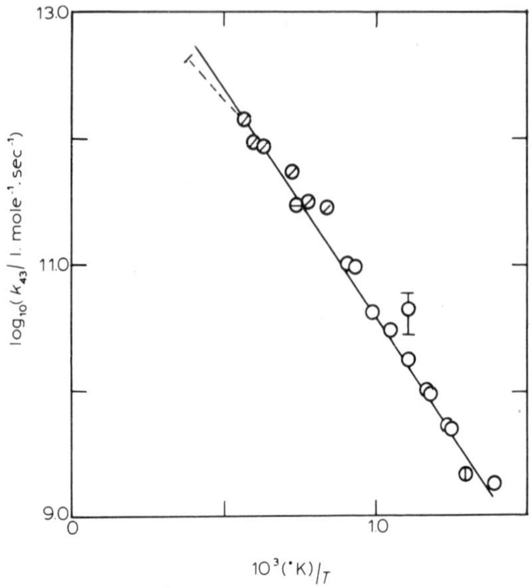

Fig. 48. Arrhenius plot of k_{43}. ⊘, Fenimore and Jones [167]; ↨, Dixon-Lewis et al. [170]; ⊖, Dixon-Lewis et al. [171]; ○, Albers et al. [292]; ⦶, Baldwin et al. [289]; ⊢ - ⊣, Henrici and Bauer [290].

Data on reaction (xliii) has also been obtained from studies with flames [167, 170, 171], shock tubes [290, 291] and discharge-flow systems [186, 292]. The data of Fenimore and Jones [167] from flame systems depend on a calibration of the H atom concentrations in the flames by means of reaction (—iDe), for which they assumed $k_{-1De} = 10^{12}$ exp $(-12750/T)$. Substitution of A_{1De} and E_{1De} from Table 46 reduces their values of k_{43} by factors of between 1.5 and 3 depending on the temperature. All the data, including values from Fenimore and Jones [167] corrected in this way, are plotted in Fig. 48, and lead to

$$k_{43} = 1.6 \times 10^{11} \exp(-8350/T) \tag{105}$$

in the temperature range 700—2500 K, with an estimated error in $\log_{10} k_{43}$ of ±0.2. This shows a slightly steeper temperature dependence than the expression suggested by Baldwin et al. [289].

Although reaction (xliii) appears to be the major reaction between H atoms and nitrous oxide, this does not exclude the occurrence of the alternative reaction (xliv). Indeed reaction (xliv) has been specifically suggested to explain (a) the formation of the nitric oxide responsible for the sensitizing effect of N_2O on $H_2 + O_2$ explosions [268] (cf. Sect. 8.3), and (b) the formation of nitric oxide in $H_2 + N_2O$ flames at 1500—2000 K [293] and shocked gases at 1900—2800 K [294]. More recent investigations of the thermal $H_2 + N_2O$ reaction have also shown that the mechanism is more complex than that suggested by Melville [287, 288].

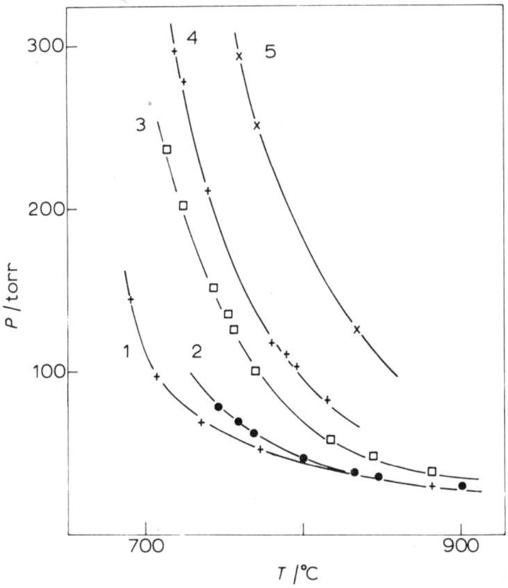

Fig. 49. Explosion limits of mixtures of hydrogen and nitrous oxide (after Navailles and Destriau [295]). Curve 1, p_{H_2}/p_{N_2O} = 1/5; curve 2, p_{H_2}/p_{N_2O} = 1/3; curve 3, p_{H_2}/p_{N_2O} = 2/1; curve 4, p_{H_2}/p_{N_2O} = 3/1; curve 5, p_{H_2}/p_{N_2O} = 10/1. Silica vessel, 1.8 cm diameter. (By courtesy of La Société Chimique de France.)

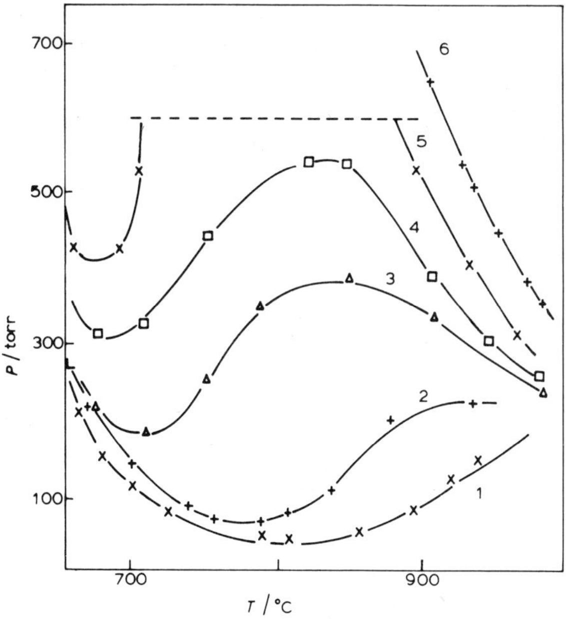

Fig. 50. Explosion limits of mixtures of hydrogen and nitrous oxide (after Navailles and Destriau [295]). Curve 1, p_{H_2}/p_{N_2O} = 1/7; curve 2, p_{H_2}/p_{N_2O} = 1/10; curve 3, p_{H_2}/p_{N_2O} = 1/15; curve 4, p_{H_2}/p_{N_2O} = 1/18; curve 5, p_{H_2}/p_{N_2O} = 1/20; curve 6, pure N_2O. Silica vessel, 1.8 cm diameter. (By courtesy of La Société Chimique de France.)

Thus Figs. 49 and 50 show the (thermal) explosion limits for both rich and lean mixtures, measured by Navailles and Destriau [295]. In the temperature range 910—1080 K there is a region of negative temperature coefficient in these limits, and also in the slow reaction, of certain lean mixtures. In Figs. 49 and 50, the mixtures showing this effect are bounded on the one side by the mixture having $H_2/N_2O \simeq \frac{1}{5}$, and on the other by pure N_2O. In addition, Holliday and Reuben [296] have measured reaction rates in $H_2 + N_2O$ mixtures at 810—870 K by following the UV absorption of N_2O at 2200 Å, and found an overall activation energy of 62.5 kcal . mole^{-1}. This is considerably larger than Melville's result [287, 288], and Holliday and Reuben maintain that Melville's method of rate measurement was probably inadequate for higher rates. Baldwin et al. [297] measured the N_2 formation in the overall reaction, and found a still higher activation energy of 71.5 kcal . mole^{-1}. Holliday and Reuben also observed that the slow reaction was strongly inhibited by addition of small quantities of NO.

Baldwin et al. [297] have recently re-examined the overall reaction at 813 and 873 K, and they confirm Melville's result that it is effectively zero order in H_2 and approximately first order in N_2O (more precisely, the order in N_2O is 1.1 ± 0.2 at 813 K and 1.2 ± 0.2 at 873 K). The rates were effectively independent of vessel surface (B_2O_3, uncoated Pyrex or uncoated silica). A small influence of helium addition was noticed at 873 K. Thus, for $p_{H_2} = p_{N_2O} = 25$ torr, the addition of 75, 200 and 400 torr He produced a gradual increase in rate to the extent that the initial rate was almost doubled at the highest pressure. For $p_{H_2} = p_{N_2O} = 10$ torr, the addition of 480 torr increased the rate approximately five-fold.

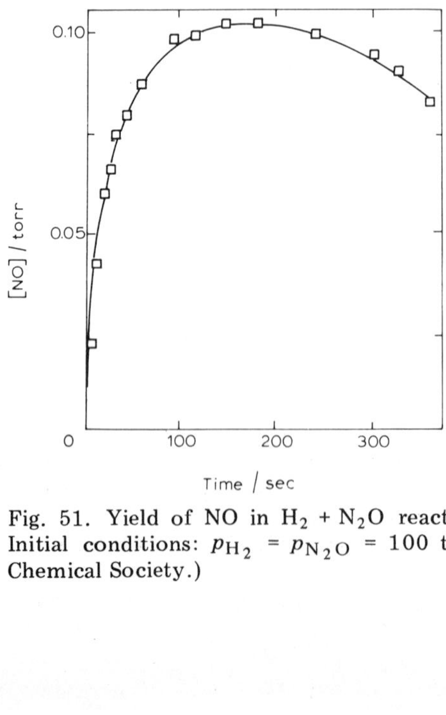

Fig. 51. Yield of NO in $H_2 + N_2O$ reaction at 873 K (after Baldwin et al. [297]). Initial conditions: $p_{H_2} = p_{N_2O} = 100$ torr; $p_{He} = 300$ torr. (By courtesy of The Chemical Society.)

At 813 K comparison of the initial rates at 250 and 500 torr, respectively, for a mixture with $H_2/N_2O/He$ = 0.2/0.2/0.6, showed an order almost exactly unity in total pressure. At 873 K, similar mixtures at 125, 250 and 500 torr gave a log (rate) versus log (total pressure) plot with a gradient of 1.5, while for an equimolar mixture of H_2 and N_2O at total pressures of 25, 50 and 200 torr, the mean gradient was 2.0 ± 0.4. Small amounts of NO are formed during the initial stages of the reaction, but its net rate of production decreases sharply as the reaction proceeds, and its concentration also passes through a maximum and decreases. A profile of [NO] versus time is shown in Fig. 51.

As a result of a preliminary analysis of the rates of chain initiation and termination in the scheme proposed by Melville [287], Baldwin et al. [297] concluded that neither reaction (xvii), nor the alternative destruction of H atoms at the vessel surface, could be the effective chain terminating step. The only reasonable alternative, confirmed by the detection of NO in the products, appeared to be reaction (xliv), and if all the NH radicals undergo termination, the rate expression becomes

$$d[N_2]/dt = 2k_{42}k_{43}[N_2O][M]/k_{44} \tag{106}$$

With the exception of the order of approximately unity in total pressure found at 813 K, this is in essential agreement with the experimental findings. Additional evidence in favour of reaction (xliv) leading to the terminating step comes from a comparison of the rates of initiation and termination for the initial conditions p_{H_2} = p_{N_2O} = 100 torr and p_{He} = 300 torr at 873 K. The calculated rates of reactions (xlii) and (xliv) agree to within a factor of around two.

Regarding the further mechanism of termination, it seems most likely [297, 298] that NH reacts with N_2O to form HNO

$$NH + N_2O = HNO + N_2 \tag{xlv}$$

The HNO may then undergo one of the reactions (xxix), (xxx) or (xlvi)

$$HNO + HNO = H_2O + N_2O \tag{xlvi}$$

The concentration of HNO in the system may be estimated from the maximum in the [NO] versus time profile, and comparative rate calculations using k_{29}, k_{30} and k_{46} from Baulch et al. [257] suggested that reaction (xlvi) is likely to be the most important. However, the story does not end there. If HNO participates in the manner suggested, its formation and decay by reaction (xxviii) must be considered, and the forward reaction (xxviii) should also lead to an inhibitory effect of added NO. Such an effect had already been observed by Holliday and Reuben [296]. Further investigation by Baldwin et al. [297] showed that the inhibition was large for small quantities of added NO (<1 torr NO in 500 torr mixture), but that it fell away at larger additions. The fall-off was attributed by them principally to the occurrence of reaction (xlvii)

$$HNO + NO = OH + N_2O \qquad \text{(xlvii)}$$

A complete numerical analysis of the system appeared to confirm the necessity for the inclusion of reaction (xlvii), inasmuch as, with $k_{47} = 0$, it was impossible satisfactorily to interpret either the rate of formation of NO during the early stages of the (initially) uninhibited $H_2 + N_2O$ reaction, or the rates inhibited by NO.

The principal reactions in the thermal oxidation of H_2 by N_2O thus appear to be

$$N_2O + M = N_2 + O + M \qquad \text{(xlii)}$$

$$O + H_2 = OH + H \qquad \text{(iii)}$$

$$H + N_2O = N_2 + OH \qquad \text{(xliii)}$$

$$OH + H_2 = H_2O + H \qquad \text{(i)}$$

$$H + N_2O = NH + NO \qquad \text{(xliv)}$$

$$NH + N_2O = HNO + N_2 \qquad \text{(xlv)}$$

$$H + NO + M' \rightleftharpoons HNO + M' \qquad \text{(xxviii)}$$

$$HNO + HNO = H_2O + N_2O \qquad \text{(xlvi)}$$

$$HNO + NO = OH + N_2O \qquad \text{(xlvii)}$$

In order fully to interpret the slope of the [NO] versus time profile in the initially uninhibited reaction, some contribution was also necessary from one or other of the steps

$$H + HNO = H_2 + NO \qquad \text{(xxix)}$$

$$OH + HNO = H_2O + NO \qquad \text{(xxx)}$$

though the absence of any marked decrease in rate as $[H_2]$ is reduced (thus allowing [OH] to increase) suggests that reaction (xxx) may be neglected.

The equations describing the progress of the overall reaction (stationary state equations for H, O, OH and HNO; differential equations for H_2O, N_2, NO) now involve the parameters k^0_{42}, k_{44}/k_{43}, k_{28}/k_{43}, k_{46} and k_{47}, together with k_{-28}, or k_{29}, or both. Attempts at optimizing the first five of these parameters showed very close agreement between observed and calculated overall reaction rates, and between observed and calculated [NO] versus time profiles, for a wide range of values of both k_{-28} and k_{29}. Thus no overall solution could be found. Nevertheless, the parameters $k^0_{42} = (2.6 \pm 0.7) \times 10^{-3}$, $k_{44}/k_{43} = (4.1 \pm 0.5) \times 10^{-3}$ and $k_{28}/k_{43} = 480 \pm 30$ at 873 K were fairly closely defined and essentially independent of the values assumed for k_{-28} and k_{29}. Using eqn. (105), the two ratios give $k_{44} = 5.2 \times 10^4$ l . mole^{-1} . sec^{-1} and $k_{28} = 6.0 \times$

10^9 l^2 . mole^{-2} . sec^{-1} at 873 K. The last figure is very close to the prediction of k_{28,H_2} = 7.6 x 10^9 l^2 . mole^{-2} . sec^{-1} from eqn. (100). Reasonably precise values of k_{46} and k_{47} were not available from the analysis, since they depended on the values assumed for k_{-28} and k_{29}. However, Baldwin et al. do draw attention to the fact that the range of values which they find for k_{47} are about ten times higher than the value suggested by Wilde [299] from his computer analysis of the H_2 + NO system (cf. Sect. 8.4.2).

In conclusion, the region of negative temperature coefficients already mentioned for lean mixtures (cf. Fig. 49) probably reflects an increasing rate of NO formation with increasing temperature, with a resulting increase in termination rate in a situation where low H atom concentrations will give low propagation rates by reaction (xliii). Reactions (xl) and (xli) of O atoms with N_2O will also need consideration in this context, and the overall result will be a reduction in chain length at higher temperatures. The negative temperature coefficient is evidence of this transfer to chain lengths characteristic of N_2O decomposition.

8.4.2 *Reaction of hydrogen with nitric oxide*

Early investigations of the reaction between hydrogen and nitric oxide in the temperature range 970—1100 K were made by Hinshelwood and Green [300] and Hinshelwood and Mitchell [301], who followed the pressure decrease in a static system. The observed order was somewhat variable, but they concluded that the reaction was essentially third order — second order in NO and first order in H_2. Later studies were by Graven [302], and Kaufman and Decker [303]. Graven studied the reaction in a

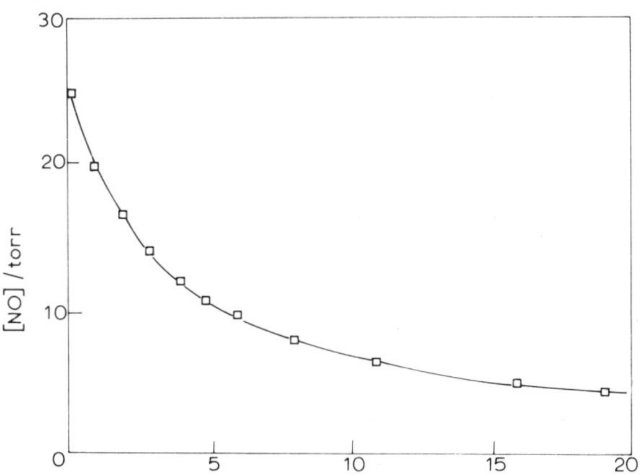

Fig. 52. Decay of NO in H_2 + NO reaction at 1323 K (after Kaufman and Decker [303]). Initial conditions: p_{NO} = 25 torr; p_{H_2} = 25 torr. Quartz vessel, 6 cm diameter. (By courtesy of The Combustion Institute.)

flow system at 1120–1330 K, following the amount of water formed at low extents of reaction. He found the initial rate to be largely second order in NO and one half order in H_2. Kaufman and Decker [303] again used a static system, and followed the NO photometrically in the temperature range 1170–1420 K. Figure 52 shows an NO decay curve for a particular set of conditions. At around 1370 K the rate of the overall reaction of the stoichiometric mixture with p_{NO} = 25 torr is only about fifty times the rate of the homogeneous NO decomposition, and because of this Kaufman and Decker limited their observations to stoichiometric and rich mixtures. They found non-integral orders in both NO and H_2. Over the temperature range 1170–1420 K the order in NO appeared to decrease from about 1.7 to 1.4, while the order in H_2 increased from about 0.5 to 0.67. The rates were quite reproducible, and apparently quite free from wall effects. The overall activation energy appeared to increase from about 36 to 50 kcal . mole^{-1} over the temperature range, but because of the changing orders with respect to the reactants, it is not a very meaningful quantity.

Kaufman and Decker [303] considered the rate controlling step in the reaction to be the reduction of NO to N_2O, after which the more rapid steps associated with the propagation mechanism in the $H_2 + N_2O$ system could take place. The mechanism has been further investigated by Wilde [299] using computer modelling techniques. The principal elementary steps contributing to the first stage, i.e. the removal of NO, are probably reactions (—xxix), (xxviii), (xlvii) and (i), with contributions also from reaction (xlvi) and the forward reaction (xxix), viz.

$$H + HNO \rightleftharpoons H_2 + NO \qquad \text{(xxix)}$$

$$H + NO + M \rightleftharpoons HNO + M \qquad \text{(xxviii)}$$

$$HNO + NO = OH + N_2O \qquad \text{(xlvii)}$$

$$OH + H_2 = H_2O + H \qquad \text{(i)}$$

$$HNO + HNO = H_2O + N_2O \qquad \text{(xlvi)}$$

Omitting reaction (xlvi), a conventional steady state treatment, with reactions (xxviii) and (xxxix) equilibrated, leads to

$$-\mathrm{d}[NO]/\mathrm{d}t = k_{47}[NO]^2[H_2]^{1/2}(K_{28}/K_{29})^{1/2} \qquad (107)$$

This approximates to the overall orders in NO and H_2 observed under some conditions. By including reaction (xlvi) with, somewhat arbitrarily, k_{46} = 3 x 10^8 exp (—1750/T), Wilde [299] found slightly greater than one half order in H_2. For the temperature range 1100–1360 K, the mean value of k_{47} giving the best overall agreement with experiment was k_{47} = 2 x 10^9 exp (—13,000/T), to within a factor of two (cf. analysis of $H_2 + N_2O$ system by Baldwin et al. [297] in Sect. 8.4.1).

The involvement of nitrogen oxides in atmospheric pollution from combustion systems has recently generated considerable interest in their higher temperature reactions, and has led to several shock tube studies of the reaction between hydrogen and nitric oxide at temperatures above 2000 K [304—307]. Bradley and Craggs [306] found that, in the absence of hydrogen, there was no measurable decomposition of nitric oxide in their shocked gases until above 3000 K, whereas in the presence of hydrogen a corresponding degree of reaction was observed at 2600 K. Experiments in which 5 % NO + 5 % H_2 + 90 % Ar were heated in a single pulse shock tube to 2700 K for 0.4 msec showed the products to be N_2 and H_2O, with possible traces of N_2O and O_2. The addition of 0.1 % O_2 to the reactants had no measurable effect on the rates. The rates were found to be about three orders of magnitude above those predicted by the mechanism discussed above for the temperature range 1000—1400 K, when reasonable rate coefficients were used in the latter. Ultra-violet absorption measurements of NO and OH indicated a single, common induction period τ for both the onset of NO removal and the appearance of OH [306, 307]. Duxbury and Pratt [307] found that a least squares fit of the gradient of a plot of $\tau\{[NO]_0[H_2]_0\}^{1/2}$ versus T^{-1} corresponded with an overall activation energy of 49 kcal . mole^{-1}. This is very close to the enthalpy change, $\Delta H^0_{298} \simeq 48.7$ kcal . mole^{-1}, of reaction (xlviii)

$$H + NO = N + OH \qquad \text{(xlviii)}$$

Attempts at computer simulation of the observed shock tube data in the initial stages of the reaction have led in all the investigations [304—307] to the conclusion that the rate-determining step at temperatures above 2400 K is reaction (xlviii); and it is worthy of note that Magnus et al. [308] have also concluded that reaction (xlviii) is important in H_2 + NO flames at similar temperatures (cf. earlier conclusions of Adams et al. [284]). The principal steps in the initial stages of the shock tube combustion are thus likely to be

$$H_2 + M = H + H + M \; (M = H_2, NO, Ar) \qquad \text{(—xvii)}$$

$$H + NO = N + OH \qquad \text{(xlviii)}$$

$$N + NO = N_2 + O \qquad \text{(xlix)}$$

$$O + H_2 = OH + H \qquad \text{(iii)}$$

$$OH + H_2 = H_2O + H \qquad \text{(i)}$$

The optimized Arrhenius parameters for reaction (xlviii) which emerge from the computer simulations, are given in Table 48, together with derived values of k_{48} at 2850 K.

To conclude, it should be added that Koshi et al. [304] found that the temperature dependence of the rate of disappearance of NO, which in their experiments corresponded to an overall activation energy of 40 ± 10

TABLE 48
Kinetic parameters for reaction (xlviii), based on $k_{48} = A \exp(-C/T)$ l . mole^{-1} . sec^{-1}

A	$C/10^3$ (K)	Temp. range (K)	k_{48} at 2850 K	Ref.
$(4.0 \pm 2.0) \times 10^{10}$	24.0 ± 0.15	2400—4000	8.7×10^6	304
1.34×10^{11}	24.6	2400—4500	2.4×10^7	305
3.5×10^{11}	23.7	2500—3020	1.0×10^8	306
2.6×10^{11}	24.4	2200—3250	4.6×10^7	307

kcal . mole^{-1} above 2400 K, decreased to almost zero below this temperature. They interpreted the change in activation energy as being associated with the change from the high temperature mechanism via reaction (xlviii) to the lower temperature mechanism via reaction (xxviii) and further reactions of HNO, but they encountered difficulty in matching the observed rates with the kinetic parameters available for the low temperature mechanism. Their experimental observation has not been confirmed.

8.5 RATE COEFFICIENTS OF ELEMENTARY PROCESSES IN THE HYDROGEN—NITROGEN OXIDE SYSTEMS

Data on the elementary steps in the H—N—O system is much less plentiful than in the case of the hydrogen—oxygen system, and for many reactions there is no reliable measurement available. Table 49 summarizes the data which are available.

9. Hydrocarbon addition to the hydrogen—oxygen system

When small quantities of hydrocarbons and related compounds are added to hydrogen—oxygen mixtures at around 500 °C, one or more of three effects may be observed.

First, there is an inhibiting effect of the additive on the low pressure explosions, so that the second limit pressure is reduced and the first limit is raised on addition of the hydrocarbon [329—331]. Secondly, there may be an increase in the maximum rate (of decrease of pressure) in the slow reaction [330]. Thirdly, induced explosions may occur in some cases (not with methane) at pressures outside the $H_2 + O_2$ explosion peninsula. In most cases such induced explosions appear as one sharp explosive reaction. However, they are sometimes characterized (e.g. with C_3H_8 at 560 °C) by an induction period during which there is a rapid pressure increase, and this is followed immediately by a very rapid pressure decrease in the system. It is probable that all the induced explosions follow this two-stage pattern. This type of explosion does not occur in H_2—O_2—CH_4 mixtures because methane is not as reactive as propane in

TABLE 49
Additional rate coefficients in H—N—O system, expressed as $k = AT^B \exp(-C/T)$, l . mole . sec units

	Reaction		A	B	C (K)	Temp. range (K)	Error in log k	Ref.
(xxviii)	$H + NO + H_2$	$= HNO + H_2$	5.4×10^9	0	—300	230—700	±0.2	See Sect. 8.1
	$H + NO + Ar$	$= HNO + Ar$	1.0×10^{10}			298	±0.05	239, 254, 256
	$H + NO + H_2O$	$= HNO + H_2O$	6.8×10^{10}			298	±0.1	256
(xxix)	$H + HNO$	$= H_2 + NO$	4.8×10^9			2000	±0.2	See Sect. 8.1
(xxx)	$OH + HNO$	$= H_2O + NO$	3.6×10^{10}			2000	±0.2	See Sect. 8.1
(xxxi)	$H_2 + NO_2$	$= H + HNO_2$	No recommendation					
(xxxii)	$H + NO_2$	$= OH + NO$	3.5×10^{11}	0	740	298—630	±0.2	257, See Sect. 8.3
(xxxiii)	$OH + NO_2 + He$	$= HNO_3 + He$	1.9×10^{10}	0	—900	230—450	±0.2	310—314
	$OH + NO_2 + Ar$	$= HNO_3 + Ar$	1.7×10^{10}	0	—900	230—450	±0.2	312—314, 317
	$OH + NO_2 + N_2$	$= HNO_3 + N_2$	4.3×10^{10}	0	—900	230—450	±0.2	314, 315, 317
	$OH + NO_2 + H_2O$	$= HNO_3 + H_2O$	4.0×10^{12}			295		311
(xxxiv)	$OH + NO + He, Ar$	$= HNO_2 + He, Ar$	7.9×10^9	0	—850	230—450	±0.2	310, 312—317
	$OH + NO + N_2$	$= HNO_2 + N_2$	1.35×10^{10}	0	—850	230—450	±0.2	314—317
(xxxvii)	$HO_2 + NO$	$= OH + NO_2$	2×10^{10}	0	1430	298—670	±0.25	281, 328
(xxxix)[a]	$O + NO_2$	$= NO + O_2$	5.7×10^9			298	±0.05	318—320
			1.05×10^{11}	—0.53	0	298—1055	±0.15	320
(xl)	$O + N_2O$	$= N_2 + O_2$	1.0×10^{11}	0	14100	1200—2000	±0.4	257
(xli)	$O + N_2O$	$= 2NO$	1.0×10^{11}	0	14100	1200—2000	±0.3	257
(xlii)[b]	$N_2O + Ar$	$= N_2 + O + Ar$	k^0 5.0×10^{11}	0	29000	1300—2500	±0.2	324
			k^∞ 1.3×10^{11}	0	30000	900—2100	±0.2	
	$N_2O + N_2O$	$= N_2 + O + N_2O$	k^0 4.9×10^{12}	0	29800	850—900	±0.2	297
			k^∞ 8.3×10^{10}	0	29800			

TABLE 49—*continued*

	Reaction		A	B	C (K)	Temp. range (K)	Error in log k	Ref.
(xliii)	$H + N_2O$	$= N_2 + OH$	1.6×10^{11}	0	8350	700—2500	±0.2	See Sect. 8.4.1
(xliv)	$H + N_2O$	$= NH + OH$	5.2×10^{4}			873		297
(xlv)	$NH + N_2O$	$= HNO + N_2$	No recommendation					
(xlvi)	HNO + HNO	$= H_2O + N_2O$	No recommendation					
	DNO + DNO	$= D_2O + N_2O$	4×10^{5}			300		325
(xlvii)	HNO + NO	$= OH + N_2O$	2×10^{9}	0	13000	1100—1360	±0.3	299 (See Sect. 8.4.1 and 2)
(xlviii)	H + NO	= N + OH	2.6×10^{11}	0	24400	2200—4000	±0.3	See Table 48
(—xlviii)	N + OH	= H + NO	3.2×10^{10}			300	±0.2	226[c]
(xlix)	N + NO	$= N_2 + O$	1.6×10^{10}	0	0	300—5000	±0.1[d]	257

[a] These values now supersede those given by Baulch et al. [257], which were based on refs. 321—323. For discussion see ref. 320.
[b] $k = k^{\infty} [1 + k^{\infty}/(k^{0}[M])]^{-1}$.
[c] From $k_{-48}/k_{-2} = 1.4 \pm 0.1$ [226].
[d] Increasing to ±0.3 above 2000 K.

the same temperature range. The slow reaction and induced explosion phenomena are discussed by Levy [330].

Of greater interest here are the inhibition phenomena at the first and second explosion limits, and certain other experiments in which traces of hydrocarbons are added to slowly reacting $H_2 + N_2 + O_2$ mixtures at around 773 K in aged boric acid coated vessels. Both types of measurement open up the possibility of examining the reactions of the radicals of the $H_2 + O_2$ system with the additive.

9.1 INHIBITION OF EXPLOSION LIMITS BY HYDROCARBONS

All the simple hydrocarbons are able to suppress the low pressure ignition of the $H_2 + O_2$ system. However, there are major differences of behaviour between methane and neopentane on the one hand, and most other hydrocarbons and related materials on the other [329—332]. With formaldehyde [333], ethane [334—336], propane [329, 337], and *n*- and *i*-butane [338] the second limit in KCl coated vessels falls more or less linearly with increasing partial pressure of additive. In the experiments of Baldwin et al. [333—338], the mole fractions, x and y, of H_2 and O_2, respectively, could be varied independently of each other by working with $H_2 + N_2 + O_2$ mixtures and adjusting the nitrogen content appropriately. The rate of fall of the second limit at constant x was almost inversely proportional to y: while at constant y and not too small x, it was almost independent of x. The limit did not change much with vessel size. The observations may be accounted for by adding reactions (l)—(lii)

$$O + RH = OH + R \tag{l}$$

$$H + RH = H_2 + R \tag{li}$$

$$OH + RH = H_2O + R \tag{lii}$$

to the basic steps (i)—(iv) of the hydrogen—oxygen mechanism

It appears that the alkyl radicals formed react predominantly with O_2 to form HO_2 and an olefin (or CO in the case of formaldehyde). HO_2 formation at second limit pressures in a KCl coated vessel is essentially a chain terminating step (cf. Sect. 3.6). If $i_{1/2}$ denotes the inhibitor concentration required to halve the second limit, then $i_{1/2}$ is inversely proportional to the rate of fall of the limit, and the scheme leads to

$$\frac{1}{i_{1/2}} = \frac{k_{51}}{k_2 y} + \frac{2k_{52}}{k_1 x + k_{52} i_{1/2}} + \frac{k_{50}}{k_3 x + k_{50} i_{1/2}} \tag{108}$$

Thus the observed marked dependence of $i_{1/2}$ on y indicates uniquely the importance of reaction (li). The small variation of $i_{1/2}$ with x, observed principally at low x, is associated with contributions from reactions (l) and (lii). These are difficult to separate, but their inclusion under the

TABLE 50
Ratios of rate coefficients for hydrocarbon inhibition of second limits [70]

Hydrocarbon	Temp (K)	$k_{50} = 0$			$k_{52} = 0$		
		k_{51}/k_2	k_{52}/k_1	R.m.s. % deviation	k_{51}/k_2	k_{50}/k_3	R.m.s. % deviation
Ethane	813	38	12	2.6	32	54	4.2
Propane	793	77	27	4.9	62	124	5.4
n-Butane	793	83	36	6.6	53	250	4.7
i-Butane	793	153	20	2.2	147	56	2.1
Formaldehyde	813	326	42	3.8	309	130	3.6
Tetraethylsilane	793	374	74	2.6	341	254	4.0

heading of one or the other (by putting $k_{50} = 0$ or $k_{52} = 0$) allows a precise prediction of the variation of $i_{1/2}$ with both x and y over a wide range [70]. The inhibition of the first limit also agrees with the proposed mechanism [333, 334, 336]. Values of the ratios k_{51}/k_2 and k_{52}/k_1 or k_{50}/k_3 for several additives are given in Table 50.

In contrast to the behaviour just described, increasing concentrations of methane [329—332, 339, 340] and neopentane [332, 341] lower the second limit only slightly, until a critical concentration is reached at which explosion is suddenly completely suppressed. In addition, there is a pronounced effect of vessel diameter, though not of surface, on the critical mole fraction. The formation of an olefin + HO_2 by the attack of molecular oxygen on either the methyl or neopentyl radical is not possible in the same way as with the other hydrocarbons considered above, and Baldwin et al. [331, 338—341] concluded that the added hydrocarbon is not directly the cause of the abrupt suppression. In the case of methane they consider that the inhibition is due to the intermediate oxidation product, formaldehyde. Formaldehyde itself is a powerful inhibitor of the type already discussed (cf. Table 50). It is formed in the system right at the commencement of reaction, when the net branching factor first becomes positive, and its formation causes a reduction in the net branching factor. However, if insufficient formaldehyde is present, the concentration of chain centres will continue to rise and to give a branched chain thermal ignition. On the other hand, if sufficient formaldehyde is formed to reduce the net branching factor below zero, the concentration of centres will rise to a maximum and then decrease. If the maximum concentration is insufficient to produce a thermal type of explosion, then the short initial burst of reaction will be suppressed by the further formation of formaldehyde (cf. ref. 339). In the case of neopentane, formaldehyde is also formed (amongst other species) as an early intermediate [312].

Of the alkyl silanes, tetraethylsilane behaves like ethane, propane and butane, but tetramethylsilane behaves like methane [343].

9.2 ADDITIVES IN SLOWLY REACTING MIXTURES OF HYDROGEN AND OXYGEN

The investigations outlined in Sect. 4, and particularly in Sect. 4.2, have shown that the slow reaction in $H_2 + N_2 + O_2$ mixtures in aged boric acid coated vessels at around 773 K provides an extremely reproducible and controllable source of the radicals H, O, OH and HO_2. Following the establishment of a detailed mechanism and the evaluation of the rate coefficients of the individual steps, the system has recently been exploited, particularly by Baldwin and co-workers, in order to examine the reactions of the radicals with small additions of foreign materials. With say 1 % of a hydrocarbon additive, the technique is to follow the rate of the overall (H_2 + O_2) reaction by means of the pressure change, and to follow the rate of disappearance of the additive and the yields of products from it by analytical methods. If R represents the additive or one of its oxidation products, measurement of the changes in the ratios $\Delta[R]/\Delta[H_2]$ with initial reaction conditions like H_2 or O_2 mole fraction allows analysis and testing of kinetic effects associated with assumed additive reactions inserted into the $H_2 + O_2$ mechanism, and eventually the derivation of rate coefficients relative to those of the $H_2 + O_2$ system. Further details are beyond the scope of this discussion. The method has been discussed by Baldwin et al. [344], and has been used for studies involving methane [345, 346], ethane [347], propane [348], *n*- and *i*-butane [348–350], formaldehyde [351] and neopentane [352, 353].

10. The oxidation of carbon monoxide and hydrogen—carbon monoxide mixtures

Despite research dating back to the 19th century, the combustion or oxidation of CO is less well defined both mechanistically and kinetically than that of hydrogen. This section, which reviews relevant research in this field, is divided into five sub-sections. The first three of these consider the reaction with oxygen, dealing with explosion limits and slow oxidation; oxidation in flames and other high temperature systems; and elementary reactions. The fourth sub-section deals with several other oxidation reactions in homogeneous systems, and the fifth will be introduced below.

Carbon monoxide—oxygen mixtures exhibit explosion phenomena similar to the hydrogen—oxygen system. However, there is an additional, and fairly extensive, adjoining region of pressure and temperature in which a slower reaction occurs, emitting a weak blue chemiluminescence rather like a feeble flame. This glow reaction can make determination of the explosion limits inaccurate owing to appreciable chemical change occurring before explosion takes place. Also, some authors have reported the limits of this glow reaction rather than the true explosion limits. The glow reaction and certain other phenomena associated with it will be discussed in Sect. 10.5.

Nearly all workers detect the first and second limits of explosion, and the third limit has been detected in the presence of a trace of moisture or hydrogen. However, it is noticeable that the positions of these limits may vary considerably from one group of researchers to another; indeed, in the case of a dry CO/O_2 mixture, the limits sometimes differ from one similar vessel to another, despite identical experimental technique. It emerges that the surface can play a complex and confusing role in the combustion of CO, and that the system is very sensitive to traces of hydrogenous impurities which can make experimental work both arduous and tedious.

Practically, one of the most important reactions of CO is its exothermic oxidation to CO_2, from which a very large proportion of the world's useful energy is derived. It has been known for many years that dry CO/O_2 flames are difficult to ignite and have lower burning velocities than similar flames derived from moist CO [354]. In this particular, the explosion and flame systems behave analogously.

Over the past decade or so, there have been a number of direct kinetic investigations of the elementary reactions involved in the oxidation of CO, particularly the reactions with OH and with oxygen atoms in various energy states. Although there has been some scatter in the reported results, the latest values, including some older data, show some measure of agreement. There have been a number of shock tube investigations of the dissociation of CO_2 in the temperature range 1500—4000 K, and these give fairly consistent high temperature values for the reverse association process. However, the activation energy at the high temperatures is uncertain.

10.1 THE EXPLOSION LIMITS AND THE SLOW COMBUSTION OF CARBON MONOXIDE—OXYGEN MIXTURES

Because of the extreme sensitivity of the position of the explosion peninsula to small amounts of hydrogenous impurity, it is not possible to separate the attempts to measure the position of the explosion region for dry mixtures on a P-T diagram from those in which impurities or additives were definitely present.

10.1.1 The explosion limits

(*a*) *The first limit.* Hadman et al. [355], in an attempt to obtain the limits of the "dry" reaction, studied the explosive oxidation of CO in quartz vessels which had been heated for lengthy periods to about 900 °C whilst under vacuum. The reactant gases were also dried by prolonged storage over P_2O_5. It was found that the lower limit pressure depended principally on the condition of the vessel surface, and this masked any systematic variation of the limit due to vessel size. Further, exposure of the surface to any considerable partial pressure of CO tended permanently to inhibit explosion. However, some reproducibility was obtainable. Hadman et al. were able to show that the limit pressure did not vary much with temperature in the range 650—700 °C, and that addition of inert gas lowered the partial pressure of the combustible mixture at the limit (see Table 51). This agrees with the earlier observations of Kopp et al. [11] and Cosslett and Garner [356] that increasing the oxygen content of the reacting mixture also expands the explosion peninsula (see Fig. 53).

TABLE 51
Effect of inert gas on lower explosion limit of $CO + O_2$ mixture at 600 °C [355]
(Pressures in torr)

Inert	P_{CO+O_2}	P_{inert}	$P_{tot.}$
N_2	19	0	19
	16.5	16.5	33
	13	26	39
	10	43	53
Ar	19	0	19
	13	13	26
	10	20	30
	7.4	30	37

Hoare and Walsh [357] attempted like Hadman et al. to work with dry gases, and tried to distinguish where possible between the onset of the glow reaction and the true explosion limits for a $2CO + O_2$ mixture. Although the lower explosion limit could not be determined owing to the extent of reaction occurring in the glow region before explosive conditions were realized, they were able to show that the glow and explosion limits were well separated.

Perhaps the closest approach to measuring the explosion limits of a pure CO/O_2 mixture was made by Dickens et al. [358]. They went to considerable lengths to exclude water and other impurities by fractional distillation and storage of the reactants at liquid oxygen temperatures. Five different cylindrical quartz vessels were used, and reproducible results for a particular reaction vessel were only obtained after some weeks of experimentation. However, there were marked differences in the behaviour of the gases in the different reactors. Experiments carried out

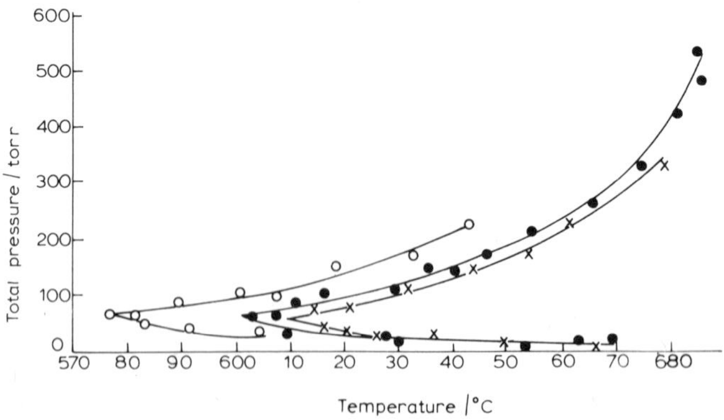

Fig. 53. Explosion limits for carbon monoxide–oxygen mixtures in quartz vessel (after Kopp et al. [11]). ○, $CO + 9O_2$; ●, $2CO + O_2$; x, $6CO + O_2$. (By courtesy of Zeitschrift für physikalische Chemie.)

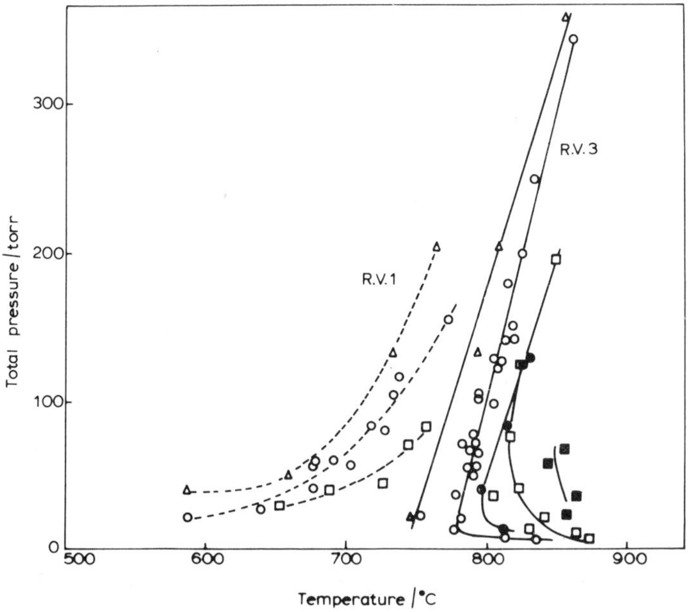

Fig. 54. Variation of explosion limits in two cylindrical quartz vessels (after Dickens et al. [358]). △, CO + 9O_2 ; ○, CO + 2O_2 ; ●, CO + O_2 ; □, 2CO + O_2 ; ■, 9CO + O_2. RV1, 8.2 cm long, 6.0 cm diameter; RV3, 9.9 cm long, 8.0 cm diameter. (By courtesy of The Faraday Society.)

with two reaction vessels gave no indication of a first limit, even at temperatures around 530 °C; whilst with another vessel, the normal explosion peninsula was obtained (Fig. 54). It seems evident that the condition of the quartz surface can be quite variable, despite identical treatments.

Because of the sensitivity of the CO/O_2 explosions to impurities, there has been a tendency to avoid coating the reaction vessels. However, some useful results have been obtained with treated surfaces. As with the H_2/O_2 system, boric acid coating lowers the first limit but inevitably introduces water [359]. By contrast, Al_2O_3 [359] and PbO [357] are active in contracting the explosion regime. The only systematic and reproducible measurements of the effect of additives on the first explosion limit of CO have been made by Nalbandjan and co-workers [360—363], who observed the lower explosion limits for CO/O_2 mixtures containing various amounts of hydrogen using a diaphragm manometer and quartz vessels coated with MgO, NaCl and KCl. Although MgO, like PbO, is an active surface, reproducible results were obtained. The first ignition limits were lowered on increasing the hydrogen content of the gases, until they approached those for a stoichiometric H_2/O_2 mixture (Fig. 55). Similarly, small amounts of ethane decreased the lower limit pressure, although above about 0.15 % C_2H_6 additive the trend reversed.

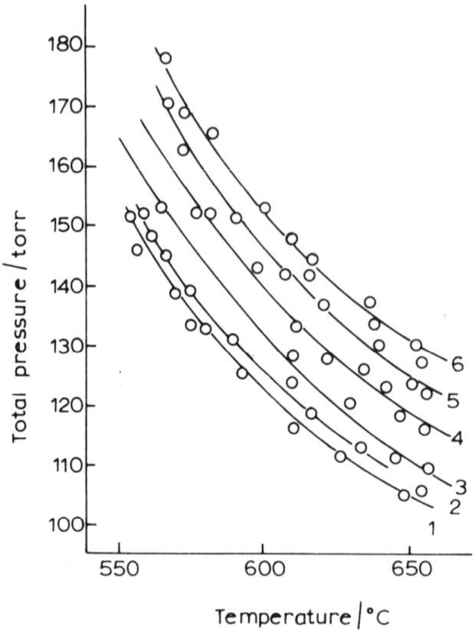

Fig. 55. Effect of added hydrogen on the lower explosion limit of $2CO + O_2$ mixture (after Azatjan et al. [360]). Percent hydrogen: (1) 3.79; (2) 3.28; (3) 2.14; (4) 1.48; (5) 1.06; (6) 0.78. (By courtesy of Kinet. Katal.)

The effect of ethane and the temperature dependence of the limits are shown in Fig. 56.

Similar promoting and inhibiting effects have been observed by Hoare and Walsh [357], who found that the addition of 6 % methane considerably expanded the explosion peninsula, but that with 10 % methane the explosion regime was less extensive. Ammonia had a similar effect, with additions above 6 % causing the first limit pressure to increase. Methanol also lowers the first limit [364], as shown in Fig. 57. ESR studies of CO/O_2 flames containing deuterated methanol indicated that hydrogen

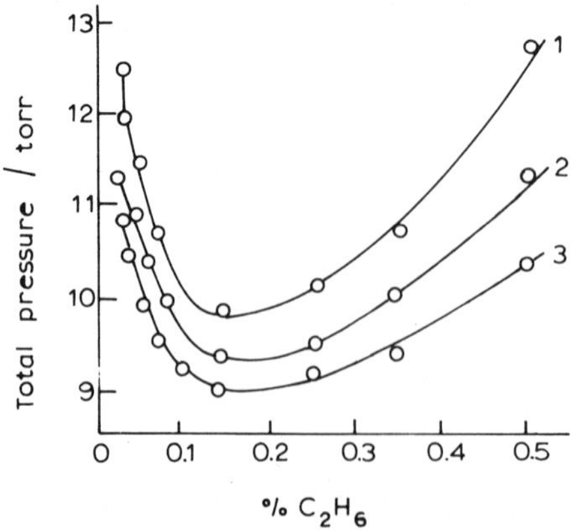

Fig. 56. Effect of added ethane on the lower explosion limit of $2CO + O_2$ mixture (after Azatjan et al. [436]). Temperature: (1) 610 °C; (2) 630 °C; (3) 650 °C. (By courtesy of Dokl. Akad. Nauk SSSR.)

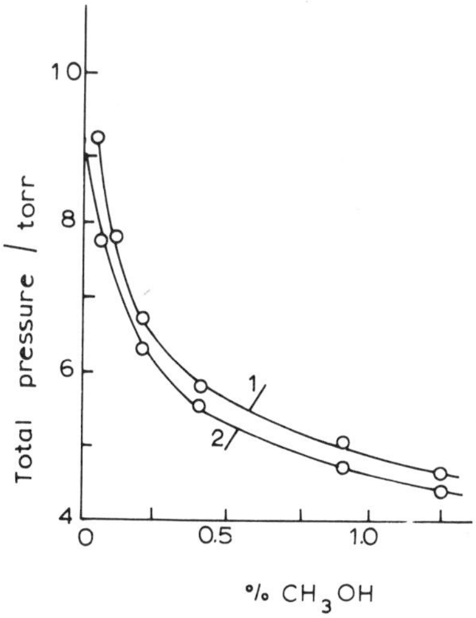

Fig. 57. Effect of added methanol on the lower explosion limit of $2CO + O_2$ mixture in quartz vessel (after Azatjan et al. [363]). Temperature: (1) 670 °C (2) 690 °C. (By courtesy of Dokl. Akad. Nauk SSSR.)

abstraction occurred more readily from the methyl group than from the hydroxyl part of the molecule. Such studies have also shown [364] that when the first limit pressure is raised again above its minimum value (by the addition of more than the optimum amount of an additive) the inhibition is accompanied by a sharp fall in the concentrations of H, O and OH.

(*b*) *The second limit.* Because the second limit largely reflects processes occurring in the gas phase, the limits should be much less dependent on the condition of the surface of the reaction vessel. This is borne out by the results of Warren [359], who showed that for a B_2O_3 surface or a vessel coated with alumina, the second limit remained the same, except of course near the tip of the explosion peninsula. Similarly, active PbO had little effect on the position of the second limit [357]. By way of contrast, however, Dickens et al. [358] found that the limit was moved towards higher temperatures when the surface/volume ratio was increased (Fig. 58). A similar result was obtained by Gordon and Knipe [365] with both "dry" and wet gases, though to a smaller extent with the latter. Generally, it is a little easier to obtain consistent results at the second limit, except with the "dry" gases. Although at one stage it was thought that the presence of a little water had no effect on the second limit, Dickens et al. [358] and Gordon [366] have shown that, as it becomes more difficult to ignite CO/O_2 mixtures the drier they become, so the explosion regime is shifted to higher temperatures. Figure 59 depicts the second limits of "dry" mixtures obtained by various workers.

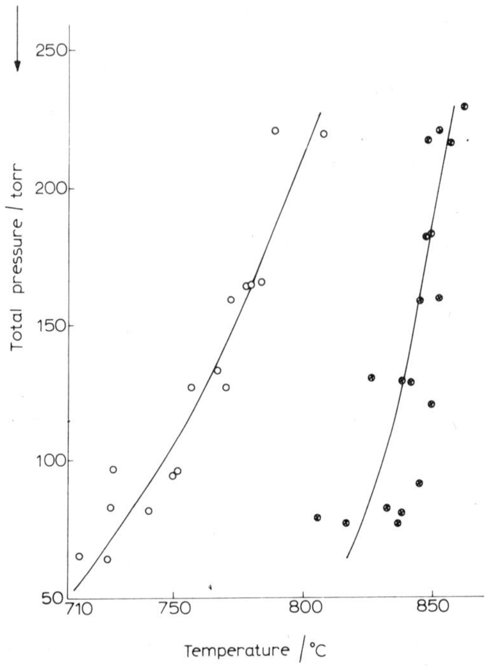

Fig. 58. Second explosion limits of CO + $2O_2$ mixture in packed and unpacked cylindrical quartz vessels (after Dickens et al. [358]). ○, Unpacked vessel, 9.9 cm long, 8.0 cm diameter; ⊗, packed vessel, 8.0 cm long, 6.0 cm diameter. (By courtesy of The Faraday Society.)

There is general agreement that increasing the O_2/CO ratio above stoichiometric raises the second limit pressure, whilst the addition of the inert gases helium, argon, or nitrogen, lowers the partial pressure of combustible gas at the limit. Von Elbe et al. [367] found that the

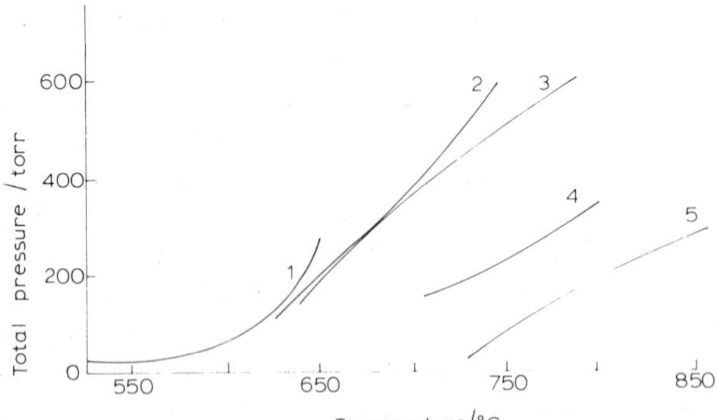

Fig. 59. Second explosion limits of "dry" CO + $2O_2$ mixtures, as measured by various workers (after Dickens et al. [358]). (1) Hoare and Walsh [357]; (2) Hadman et al. [355]; (3) von Elbe et al. [367]; (4) Gordon and Knipe [365]; (5) Dickens et al. [358]. (By courtesy of The Faraday Society.)

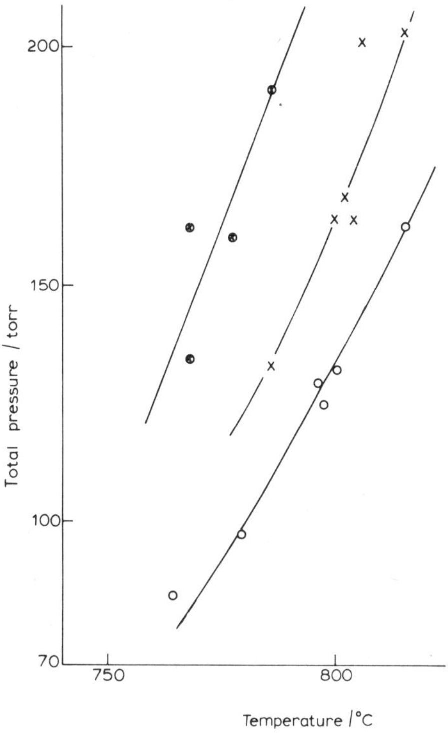

Fig. 60. Effect of added inert gas on the second explosion limit in cylindrical quartz vessel, 9.9 cm long, 8.0 cm diameter (after Dickens et al. [358]). ○, $CO + 2O_2$; x, $CO_2 + 2O_2 + 3He$; ⊗, $CO + 2O_2 + 3Ar$. (By courtesy of The Faraday Society.)

addition of nitrogen to a 2:1 and a 1:2 CO/O_2 mixture did not affect the total pressure at the limit, confirming the earlier work of Kopp et al. [11]. Dickens et al. [358] found that He and Ar raised the total pressure (see Fig. 60), while CO_2 caused a small decrease. Contrary to the observation above about the effect of oxygen, von Elbe et al. [367] observed that the limit passed through a minimum at an O_2/CO ratio of 1 : 2. However, their further observation that exposure of the surface to CO_2 tended to inhibit the explosions may not be unconnected with this.

The effects of inert gases on the second ignition limits of $2CO + O_2$ mixtures containing a little hydrogen have been determined by Buckler and Norrish [369]. They added sufficient inert gas to 50 torr of $2CO + O_2$ to approximate to the limit pressure in the reaction vessel; and then added a little extra inert gas containing the hydrogen, so that the final hydrogen pressure in the vessel was 2 torr. Depending on the total pressure of inert gas, the mixture underwent either slow reaction or ignition, and by repeated experiment the partial pressure of inert gas at the limit could be found. Such partial pressures are given for a range of temperatures in Table 52.

TABLE 52
Effect of inert gases on the second ignition limits of $2CO + O_2$ containing small amount of hydrogen [369]. (Pyrex reaction vessel, 27.0 mm. diameter; concentration of $2CO + O_2$ = 50.0 torr; concentration of hydrogen = 2.0 torr.)

Temperature (°C)	Partial pressures of inert gases at limit (torr)			
	Argon	Helium	Nitrogen	Carbon dioxide
585	145.0	106.2	103.2	60.1
575	132.8	99.5	94.8	53.9
565	117.5	90.7	86.5	48.3
555	105.1	80.6	76.7	42.9
545	92.6	69.8	66.4	38.2
535	77.4	58.8	54.9	30.9
525	58.3	48.8	42.2	24.0

It has already been noted that the presence of small quantities of hydrogenous impurities expands the explosion peninsula. Such sensitization allows easier experimentation and provides for more reproducible results. The effect of hydrogen on the ignition limits is shown in Fig. 61. As was observed when considering the effect on the first limit, addition of sufficient hydrogen causes the reaction system to behave in essentially the same way as the H_2/O_2 reaction. Dixon-Lewis and Linnett [30] found that, on replacing more than about 10 % of the CO by H_2 in a KCl coated vessel at 510—570 °C, the second limit pressure could be extrapolated

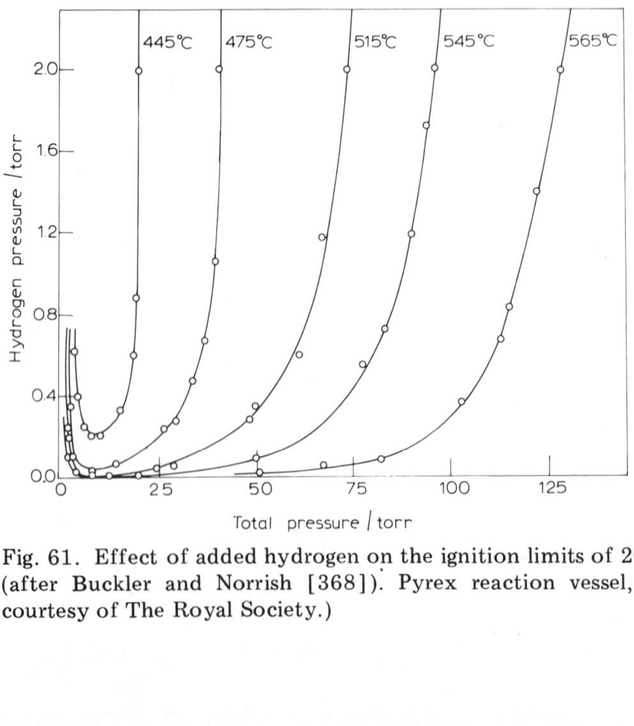

Fig. 61. Effect of added hydrogen on the ignition limits of $2CO + O_2$ in a flow system (after Buckler and Norrish [368]). Pyrex reaction vessel, diameter 27.0 mm. (By courtesy of The Royal Society.)

from the H_2/O_2 system rather than the CO/O_2 system. The presence of 1 % methane similarly expands the explosion regime, although a large quantity (10 %) causes a contraction and a delay before explosion sets in [357]. A similar inhibiting action is found on addition of 1 % CH_2O or 1 % HCl [357]. Retardation of ignition of a CO/air mixture by formaldehyde, methane and ethane was observed by Burgoyne and Hirsch [370] in a flow system employing short reaction times of the order of a few millisec. Here the ignition temperatures are considerably higher than those encountered with static systems. These higher temperatures may be related to data of Jono [371], who observed an induction period of about 1.3 sec before true low pressure explosion occurred. In the first half of this induction period the reaction rate rose to a maximum and fell rapidly, and then reaction proceeded steadily until explosion resulted.

Tipper and Williams [372], using a boric acid coated vessel, found the explosion region of a wet stoichiometric mixture to contract on addition of 0.5 % SO_2, the lower limit being unaffected whilst the upper limit was reduced and the tip of the peninsula displaced to higher temperatures. Iodine is a very powerful inhibitor, minute traces being sufficient to prevent explosion [355].

The effect of temperature on the second limit is difficult to ascertain quantitatively, the values of the "activation energy" ranging between 10 kcal . $mole^{-1}$ and infinity (assuming an Arrhenius type relation between the P_2 and temperature, which not all experimenters record). The most recent data on the explosion limits of "dry" CO/O_2 mixtures [358] suggest a value of about 21 kcal . $mole^{-1}$ above 300 torr, but non-linear Arrhenius plots were also obtained. In the presence of small quantities of H_2, Buckler and Norrish [368] found the activation energy to be about 13 kcal . $mole^{-1}$, though with increasing hydrogen content there was not a linear relationship between log P_2 and T^{-1}. Dixon-Lewis and Linnett [30] did not find quite such a low activation energy for P_2. They found 19.3 kcal . $mole^{-1}$ when $[CO]/[H_2]$ was equal to 200, as opposed to 21.4 kcal . $mole^{-1}$ for pure hydrogen.

(c) *The third limit.* Evidence of the existence of a third explosion limit has come from a study of the influence of small concentrations of hydrogen on the combustion of CO. Gaillard-Cusin and James [373], using the static method, recorded the onset of explosion from the rate of change of the chemiluminescent intensity as monitored by a photomultiplier. Using 10 % CO/air mixtures containing 0.02—1 % hydrogen, they detected a pressure above the second limit at which there occurred a sharp flash of light. The locus of these pressures is shown in Fig. 62. An Arrhenius plot yields an apparent activation energy of approximately 100 kcal . $mole^{-1}$.

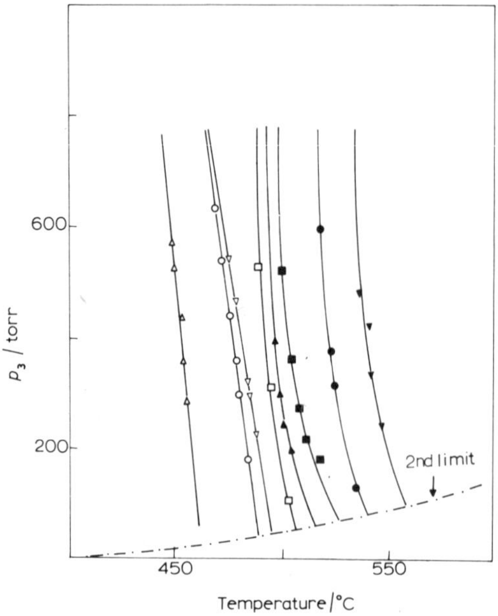

Fig. 62. The third explosion limits of 10 % CO-air mixtures containing a little hydrogen (after Gaillard-Cusin and James [373]). H_2 : CO ratios: △, 1:10; ○, 1:30; ▽, 1:50; □, 1:70; ▲, 1:80; ■, 1:100; ●, 1:200; ▼, 1:500. (By courtesy of La Société de Chimie Physique.)

10.1.2 The slow oxidation of carbon monoxide

Outside the explosion peninsula, particularly in the region above the second limit, the oxidation of CO can take place at a speed convenient for normal kinetic measurements. As with the explosive combustion, the kinetics are very sensitive to the surface and to the purity of the reacting gases.

Cosslett and Garner [356] found that, although the rate of reaction (as measured by pressure change) was proportional to the initial pressure of the reactant mixture, the order of the reaction was dependent on the state of the surface. The more active the surface, the smaller the reaction order. Hadman et al. [374] noted a similar effect in that, for the "dry" gases at a temperature close to 700 °C, the rate was proportional to $[CO]^{\alpha}[O_2]^{\beta}$, where α and β are less than unity, with β decreasing as the surface/volume ratio increased. A study of isotopic scrambling and exchange in "pure" CO/O_2 mixtures enriched in $O^{18}O^{18}$ at lower temperatures (500 °C) by Verdurmen [375] showed that the rate of production of CO_2 in quartz vessels at these temperatures was proportional to the surface area.

The reaction is promoted by small additions of hydrogenous materials, the kinetics being radically altered with the rate of oxidation. Near the tip of the explosion peninsula [374] the rate is proportional to $[CO][H_2]/$

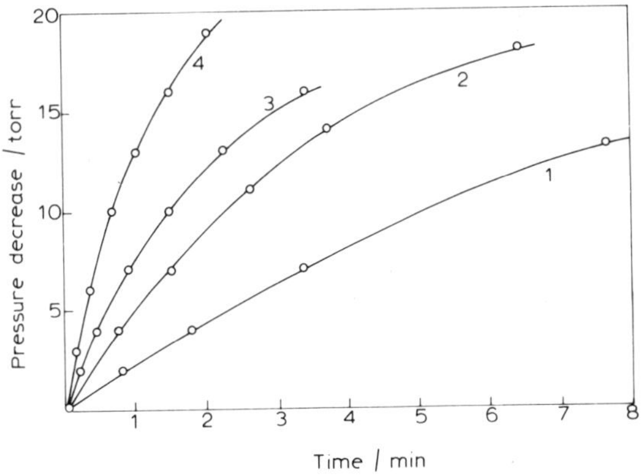

Fig. 63. Rate of slow oxidation of wet CO/O_2 mixtures at 560 °C in cylindrical quartz vessel (after Hadman et al. [374]). Initial pressure: p_{CO} = 50 torr, p_{H_2O} = 10 torr. p_{O_2}: (1) 250 torr; (2) 150 torr; (3) 100 torr; (4) 50 torr. (By courtesy of The Royal Society.)

$[O_2]$. However, with packed vessels the rate does not extrapolate back to zero as the water content is reduced, thus suggesting a concurrent surface reaction.

The results of Hadman et al. [374], except when iodine was present, are somewhat contrary to a more detailed investigation of Tsvetkova et al. [376], who found the reaction to be first order with respect to H_2O (for small additions of water) and also to increase with both [CO] and $[O_2]$, though in a complex manner. Tipper and Williams [372] also found that oxygen may have had a retarding effect on the rate of oxidation of wet

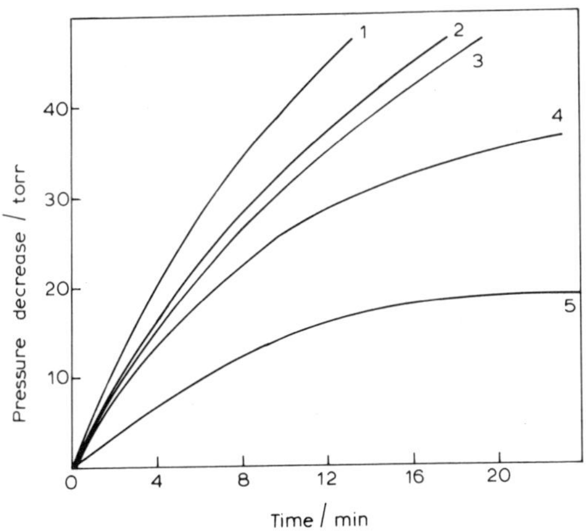

Fig. 64. Rate of slow oxidation of wet CO/O_2 mixtures at 556 °C (after Tsvetkova et al. [376]). Initial pressures: p_{CO} = 200 torr, p_{H_2O} = 20 torr, p_{O_2}: (1) 200 torr; (2) 117 torr; (3) 100 torr; (4) 50 torr; (5) 25 torr. (By courtesy of The Academy of Sciences of the USSR.)

CO, but it appeared that the oxygen could have changed the nature of the vessel surface, which was coated with boric acid. They finally concluded that the addition of CO, O_2 or N_2 had little effect on the initial oxidation rate. On the other hand, Hoare and Walsh [357] found an increase in the oxidation rate when the concentration of either the carbon monoxide or oxygen was raised. The contrasting results on the effect of oxygen are shown in Fig. 63 and 64.

Tsvetkova et al. [376] found the initial rate to be proportional to $a[H_2O]/(1 + b[H_2O])$. The activation energy was independent of the initial composition of the reaction mixture, but increased with increasing initial pressure. Very similar results were obtained when water was replaced by hydrogen.

The addition of hydrogen increases the rate of oxidation of stoichiometric CO/O_2 mixtures containing about 5 % water; indeed, sufficient hydrogen leads to explosion. Small quantities of methane have a similar accelerating effect, but larger amounts reduce the rate to close to its original value again [357].

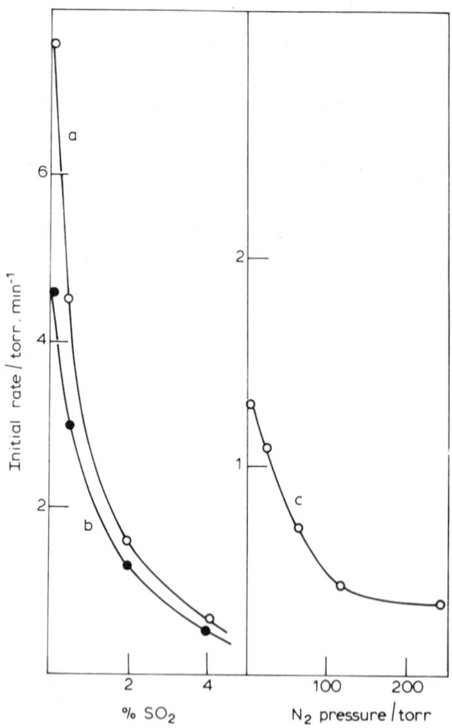

Fig. 65. Effect of added SO_2 and N_2 on the initial rate of oxidation of a wet $2CO + O_2$ mixture at 540 °C in cylindrical B_2O_3 coated vessel, volume 125 cm^3 (after Tipper and Williams [372]). Initial pressures: (a) and (b), $p_{CO} + p_{O_2} = 365$ torr, $p_{H_2O} = 8$ torr; (c) $p_{CO} + p_{O_2} = 90$ torr, $p_{H_2O} = 2$ torr. Note that for (b), the ordinate scale should be halved. (By courtesy of The Faraday Society.)

Addition of SO_2 retards the oxidation of moist stoichiometric CO/O_2 mixtures (Fig. 65), but at the same time the rate becomes more sensitive to increases in the CO or O_2 concentration, which accelerate the oxidation. Also, when SO_2 is present, addition of N_2 causes a further retardation. The effects of SO_2 appear to be independent of the vessel surface [372]. HCl and iodine are effective inhibitors of the wet reaction [357]. In the presence of iodine the influence of the surface largely disappears [374].

Knipe and Gordon [377] studied the slow oxidation of rich, "dry" CO/O_2 mixtures at 600 °C, and observed that the rate of pressure change fell to a steady value after an initially faster reaction. Similar effects have been observed in the oxidation of CO by mixtures of O_2 and O_3 [378]. Although inhibition by product CO_2 seemed to offer the obvious explanation, premixing a little CO_2 with the reactants had no effect. During the reaction of large amounts of CO with oxygen and ozone, small particles of graphite were formed which led Hartech and Dondes [378] to suggest carbon suboxide as the inhibitor. Inhibition of CO oxidation by C_3O_2 has been inferred from a study [394] of the oxidation of C_3O_2.

10.1.3 Reaction mechanism

(*a*) *General discussion.* The existence of first and second explosion limits implies a chain branching reaction. Generally the first limit is considered to arise from the diffusion of chain carriers to the wall, where they are destroyed faster than they can be formed in the gas phase, while the second limit occurs at pressures at which gas phase termolecular chain breaking reactions proceed fast enough to balance the rates of chain creation. When pure CO and O_2 react, the number of possible chain branching reactions is small. Semenov [2] has postulated reactions (liii) and (liv) as the branching steps, while reactions (lv) and (lvi) have been suggested by Lewis and von Elbe [4]. There have been other variations on this theme.

$$CO + O \rightarrow CO_2^* \qquad \text{(liii)}$$

$$CO_2^* + O_2 \rightarrow CO_2 + 2O \qquad \text{(liv)}$$

$$O + O_2 + M' \rightarrow O_3 + M' \qquad \text{(lv)}$$

$$CO + O_3 \rightarrow CO_2 + 2O \qquad \text{(lvi)}$$

The pair of reactions (lv) and (lvi) present some difficulties as a branching process. If the termolecular chain breaking at the second limit is due to reaction (lvii)

$$O + CO + M'' \rightarrow CO_2 + M'', \qquad \text{(lvii)}$$

and the thermal decomposition of O_3 (reaction (—lv)) is included in the mechanism, the expression for the second limit becomes

$$k_{55}k_{56}[O_2][M'] = k_{-55}k_{57}[M'][M''] + k_{56}k_{57}[CO][M''] \quad (109)$$

or, assuming $[M'] = [M'']$

$$k_{55}k_{56}[O_2] = k_{-55}k_{57}[M'] + k_{56}k_{57}[CO] \quad (109a)$$

According to eqn. (109a), only CO may be replaced by inert gas without lowering the overall second limit pressure, and this is contrary to experiment. The situation can be modified by inclusion of further reactions, so that some semblance of agreement is obtained. However, there is a stronger objection. There is no evidence that CO reacts directly with O_3 at 350 °C, the formation of CO_2 in these conditions being due to reaction with oxygen atoms which arise from the thermal decomposition of the ozone [379, 380].

If reactions (liii), (liv) and (lvii) alone are controlling, the second limit condition becomes

$$k_{57}[M''] = k_{53} \quad (110)$$

However, this expression does not allow for the possibility of collisional quenching of the CO_2^* formed in reaction (liii). Such a possibility may be included by introducing, for example, reactions (lx) and (lxi) into the chain [5, 381], viz.

$$CO_2^* + CO \rightarrow CO^* + CO_2 \quad (lx)$$

$$CO^* + O_2 \rightarrow CO_2 + O \quad (lxi)$$

In this case eqn. (110) becomes

$$k_{57}[M''] = k_{53}/(1 + k_{60}[CO]/k_{54}[O_2]) \quad (110a)$$

In either event, it seems doubtful whether the probability of reaction (liii) is sufficiently high to produce the observed effects. The reaction of O atoms with CO will be discussed in Sect. 10.3.

The great sensitivity of the oxidation towards hydrogen, water and other hydrogenous materials, all of which have been found to expand the explosion regime, indicates a drastic influence on the nature of the chain branching step; and, indeed, both the first and second limits occur at pressures and temperatures characteristic of the H_2/O_2 system [30, 368, 369]. Also, in the presence of hydrogen, the CO/O_2 reaction becomes sensitive to NO_2 [382], again like H_2/O_2 mixtures. Reaction (xxiii) is fairly rapid, and is analogous to reaction (i) of the hydrogen—oxygen system. When large amounts of hydrogen are present, the dominant reactions are therefore probably those of the H_2/O_2 system, together with reaction (xxiii)

$$OH + CO \rightarrow CO_2 + H \tag{xxiii}$$

and in this case the situation would be expected to be almost identical with that when CO is absent (or better, replaced by N_2 which has similar physical properties to CO). It turns out that this expectation is fulfilled in principle, although there are chemical effects due to several other reactions which may be superimposed on this simple mechanism. These will be discussed more fully in Sect. 10.1.3(*b*).

The rate of oxidation of moist CO on the basis just discussed is given by

$$d[CO_2]/dt \propto [CO][H_2O]^{0.5}[O_2]^{0.25} \tag{111}$$

and Kozlov [383], Hottel et al. [384], Williams et al. [422], Dryer and Glassman [455] and others have obtained kinetic expressions very close to this from their studies of CO oxidation in flow reactors in the presence of large quantities of water. Just how much hydrogen is needed for the branching reactions (ii) and (iii)

$$H + O_2 \rightarrow OH + O \tag{ii}$$

$$O + H_2 \rightarrow OH + H \tag{iii}$$

to become dominant is obviously crucial, and it is perhaps at this stage that a number of high temperature shock tube results should be considered.

Sulzmann et al. [386] have studied the onset of CO_2 formation behind incident shock waves in $CO/O_2/Ar$ mixtures containing about 80 % argon by means of emission intensity measurements at 3064 and 4470 Å, and also at 4.25 and 5.07 μm. They also investigated the influence of hydrogen on the induction period [387]. Temperatures were in the range 1500—3000 K. In comparing the results of the experiments with an analytical expression for the induction times, they considered reaction (lxii) to be the initiating step in both cases, viz.

$$CO + O_2 \rightarrow CO_2 + O \tag{lxii}$$

In the hydrogen containing system the H_2/O_2 branching cycle of reactions (i), (ii) and (iii) was then invoked together with reactions (xxiii), (liii) and (liv) of CO. The "dry" gases were reported to contain less than 1 p.p.m. hydrogenous impurity, and for these Sulzmann et al. assumed a kinetic scheme consisting of reactions (lxii), (liii), (liv) and (lvii). However, again by obtaining an analytical solution for the "dry" system, Brokaw [388] has shown that the observed induction times cannot be explained by this mechanism, and he has suggested therefore that there is little or no experimental evidence for reaction (liv). The data of Sulzmann et al. may nevertheless be explained using known rate coefficients for the

References pp. 234—248

H_2/O_2 system, if it is assumed that their reportedly "dry" mixtures in fact contained about 20 p.p.m. water vapour. Following the initiating step (lxii), the reaction mechanism in the presence of either water or hydrogen then becomes

$$O + H_2O \rightarrow OH + OH \quad (-xvi)$$

$$OH + H_2 \rightarrow H_2O + H \quad (i)$$

$$H + O_2 \rightarrow OH + O \quad (ii)$$

$$O + H_2 \rightarrow OH + H \quad (iii)$$

$$OH + CO \rightarrow CO_2 + H \quad (xxiii)$$

From a detailed analysis of the results of Sulzmann et al. [386, 387] on this basis, Brokaw was able to estimate k_{-16} by two methods as 8×10^8 and 1.5×10^9 l . mole^{-1} . sec^{-1} near 1600 K.

Dean and Kistiakowsky [230] made a further attempt to see if they could be more definite in selecting a reaction mechanism, by not only

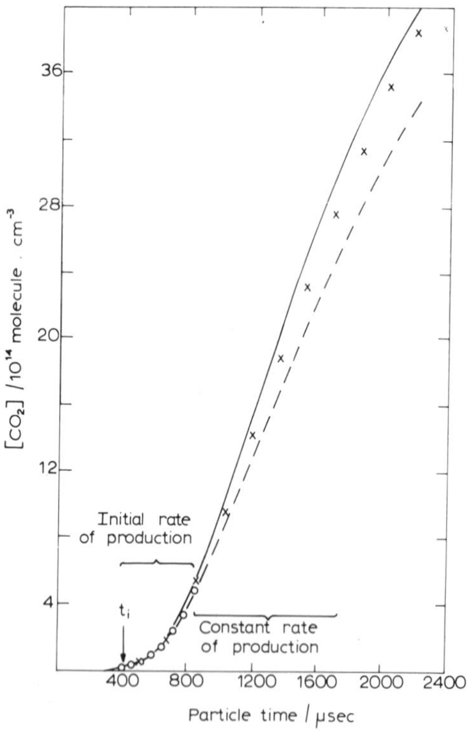

Fig. 66. Initial formation of CO_2 in a shocked $4CO/2O_2/96$ Ar mixture (after Dean and Kistiakowsky [230]). Temperature: 2155 K; [CO] = 2.32×10^{16} molecule . cm^{-3}. Lines are from Brokaw's mechanism. Solid line, $k_{23} = 1.3 \times 10^9 \exp(-500/T)$, $k_3 = 8.0 \times 10^{10} \exp(-5{,}000/T)$; broken line, $k_{23} = 9.9 \times 10^8 \exp(-500/T)$, $k_3 = 1.0 \times 10^{11} \exp(-5{,}000/T)$. (By courtesy of J. Chem. Phys.)

studying the induction period theoretically, but also calculating the rates of CO_2 formation and CO depletion during the induction period by numerical solution of the governing differential equations. They computed profiles both using the Brokaw mechanism and that involving branching via CO_2^*, and compared the results both with their own shock tube measurements using CO/O_2 mixtures diluted with argon, and with the measurements of Sulzmann et al. An example using the Brokaw mechanism is shown in Fig. 66. If, on the other hand, the "dry" mechanism was used, either the exponential growth constant should have been dependent on the product $[CO][O_2]$, which was not observed, or alternatively the pre-exponential part of the rate coefficient for the branching reaction had to be several orders of magnitude above the collision frequency. These workers considered that the likely impurity levels in their reacting mixtures could well have been sufficient to invoke the well known H_2/O_2 branching reactions.

In similar vein, Fishburne et al. [389] calculated that only 7 p.p.m. of water were required to explain their data on the basis of the Brokaw mechanism.

It would appear that in the shock tube experiments there is no real evidence for a significant contribution from the branching reaction (liv) to the overall CO_2 production. However, it is not possible to exclude it altogether. It may be that with sufficiently pure reactants the oxidation would proceed through an excited CO_2 molecule, but as yet these conditions have not been definitely established in shock tube experiments. It is natural to ask if such conditions have been obtained in explosion and slow oxidation studies, or if a Brokaw type mechanism should be applied to these as well. Although surface effects become more severe in such experiments, it also should be easier to attain low impurity levels than using shock tubes. Brokaw [388] has shown that on adding the chain breaking steps (iv) and (lvii) to his reaction mechanism above, the second limit condition becomes

$$2k_2 = \Sigma k_4[M]\{1 + \Sigma k_{57}[CO][M'']/(k_3[H_2] + k_{-16}[H_2O])\} \quad (112)$$

In the presence of trace amounts of water and no hydrogen, i.e. $k_3[H_2] \ll k_{-16}[H_2O]$, eqn. (112) simplifies to

$$[M][M''] = 2k_2k_{-16}[H_2O]/k_4k_{57}[CO] \quad (112a)$$

The data of Dickens et al. [358], who have probably achieved the closest approach to "dry" mixtures, still do not exclude such a relationship; and indeed, their reported activation energy supports it. Nevertheless, assuming the best reported values of the rate coefficients k_2, k_4, k_{-16} and k_{57} at 1000 K, the impurity levels required to reproduce the results of Dickens et al. are rather higher than would be expected, being of the order of a few parts per million. If methane were responsible for the production of OH by reaction with O atoms, then only about 0.3 p.p.m. would suffice, and methane as an impurity would be difficult to eliminate completely. On balance it would still seem that all the explosion limit data for "pure" CO/O_2 mixtures, and those containing a little H_2 or H_2O by design, can be explained on the basis of the branching reaction between H atoms and molecular oxygen.

A corollary of this hypothesis, which implies the extreme difficulty of removing or even measuring the last traces of hydrogen or water in the "dry" systems, is that the effect of additives on the limit systems is essentially due to modification of the radical concentrations produced by the H_2/O_2 branching cycle. Such effects should therefore be similar to those observed in H_2/O_2 systems. Small quantities (ca. 1—2 %) of hydrocarbons, methanol and ammonia lower the pressure of the first limit and raise that of the second for CO/O_2 explosions; that is, they expand the explosion regime. Additions of larger amounts (ca. 10 %) of these compounds cause contraction of the explosion peninsula rather than expansion, and in this they are exerting an influence similar to that in H_2/O_2 explosions, except that the concentrations are higher (see Sect. 10.1.1). Sensitization by small amounts of additives can readily be explained on the basis of hydrogen atom abstraction reactions, which thereby raise the H_2 concentration available for the branching reactions (ii) and (iii), e.g.

$$RH + O \rightarrow R. + OH \qquad (l)$$

or

$$RH + H \rightarrow R. + H_2 \qquad (li)$$

On the other hand, when substantial additions are made, chain termination reactions such as (lxiii) can take place at appreciable speed (see Sect. 9.1).

$$C_2H_5 + O_2 \rightarrow C_2H_4 + HO_2 \qquad (lxiii)$$

ESR observations have confirmed that hydrocarbons, methanol and ammonia derivatives initially cause large increases in the amounts of H, OH and O found in a "pure" CO/O_2 flame, but that when additive is present in an amount sufficient to cause contraction of the explosion regime, there is a sharp drop in the radical concentration [364]. Similar observations to the latter have been recorded in the inhibition of hydrogen flames. Here, of course, the initial accelerating effect is negligible, since there are already large amounts of H atoms present.

Iodine inhibits both CO and H_2 oxidation by removing hydrogen atoms via reaction (lxiv)

$$H + I_2 \rightarrow HI + I \qquad (lxiv)$$

In the case of sulphur dioxide addition only the second limit is affected, so that a gas phase, termolecular chain breaking reaction must be involved. An obvious candidate is reaction (lxv), which has been invoked together with reaction (lxvi) to account for similar inhibition of the limit in H_2/O_2 mixtures [390], viz.

$$O + SO_2 + M \rightarrow SO_3 + M \quad \text{(lxv)}$$

$$H + SO_2 + M \rightarrow HSO_2 + M \quad \text{(lxvi)}$$

$$H + HSO_2 \rightarrow H_2 + SO_2 \quad \text{(lxvii)}$$

$$OH + SO_2 + M \rightarrow HSO_3 + M \quad \text{(lxviii)}$$

Recent flame studies [391, 392] have discounted a suggestion that reaction (lxviii) is important at high temperatures, but it could be an inhibiting reaction at the temperatures of the second limit. These and other studies [390—393] have shown that SO_2 catalyzes the recombination of H atoms by reactions (lxvi) and (lxvii). SO_3 might also be expected to be an intermediate in a similar catalytic sequence, but its drastic effect on the vessel surface would predominate over such considerations.

The rate of oxidation of CO outside the explosion region is accelerated by the presence of water, hydrogen, and small quantities of hydrocarbon, with larger quantities of hydrocarbon decreasing the rate in a manner analogous to the effect on the explosions. Although most workers report that packing the vessel decreases the rate, Verdurmen [375] found in his isotopic scrambling experiments that the rate of formation of CO_2 is proportional to the surface area. Another contrast is that for the "dry" mixtures oxygen has been found by different groups both to increase [357, 376] and to decrease [372, 374] the reaction velocity. The existing information here allows little but speculation. Most of the factors can be accommodated if it is assumed that for "dry" CO/O_2 mixtures below the temperature of the explosion peninsula, CO_2 forms mainly at the surface with an activation energy of ca. 30—35 kcal . mole^{-1}, whereas, in the region above the second limit and with more substantial amounts of hydrogen or water present, a gas phase process occurs. A quantitative treatment of the latter is given in Sect. 10.1.3(*b*). The effects of additives on the rate of the slow reaction may be discussed in terms similar to those already presented for the explosions.

(*b*) *Quantitative treatment of the hydrogen—carbon monoxide—oxygen system.* In order to include the whole range of possibilities in hydrogen—carbon monoxide—oxygen mixtures (or moist carbon monoxide—oxygen mixtures) it is necessary to add to the H_2/O_2 reaction mechanism not only the reactions of CO with OH and O, but also with H and HO_2. Including the possibilities of both termolecular and bimolecular association of O atoms with CO, the most likely reactions to be appended to the H_2/O_2 scheme therefore become

$$OH + CO = CO_2 + H \quad \text{(xxiii)}$$

$$O + CO + M'' = CO_2 + M'' \quad \text{(lvii)}$$

$$O + CO = CO_2 + h\nu \quad \text{(lviia)}$$

$$H + CO + M''' = HCO + M''' \quad \text{(lxxiii)}$$

$$HCO + O_2 = HO_2 + CO \quad \text{(lxxiv)}$$

$$HO_2 + CO = CO_2 + OH \quad \text{(lxxv)}$$

In addition, the explosion limits and reaction rates when large quantities of CO are present will be influenced by the "chaperon" or "third body" coefficients $k_{CO} = k_{4,CO}/k_{4,H_2}$ in reaction (iv) and, if quadratic branching occurs, the similar coefficients k'_{CO} in reaction (vii). The appropriate elementary processes in the H_2/O_2 system are

$$OH + H_2 = H_2O + H \quad \text{(i)}$$

$$H + O_2 = OH + O \quad \text{(ii)}$$

$$O + H_2 = OH + H \quad \text{(iii)}$$

$$H + O_2 + M = HO_2 + M \quad \text{(iv)}$$

$$HO_2 \xrightarrow{\text{surface}} \text{destruction} \quad \text{(v)}$$

$$H_2O_2 + M' = OH + OH + M' \quad \text{(vii)}$$

$$H + HO_2 = OH + OH \quad \text{(viii)}$$

$$HO_2 + HO_2 = H_2O_2 + O_2 \quad \text{(x)}$$

$$HO_2 + H_2 = H_2O_2 + H \quad \text{(xi)}$$

$$O + H_2O_2 = H_2O + O_2 \quad \text{(xiii)}$$

$$H + H_2O_2 = H_2O + OH \quad \text{(xiv)}$$

$$H + H_2O_2 = H_2 + HO_2 \quad \text{(xiva)}$$

$$OH + H_2O_2 = H_2O + HO_2 \quad \text{(xv)}$$

To investigate the new reactions, both slow reaction rates and explosion limits have been examined.

(*i*) *Addition of small amounts of carbon monoxide to slowly reacting mixtures of hydrogen and oxygen.* Baldwin et al. [70] have studied the relative rates of formation of CO_2 and H_2O when small amounts (ca. 1 %) of carbon monoxide are added to slowly reacting hydrogen—oxygen mixtures in a 51 mm diameter, aged, B_2O_3 coated vessel at 500 °C. Figure 67 shows a typical set of data for the changes in total pressure, the partial

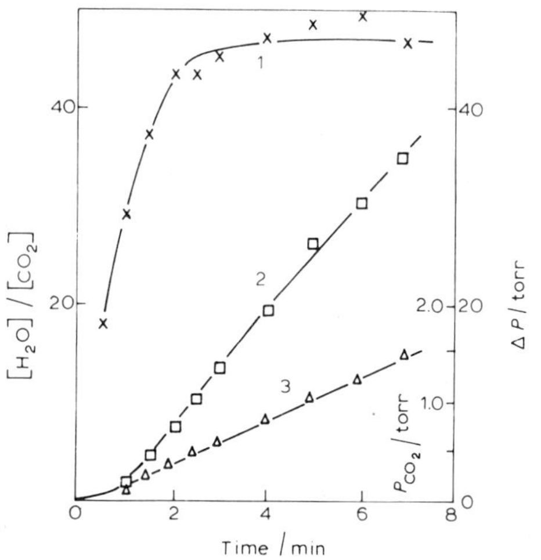

Fig. 67. CO_2 yields and pressure changes in $H_2/CO/O_2$ mixture in 51 mm diameter, aged, boric acid coated vessel at 500 °C (after Baldwin et al. [70]). x, $[H_2O]/[CO_2]$; □, ΔP; △, p_{CO_2}. Total pressure: 500 torr. Initial mole fractions: H_2, 0.28; O_2, 0.14; CO, 0.01; N_2, 0.57. (By courtesy of The Combustion Institute.)

pressure of CO_2 and the ratio $[H_2O]/[CO_2]$. This ratio is low during the induction period, when $[H_2O_2]$ is building up to its stationary concentration, but after the induction period it stays approximately constant. The limiting (constant) ratio is effectively independent of total pressure, inversely proportional to the CO mole fraction, slightly dependent on the O_2 mole fraction, and proportional to the H_2 mole fraction to a power less than one.

Using a simplified H_2/O_2 mechanism in which reactions (viii), (xi), (xiii) and (xv) are neglected, and then adding reactions (xxiii), (lviia) and (lxxv) in turn, gives the expressions [70]:
for reaction (xxiii) OH + CO,

$$\frac{d[H_2O]}{d[CO_2]} = \frac{k_1[H_2](1 + 2k_2/k_4[M] + k_2k_{23}[CO]/k_1k_4[H_2][M])}{k_{23}[CO](1 + k_2/k_4[M])}$$

$$\approx k_1[H_2]/k_{23}[CO] \qquad (113)$$

(the slow reaction conditions studied being such that $k_2/k_4[M] \approx 0.05$);
for reaction (lviia) O + CO,

$$\frac{d[H_2O]}{d[CO_2]} = \frac{(k_3[H_2] + k_{57a}[CO])(k_4[M] + k_2)}{k_2k_{57a}[CO]}$$

$$\approx \frac{k_3[H_2]k_4[M]}{k_2k_{57a}[CO]} \qquad (114)$$

and for reaction (lxxv) HO_2 + CO

$$\frac{d[H_2O]}{d[CO_2]} \approx \frac{\{4k_2k_7k_{10}k_{14}[O_2][M'](1-2k_2/k_4[M])\}^{1/2}(k_4[M]+2k_2)}{k_{14}k_{75}[CO](k_4[M]-2k_2)}$$

$$\approx \frac{(4k_2k_7k_{10}[O_2][M']/k_{14})^{1/2}}{k_{75}[CO]} \qquad (115)$$

The experimental observation that $[H_2O]/[CO_2]$ is effectively independent of the total pressure is completely inconsistent with eqn. (114), so that the bimolecular reaction (lviia) can be ruled out. Reaction (lvii) instead of (lviia) would give the observed independence of pressure, but, if it were the main source of CO_2, the magnitude of the ratio k_{57}/k_3 would be such that, unless the CO_2 were able to continue the reaction chain, CO would be a powerful inhibitor of the H_2/O_2 system. Second limit studies [30] have shown this not to be the case (see Sect. 10.1.1(*b*)). Effectively, the observations are consistent with contributions from both reactions (xxiii) and (lxxv), which result in observed concentration dependences lying between those in eqns. (113) and (115). The contribution from reaction (lxxv) also explains the relatively high yields of CO_2 during the induction period. During this period the radical concentration is comparatively small, and reaction (lxxv) is favoured relative to the radical—radical reaction (x).

Quantitative interpretation of the situation involves the inclusion of reactions (xi) and (xv) in the analysis, and then the determination of the unknown ratios k_{23}/k_1 and $k_{75}/k_{10}^{1/2}$. Because of the complexity of the system, this is best done using computerized optimization techniques (cf. Sect. 4.3.3). The optimization is here assisted by independent measurements of the rate of decomposition of H_2O_2 in the presence of carbon monoxide [211]. These will be discussed below. With the help of these additional measurements, which contribute towards the expression given in Sect. 6.1 for the ratio k_1/k_{23}, an optimum value of $k_{75}/k_{10}^{1/2}$ = 0.4 ± 0.04 at 773 K is suggested by the effect of the small additions of CO on the induction period and maximum rate of the slow reaction in the H_2/O_2 system [395].

(*ii*) *Decomposition of hydrogen peroxide in the presence of carbon monoxide at 713 K.* At low concentrations of carbon monoxide, the rate of decomposition increases linearly with [CO]. However, in contrast with the H_2-sensitized decomposition which reaches a limiting rate after a certain concentration of hydrogen has been added (cf. Sect. 6.8), the CO-sensitized rate continues to increase even at higher CO concentrations [211]. The different behaviour is attributed by Baldwin et al. [211] to a contribution to the CO sensitization by reaction (lxxv). This occurs because the reaction of HO_2 with CO at 713 K is about ten times faster than its reaction with H_2. Including reaction (lxxv) in a mechanism consisting also of reactions (vii), (x), (xiv), (xiva) and (xv) gives the expression

$$-\frac{d[H_2O_2]}{dt} = 2k_7[H_2O_2][M']\left\{1 + \frac{k_{75}[CO]}{(k_{10}k_7[H_2O_2][M'])^{1/2}}\right\} \times \left\{\frac{1 + k_{23}[CO]/k_{15}[H_2O_2]}{1 + \dfrac{k_{23}[CO]}{k_{15}[H_2O_2]}\left(\dfrac{k_{14a}}{k_{14} + k_{14a}}\right)}\right\} \tag{116}$$

in which the ratios k_{15}/k_{23} and $k_{75}/k_{10}^{1/2}$ appear in addition to the parameters k_7 and $k_{14a}/(k_{14} + k_{14a})$ which are known from studies of the H_2/H_2O_2 system (cf. Sect. 4.3.3 and Table 18). Optimization of the agreement between theory and experiment led to $k_{15}/k_{23} = 21.3 \pm 1.0$ and $k_{75}/k_{10}^{1/2} = 0.12 \pm 0.03$ at 713 K, with an r.m.s. deviation of 5.3 %.

Combining k_{15}/k_{23} with k_{15}/k_1 from the H_2-sensitized decomposition [68, 69] leads to $k_1/k_{23} = 4.3 \pm 0.9$ at 713 K. This has been plotted in Fig. 37, and contributes towards the equation given in Sect. 6.1,

$$k_1/k_{23} = 77.5 \exp(-2{,}210/T) \tag{117}$$

(*iii*) *Second limits in hydrogen—carbon monoxide—oxygen mixtures.* Figure 61 shows that at large $[CO]/[H_2]$ ratios the second limit of $2X + O_2$ mixtures, where X represents the mixture of $H_2 + CO$, decreases markedly with decreasing percentage of hydrogen. Similar effects are observed over a range of $[X]:[O_2]$ ratios extending around stoichiometric from at least 4 : 1 to 1 : 2. Such behaviour indicates that some radical, which normally reacts with H_2 in the H_2/O_2 system, begins to undergo a chain terminating reaction with CO. The effect can only be attributed to reaction (lvii). It does not become important until $[CO]/[H_2] >$ ca. 10. Inclusion of reaction (lvii) together with reactions (i)—(v) at the second limit in vessels with surfaces of high destruction efficiency for HO_2 gives

$$[M] = \frac{2k_2}{k_4} - \frac{k_{57}[CO][M][M'']}{k_3[H_2]} \tag{118}$$

so that the derivation of k_{57}/k_3· appears straightforward. However, Baldwin et al. [395] have drawn attention to two difficulties in the analysis of the second limit results of Buckler and Norrish [368, 369] with Pyrex reaction vessels, and Dixon-Lewis and Linnett [30] with KCl coated vessels, both at high $[CO]/[H_2]$ ratios.

The first difficulty is that neither clean Pyrex nor KCl coated surfaces are of the highest efficiency for removal of HO_2 (cf. Sect. 3.6.4 and Fig. 11), so that there may be a variable contribution from the regeneration term (cf. Sect. 3.6.2 and 4) or quadratic branching as the $[CO]/[H_2]$ ratio is changed. However, since it is also found that the limits at high

$[CO]/[H_2]$ do not increase markedly with decreasing oxygen, Baldwin et al. assume as a first approximation that

$$[M] = K - \frac{k_{57}[CO][M][M'']}{k_3[H_2]} \tag{118a}$$

where now K is a constant, at a given temperature, which is greater than $2k_2/k_4$. This is tantamount to assuming a constant contribution from quadratic branching in all the high $[CO]/[H_2]$ mixtures. Plotting [M] against $[CO][M]^2/[H_2]$ should then give straight lines of gradient k_{57}/k_3.

The calculation of the [M] leads to discussion of the second difficulty, which is that neither the "chaperon" coefficient of CO relative to H_2 in reaction (iv), nor the coefficients for CO and O_2 relative to H_2 in reaction (lvii) are known. Anticipating the results of the discussion immediately below, k_{CO} is given the value 0.74. It is further assumed that $k''_{CO} = k_{CO}$ and $k''_{O_2} = k_{O_2}$. With these assumptions both the results of Buckler and Norrish [368, 369] and Dixon-Lewis and Linnett [30] give values of k_{57}/k_3 ranging from about 12 l . mole^{-1} at 500 °C to 6 l . mole^{-1} at 570 °C.

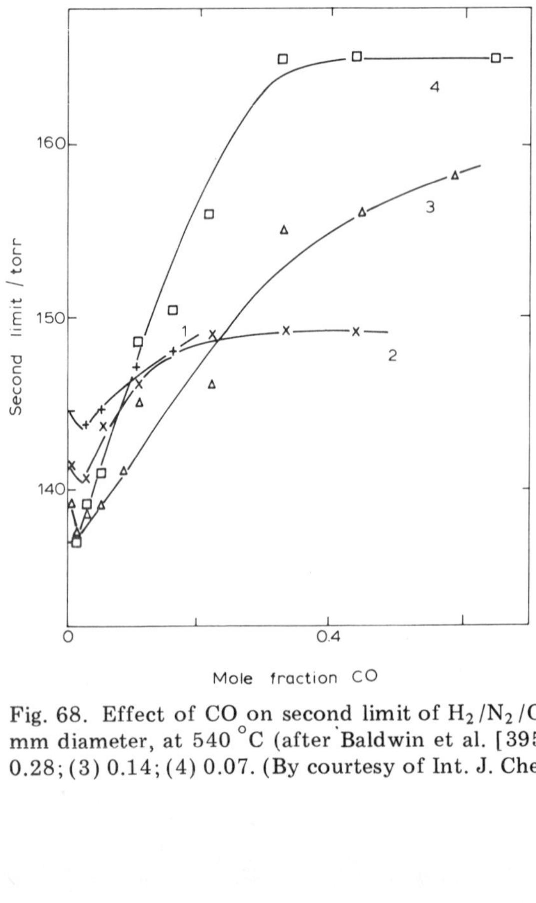

Fig. 68. Effect of CO on second limit of $H_2/N_2/O_2$ mixtures in KCl coated vessel, 51 mm diameter, at 540 °C (after Baldwin et al. [395]). $X_{H_2} = 0.28$; X_{O_2}: (1) 0.56; (2) 0.28; (3) 0.14; (4) 0.07. (By courtesy of Int. J. Chem. Kinet.)

The remaining stage of the analysis is to consider the limits for mixtures with lower $[CO]/[H_2]$ ratios. In contrast with the earlier measurements of Dixon-Lewis and Linnett [30] who simply studied $H_2/CO/O_2$ mixtures and assumed $k_{CO} = k_{N_2}$ in reaction (iv), Baldwin et al. [395] have directly replaced N_2 by CO in $H_2/N_2/O_2$ mixtures with constant mole fractions of H_2 and O_2. With both KCl and CsCl coated vessels, but particularly with KCl coated vessels, they obtained results at 813 K, and for CO mole fractions up to about 0.6, which they could only attribute to vessel surface changes with increasing concentration of CO, so that there was an increasing contribution due to the regeneration term, or to quadratic branching. For KCl vessels, for example, the limit was usually depressed slightly on addition of the first small amount of CO (up to a mole fraction of about 0.02), but then, with increasing addition, it rose rather sharply to an almost constant value. The rise in the limit increased as the O_2 mole fraction decreased, in line with the quadratic branching ideas, and gave results as shown in Fig. 68. With CsCl coated vessels the behaviour was not so pronounced; the limit decreased continuously with increasing CO concentration, but, after an initial sharp fall, the rate of decrease became less for CO mole fractions greater than about 0.05. If the initial inhibiting effect of CO is due only to reaction (lxxiii) and to the chaperon effect of CO in reaction (iv), then using reactions (i)—(iv) and (lxxiii) leads to the limit expression

$$[M] = \frac{2k_2}{k_4} - \frac{k_{73}[CO][M'']}{k_4[O_2]} \tag{119}$$

Such analysis as was possible on the initial steeper regions (where quadratic branching or regeneration effects were presumed to be negligible) led to $k_{CO} \nless 0.6$ and $k_{73}/k_4 \ngtr 0.07$ at 813 K. However, more precise studies of these parameters is possible using aged B_2O_3 coated vessels [395], particularly since, in such vessels, the limit can be investigated at low O_2 mole fractions where reaction (lxxiii) becomes important. Computer analysis to fit the results for the boric acid coated vessel requires the assignment of values to the ratios k_{23}/k_1, k_{57}/k_3, k_{73}/k_4 and $k_{75}/k_{10}^{1/2}$, together with values for all the chaperon efficiencies relative to $H_2 = 1$. Assuming the chaperon efficiencies in reactions (lvii) and (lxxiii) to be the same as in reaction (iv), it turns out that only two of these parameters, k_{73}/k_4 and k_{CO}, have a marked effect on the limits for the range of compositions under consideration. The remaining ratios and efficiencies were therefore given values already determined independently in the studies of the $CO + H_2O_2$ reaction (Sect. 10.1.3 (*b*) (*ii*)), the second limits at high $[CO]/[H_2]$ ratios (see above), and the addition of CO to slowly reacting mixtures of H_2 and O_2 (Sect. 10.1.3 (*b*) (*i*)), together with previous studies of the $H_2/N_2/O_2$ system. For the most probable values of all the independent parameters, the optimization procedure gave $k_{CO} = 0.74 \pm 0.04$ and $k_{73}/k_4 = 0.022$ at 773 K, with an r.m.s. deviation of 1.1 %.

10.2 OXIDATION OF CARBON MONOXIDE IN FLAMES AND OTHER HIGH TEMPERATURE FLOW SYSTEMS

In this section the nature of the light emitted in the flames, explosions, and the oxidations in general will first be briefly discussed and the more general properties of the flames and other high temperature flow systems will then be described.

10.2.1 The nature of the light emission

The emission from the flames has received most attention. A carbon monoxide flame burning in air or oxygen is bright blue in colour, and spectroscopic examination shows that the light consists of a strong continuum with numerous diffuse bands. Gaydon [396] has dealt thoroughly with earlier investigations, and detailed spectroscopic considerations are outside the scope of this review.

Most of the banded spectrum comprises what are called the "CO flame bands". However, emission from molecular oxygen has also been detected, particularly in lean, hot flames: these are the "Schumann-Runge" and "Atmospheric" bands [397]. The CO flame bands are also present in the emission from $CO—N_2O$ flames. They do not resemble spectra obtained from O_3 decomposition flames, and neither do they correlate with the band spectra of O_2, C_2 or CO [398]. Fowler and Gaydon [399] tentatively assigned the emission to excited CO_2 molecules, but detailed analysis presented some difficulties. Gaydon [400] has suggested that the emission may be connected with transitions which give rise to a weak banded absorption by gaseous CO_2 below 1700Å, with the large difference in wavelength between this absorption and the emission of CO_2 in discharge tubes being due to a severe change in the shape of the molecule undergoing the absorption (or the emission in the flame). Gaydon also pointed out the analogy between the SO_2 and CO_2 afterglows, the former of which involves an excited singlet state. Walsh [401], on the other hand, favoured a triplet—triplet transition in the case of CO_2, on the pragmatic basis that it allows for easier frequency assignment. However, Walsh's proposal is not supported by absorption measurements.

Dixon [402] has analyzed the spectrum of the CO_2 afterglow photographed under high resolution, and has shown that Gaydon's views were essentially correct. The spectrum arises from CO_2 molecules at the lowest vibrational level of the $^1B_2$ state radiating to high vibrational levels of the ground electronic state ($^1\Sigma_g^+$). These transitions are associated with the weak absorption system of CO_2 at 1475 Å.

The continuum accounts for most of the emission from CO flames and other high temperature sources; whereas in low pressure systems which emit the CO_2 afterglow, the continuum is not detected. Studies of preheated CO diffusion flames [403] show that hydrogen reduces the intensity of both the continuum and the band system equally. Increasing the flame temperature both increases the intensity of the continuum and enhances the blue end. On the other hand, the band system is unaffected by the temperature. The bands are thus not thermal in origin, and they also appear not to depend on the equilibrium concentration of O atoms, which rises considerably on preheating. Although the continuum was interpreted by Gaydon [396] in terms of an association process such as reaction (lviia)

$$O + CO \rightarrow CO_2 + h\nu \qquad \text{(lviia)}$$

the data do not completely preclude Kondratiev's view that it consists of unresolved band spectra. This idea is supported by Clyne and Thrush [404], on the basis that the continuum and the flame bands have similar spectral distributions. The flame bands may then broaden into the continuum as the temperature is raised.

In connection with the origin of the continuum, Kaskan [405] has shown, by additional measurement of [OH] and calculation of partial equilibrium [O], [CO] and $[H_2O]$ in the recombination zones of some rich CO/H_2/air flames, that its intensity corresponds with its production by a reaction between O atoms and CO. He found $I/[CO] \propto [OH]^2/[H_2O]$, and hence $I \propto [O][CO]$ if reaction (xvi) is equilibrated. However, Kaskan's calculation of [CO] depends also on the assumption that the water gas equilibrium is maintained in the measurement region. This will be discussed in Sect. 10.2.3. Kaskan [406] also found the intensity of the continuum in lean CO/H_2/air flames to be given by $I \propto [OH]^4[CO_2]/[H_2O][O_2]$ at temperatures above 1500 K and for a range of unburnt $[CO]/[H_2]$ ratios. This again implies $I \propto [O][CO]$ if all the partial equilibrium assumptions are valid. It is likely that the partial equilibria are all maintained in both the rich and lean flames at temperatures above about 1500 K. Clyne and Thrush [404] found the (banded) emission in discharge-flow experiments to be proportional to [O][CO].

10.2.2 Burning velocities

Jahn [407] has measured the burning velocities of flames at atmospheric pressure formed from a range of $CO/O_2/N_2$ and $CO/O_2/CO_2$ mixtures, containing a little water vapour or hydrogen. The data are reproduced by Lewis and von Elbe [4]. They refer to an average burning velocity over the surface of the inner cone of a Bunsen type flame. The difference between the flames with H_2 and CO_2 as diluent was small, and probably reflects the differences in transport properties. Fiock and Röder [408] used the soap bubble technique (for details see e.g. Fristrom and

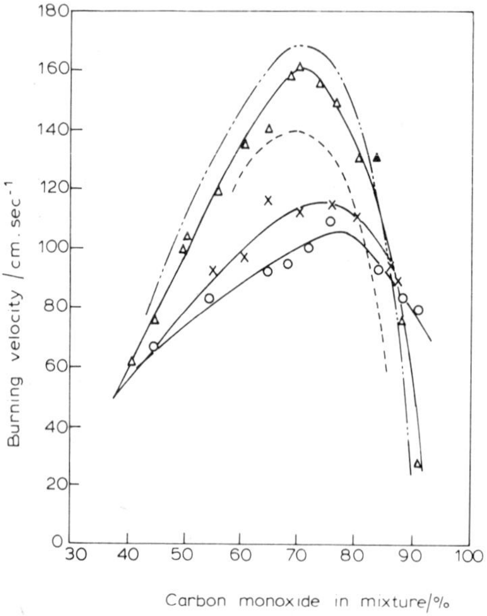

Fig. 69. Effect of pressure on burning velocities of carbon monoxide—air mixtures (after Strauss and Edse [409]). ○, 1 atm; x, 5.1 atm; △, 21.4 atm; —·—·— 52 atm; - - - -, 90 atm. (By courtesy of The Combustion Institute.)

Westenberg [120]) to measure the burning velocities of moist CO/O_2 mixtures. Their results agree reasonably well with Jahn's.

Lewis and von Elbe have ignited CO/O_2 mixtures in the centre of spherical reaction vessels, and have obtained data on the rate of propagation of the flame and the rate of pressure change in the vessel. From such results the burning velocity S_u can be calculated [4], according to the thin flame approximation, from eqn. (120),

$$S_u = \frac{dr}{dt} - \frac{1-\epsilon}{3p\gamma_u\epsilon/r}\frac{dp}{dt} \tag{120}$$

where r is the radius of the flame at time t, p is the pressure, γ_u is the ratio of specific heats of the unburnt gas, and $\epsilon = r^3/R^3$, R being the radius of the vessel. The accelerating effect of water was clearly shown.

Photographs of the progress of a flame through CO/O_2 mixtures ignited at the centre of a soap bubble placed in a constant pressure bomb enabled Strauss and Edse [409] to investigate the effect of pressure on the burning velocity. Their results are shown in Fig. 69. Watermeier [410] has also used the constant pressure bomb technique to obtain the velocities of CO/O_2 flames containing traces of H_2 or D_2. His results are lower than those reported by Strauss and Edse, but are internally consistent in that the ratio $S_{u,H_2}/S_{u,D_2}$ is fairly constant at about 1.3. It is noteworthy that very pure CO/O_2 mixtures would not ignite, and this condition was also used as a criterion for purity. In a later paper, Wires et al. [411]

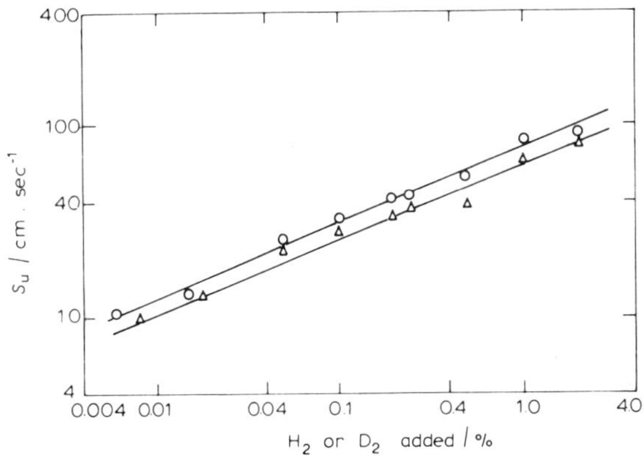

Fig. 70. Effect of H_2 and D_2 on the burning velocities of $2CO + O_2$ mixtures (after Wires et al. [411]). ○, Hydrogen; △, deuterium. (By courtesy of J. Phys. Chem.)

determined flame velocities, quenching distances and minimum ignition energies for "dry" $2CO + O_2$ mixtures containing some H_2 or D_2. The effects of the hydrogen and deuterium on the burning velocity are shown in Fig. 70. The ignition energy is inversely related to the burning velocity, and hence it increases with decreasing H_2 content. For the "pure" $2CO + O_2$ mixtures the ignition energy was extremely high (>500 mJ), and the burning velocity, obtained by backward extrapolation to zero hydrogen content, was less than 3 cm . sec^{-1}. The isotopic burning velocity ratio $S_{u,H_2}/S_{u,D_2}$ was 1.22 and, rather interestingly, there was no isotopic effect on either the ignition energies or quenching distances when mixtures of the same burning velocity were compared.

Friedman and Cyphers [412] attempted to correlate the burning velocity of CO/O_2 flames with the flame temperature and the initial mixture composition. Using a flat flame burner, they measured the burning velocities of a number of $CO/O_2/H_2O/N_2$ mixtures under conditions of varying pressure, initial $[O_2]/[CO]$ ratio and amount of added water, at the same time adjusting the flame temperature by altering the amount of diluent nitrogen. Their results are illustrated in Fig. 71. On the assumption of a first order loss of CO which they had observed in the post-flame gases of a propane flame [413], they found the CO flame data to correlate empirically with the equation

$$S_u^2 = 3.8 \times 10^6\ X_{CO,u}\ X_{H_2O,u}^{0.5}\ (P/P_{at})^{-0.24} \exp(-11{,}130/T_b) \quad (121)$$

where X_u denotes initial mole fraction, P the pressure and T_b the flame temperature.

Friedman and Cyphers also developed a correlation between S_u and the term $([H] + 0.15[OH])^{1/2}$, where [H] and [OH] denote equilibrium concentrations in the burnt gas. Using other workers' measurements, they further showed that a plot of log $(S_u^2/X_{CO,u}X_{H_2O,u}^{1/2})$ versus $1/T$ was linear within the scatter which would be expected for measurements from a

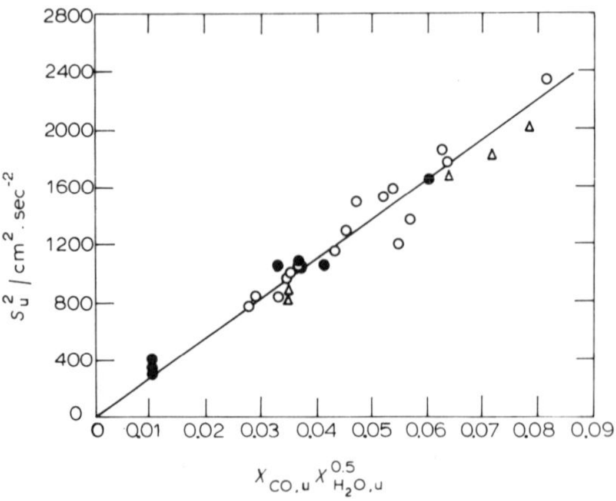

Fig. 71. Correlation of burning velocities (S_u) of $CO/O_2/N_2/H_2O$ mixtures with unburnt gas composition (after Friedman and Cyphers [412]). Pressure 60 torr; flame temperature T_b = 2010 K. Equivalence ratios: ●, 0.62—0.71; ○, 1.0; △, 2.4—2.5. (By courtesy of J. Chem. Phys.)

number of sources. The slope of the line corresponded with an apparent activation energy of about 23 kcal . mole^{-1}.

Taking into consideration the far-reaching effects of hydrogenous impurities, the data on the burning velocities of moist CO/O_2 flames are in fair agreement. Particularly interesting is the low value of the "burning-velocity" of the dry stoichiometric mixture obtained by back extrapolation to zero H_2 or H_2O content [409]. This is less than 3 cm . sec^{-1}. The low value supports the view that the branching reaction sequence (liii) and (liv) is of little importance, and that CO/O_2 mixtures are unreactive when sufficiently pure.

$$O + CO \rightarrow CO_2^* \qquad \text{(liii)}$$

$$CO_2^* + O_2 \rightarrow CO_2 + 2O \qquad \text{(liv)}$$

10.2.3 Flame profiles

Species profiles have not been measured directly for dry CO/air or CO/O_2 flames in the same way as they have for hydrogen flames. Several investigations, however, have been concerned with the oxidation of carbon monoxide in lean hydrocarbon flames (e.g. refs. 406, 413, 417, 429) or in moist CO flames flames of H_2/CO mixtures in air [167, 406, 414, 418] or O_2 [523]. The interest in the oxidation in hydrocarbon flames has arisen since the overall reaction in such flames is a two stage process. In the first rapid stage (the main flame reaction zone) the hydrocarbon is essentially converted to CO and water, with traces of hydrogen also appearing. The second, more extended, stage is devoted to radical recombination and to the slower oxidation of CO, predominantly by reaction (xxiii).

$$OH + CO \rightleftharpoons CO_2 + H \qquad \text{(xxiii)}$$

Both Friedman and Cyphers [413] and Fenimore and Jones [414] obtained a first order decrease of [CO] in specific flames, while for a series of flames, Fenimore and Jones [414] found that the gradient d(ln[CO])/dt was proportional to [O_2] in flames containing ample water, but was proportional to [H_2O] and independent of [O_2] in flames containing little water.

The question whether reaction (xxiii) is fairly rapidly equilibrated, i.e. whether the water gas equilibrium is rapidly established, in the recombination zones of any flames has received considerable attention. The equilibrium is certainly not rapidly established in lower temperature ($T_b \simeq 1100$ K), fuel-rich $H_2/N_2/CO_2/O_2$ or $H_2/N_2/CO/O_2$ flames at atmospheric pressure. In such $H_2/N_2/O_2$ flames with a trace of added CO_2, Dixon-Lewis et al. [169] found that only about 12 % of the CO_2 reacted in the flame, whereas for equilibration the final [CO]/[CO_2] ratio should have been about unity. Further, in a similar $H_2/N_2CO/O_2$ flame which contained about 20 % H_2 and 8 % CO initially, Dixon-Lewis et al. [419] found the [CO_2]/[CO] ratio in the effluent gas to be only about $\frac{1}{8}$. In higher temperature flames the equilibrium is approached more rapidly. Fenimore and Jones [209], similarly using mass spectrometric probing of atmospheric pressure H_2/O_2/Ar flames containing a little CO_2, found that at 1345 K the rate of approach to equilibrium was "as rapid as could be followed conveniently". However, at 1605 K with a very lean, low pressure propane-air flame, Friedman and Cyphers [413] found the ratio [CO]/[H_2] at distances up to 1.5 cm above the luminous zone to be two to three times higher than would be expected on the basis of equilibration.

In connection with the same problem, Jost et al. [420] have added CO and H_2 to the burnt gases of a hydrocarbon flame, which flowed isothermally at atmospheric pressure along a heated ceramic tube. CO profiles were obtained by sampling at various points downstream. Their results indicate that equilibrium is rapidly attained at 1700 K, but that at 1400 K under their conditions some seconds are required to attain better than 90 % equilibrium. The rates corresponded with an overall activation energy of 72 kcal . mole^{-1}. These results are more in line with those of Fristrom et al. [415, 416], who measured profiles in the burnt gas of low pressure (1/10 and 1/20 atm) methane-air flames. They found that equilibration was not attained at the low pressures until the final flame temperature of about 2000 K was reached. The results overall, including those of Kaskan [406] referred to in Sect. 10.2.1, suggest that equilibration is rapid in atmospheric pressure flames at temperatures above about 1500—1600 K. Measurements of the rates of reaction of traces of CO_2 and D_2O added to fuel-rich $H_2/N_2/O_2$ flames at lower temperatures than this have led [172, 209] to values of the ratio k_{H+D_2O}/k_{H+CO_2}, and then by further analysis to two of the values of k_1/k_{23} given in Table 36. The results of Jost et al. [420] lead to $k_{23} = 10^9 \exp(-2350/T)$ between

1380 and 1720 K, assuming $\Delta H^0_{f,OH} = 9.33$ kcal . mole^{-1}. Singh and Sawyer [417] find $k_{23} \approx 8 \times 10^8$ at 1600 K, the mid-point of their temperature range of 1500—1720 K.

10.2.4 Studies using high temperature flow reactors

In addition to the flame studies there have been several investigations of CO oxidation in other flow reactors.

Kozlov [383] investigated the rate of burning of CO in the temperature range 700—1100 °C by this means, determining the CO and CO_2 profiles by sampling and IR gas analysis. He was able to show that the rate of disappearance of CO for lean conditions was proportional to [CO] and to $[H_2O]^{0.5}$. The order with respect to oxygen was 1.0 or 0.25, respectively, depending on whether the mixture contained more or less than 5 % oxygen. The overall activation energy was 32 kcal . mole^{-1}.

Very similar dependences of the rate on the reactant concentrations were obtained by Longwell and Weiss [421], Hottel et al. [384], Williams et al. [422] and Dryer and Glassman [455] using stirred flow reactors. Hottel et al. [384] found

$$d[CO_2]/dt = 1.2 \times 10^8 \exp(-8{,}000/T) X_{CO} X_{H_2O}^{0.5} X_{O_2}^{0.3} (P/RT)^{1.8} \tag{122}$$

while Williams et al. [422] obtained

$$d[CO_2]/dt = 1.8 \times 10^{10} \exp(-12{,}500/T) X_{CO} X_{H_2O}^{0.5} X_{O_2}^{0.5} (P/RT)^2 \tag{123}$$

where the rates are in mole . l^{-1} . sec^{-1}, P is in atmospheres, T in K, and R in l . atm . mole^{-1} . K^{-1}. Williams et al. also found that a lower oxygen exponent e.g. 0.25, gave a better fit at higher oxygen mole fractions. Remembering the flame theory result that $S_u^2 \propto$ reaction velocity, the reaction orders with respect to [CO] and $[O_2]$ agree with the relation (121) found by Friedman and Cyphers [412]. Similar relationships were also found by Sobolev [423]. A formulation approximating to that of Hottel et al. has been recommended for interpreting studies on pulverized coal combustion [424].

The mechanism of oxidation in all these systems containing hydrogen or water vapour will consist of the addition of reaction (xxiii), and to a lesser extent reactions (lvii), (lxxiii) and (lxxv), to the hydrogen—oxygen mechanism. The experimental findings are in accord with the theoretical eqn. (111).

10.3 ELEMENTARY REACTIONS IN THE HYDROGEN—CARBON MONOXIDE—OXYGEN SYSTEM

The rate coefficients of the hydrogen—oxygen system have already been discussed in Sect. 6. It remains to consider reactions (xxiii), (lvii), (lxii), (lxxiii) and (lxxv).

10.3.1 Reaction (xxiii) $OH + CO \rightarrow CO_2 + H$

The data for this reaction has recently been thoroughly reviewed by Baulch and Drysdale [178]. Not only is it an important reaction in relation to exhaust emission and air pollution studies (cf. slowness of CO oxidation stage in flames, Sect. 10.2.3), but it is also a useful reference reaction for the measurement of OH reaction rates by competitive methods [213]. An example of this approach is the use of the ratios k_1/k_{23} given in Table 36 and Fig. 37 in order to deduce eqn. (71) for k_1.

Absolute measurements of k_{23} have been made by a variety of methods similar to those already briefly described in Table 35 and Sect. 6. Results are summarized in Table 53. In evaluating results, Baulch and Drysdale have drawn attention to the need (i) to allow for first order surface decay of OH in discharge-flow systems (cf. Sect. 6.4), and (ii) to measure OH concentrations at the "hot boundaries" of flame reaction zones rather than using calculated full equilibrium concentrations. Results which do not conform with these requirements are excluded from the table, or an appropriate comment is made. Shock tube results which are uncorrected for boundary layer effects (cf. Sect. 5.1) are also excluded. Neither Dean and Kistiakowsky [230] nor Izod et al. [432] reported that they had made this correction.

Studies of the rate of approach to the water gas equilibrium from the $H_2 + CO_2$ side have been made by Tingey [433], Kochubei and Moin [434] and others using tubular flow reactors at temperatures around 1000 K. Such measurements rely on the thermal dissociation of hydrogen for their radical concentrations, and in the absence of measurements of these the systems are not regarded as sufficiently well defined for a valid determination of k_{-23}.

Considering the results in Table 53, there seems to be very close agreement between measurements of k_{23} at 300 K from several laboratories. As a result, Smith and Zellner [214] recommend $k_{23} = 8.7 \times 10^7$ at this temperature. The whole series of results in the table is plotted in Arrhenius form in Fig. 72, which shows that the rate coefficient has only a very small temperature dependence, at least up to 500 K. Baulch and Drysdale [178] found that the simplest expression to fit the reliable data adequately over the temperature range 250—2500 K was

$$\log (k_{23}/\text{l.mole}^{-1}\,.\text{sec}^{-1}) = 7.83 + 3.9 \times 10^{-4}\, T \qquad (124)$$

The Arrhenius type expression

$$k_{23} = 1.5 \times 10^4\, T^{1.3} \exp(+385/T) \qquad (124a)$$

is in close agreement with eqn. (124) up to 2000 K. The line in Fig. 72 corresponds with eqn. (124). Its curvature has been discussed by Dryer et al. [196] and by Smith and Zellner [214].

TABLE 53
Absolute measurements of k_{23}

k_{23} (l . mole^{-1} . sec^{-1})	Temp. (K)	Method and comments[a]	Ref.
$6.6 \times 10^7\ T^{1/2} \exp(-2{,}500/T)$	348—520	As ref. 198, Table 35[b]. Results invalid.	425
$6 \times 10^8\ T^{1/2} \exp(-3{,}500/T)$	348—520	As ref. 198, Table 35[b]. Results invalid.	198
2.2×10^8	1400	Stirred flow reactor	384
$1.0 \times 10^9 \exp(-2{,}350/T)$	1380—1720	See Section 10.2.3.	420
$(1.15 \pm 0.05) \times 10^8$	300	D.F. As ref. 202, Table 35[b]. No correction for surface loss of OH.	202
$(5.1 \pm 2.0) \times 10^7$	300	D.F. OH from H + NO_2. [CO_2] by mass spectrometry. No correction for surface loss of OH. Criticized in ref. 231.	426
$(8.9 \pm 0.9) \times 10^7$	301	As ref. 205, Table 35[b].	205
1.9×10^8	1600	As ref. 203, Table 35, but without added H_2. CO produced within flame.	427
$(9.0 \pm 0.3) \times 10^7$	300	As ref. 426, but with correction for surface loss of OH.	231
5.8×10^7	310	As ref. 206, Table 35[b].	428
10.6×10^7	440		
15.3×10^7	610		
$(8.15 \pm 0.43) \times 10^7$	300	As ref. 205, Table 35, but using $H_2O/CO/Ar$ mixture.	205
$(8.84 \pm 0.54) \times 10^7$	300		
$(8.83 \pm 0.39) \times 10^7$	300		
$(8.40 \pm 0.37) \times 10^7$	305		
$(9.80 \pm 0.53) \times 10^7$	334		
$(8.84 \pm 0.27) \times 10^7$	373		
$(8.43 \pm 0.24) \times 10^7$	421		
$(9.97 \pm 0.20) \times 10^7$	495		
$(9.90 \pm 0.11) \times 10^7$	495		
$(9.86 \pm 0.54) \times 10^7$	498		
$(1.00 \pm 0.05) \times 10^8$	495		
5.6×10^7	310	As ref. 206, Table 35[b].	206
10.1×10^7	440		
13.6×10^7	610		
2.31×10^8	1297	Shock tube. See Section 5.2 and Table 23.	92
2.07×10^8	1342		
3.46×10^8	1369		
2.27×10^8	1372		
2.12×10^8	1445		
3.54×10^8	1521		

TABLE 53—*continued*

k_{23} (l. mole^{-1} . sec^{-1})	Temp. (K)	Method and comments[a]	Ref.
2.69×10^8	1535		
5.52×10^8	1626		
3.70×10^8	1777		
4.10×10^8	1843		
3.73×10^8	1896		
2.41×10^8	1899		
8×10^8	1600	C_2H_4/O_2 and C_2H_6/O_2 flames at atmospheric pressure. [CO] and [CO_2] by mass spectrometry. [OH] by UV absorption. Requires absolute [OH].	417
$(1.0 \pm 0.3) \times 10^8$	300	D.F. OH from H + NO_2. Stable products by mass spectrometry. Additional reaction between OH and NO_2 proposed to account for stoichiometry. (See Sect. 6.4.)	232
$(1.65 \pm 0.1) \times 10^8$	1050	Flame study. $H_2/N_2/O_2$ + trace CO_2 at atmospheric pressure. For analysis see Sect. 5.4.2.	172
8.1×10^7	298	As ref. 207, Table 35[b] using H_2O/CO mixtures in He.	207
$4 \times 10^9 \exp(-4{,}000/T)$ (±25 %)	1500—2000	Shock tube. Relative [O][CO] in "oxidizing" ($H_2/5O_2/3CO/91$ Ar) and "reducing" ($5H_2/O_2/4CO_2/90$Ar) mixtures from CO flame spectrum. Absolute [CO_2] by calculated infrared emission. Optimized fit to rates of CO_2 formation or removal and [O][CO] intensities using "best" H_2/O_2 rate parameters (see under ref. 111 in Table 35).	111
$1.4 \times 10^9 \exp(-2{,}770/T)$	1500—1900	Low pressure CH_4/O_2 flames. All species, including OH, by mass spectrometry with molecular beam sampling.	429
8.67×10^7	216	F.P. (a) H_2O/CO mixtures. (b) $N_2O/H_2/CO$ mixtures. As ref. 197, Table 35.	214
9.28×10^7	229		
8.91×10^7	262		
8.49×10^7	300		
9.09×10^7 (a)	300		
9.34×10^7 (a)	300		
1.00×10^8	333		

TABLE 53—*continued*

k_{23} (l.mole^{-1}.sec^{-1})	Temp. (K)	Method and comments[a]	Ref.
1.10×10^8	357		
1.15×10^8	424		
1.26×10^8	459		
8.49×10^7	224		
9.09×10^7	248		
8.49×10^7	248		
9.09×10^7 (b)	253		
8.13×10^7	262		
8.67×10^7	275		
8.91×10^7	300		
1.12×10^8	380		
8.0×10^7	298	As ref. 208, Table 35[b].	208
8.3×10^7	396		
8.7×10^7	523		
1.02×10^8	707		
1.31×10^8	915		
8.79×10^7	220	F.P. $H_2O/CO/He$ mixtures. [OH] by resonance fluorescence. Relative values only needed (effective 1st order decay of OH in presence of large excess CO).	430
8.97×10^7	240		
9.28×10^7	273		
9.58×10^7	300		
1.04×10^8	373		
9.36×10^7	298	D.F. [OH] by laser magnetic resonance. Relative values only needed (1st order decay of OH).	431
8.0×10^7	400	Low pressure $H_2/CO/O_2$ flame with 9.4% CO, 11.4% H_2, 79.2% O_2 initially. All species, including radicals, by mass spectrometry with molecular beam sampling.	
$2.3 \times 10^9 \exp(-2{,}850/T)$	1000–1800		

[a] D.F., discharge-flow method; F.P., flash photolysis. [b] With CO as reactant instead of H_2.

10.3.2 *Recombination reaction between O atoms and CO*

The reaction between oxygen atoms and carbon monoxide produces visible evidence of its occurrence in that it is accompanied by the emission of a blue chemiluminescence in the visible and near UV. The emission is particularly noticeable from CO/O_2 flames, and has been discussed in Sect. 10.2.1.

Clyne and Thrush [404] have examined the intensity I of the chemiluminescent emission between 200 and 300 K in a fast flow system. Oxygen atoms were generated either by dissociation of pure oxygen or

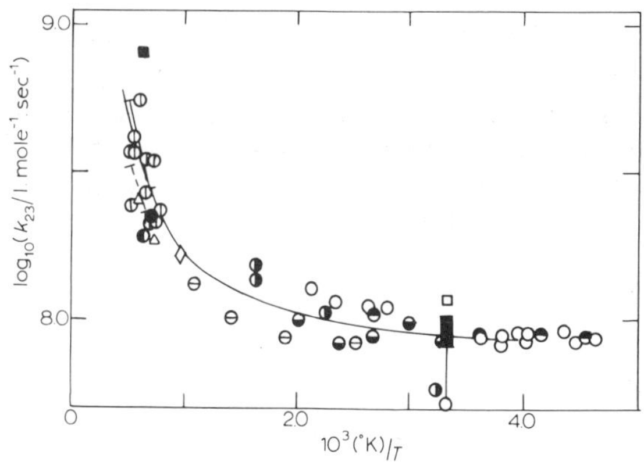

Fig. 72. Arrhenius plot of k_{23}. ▮, Greiner [205], Stuhl and Niki [207], Westenberg and de Haas [208], Smith and Zellner [214], Wilson and O'Donovan [231], Mulcahy and Smith [232], Davis et al. [430], Howard and Evenson [431]; □, Dixon-Lewis et al. [202]; ⌽, Herron [426]; ○, Smith and Zellner [214]; ◓, Davis et al. [430]; ◑, Wong et al. [206, 428]; ◒, Greiner [205]; ⊖, Westenberg and de Haas [208]; ◇, Dixon-Lewis (see Sect. 5.4.2); ⦶, Brabbs et al. [92]; △, Jost et al. [420]; ●, Hottel et al. [384]; ◐, Porter et al. [427]; ■, Singh and Sawyer [417]; ⊢⊣, Gardiner et al. [111]; ⊢ - ⊣, Peeters and Mahnen [429].

mixtures of about 1 % oxygen in Ar, He or Ne in an electrodeless discharge, or by addition of the stoichiometric quantity of NO to a stream of N atoms produced by a discharge in pure nitrogen. Pressures of up to 0.1 torr of CO were admitted to the stream containing O atoms (total pressure ca. 1.7 torr) at one of four subsequent inlets to the flow tube, so that the kinetics of the emission could be studied by observation at a fixed downstream position. The intensity I was found to be given by

$$I = I_0 [O][CO] \tag{125}$$

where I_0 was independent of the total pressure, but depended on the nature of the inert carrier gas M″. In this the behaviour is similar to that in the O + NO and O + SO reactions [404, 435]. For the O + NO reaction, Clyne and Thrush found that the recombination rate coefficient k also depends on the nature of the carrier gas, but k depends on the pressure as well. For O + CO, I_0 is found to increase with temperature in a manner fitting the expression

$$I_0 = 6 \times 10^3 \exp\{(-1{,}850 \pm 250)/T\} \tag{126}$$

At 273 K it is about 2000 times less (for M = O_2) than the proportionality constant for O + NO. The latter has a small negative temperature coefficient, and plots of log $I_{0,CO}$ and log $I_{0,NO}$ versus T^{-1} show that the two pre-exponential factors are similar. The rate coefficient k_{57} has also been found to be very much smaller than the corresponding coefficient for O + NO at 300 K, and the evaluation to be given below indicates that k_{57} also has an activation energy of about 4 kcal . mole^{-1}. The mechanism of light emission is therefore closely related to the mechanism of combination.

The fact that I_0 depends on the nature of the carrier gas indicates that the chemiluminescent reactions take place in three body processes. The whole range of phenomena may then be explained by postulating an initial termolecular combination to an excited state of CO_2, followed at a later stage by the emission of radiation or by collisional quenching to form CO_2 in the ground electronic state, viz.

$$O + CO + M'' \rightarrow CO_2^* + M'' \tag{lvii}$$

$$CO_2^* \rightarrow CO_2 + h\nu \tag{lviii}$$

$$CO_2^* + M'' \rightarrow CO_2 + M'' \tag{lix}$$

Using a stationary state treatment for CO_2^* then gives

$$[CO_2^*] = \frac{k_{57}[O][CO][M'']}{k_{58} + k_{59}[M'']} \tag{127}$$

If $k_{59}[M''] \gg k_{58}$, then for appropriate units of I_0,

$$I_0 = k_{58}[CO_2^*]$$

$$= \frac{k_{57}k_{58}}{k_{59}}[O][CO] \tag{128}$$

where k_{57} and k_{59} depend on the nature of M″, and

$$d[CO_2]/dt = (k_{58} + k_{59}[M''])[CO_2^*]$$

$$= k_{57}[O][CO][M''] \tag{129}$$

There is still one further complexity, however, in that the overall recombination

$$O(^3P) + CO(^1\Sigma^+) \rightarrow CO_2(^1\Sigma_g^+)$$

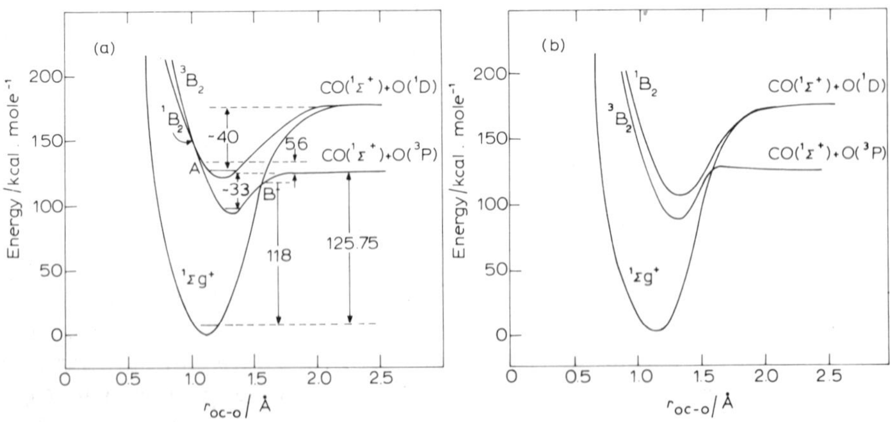

Fig. 73. Schematic energy diagram for O + CO⇄, CO_2 system as suggested by (a) Lin and Bauer [439], and (b) Clyne and Thrush [404] (after Lin and Bauer [439]). (By courtesy of J. Chem. Phys.)

is spin forbidden. On the basis of spectroscopic and molecular orbital considerations, Clyne and Thrush [404] proposed that the overall process could be described by Fig. 73b. CO_2 is first formed in a $^3B_2$ state, with the observed activation energy of the overall process corresponding with the height of the energy barrier over which newly formed CO_2 molecules must pass to reach the stable $^3B_2$ state. The triplet molecule then passes to an upper singlet level by a radiationless transition, and it is this singlet molecule which is either quenched or radiates to the $^1\Sigma_g^+$ ground state of CO_2. From a detailed investigation of the CO flame bands, Dixon [402] concluded that the upper singlet state was the $^1B_2$ molecule (see also Sect. 10.2.1). Studies of the reverse process of dissociation of CO_2 in both the high and low pressure limits of the unimolecular dissociation reaction [437,438] support the broad lines of the reasoning, and suggest independently that the crossing point of the singlet-triplet transition is at an energy of some 115 kcal . mole^{-1}, or a little higher, above the ground state.

An alternative detailed interpretation is that of Lin and Bauer [439], who investigated the reaction between CO and N_2O in a single pulse shock tube at temperatures between 1320 and 2280 K. At the lower end of the temperature range the direct bimolecular reaction between CO and N_2O was important, but above 1600 K the dominant reaction path was the dissociation of N_2O followed by reactions of O atoms. In the analysis of their results, Lin and Bauer used the rate coefficient from Olschewski et al. [324] for the primary N_2O dissociation step, and they obtained an apparent negative activation energy of -23.4 kcal . mole^{-1} for reaction (lvii)

$$O + CO + M'' \rightarrow CO_2 + M'' \qquad \text{(lvii)}$$

They therefore visualized the reaction as in Fig. 73a. Here the positive activation energy of the chemiluminescent association is explained in terms of the height of the crossing point A above the dissociation energy, while the negative activation energy of the overall reaction is indicated by the depth of B below the dissociation limit. However, there remains some considerable doubt about Lin and Bauer's expression for k_{57}, and about the large negative activation energy, as will be discussed below.

Until recently, it had not been established whether the association of O atoms with CO was bi- or termolecular. Although Dixon-Lewis and Linnett [30] and Buckler and Norrish [368] considered their results to be more consistent with a bimolecular association, Baldwin et al. [395] have pointed out that their interpretation was based on too simple a mechanism for data obtained with KCl coated vessels (see Sect. 10.1.3(*b*)(*iii*) and Fig. 68). Shock tube studies definitely indicate that the dissociation is second order at lower pressures, and this implies by microscopic reversibility that the reverse association reaction is termolecular. Kondratiev and Intezarova

[440, 441] have obtained data on the decomposition of O_3 in the presence of CO at atmospheric pressure which showed a negative activation energy for the O + CO combination, and they considered that this was only meaningful in the context of third order kinetics. More directly, Simonaitis and Heicklen [442] find a distinct pressure effect in their study of the competition of CO and 2-trifluoromethylpropene for O atoms produced by the mercury photosensitized decomposition of N_2O. Simonaitis and Heicklen worked at two pressures — the lower at about $\frac{1}{3}$ atm and the higher at 1 atm. In this pressure range they found the O + CO reaction to be intermediate between second and third order, and they were able to obtain from their results values of both the limiting second and third order rate coefficients. Their general findings at pressure just below atmospheric have since been confirmed by de Mòre [450]. These data suggested that the O + CO reaction should have been at least of intermediate order under the conditions of Kondratiev and Intezarova [440, 441], and recomputation of the Russian data on a second order basis gave a rate coefficient in agreement with their own (Simonaitis and Heicklen's) limiting second order value. Nevertheless, a pure second order process is incompatible with the reported [441] negative activation energy, and it was also found [441] that the ratio of the second order rate coefficients of the reactions of O atoms with O_3 and CO varies with the composition of the mixture. From the experimental conditions described in the Russian work, there may have been problems connected with heat transfer, and consequently non-isothermal conditions. It seems most likely that atmospheric pressure lies within the transition region.

The kinetic data for the termolecular association reaction were reviewed by Baulch et al. [443] in 1968. The data, together with more recent determinations, are summarized in Table 54, and are plotted in Arrhenius form in Fig. 74 for Ar, CO, CO_2 and N_2O as third bodies. It is only in these cases that measurements have been made over a temperature range.

It is immediately clear from Fig. 74 that the large negative activation energy reported by Lin and Bauer [439] (see above) is quite inconsistent with the results at lower temperatures. In this connection it is noteworthy that Clark et al. [453] also used the decomposition of N_2O as a source of O atoms, in a shock tube study of the exchange reaction between ^{18}O and $S^{16}O$. They found that the use of the rate coefficient of Olschewski et al. [324] for the primary dissociation step of N_2O, namely $k_{N_2O} = 10^{11.7}$ exp (—29,000/T), led to an extremely improbable, large negative activation energy (—23 kcal . mole^{-1}) for the exchange reaction. On the other hand, if they used $k_{N_2O} = 10^{9.3}$ exp (—20,500/T), in agreement with much shock tube work (e.g. ref. 454), their analysis gave a much lower negative value of —4.5 kcal . mole^{-1}. The rate coefficient k_{N_2O} is important in estimating the O atom concentrations in the systems, and it seems highly probable that Lin and Bauer's result is connected with uncertainties here.

TABLE 54
Measurements of k_{57}

k_{57} ($l^2 . mole^{-2} . sec^{-1}$)	M	Temp. (K)	Method and comment[a]	Ref.
$<5 \times 10^7$	Ar	293	D.F. See text.	404
$2 \times 10^7 < k_{57} < 10^8$	Ar	2800—3600	Shock tube. Recombination followed in expansion wave after shock dissociation of CO_2 in Ar. Measured CO flame band emission.	385
10^8	Ar	3500	Shock tube decomposition of CO_2 measured to give k_{-57}. Equilibrium constant from JANAF Tables [174].	437, 438
$<4 \times 10^7$	O_2 and N_2	500		428
8×10^7	Ar	456	Flow system. O by pyrolysis of O_3 at 1300 K. 0.1—1.2 torr CO in total pressure of 2—4 torr. Relative [O] by total chemiluminescence from O + CO.	444
$2.8 \times 10^5 \exp(+11{,}900/T)$	Ar	1500—3000	Shock tube study of CO + N_2O. See text.	439
$6.8 \times 10^5 \exp(+1{,}490/T)$	CO	430—500	Measured yields of CO_2 in decomposing CO + O_3 at atm pressure. Gives k_{O+O_3}/k_{57}. Took $k_{O+O_3} = 2.9 \times 10^9 \exp(-1{,}850/T)$. Also concluded $k_{57,O_2} = 4k_{57,CO}$. See text.	441
$(2.2 \pm 0.5) \times 10^6$ $(2.5 \pm 1.2) \times 10^6$ $(5.1 \pm 1.5) \times 10^6$	He Ar N_2	300	F.P. 10 % CO in He + 0.1 torr O_2. [O] measured by resonance fluorescence. Relative [O] only needed, but measurement difficult due to slowness of reaction.	445
$<1.7 \times 10^7$	He	300	F.P. of O_2 + kinetic spectroscopy in vacuum UV. Relative [O] only needed.	446
3.6×10^7 6.5×10^7	CO Ar	300	D.F. O by R.F. discharge in O_2/Ar, 0.4—1.8 % O_2. Total pressure 2.8—4.4 torr. Excess CO (0.5—2.5 times [Ar]) added downstream. ESR measurement of [O].	447

TABLE 54—*continued*

k_{57} ($l^2 . mole^{-2} . sec^{-1}$)	M	Temp. (K)	Method and comment[a]	Ref.
1.15 x 10^6 (±25 %) 7.9 x 10^5 (±25 %) 6.1 x 10^5 (±25 %)	CO N_2 He	300	F.P. O atoms by vacuum UV photolysis of 1 torr CO_2, 0.3 torr N_2O or 0.1 torr O_2 in 13—165 torr CO or 19 torr CO with up to 350 torr N_2 or 430 torr He. Monitored by O + CO chemiluminescence. Relative [O] only required.	448
2.9 x 10^8 2.6 x 10^8	CO CO	773 833	Analysis of $H_2/CO/O_2$ second limits with large $[CO]/[H_2]$ from refs. 30 and 368.	395
2.34 x 10^9 exp {(−2,170 ± 275)/T} 8.3 x 10^5 2.2 x 10^6	CO N_2 CO_2	250—370 296 296	As ref. 445, but much greater care to purify CO.	449
6.0 x 10^6 3.5 x 10^7	N_2O N_2O	298 383	O by Hg photo-sensitized decomposition of N_2O. Total pressure between 1/3 and 1 atm. Measurements indicate this is transition region between 2nd and 3rd order (see text). Limiting low pressure k_{57} is quoted.	442
	CO_2 or N_2	293	O by photolysis of CO_2 at 1849 Å. Total pressure varied between 0.74 and 40 atm. Rate measured relative to $O + O_2 + M \rightarrow O_3 + M$ Intermediate between 2nd and 3rd order below atm pressure (cf. ref. 442). If calculated as 3rd order from results at 0.74 atm, gives good agreement with ref. 449.	450
(2.0 ± 0.25) x 10^6 (1.6 ± 0.4) x 10^6 (3.5 ± 0.7) x 10^6	CO_2 CO Ar	296	As ref. 448.	451
8.0 x 10^8 exp {(−1,770 ± 400)/T}	CO_2	257—296	As ref. 451.	452

[a] D.F. discharge flow method. F.P., flash photolysis.

The points in Fig. 74 show considerable scatter, but they all indicate a small activation energy for reaction (lvii). The points for M″ = CO form

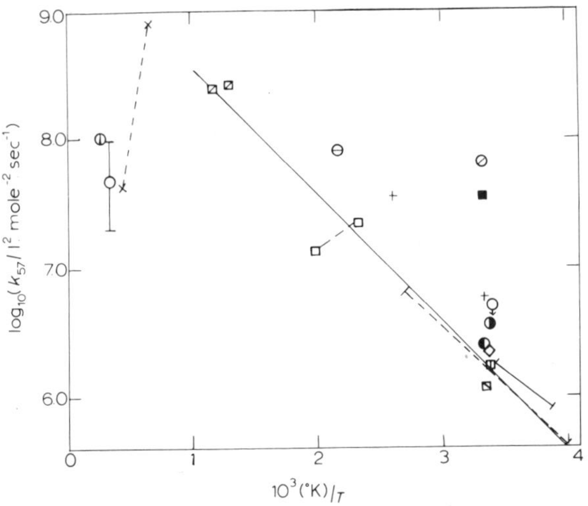

Fig. 74. Arrhenius plot of k_{57}. $M = Ar$: ♀, Clyne and Thrush [404]; Brabbs and Belles [385]; ⦶, Olschewski et al. [437, 438]; ⊖, Mulcahy and Williams [444]; ◐, Slanger and Black [445]; ⊘, Azatjan et al. [447]; ◑, Inn [451]; x- - -x, Lin and Bauer [439]. $M = CO$: □- - - □, Kondratiev and Intezarova [441]; ■, Azatjan et al. [447]; ⧅, Stuhl and Niki [448]; ⧄, Baldwin et al. [395] (see text); ⊢ - ⊣, Slanger et al. [449]; □, Inn [451]. $M = CO_2$: ◇, Slanger et al. [449]; ⊢—⊣, Inn [452]. $M = N_2O$: +, Simonaitis and Heicklen [442].

the most coherent set, particularly if the second limit analysis of Baldwin et al. [395] is expressed on the basis of M ≡ CO. The line drawn gives

$$k_{57,CO} = 4 \times 10^9 \exp(-2{,}300/T) \quad (130)$$

in the temperature range 250—800 K. The activation energy of 4.6 kcal . mole^{-1} may be an upper limit, since the points at 773 and 833 K were calculated [395] for $2CO + O_2$ mixtures on the basis that $k_{57,O_2} \simeq 0.5k_{57,CO}$. Kondratiev and Intezarova [441], on the other hand, find $k_{57,O_2} = 4k_{57,CO}$.

The reaction of $O(^1D)$ with carbon monoxide is also of considerable interest in relation to the foregoing discussion. The most reliable measurements of the rate of disappearance of $O(^1D)$ are probably those of Heidner et al. [456], who generated $O(2^1D_2)$ by the pulsed irradiation of O_3 in the Hartley band continuum (2000—3000 Å), and then measured the absorption of the atomic resonance line at λ = 1152 Å. At total pressures of the order of 20 torr (principally He buffer gas containing up to 1000 p.p.m. CO) they found the pseudo-first order decay rates to increase linearly with [CO], giving a second order decay constant $k_{CO} = (4.4 \pm 0.4) \times 10^{10}$ l . mole^{-1} . sec^{-1}. This value is slightly higher than the upper limit of 3×10^{10} previously found by Noxon [457], and Clark and Noxon [458]. The removal is clearly too rapid at the overall pressures employed for its rate to be that of a three body recombination leading to CO_2. Since chemical reaction leading to $C + O_2$ is highly endothermic [459], the removal process must therefore be a non-adiabatic transition with change of spin to yield $O(^3P)$. The observed reaction is thus essentially a pure quenching of the singlet oxygen atoms, with possible assistance from energetically accessible electronic states of CO_2 which correlate with $O(2^1D_2)$ and $CO(X^1\Sigma^+)$ to give the high rate.

10.3.3 Reaction (lxii) $CO + O_2 = CO_2 + O$

Reaction (lxii) is a chain initiating step at shock tube temperatures. By measuring the initial slopes of the CO_2–time histories in shocked $CO/O_2/Ar$ mixtures (see Sect. 10.1.3(a)), Sulzmann et al. [386] obtained $k_{62} = (3.5 \pm 1.6) \times 10^9 \exp\{(-25{,}500 \pm 3500)/T\}$ in the temperature range 2400–3000 K. This particular result is independent of the rest of the detailed mechanism assumed for the "dry" CO/O_2 reaction, and at the shock tube temperatures it is in close agreement with the expression $k_{62} = 2.5 \times 10^9 \exp(-24{,}000/T)$ which Brokaw [388] chose to fit the overall induction period data of Sulzmann et al. [386, 387]. It will be recalled that Brokaw's detailed mechanism differed from that of Sulzmann et al. in that even in supposedly "dry" mixtures the moist CO/O_2 branching steps were introduced to replace those involving only CO and O_2 (see Sect. 10.1.3(*a*)). A further value of k_{62} in agreement with the above expressions was measured by Dean and Kistiakowsky [230], again from initial rates of CO_2 production in the shock tube environment. They found $k_{62} = 1.8 \times 10^9 \exp(-22{,}800/T)$. The three expressions give values agreeing to within 30 % at 3000 K.

On the other hand, examination of the very early O + CO light emission (immediately after the shock front) in mixtures 3 and 5 of Table 23 allowed Brabbs et al. [92] to deduce $k_{62} = 1.6 \times 10^{10} \exp(-20{,}500/T)$, with a standard deviation in $\ln k_{62}$ of ± 0.54. This is more than an order of magnitude larger than the above estimates. Applying detailed balancing, the high value was also found to be in close agreement (only 30 % lower) with quite independent measurements of k_{-62} by Clark et al. [460] and Garnett et al. [461], and, with some reservations about its accuracy, Brabbs et al. believed the higher value to be an improved estimate. However, a still more recent determination by Rawlins and Gardiner [462], based on OH induction times in $H_2/CO/O_2/Ar$ mixtures measured by Gardiner et al. [463], leads to $k_{62} = 1.2 \times 10^8 \exp(-17{,}500/T)$ between 1500 and 2500 K. At 2500 K this expression gives a rate coefficient some 50 % below those predicted by the earlier, lower estimates. Also in agreement with the lower estimates are the results of Drummond [464], and Sulzmann et al. [465]. Since the observations are made immediately after the passage of the shock fronts, it is possible [92, 463] that varying rates of vibrational relaxation of CO in the different experiments may be responsible for the discordant results. However, Gardiner et al. [463] also suggest that the high result of Brabbs et al. may more likely be due to their measurements being affected by scattered light within the shock tube. Indeed, it seems the more likely that the result of Brabbs et al. is high, since Clark et al. [466] have also since shown that small amounts of hydrogenous organic impurities may have been responsible for the high values of Clark et al. [460] and Garnett et al. [461]. In particular, it was found that the observations of Clark et al. [460] become consistent with the lower estimates of k_{62} if the presence of 400 p.p.m. atomic hydrogen is assumed in the experiment. The lower values of k_{62} are thus to be preferred, probably with a curved Arrhenius plot [462].

10.3.4 Reaction (lxxiii) $H + CO + M''' = HCO + M'''$

Apart from the derivation of a value for the ratio $k_{73,H_2}/k_{4,H_2}$ at 773 K by Baldwin et al. [395] (see Sect. 10.1.3(*b*)), there have been a few direct determinations of k_{73} at or near room temperature. The data are summarized in Table 55. Wang et al. [467] find the reaction to have a small positive activation energy between 298 and 373 K. The results of Baldwin et al. [395] confirm that this trend is continued as the temperature is increased further. Combining the results of Hikida et al. [236] and Baldwin et al. [395] for M= H_2 leads to

$$k_{73,H_2} = 5 \times 10^8 \exp(-755/T) \qquad (131)$$

in the temperature range 300--800 K.

TABLE 55
Measurements of k_{73}

k_{73} ($l^2 . mole^{-2} . sec^{-1}$)	M	Temp. (K)	Method and comment	Ref.
$(4.0 \pm 0.6) \times 10^7$	H_2	298	Pulse radiolysis and measurement of [H] by Lyman-α absorption (see also Section 6.5.1).	236
$(2.6 \pm 0.4) \times 10^7$	Ar	298		
$(2.9 \pm 0.3) \times 10^7$	H_2	298	Hg photo-sensitized production of H atoms and measurement of [H] by Lyman-α absorption. Results may be low (cf. Sect. 6.5.1 and Table 41).	239
$(8.0 \pm 0.4) \times 10^6$	N_2	298		
$(6.9 \pm 0.3) \times 10^6$	Kr	298		
$(6.2 \pm 0.7) \times 10^6$	Ar	298		
$(6.0 \pm 0.7) \times 10^6$	He	298		
$(4.8 \pm 0.5) \times 10^6$	Ne	298		
	H_2	298—373	Method as in ref. 236. Activation energy = 2.0 ± 0.4 kcal . $mole^{-1}$ in temperature range.	467
$(4.0 \pm 2.0) \times 10^7$	Ar	293	H_2/Ar D.F. system with CO added downstream. [H] by ESR.	468
$(8.0 \pm 2.0) \times 10^7$	CO	293		
1.9×10^8	H_2	773	Second limits of $H_2/CO/O_2/N_2$ mixtures with low $[CO]/[H_2]$. Gives k_{73}/k_4. Combined with k_{4,H_2} from eqn. (81).	395

The size of E_{73} is of considerable interest in relation to the shape of the H + CO potential energy surface. Both the $^2A'$ ground state and the $^2A''$ first excited state of HCO correlate with $H(^2S)$ + $CO(^3\Pi)$. The electronic state of HCO which correlates with $H(^2S) + CO(^1\Sigma^+)$ is the $^2A'$ repulsive state, and the formation of HCO from this involves an avoided potential crossing into the $^2A'$ ground state [470]. Now a predissociation level in the $^2A''$ first excited state is observed 35.4 kcal . mole^{-1} above the $^2A'$ ground state [471], and the interesting question then arises [239, 467] whether the level crossing from the repulsive $^2A'$ state to the ground state also occurs near to this predissociation level. The situation is confused, since the heat of dissociation of HCO is still a matter of controversy [472–474], with values of 28 and 17 kcal . mole^{-1} being canvassed. However, even with the higher of these values, E_{73} would need to be around 7 kcal . mole^{-1} for the predissociation level to be approached. The observed activation energy of only 1.5—2 kcal . mole^{-1} thus indicates that the crossing is unlikely to be near the predissociation level. Indeed it may be very considerably less — at 19 or 30 kcal . mole^{-1} depending on the heat of dissociation. The low activation energy thus brings to light considerable complexity in the potential surface, while the low pre-exponential factor in eqn. (131) indicates a small cross-over probability.

10.3.5 *Reaction (lxxv)* $HO_2 + CO = CO_2 + OH$

A number of recent measurements of k_{75} at around room temperature and at 700—905 K are summarized in Table 56. With the exception of the results of Westenberg and de Haas [234], the findings all point to the conclusion that reaction (lxxv) is slow; indeed, the small effect on the $H_2/N_2/O_2$ second limits of replacing N_2 by CO leads directly to the same

TABLE 56
Measurements of k_{75}

k_{75} (l . mole^{-1} . sec^{-1})	Temp. (K)	Method and comments	Ref.
5.4×10^3	713	See Sect. 10.1.3(*b*). Value of $k_{75}/k_{10}^{1/2}$ combined with $k_{10} = 2.0 \times 10^9$ (cf. Table 43 and Sect. 6.6).	211
$<6 \times 10^5$	853—945	Second limits of $2H_2 + O_2 + 0.6$ % C_2H_6 containing 21 % N_2 or 21 % CO found to be identical. If inhibitory effect of C_2H_6 is described by mechanism of Sect. 9.1, then reaction (lxxv) cannot be important. Upper limit quoted for k_{75}.	476

TABLE 56—*continued*

k_{75} (l.mole^{-1}.sec^{-1})	Temp. (K)	Method and comments	Ref.
1.8×10^4	773	See Sect. 10.1.3(*b*). Value of $k_{75}/k_{10}^{1/2}$ combined with $k_{10} = 2.0 \times 10^9$ as above.	70, 395
1×10^9	300	D.F. Excess CO (compared with H) added to $H + O_2 + M$ system (M = Ar, He) in B_2O_3 coated tube. [H] and [OH] measured by ESR. Steady state assumed for [HO_2]. Change in measured ratio ([H] + [OH])/[O_2] along tube depends, inter alia, on k_{75} [CO]/k_8[H]. Gives $k_{75}/k_8 = 0.06$. Combined with $k_8 = 2 \times 10^{10}$ from Table 43, leads to $k_{75}/k_{23} > 1$.	234 (See also Sect. 6.5.1)
	298	Irradiated H_2O_2—CO—O_2 mixtures at λ = 2540 Å. Measured d[CO_2]/d*t*, and found $k_{75}/k_{23} < 6 \times 10^{-3}$.	477
<6	300	Low intensity photolysis of CO^{16}—H_2O^{16}—Ar—$O_2^{18,18}$. Measured rates of production of $CO_2^{16,18}$ and $CO_2^{16,16}$.	478
$<2 \times 10^3$	373—473	HO_2 by photolysis of N_2O at 2139 Å in presence of excess H_2O or H_2. Smaller amounts of CO and O_2 present. Products analyzed for CO_2. Gives $k_{75}/k_{10}^{1/2} < 0.046$. Combined with $k_{10} = 2.0 \times 10^9$ as above.	278
2.5×10^5 5.8×10^5 7.2×10^5	878 927 952	Initial stages of high temperature reaction in 1 % CH_2O + 39 % CO + 40 % N_2 + 20 % O_2 investigated in flow system at atmos. pressure. Flow tube coated with B_2O_3. d[CO_2]/dt measured. [HO_2] determined by freezing out on cold finger, and ESR. Theoretical analysis based on complex mechanism shows CO_2 formed chiefly by reaction (lxxv) in initial stages.	479, 480

D.F., discharge-flow method.

conclusion. The results of Westenberg and de Haas [234] must therefore be regarded as erroneous. Combination of the results of Baldwin et al. [70, 211, 395] and Vardanjan et al. [479, 480] at the higher temperatures leads to

$$k_{75} = (8 \pm 4) \times 10^{10} \exp\{-(11{,}500 \pm 1500)/T\} \quad (132)$$

in the temperature range 700—950 K. At 300 K eqn. (132) gives $k_{75} = 1.8 \times 10^{-6}$ l. mole^{-1}. sec^{-1}.

10.3.6 *Reaction (xxiiiD)* $OD + CO = CO_2 + D$

A single measurement of k_{23D} has been made by Westenberg and Wilson [251] at room temperature, using a discharge-flow system (cf. ref. 202, Tables 35 and 53) with OD prepared from $D + NO_2$ and measured by ESR. The experiment led to $k_{23D} = (3.3 \pm 0.1) \times 10^7$ l. mole^{-1}. sec^{-1} at 300 K. Comparison with k_{23} gives $k_{OH + CO}/k_{OD + CO} \simeq 2.7$ at this temperature.

10.4 FURTHER OXIDATION REACTIONS OF CARBON MONOXIDE IN HOMOGENEOUS SYSTEMS

In this sub-section it is proposed to deal first with effects of oxides of nitrogen in the oxidation of carbon monoxide, in a manner similar to that adopted for the hydrogen—oxygen system. Following this the reactions with fluorine, fluorine monoxide and sulphur dioxide will be considered.

10.4.1 *Nitrogen oxides and carbon monoxide oxidation*

(*a*) *Sensitization of the carbon monoxide—oxygen system and the reaction between carbon monoxide and nitrogen dioxide.* Trace additions (<0.1 %) of NO_2 produce sensitized low pressure explosions of CO/O_2 mixtures containing water vapour or hydrogen at much reduced temperatures compared with unsensitized mixtures [11]. Ammonia is also able to promote ignition outside the normal explosion region [481]. With NO_2 as sensitizer, the ignition with hydrogen present enters the temperature region of the NO_2-sensitized H_2/O_2 system. Ignition with water vapour present occurs in a temperature range between this and the unsensitized ignitions. The phenomenon is exactly similar to the sensitized H_2/O_2 ignitions (cf. Sect. 8.3), and can be explained similarly if reaction (xxiii), and possibly reaction (lxxvi)

$$CO + NO_2 = CO_2 + NO \quad \text{(lxxvi)}$$

are added to the sensitized H_2/O_2 mechanism.

TABLE 57
Measurements of k_{76}

k_{76} (l.mole^{-1}.sec^{-1})	Temp. (K)	Method and comment	Ref.
0.276 ± 0.017	658	See text	484
2.38 ± 0.05	718		
8.80 ± 0.47	763		
14.5 ± 0.3	783		
21.6 ± 0.7	800		
$(3.12 \pm 0.04) \times 10^{-4}$	498	See text	485
$(6.75 \pm 0.75) \times 10^{-4}$	510		
$(1.20 \pm 0.03) \times 10^{-3}$	522		
$(3.08 \pm 0.01) \times 10^{-3}$	536		
$(1.35 \pm 0.12) \times 10^{-2}$	563		
1.9×10^{-3}	540	Vycor reaction vessel. Technique similar to ref. 485, with 20-fold excess CO in reactants to avoid errors due to decomp. of NO_2. Total pressure between 1 and 20 torr. Reaction stopped at <1 % conversion of CO. CO_2 removed from other products by repeated distillation from -135 °C to liquid N_2 temp. Also studied C isotope effect. Found $k_{(12)}/k_{(13)} = 1.022$, 1.019 and 1.016 at 540, 638 and 727 K.	486
1.6×10^{-3}	541		
0.11	638		
0.12	638		
2.7	727		
3.2	727		
$10^{8.8} \exp\{-(13{,}850 \pm 200)/T\}$	393—473	Cylindrical Pyrex reaction vessel, 45 mm diam., 278 ml vol. NO_2 in equilibrium with dissociation products. Zero rate of pressure change initially; hence initial rate due to reaction (lxxvi) alone. $d[NO_2]/dt$ measured photometrically. Found $-d[NO_2]/dt \propto [CO]^{1.08 \pm 0.02} \times [NO_2]^{0.96 \pm 0.02}$ Rate coefficient in good agreement with ref. 485.	487
$10^{9.1} \exp(-13{,}800/T)$	1050—1500	Single pulse shock tube. Mixtures in range 0.74—4.55 % NO_2 + 1.25—5.0 % CO + Ar. Product analysis by gas solid chromatography. Found $d[CO_2]/dt \propto [NO_2]^{0.77}[CO]^{1.13} \times [Ar]^{0.1}$. No correction for boundary layer effects.	488

There are no sensitized ignitions if the CO/O_2 mixtures are sufficiently dry; indeed chloropicrin, which acts as a sensitizer in the H_2/O_2 system, then acts as an inhibitor, and completely eliminates the low pressure explosion region if present in any appreciable quantity [510]. On the other hand, NO_2 does catalyze the slow oxidation of dry, as well as moist mixtures [4, 382, 482—484]. The catalysis for the moist mixtures clearly involves reactions like (xxiii) and (xxxii). For the dry mixtures, the catalyzed reaction is surface dependent at pressures of NO_2 below 10 torr, and homogeneous for pressures above this [482—484]. The velocity of the heterogeneous reaction is markedly reduced by KCl coating of the reaction vessel. By assuming that at 658—800 K the NO_2, NO and O_2 in the system remained in equilibrium if the reacting mixtures contained a large excess of oxygen, Calhoun and Crist [484] were able to derive apparent second order rate coefficients for reaction (lxxvi) at five temperatures. Mean values are given in Table 57.

A more direct study of the reaction between CO and NO_2 was made by Brown and Crist [485], who used a KCl coated Pyrex reaction vessel fitted with a greaseless valve to avoid decomposition of the NO_2. In order also to avoid complications due to gas phase dissociation of the NO_2, its pressure was kept very low (<0.5 torr), and the reaction times were kept comparatively short. Amounts of reaction were measured by freezing and then analyzing for the product CO_2 by vacuum sublimation from the nitrogen oxides. In order to obtain measurable amounts of reaction under the conditions stated, it was necessary to employ high concentrations of CO. Even then the partial pressures of CO_2 in the products were less than 30 microns, and often as little as 5 microns, so that good experimental technique was required. It was confirmed that the reaction was second order over some two- to three-fold variation of the partial pressures of CO and NO_2. Mean rate coefficients between 500 and 563 K are given in Table 57.

Table 57 also summarizes the results of three more recent measurements of k_{76}. All five sets of results are in reasonable agreement in the temperature range 448—1500 K, and lead to a mean Arrhenius relation

$$k_{76} = 3.2 \times 10^9 \exp(-14{,}800/T) \tag{133}$$

(*b*) *Reaction of carbon monoxide with nitrous oxide.* The thermal reaction between carbon monoxide and nitrous oxide was studied by Bawn [489]. In quartz vessels at temperatures in the neighbourhood of 820 K and at pressures below about 200 torr the reaction was heterogeneous in character, with its rate being directly proportional to $[N_2O]$ and inversely proportional to [CO]. The reaction was not influenced by addition of inert gases, and nitric oxide had only a slight retarding action. Added carbon dioxide markedly accelerated the initial rate, and carbon dioxide formed during the reaction also exerted a catalytic effect. There is no volume change during the heterogeneous reaction, and the overall change is represented by

$$CO + N_2O = CO_2 + N_2$$

At higher temperatures and pressures the reaction passes into explosion [489], during which other oxides of nitrogen are also formed and the unimolecular decomposition of the N_2O contributes at the higher temperatures involved. Small amounts of added NO inhibit the explosion.

During a study of the oxidation of carbon at lower temperatures, Strickland-Constable [490], and Madley and Strickland-Constable [491] found in subsidiary experiments that CO and N_2O did not react together appreciably in Pyrex or silica vessels at around 620 K. Oxidation of CO did, however, occur in a vessel in which N_2O was already reacting with charcoal at 593 K. The latter reaction (without added CO) gave CO_2 as the principal gaseous product, with very little CO formed. Thus in the presence of added CO two reactions occur simultaneously

$$C + 2N_2O = CO_2 + 2N_2 \qquad \text{(lxxvii)}$$

$$CO + N_2O = CO_2 + N_2 \qquad \text{(lxxviii)}$$

At rather lower temperatures (just below 573 K) reaction (lxxvii) becomes quite slow, and it is possible to study reaction (lxxviii) independently under these conditions. The heterogeneous reaction was found to be first order in [CO], zero order in $[N_2O]$, and to have an apparent activation energy of 12 kcal . mole^{-1}.

The homogeneous reaction between CO and N_2O in a single pulse shock tube in reaction mixtures heavily diluted with argon, and in the temperature range 1320—2280 K, was studied by Lin and Bauer [439] (cf. Sect. 10.3.2). Amounts of conversion for various computed shock conditions and residence times were obtained by analysis of the concentrations of N_2O and CO_2 in samples abstracted from near the reflecting end of the shock tube immediately after the reaction. At the higher end of the temperature range the reaction proceeded chiefly by the unimolecular dissociation of N_2O (reaction (xlii)), followed by reactions of O atoms. However, Lin and Bauer found that the measured conversions below about 1600 K were too large to be accounted for in this way, and they concluded that at the lower temperatures the bimolecular reaction (lxxviii) is dominant. They deduced $k_{78} = 1.1 \times 10^8 \exp(-11{,}500/T)$ in the temperature range 1317—1908 K, though no corrections were applied for boundary layer effects in the shock tube (cf. Sect. 5.1). A similar shock tube study with krypton as diluent, and with corrections for boundary layer effects using the formula of Emrich and Wheeler [492]. has been made more recently by Milks and Matula [493]. Their analysis again led to the conclusion that reaction (lxxviii) is dominant below about 1600 K, and they obtained $k_{78} = (3 \pm 2) \times 10^8 \exp\{-(8{,}650 \pm$

1,150)/T} between 1169 and 1655 K. Their values in this temperature range are about 10 to 15 times higher than those of Lin and Bauer [439]. As in the case of the CO/O_2 system, catalytic effects of traces of hydrogenous impurities may account for some of the discrepancy. The effect of added hydrogen on the "induction times before explosion" in shocked $CO/N_2O/Ar$ mixtures at temperatures around 1300—1700 K has been measured by Drummond [464]. The induction periods preceding the rapid consumption of N_2O were determined from oscilloscope traces of its absorption at 2500 or 2590 Å. They amounted to some 150—200 μsec in "pure" $CO/N_2O/Ar$ mixtures even at 1350—1400 K where reaction (lxxviii) is supposed to be dominant. Similar induction periods have also been observed by Soloukhin [494, 495], and Zaslonko et al. [496]. Drummond [464] found the activation energy governing the induction lag to be 31.5 ± 1.2 kcal . mole^{-1}.

A major difficulty in the way of the interpretation of the lower temperature shock tube results in terms of the direct O atom transfer reaction (lxxviii) is that no appreciable induction time would be expected if this reaction were dominant. In a more detailed study of the system, Zaslonko et al. [496] measured N_2O concentrations from absorption in the region 2400 Å, and relative O atom concentrations from the O + CO emission at around 4350 Å. During the induction period at about 1500 K it was observed that the O atom concentration increased exponentially, and that the overall rate coefficient $k_{obs} = -[N_2O]^{-1}d[N_2O]/dt$ increased markedly during the course of the reaction. At 20 % conversion of the N_2O, the k_{obs} corresponded with an effective rate of the direct exchange reaction (lxxviii) some ten times higher than would be predicted by the Lin and Bauer expression. The activation energy of the effective exchange rate coefficient decreased from about 30 kcal . mole^{-1} below 1500 K to practically zero at temperatures above 1800 K.

A further observation was that the optical absorption by the N_2O increased somewhat during the induction period. At the same time additional measurements of the IR emission at 4.75 μm showed a growth in the radiation from the oscillators CO, $N_2O(\nu_3)$ and $CO_2(\nu_3)$ before appreciable decomposition of the N_2O had occurred. Both these effects can be associated with vibrational disequilibrium in the system (the absorption coefficient of the N_2O being directly related to its vibrational temperature), and Zaslonko et al. use this evidence to support the view that the overall $CO + N_2O$ reaction occurs even at lower temperatures via a chain process initiated by the unimolecular dissociation of N_2O. The overall $CO + N_2O$ reaction is strongly exothermic ($\Delta H \simeq -87$ kcal . mole^{-1}), as also is the recombination reaction (lvii), and the reaction product CO_2 has a similar vibration frequency to the reactants, particularly N_2O. They therefore visualize rapid vibrational energy transfer from the product CO_2 to fresh N_2O, thus enhancing its rate of dissociation above the thermal value and producing the enhanced overall rate. They estimate that approximately 50 % of the heat of the overall reaction is converted back to vibrational energy of the reagents.

Finally, it is interesting to note that Drummond [464] suggested a rather similar vibrational energy transfer process in order to explain why the apparent activation energy associated with the $CO + N_2O$ induction periods was some 35 kcal . mole^{-1} below the dissociation energy of N_2O.

(*c*) *Reaction of carbon monoxide with nitric oxide.* The direct formation of CO_2 by collisions between CO and NO is not favoured by either energy or spin conservation considerations, and Drummond [464] found no evidence for an overall CO + NO reaction in shocked 5 CO/5 NO/90 Ar mixtures at temperatures below 2916 K, either from pressure records or from optical absorption measurements at 2590 and 4457 Å. Shock tube studies of the reaction in several CO/NO/Ar mixtures at higher temperatures between 3200 and 4500 K have been carried out by Sulzmann et al [465], who measured the time variation of the radiant emission intensities at 4.25 μm (CO_2 emission) and 2312 Å (NO emission in early stages and O_2 emission in later stages) behind reflected shock waves. The initial slopes of the CO_2 intensity—time signals were zero, confirming the absence of the direct molecular rearrangement. Sulzmann et al. [465] interpret their results in terms of the mechanism

$$NO + NO \rightleftharpoons N_2O + O \quad (-xli)$$

$$O + NO \rightleftharpoons N + O_2 \quad (lxxix)$$

$$N + NO \rightleftharpoons N_2 + O \quad (xlix)$$

$$N_2O + M \rightleftharpoons N_2 + O + M \quad (xlii)$$

$$O + CO + M'' \rightleftharpoons CO_2 + M'' \quad (lvii)$$

$$CO + O_2 \rightleftharpoons CO_2 + O \quad (lxii)$$

$$CO + N_2O \rightleftharpoons CO_2 + N_2 \quad (lxxviii)$$

$$O + O + M' \rightleftharpoons O_2 + M' \quad (lxxx)$$

$$N + N + M' \rightleftharpoons N_2 + M' \quad (lxxxi)$$

in which presumably the first six reactions are the most important.

10.4.2 Reactions of carbon monoxide with fluorine, fluorine + oxygen, and fluorine monoxide

The thermal reaction in mixtures of carbon monoxide, fluorine and oxygen was first investigated by Arvia et al. [497], and Heras et al. [498] in the temperature range 288—318 K. In the presence of large amounts of oxygen it led almost quantitatively to a peroxide, $(FCO)_2O_2$; while with smaller amounts of oxygen present, CO_2 and COF_2 were formed as well.

In the later case the reaction can also become explosive. The reaction was homogeneous. Wechsberg and Cady [499], and more recently Kapralova et al. [500], have also shown the reaction between CO and F_2 to be homogeneous in glass, copper and aluminium vessels at temperatures up to 430 K. This reaction leads to COF_2 as product.

Kapralova et al. [500] found the reaction between CO and F_2 to be strongly inhibited by oxygen. Using "fluorine" containing 1.5 % O_2 as their major oxidant, they found that addition of a further 0.28 torr O_2 to a reacting mixture containing 8.2 torr CO + 28 torr "fluorine" caused approximately a five-fold diminution in rate of pressure change, but that further addition of oxygen had very much less effect on the "limiting" rate so obtained. For a constant ratio $[F_2]/[O_2]$, the reaction was first order with respect to both $[F_2]$ and [CO], in agreement with the results of Heras et al. [498] at large $[O_2]$. Kapralova et al. were able to explain their results by means of the series of reactions (lxxxii)—(lxxxvi), which involves the addition of reaction (lxxxiv) to the scheme proposed by Heras et al. viz.

$$CO + F_2 = COF + F \quad \text{(lxxxii)}$$

$$F + CO + M = COF + M \quad \text{(lxxxiii)}$$

$$COF + F_2 = COF_2 + F \quad \text{(lxxxiv)}$$

$$COF + O_2 = COF.O_2 \quad \text{(lxxxv)}$$

$$2COF.O_2 = (COF)_2O_2 + O_2 \quad \text{(lxxxvi)}$$

The scheme leads to the following rate expression for the pressure change

$$-\frac{dp}{dt} = k_{82}[CO]\,[F_2]\left\{3 + \frac{2k_{84}[F_2]}{k_{85}[O_2]}\right\} \quad (134)$$

and clearly with increasing $[O_2]$ the rate falls to the limiting value $3k_{82}[CO][F_2]$. From their results, Kapralova et al. [500] obtained $k_{82} = 0.19 \pm 0.019$ l . mole^{-1} . sec^{-1} and $k_{84}/k_{85} = 0.12 \pm 0.02$ at 294 K, with an activation energy for the overall reaction equal to 14.9 ± 0.8 kcal . mole^{-1}. From their experiments with large $[O_2]$, Heras et al. [498] find $k_{82} = 4.7 \times 10^8 \exp(-6{,}750/T)$, giving a value of 0.13 l mole^{-1} . sec^{-1} at 294 K.

If instead of a mixture of fluorine and oxygen, the reaction is carried out between CO and F_2O [501], the temperature range over which it proceeds at a conveniently measurable rate is between 420 and 470 K. Again the reaction is largely homogeneous, and is not affected by the total pressure. At temperatures below 450 K the products are CO_2 and COF_2 in equal proportions, but at higher temperatures the COF_2 reacts further to produce $(CF_3O)_2$. Oxygen addition strongly accelerates the reaction, the oxygen at the same time being itself consumed and altering the course of the process. In the absence of oxygen the overall reaction is first order in both CO and F_2O. It is a chain reaction, and the mechanism suggested by Arvia et al. [501] is

$$CO + F_2O = COF + FO \quad (lxxxvii)$$

$$COF + F_2O = COF_2 + FO \quad (lxxxviii)$$

$$CO + FO = COF + O \quad (lxxxix)$$

$$O + CO + M'' = CO_2 + M'' \quad (lvii)$$

$$COF + FO = COF_2 + O \quad (xc)$$

This leads to $d[CO_2]/dt = k_{obs}[CO][F_2O]$, where $k_{obs} = (k_{87}k_{88}k_{89}/k_{90})^{1/2} = 1.45 \times 10^{12} \exp(-12{,}700/T)$. It will be noted that reactions (lxxxix) and (lvii) may be replaced by reactions (xci) and (lxxxiii)

$$CO + FO = CO_2 + F \quad (xci)$$

$$F + CO + M = COF + M \quad (lxxxiii)$$

without altering the form of the expression for the reaction rate. However, Arvia et al. reject the latter path on the basis of a previous observation by Aymonino [502] that CF_3OF is formed in considerable quantity by prolonged irradiation of mixtures of $COF_2 + F_2$ at 308 K from a mercury-in-quartz lamp. From this it was concluded that F atoms will readily attack COF_2, and therefore cannot be present in the $CO + F_2O$ system. On the other hand, Appelman and Clyne [503] find COF_2 to be the final product from the slow, third order reaction between F atoms and CO in a discharge-flow system at room temperature.

Henrici et al. [504] carried out a shock tube study of the $CO + F_2O$ reaction in mixtures heavily diluted with argon at higher temperatures. They obtained data on overall CO_2, O_2 and COF_2 production from single pulse experiments, and they also made time-resolved optical measurements of the rate of formation of CO_2 and depletion of F_2O by studying the emission at 4.3 μm and the absorption at 2200 Å, respectively. The major path for the decomposition of F_2O was assumed to be by reactions (xcii)—(xciv)

$$F_2O + M \rightleftharpoons F + FO + M \quad (xcii)$$

$$FO + FO = O_2 + 2F \quad (xciii)$$

$$F + F + M \rightleftharpoons F_2 + M \quad (xciv)$$

rate coefficients of which had been determined in a prior investigation [505]. It was assumed that the CO_2 was formed by reaction (xci). However, the formation of COF_2 means that a rather complex mechanism similar to that of Arvia et al. [501] must be added to this scheme, and the interpretation becomes complex. Nevertheless, on the basis of a multi-parameter fit of their whole range of results, Henrici et al. [504] suggest $k_{91} \simeq$, 7.5×10^{10} l . mole^{-1} . sec^{-1} between 800 and 1400 K, believed to be correct to within a factor of two.

10.4.3 Reaction of carbon monoxide with sulphur dioxide

The only study of the thermal reaction between CO and SO_2 has been performed in a shock tube by Bauer et al. [506] using the single pulse technique with a mixture of CO and SO_2 heavily diluted with argon, over the temperature range 1770—2453 K. Reflected shock residence times were 370—540 μsec. No S_2O, SO_3 or O_2 were found in the reaction products, and only traces of COS and CS_2 were generated. The principal product was CO_2, of which two moles were formed per mole of SO_2 which disappeared. Although not precisely established, the stoichiometry is close to

$$2CO + SO_2 = 2CO_2 + S$$

The rate of production of CO_2 was found to be described by

$$d[CO_2]/dt = 2.7 \times 10^9 \exp\{-(24{,}150 \pm 600)/T\}[Ar]^{0.57}[CO]^{0.76}[SO_2]^{0.67}.$$

To obtain the observed partial orders for each reactant, the reactions postulated were

$$SO_2 + Ar \rightarrow SO_2^* + Ar \quad \text{(xcv)}$$

$$CO + Ar \rightarrow CO^* + Ar \quad \text{(xcvi)}$$

$$SO_2 + CO \rightarrow CO_2 + SO \quad \text{(xcvii)}$$

$$SO_2^* + CO \rightarrow CO_2 + SO \quad \text{(xcviii)}$$

$$SO_2 + CO^* \rightarrow CO_2 + SO \quad \text{(xcix)}$$

$$SO + SO_2 \rightarrow SO_3 + S \quad \text{(c)}$$

$$SO_3 + CO \rightarrow CO_2 + SO_2 \quad \text{(ci)}$$

with steady state conditions imposed on SO_3, SO, SO_2^* and CO*. However, the data are not sufficient to exclude other possibilities.

10.4.4 Effect of metal carbonyls

The gas phase homogeneous catalysis of the CO + O_2 reaction in shock waves by addition of chromium, iron and nickel carbonyls has been described by Izod et al. [507], and Matsuda [508, 509]. It will not be discussed further here.

10.5 THE GLOW REACTION IN THE CARBON MONOXIDE—OXYGEN SYSTEM AND ITS RELATION TO THE EXPLOSION REGION: OSCILLATORY BEHAVIOUR

Qualitative mention has already been made in the introduction to Sect. 10 of a fairly extensive region of temperature and pressure surrounding the explosion peninsula of the CO/O_2 system, in which a slower glow reaction occurs. Hoare and Walsh [357] observed the reaction visually, and reported definite pressure–temperature limits for the glow. Their upper and lower glow limits, together with the observed upper explosion limit, for "dry" $2CO + O_2$ mixtures are shown in Fig. 75. Hoare and Walsh also include a qualitative lower explosion limit line in the diagram. However, later work by Linnett et al. [511] showed that this limit does not exist as a sharp limit, and further that even the visual detection of the lower glow limit was a subjective exercise. Replacing visual observation by the use of a very sensitive photomultiplier and recording system to obtain light intensity—time graphs, they found no sign of a lower glow limit with silica or alumina vessels. The glow simply became weaker and weaker at lower pressures until it was swamped by the furnace glow.

With an alumina coated vessel (but not with silica), Linnett et al. [511] also observed oscillatory behaviour inside the glow region. Instead of a single glow corresponding with reaction, they observed a series of successive glows, and these were particularly numerous in CO-rich mixtures. Linnett et al. worked with moist gases. However, similar oscillatory

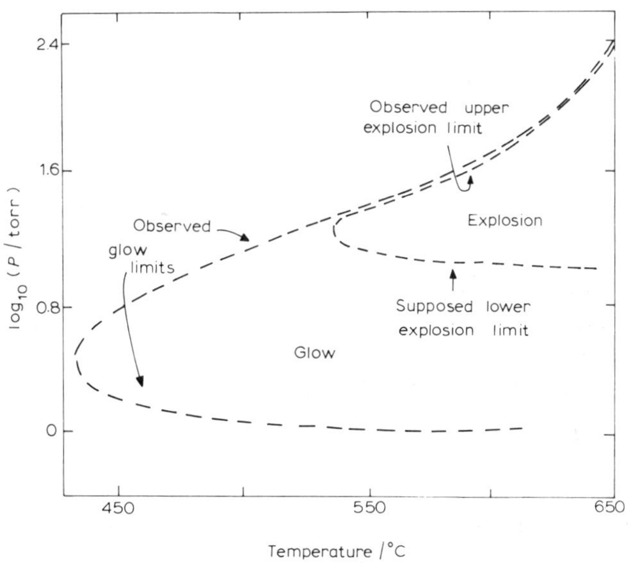

Fig. 75. Glow and explosion limit curves for "dry" $2CO + O_2$ mixtures (after Hoare and Walsh [357]). Tip and lower part of explosion limit curve are qualitative only. (By courtesy of The Faraday Society.)

behaviour had previously been observed by Ashmore and Norrish [510] when mixtures were introduced into vessels which had previously been used for experiments with chloropicrin, and by Dickens et al. [358] in "dry" mixtures. More recently, McCaffrey and Berlad are reported [512] to have observed up to two hundred oscillations in some circumstances. Dove [513] also found that oscillation was favoured by a high $[CO]/[O_2]$ ratio, with no oscillations occurring in the reverse situation. All workers are unanimous that the oscillations are isothermal, at least in "dry" mixtures.

A further observation of Linnett et al. [511], which has also since become of interest in connection with kinetic modelling of the system, is that afterglows following explosions occurred in the alumina coated vessel, lasting sometimes as long as twenty seconds.

Theoretical interpretations of the oscillatory behaviour have been offered by Gray [514], Yang [515, 516], and Yang and Berlad [517], and the overall topic of oscillatory reactions has recently been reviewed by Gray [512]. According to these interpretations the CO oxidation proceeds by an isothermal branched chain (autocatalytic) mechanism as already outlined in Sect. 10.1.3 (a), but there is an additional quadratic termination step. Simplifying to a binary (two radical) model as has been done by Gray [514] and Yang [515], then if X and Y are the two intermediates undergoing the quadratic termination, the skeleton oxidation mechanism becomes

$$\mathrm{X} \xrightarrow{k_b} 2\mathrm{X}$$

$$\mathrm{X} \xrightarrow{k_{t1}} \text{inert}$$

$$\mathrm{X} \xrightarrow{k_p} \mathrm{Y}$$

$$\mathrm{X} + \mathrm{Y} \xrightarrow{k_q} \text{inert}$$

$$\mathrm{Y} \xrightarrow{k_{t2}} \text{inert}$$

where k_b is the branching rate coefficient, k_{t1} and k_{t2} are the rate coefficients for linear termination, k_q is that for quadratic termination, and k_p is a propagation rate coefficient. Putting $\phi = k_b - k_{t1} - k_p$, the differential equations describing the system become

$$\mathrm{d}x/\mathrm{d}t = \phi x - k_q xy = P(x, y) \tag{135}$$

$$\mathrm{d}y/\mathrm{d}t = k_p x - k_{t2} y - k_q xy = Q(x, y) \tag{136}$$

where x and y denote concentrations. These lead to two singularities which are related to possible kinetic states [518, 519]

$$\left.\begin{aligned} x_1 &= 0 \\ y_1 &= 0 \end{aligned}\right\} S_1 \quad \text{and} \quad \left.\begin{aligned} x_2 &= k_{t2}\phi/k_q(k_p - \phi) \\ y_2 &= \phi/k_q \end{aligned}\right\} S_2 \tag{137}$$

and the nature of these singularities depends on the partial derivatives $\partial P/\partial x$, $\partial P/\partial y$, $\partial Q/\partial x$ and $\partial Q/\partial y$ evaluated at the singularity [515, 520]. For $\phi < 0$, S_2 represents no real physical state, since eqn. (137) shows that both x_2 and y_2 become negative. The singularity S_1 becomes a stable nodal point under this condition, and represents a state of no reaction (or normal slow reaction if a chain initiation step is included in the scheme). For $\phi > 0$, S_1 is a saddle point, which represents no stable physical state; while S_2 is stable if $\phi < k_p$. S_2 is a stable focus if $4\phi > k_{t2}\{1 + \phi/(k_p - \phi)\}^2$, and a stable node if the reverse is the case. At a stable focus the trajectories of x_2 and y_2 approach the steady state with damped oscillations. The node corresponds with a sustained reaction.

In carbon monoxide oxidation, the species X has been identified as the O atom [514, 515], and the observed glows, due to O + CO emission, are an indicator of its concentration history. The scheme proposed by Yang [515] is based on the Brokaw mechanism (see Sect. 10.1.3(*a*)), and consists of the reactions

$O + H_2O$	$\rightarrow OH + OH$	(—xvi)
$H + O_2$	$\rightarrow OH + O$	(ii)
$O + CO + M''$	$\rightarrow CO_2^*(CO_2) + M''$	(lvii)
$CO_2^* + O$	$\rightarrow CO + O_2$	(cii)
$CO_2^* + M''$	$\rightarrow CO_2 + M''$	(ciii)
$OH + CO$	$\rightarrow CO_2 + H$	(xxiii)
H	$\rightarrow$ destruction at surface	(civ)
O	$\rightarrow$ destruction at surface	(cv)
OH	$\rightarrow$ destruction at surface	(cvi)
$H + O_2 + M$	$\rightarrow HO_2 + M$	(iv)
HO_2	$\rightarrow$ destruction at surface	(v)

Here the species Y is taken to be CO_2^* and the new quadratic termination step is reaction (cii). The system is reduced to a binary one in [O] and [CO_2^*] by introducing the steady state relations for [H] and [OH] only. Analysis along the lines indicated above then predicts the three types of behaviour: (i) no reaction (or very slow reaction controlled by initiation) when $\phi < 0$; (ii) damped oscillation or a sustained glow when $\phi > 0$ but less than some critical value; and (iii) explosive behaviour when ϕ is greater than the critical value. The distinctive difference from the hydrogen oxidation system, where there is a sharp transition from slow reaction to explosion at $\phi = 0$, is that now there is a more gradual transition within the region $0 < \phi < k_p$. This is in accord with the experimental observations [511].

There is still one shortcoming in the proposed mechanism in the above form. This is that it only ever predicts damped oscillatory behaviour, whereas the large number of oscillations which have often been observed clearly indicates that they may be of a sustained character under certain conditions. Yang [515] overcame this difficulty by supposing a Langmuir-type absorption of the O atoms for reaction (cv), such that the surface becomes saturated when their concentration is high. This in turn produces a saturation effect on the termination step itself, and is equivalent to a pumping action which amplifies the carrier concentration during the rising part of its cycle. With suitable values of the associated absorption and rate parameters the pumping action is able to produce a sustained oscillation. It should be added that Yang and Berlad [517], by full numerical integration of a multi-radical model of the system, with no steady state assumptions, were able to demonstrate another possible pumping mechanism by way of the quadratic reaction (x) and the linear reaction (vii) when the following reactions of HO_2 and H_2O_2 were incorporated in the scheme

$$H_2O_2 + M' \rightarrow OH + OH + M' \qquad \text{(vii)}$$

$$HO_2 + HO_2 \rightarrow H_2O_2 + O_2 \qquad \text{(x)}$$

$$H_2O_2 \rightarrow \text{destruction at surface} \qquad \text{(cvii)}$$

$$H + H_2O_2 \rightarrow H_2O + OH \qquad \text{(xiv)}$$

The precise pumping mechanism is not yet certain.

Finally, in three series of numerical calculations, Yang [515] has shown how the CO/O_2 system may go through the whole range of kinetic states (inactivity, sustained oscillation, afterglow, sustained glow and explosion) as the reactivity ϕ increases; while Yang and Berlad [517] and Yang [516], with the additional assumption of a lower threshold intensity below which neither the human eye nor optical instruments can detect the emission, have demonstrated the passage through the range of observed phenomena as the temperature, composition and pressure of the CO/O_2 mixture is changed. Although the isothermal assumptions used in the calculations expose them to some criticism when dealing with moist mixtures, for which the system is not truly isothermal [521], the results are nevertheless most valuable. For details of these most interesting and informative contributions the reader is referred to the original publications.

REFERENCES

1 C. N. Hinshelwood and A. T. Williamson, The Reaction Between Hydrogen and Oxygen, Oxford University Press, Oxford, 1934.

2 N. Semenov, Chemical Kinetics and Chain Reactions, Oxford University Press, Oxford, 1935.

3 W. Jost, Explosion and Combustion Processes in Gases, McGraw-Hill, New York, 1939—1946.
4 B. Lewis and G. von Elbe, Combustion, Flames and Explosions of Gases, Cambridge University Press, Cambridge, 1938 and Academic Press, New York, 1951 and 1961.
5 G. J. Minkoff and C. F. H. Tipper, Chemistry of Combustion Reactions, Butterworths, London, 1962.
6 L. S. Kassel, Chem. Rev., 21 (1937) 331.
7 C. N. Hinselwood, Proc. R. Soc. London, Ser. A, 188 (1946) 1.
8 G. Williams and K. Singer, Annu. Rep. Prog. Chem., 45 (1948) 67.
9 A. B. Sagulin, Z. Phys., 48 (1928) 571.
10 A. B. Sagulin, Z. Phys. Chem., Abt. B, 1 (1928) 275.
11 D. Kopp, A. Kowalsky, A. B. Sagulin and N. Semenov, Z. Phys. Chem., Abt. B, 6 (1930) 307.
12 H. W. Thompson and C. N. Hinshelwood, Proc. R. Soc. London, Ser. A, 122 (1929) 610.
13 C. N. Hinshelwood and E. A. Moelwyn-Hughes, Proc. R. Soc. London, Ser. A, 138 (1932) 311.
14 A. A. Frost and N. H. Alyea, J. Am. Chem. Soc., 55 (1933) 3227; 56 (1934) 1251.
15 A. Kowalsky, Phys. Z. Sowjetunion, 1 (1932) 595; 4 (1933) 723.
16 N. Semenova, Acta Phys. Chim U.R.S.S., 6 (1937) 25.
17 A. Biron and A. Nalbandjan, Acta Phys. Chim U.R.S.S., 6 (1937) 43.
18 N. Semenov, Z. Phys., 46 (1927) 109.
19 R. N. Pease, J. Am. Chem. Soc., 52 (1930) 5106; 53 (1931) 3188.
20 R. R. Baldwin and P. Doran, Trans. Faraday Soc., 57 (1961) 1578.
21 R. R. Baldwin, P. Doran and L. Mayor, Trans. Faraday Soc., 58 (1962) 2410; 8th Int. Symp. Combust., Williams and Wilkins, Baltimore, 1962, p. 103.
22 R. R. Baldwin and R. M. Precious, Nature (London), 169 (1952) 290.
23 B. Lewis and G. von Elbe, J. Chem. Phys., 9 (1941) 194; 10 (1942) 366
24 A. C. Egerton and D. R. Warren, Proc. R. Soc. London, Ser. A, 204 (1951) 465.
25 D. R. Warren, Proc. R. Soc. London, Ser. A, 211 (1952) 86, 96.
26 H. P. Broida and O. Oldenberg, J. Chem. Phys., 19 (1951) 196.
27 A. H. Willbourn and C. N. Hinshelwood, Proc. R. Soc. London, Ser. A, 185 (1946) 353, 369, 376.
28 G. H. Grant and C. N. Hinshelwood, Proc. R. Soc. London, Ser. A, 141 (1933) 29.
29 A. Nalbandjan, Acta Phys. Chim. U.R.S.S., 19 (1944) 483, 497; 20 (1945) 31; J. Phys. Chem. Russ., 19 (1945) 210.
30 G. Dixon-Lewis and J. W. Linnett, Trans. Faraday Soc., 49 (1953) 756.
31 V. V. Voevodsky and V. L. Talrose, J. Phys. Chem. Russ., 22 (1948) 1192.
32 R. R. Baldwin and C. T. Brooks, Trans. Faraday Soc., 58 (1962) 1782.
33 P. G. Ashmore and B. J. Tyler, Trans. Faraday Soc., 58 (1962) 1108.
34 O. Oldenberg and H. S. Sommers, J. Chem. Phys., 8 (1940) 468; 9 (1941) 114, 432, 573.
35 H. R. Heiple and B. Lewis, J. Chem. Phys., 9 (1941) 584.
36 N. Chirkov, Acta Phys. Chim. U.R.S.S., 6 (1937) 915.
37 M. Prettre, J. Chim. Phys., 33 (1936) 189.
38 C. F. Cullis and C. N. Hinshelwood, Proc. R. Soc. London, Ser. A, 186 (1946) 462, 469.
39 P. G. Ashmore and F. S. Dainton, Nature (London), 158 (1946) 416.
40 C. H. Gibson and C. N. Hinshelwood, Proc. R. Soc. London, Ser. A, 119 (1928) 591.

41 E. A. Moelwyn-Hughes, A. C. Rolfe and C. N. Hinshelwood, Proc. R. Soc. London, Ser. A, 139 (1933) 521.
42 R. B. Holt and O. Oldenberg, J. Chem. Phys., 17 (1949) 1091.
43 W. F. Anzilotti, J. D. Rogers, G. W. Scott and V. J. Tomsic, 5th Int. Symp. Combust., Reinhold, New York, 1955, p. 356; Ind. Eng. Chem., 46 (1954) 1316.
44 J. W. Linnett and C. P. Tootal, 7th Int. Symp. Combust., Butterworths, London, 1959, p. 23.
45 R. R. Baldwin and L. Mayor, Trans. Faraday Soc., 56 (1960) 80, 103; 7th Int. Symp. Combust., Butterworths, London, 1959, p. 8; Rev. Inst. Fr. Pet. Ann. Combust. Liq., 13 (1958) 397.
46 L. S. Kassel and H. H. Storch, J. Am. Chem. Soc., 57 (1935) 672.
47 V. Bursian and V. Sorokin, Z. Phys. Chem., Abt. B, 12 (1931) 247.
48 N. Semenov, Acta Phys.-Chim. U.R.S.S., 18 (1943) 93.
49 R. R. Baldwin, Trans. Faraday Soc., 52 (1956) 1337.
50 B. Lewis and G. von Elbe, J. Am. Chem. Soc., 59 (1957) 970.
51 V. V. Voevodsky, J. Phys. Chem. Russ., 20 (1946) 1285; 7th Int. Symp. Combust., Butterworths, London, 1959, p. 34.
52 B. Lewis and G. von Elbe, 3rd. Int. Symp. Combust., Williams and Wilkins, Baltimore, 1949, p. 484.
53 D. R. Warren, Trans. Faraday Soc., 53 (1957) 199, 206.
54 R. R. Baldwin,Trans. Faraday Soc., 52 (1956) 1344.
55 D. L. Baulch, D. D. Drysdale, D. G. Horne and A. C. Lloyd, Evaluated Kinetic Data for High Temperature Reactions, Vol. 1, Butterworths, London, 1972.
56 J. W. Linnett and D. G. H. Marsden, Proc. R. Soc. London, Ser. A, 234 (1956) 504.
57 S. Weissman and E. A. Mason, J. Chem. Phys., 36 (1962) 704.
58 S. C. Kurzius and M. Boudart, Combust Flame, 12 (1968) 477.
59 N. Semenov, Acta Phys.-Chim. U.R.S.S., 20 (1945) 291.
60 N. Semenov, Some Problems in Chemical Kinetics and Reactivity, Vol. II, Pergamon, Oxford, 1958, Chap. 3.
61 L. V. Karmilova, A. Nalbandjan and N. Semenov, J. Phys. Chem. Russ., 32 (1958) 1193.
62 R. R. Baldwin, L. Mayor and P. Doran, Trans. Faraday Soc., 56 (1960) 93.
63 G. Dixon-Lewis, J. W. Linnett and D. F. Heath, Trans. Faraday Soc., 49 (1953) 766.
64 R. R. Baldwin, B. N. Rossiter and R. W. Walker, Trans. Faraday Soc., 65 (1969) 1044.
65 W. Forst, Can. J. Chem., 36 (1958) 1308.
66 D. E. Hoare, J. B. Prothero and A. D. Walsh, Nature (London), 182 (1958) 654; Trans. Faraday Soc., 55 (1959) 548.
67 R. R. Baldwin and D. Brattan, 8th Int. Symp. Combust., Williams and Wilkins, Baltimore, 1962, p. 110
68 R. R. Baldwin, D. Booth and D. Brattan, Can. J. Chem., 39 (1961) 2130.
69 R. R. Baldwin, D. Brattan, B. Tunnicliffe, R. W. Walker and S. J. Webster, Combust. Flame, 15 (1970) 133.
70 R. R. Baldwin, D. Jackson, R. W. Walker and S. J. Webster, 10th Int. Symp. Combust., Combustion Institute, Pittsburgh, 1965, p. 423.
71 W. Forst and P. A. Giguère, J. Phys. Chem., 62 (1958) 340.
72 R. R. Baldwin, D. Jackson, R. W. Walker and S. J. Webster, Trans. Faraday Soc., 63 (1967) 1665, 1676.
73 R. R. Baldwin, M. E. Fuller, J. S. Hillman, D. Jackson and R. W. Walker, J. Chem. Soc., Faraday Trans. 1, 70 (1974) 635.
74 E. A. Albers, K. Hoyermann, H. Gg. Wagner and J. Wolfrum, 13th Int. Symp. Combust., Combustion Institute, Pittsburgh, 1971, p. 81.

75 J. N. Bradley, Shock Waves in Chemistry and Physics, John Wiley and Sons, New York, 1962.
76 A. G. Gaydon and I. R. Hurle, The Shock Tube in High Temperature Chemical Physics, Reinhold, New York, 1963.
77 E. F. Greene and J. P. Toennies, Chemical Reactions in Shock Waves, Academic Press, New York, 1964.
78 G. L. Schott and R. W. Getzinger, Phys. Chem. Fast React., 1 (1973) 81.
79 H. Mirels, Phys. Fluids, 6 (1963) 1201; 9 (1966) 1907; A.I.A.A.J., 2 (1964) 84.
80 F. E. Belles and T. A. Brabbs, 13th Int. Symp. Combust., Combustion Institute, Pittsburgh, 1971, p. 165.
81 T. Asaba, W. C. Gardiner, Jr. and R. F. Stubbemann, 10th Int. Symp. Combust., Combustion Institute, Pittsburgh, 1965, p. 295.
82 W. C. Gardiner, Jr., K. Morinaga, D. L. Ripley and T. Takeyama, Phys. Fluids, 12 (1969) 1.
83 A. L. Myerson and W. S. Watt, J. Chem. Phys., 49 (1968) 425.
84 W. S. Watt and A. L. Myerson, J. Chem. Phys., 51 (1969) 1638.
85 L. S. Blair and R. W. Getzinger, Combust. Flame, 14 (1970) 5.
86 R. W. Getzinger, L. S. Blair and D. B. Olson, in I. I. Glass (Ed.), Proc. 7th Int. Shock Tube Symp., University of Toronto Press, 1970, p. 605.
87 C. W. von Rosenberg, N. H. Pratt and K. N. C. Bray, J. Quant. Spectrosc., Radiat. Transfer, 10 (1970) 1155.
88 D. Gutman and G. L. Schott, J. Chem. Phys., 46 (1967) 4576.
89 D. Gutman, E. A. Hardwidge, F. A. Dougherty and R. W. Lutz, J. Chem. Phys., 47 (1967) 4400.
90 D. Gutman, R. W. Lutz, N. Jacobs, E. A. Hardwidge and G. L. Schott, J. Chem. Phys., 48 (1968) 5689.
91 G. L. Schott, 12th Int. Symp. Combust., Combustion Institute, Pittsburgh, 1969, p. 569.
92 T. A. Brabbs, F. E. Belles and R. S. Brokaw, 13th Int. Symp. Combust., Combustion Institute, Pittsburgh, 1971, p. 129.
93 D. R. White and K. H. Cary, Phys. Fluids, 6 (1963) 749.
94 D. R. White and G. E. Moore, 10th Int. Symp. Combust., Combustion Institute, Pittsburgh, 1965, p. 785.
95 D. R. White, 11th Int. Symp. Combust., Combustion Institute, Pittsburgh, 1967, p. 147.
96 W. G. Browne, D. R. White and G. R. Smookler, 12th Int. Symp. Combust., Combustion Institute, Pittsburgh, 1969, p. 557.
97 R. A. Strehlow and A. Cohen, Phys. Fluids, 5 (1962) 97.
98 V. V. Voevodsky and R. I. Soloukhin, 10th Int. Symp. Combust., Combustion Institute, Pittsburgh, 1965, p. 279.
99 R. I. Soloukhin, Shock Waves and Detonations in Gases, Mono Book Corp., Baltimore, 1966, Chap. 4.
100 D. R. White and R. C. Millikan, J. Chem. Phys., 39 (1963) 2107.
101 F. E. Belles and M. R. Lauver, 10th Int. Symp. Combust., Combustion Institute, Pittsburgh, 1965, p. 285.
102 G. L. Schott and J. L. Kinsey, J. chem. Phys., 29 (1958) 1177.
103 H. Miyama and T. Takeyama, J. Chem. Phys., 41 (1965) 2287.
104 S. Fujimoto, Bull. Chem. Soc. Jpn., 36 (1963) 1233.
105 G. B. Skinner and G. H. Ringrose, J. Chem. Phys., 42 (1965) 2190.
106 S. G. Saytzev and R. I. Soloukhin, 8th Int. Symp. Combust., Williams and Wilkins, Baltimore, 1962, p. 344.
107 J. W. Meyer and A. K. Oppenheim, 13th Int. Symp. Combust., Combustion Institute, Pittsburgh, 1971, p. 1153; Combust. Flame, 17 (1971) 65.

108 R. S. Brokaw, 10th Int. Symp. Combust., Combustion Institute, Pittsburgh, 1965, p. 269.
109 V. N. Kondratiev, Chemical Kinetics of Gas Reactions, Pergamon Press, London, 1964, Chaps. 9 and 10.
110 C. J. Jachimowski and W. M. Houghton, Combust. Flame, 15 (1970) 125.
111 W. C. Gardiner, Jr., W. G. Mallard, M. McFarland, K. Morinaga, J. H. Owen, W. T. Rawlins, T. Takeyama and B. F. Walker, 14th Int. Symp. Combust., Combustion Institute, Pittsburgh, 1973, p. 61.
112 D. L. Ripley and W. C. Gardiner, Jr., J. Chem. Phys., 44 (1966) 2285.
113 D. L. Ripley, Dissertation, University of Texas, Austin, 1967.
114 W. C. Gardiner, Jr., K. Morinaga, D. L. Ripley and T. Takeyama, J. Chem. Phys., 48 (1968) 1665.
115 C. B. Wakefield, Dissertation, University of Texas, Austin, 1969.
116 C. B. Wakefield, D. L. Ripley and W. C. Gardiner, Jr., J. Chem. Phys. 50 (1969) 325.
117 W. C. Gardiner, Jr. and C. B. Wakefield, Astronaut. Acta, 15 (1970) 399.
118 G. Rudinger, Phys. Fluids, 4 (1961) 1463.
119 C. P. Fenimore, Chemistry in Premixed flames, Pergamon Press, Oxford, 1964.
120 R. M. Fristrom and A. A. Westenberg, Flame Structure, McGraw Hill, New York, 1965.
121 G. Dixon-Lewis and A. Williams, Q. Rev., Chem. Soc., 17 (1963) 243.
122 G. Dixon-Lewis, Proc. R. Soc. London, Ser. A, 307 (1968) 111.
123 G. Dixon-Lewis, Proc. R. Soc. London, Ser. A, 298 (1967) 495; 317 (1970) 235.
124 G. Dixon-Lewis and G. L. Isles, Proc. R. Soc. London, Ser. A, 308 (1969) 517.
125 D. B. Spalding, Philos. Trans. R. Soc. London, Ser. A, 249 (1956) 1.
126 G. K. Adams and G. B. Cook, Combust. Flame, 4 (1960) 9.
127 Y. B. Zeldovich and G. I. Barrenblatt, Combust. Flame 3 (1959) 61.
128 K. A. Wilde, Combust. Flame, 18 (1972) 43.
129 D. B. Spalding and P. L. Stephenson, Proc. R. Soc. London, Ser. A, 324 (1971) 315.
130 P. L. Stephenson and R. G. Taylor, Combust. Flame, 20 (1973) 231.
131 C. G. James and T. M. Sugden, Proc. R. Soc. London, Ser. A, 227 (1954) 312.
132 E. M. Bulewicz and T. M. Sugden, Trans. Faraday Soc., 52 (1956) 1475.
133 E. M. Bulewicz, C. G. James and T. M. Sugden, Proc. R. Soc. London, Ser. A, 235 (1956) 89.
134 P. J. Padley and T. M. Sugden, Proc. R. Soc. London, Ser. A, 248 (1958) 248; 7th Int. Symp. Combust., Butterworths, London, 1959, p. 235.
135 H. Smith and T. M. Sugden, Proc. R. Soc. London, Ser. A, 219 (1953) 204.
136 M. J. McEwan and L. F. Phillips, Combust. Flame, 9 (1965) 420.
137 W. E. Kaskan, Combust. Flame, 2 (1958) 229.
138 G. L. Schott and P. F. Bird, J. Chem. Phys., 41 (1964) 2869.
139 O. Oldenberg and F. F. Rieke, J. Chem. Phys., 6 (1938) 439.
140 R. W. Patch, J. Chem. Phys., 36 (1962) 1919.
141 J. P. Rink, J. Chem. Phys., 36 (1962) 262, 1398.
142 E. A. Sutton, J. Chem. Phys., 36 (1962) 2923.
143 I. R. Hurle, 11th Int. Symp. Combust., Combustion Institute, Pittsburgh, 1967, p. 827.
144 T. A. Jacobs, R. R. Giedt and N. Cohen, J. Chem. Phys., 43 (1965) 3688.
145 T. A. Jacobs, R. R. Giedt and N. Cohen, J. Chem. Phys., 47 (1967) 54.
146 I. R. Hurle, A. Jones and J. L. J. Rosenfeld, Proc. R. Soc. London, Ser. A, 310 (1969) 253.
147 F. S. Larkin and B. A. Thrush, Discuss. Faraday Soc., 37 (1964) 112; 10th Int. Symp. Combust., Combustion Institute, Pittsburgh, 1965, p. 397; F. S. Larkin, Can. J. Chem., 46 (1968) 1005.

148 D. O. Ham, D. W. Trainor and F. Kaufman, J. Chem. Phys., 53 (1970) 4395; 58 (1973) 4599.
149 P. Walkauskas and F. Kaufman, 15th Int. Symp. Combust., Combustion Institute, Pittsburgh, 1975, p. 691.
150 G. L. Schott, J. Chem. Phys., 32 (1960) 710.
151 R. W. Getzinger and L. S. Blair, Combust. Flame, 13 (1969) 271.
152 A. Gay and N. H. Pratt, Proc. 8th Shock Tube Symp., Chapman and Hall, London, 1971, paper 39.
153 W. G. Mallard and J. H. Owen, Int. J. Chem. Kinet., 6 (1974) 753.
154 C. J. Halstead and D. R. Jenkins, 12th Int. Symp. Combust., Combustion Institute, Pittsburgh, 1969, p. 979; Combust. Flame, 14 (1970) 321.
155 G. Dixon-Lewis and J. B. Greenberg, to be published.
156 G. Dixon-Lewis, M. M. Sutton and A. Williams, Proc. R. Soc. London, Ser. A, 317 (1970) 227.
157 K. H. Eberius, K. Hoyermann and H. Gg. Wagner, Ber. Bunsenges. Phys. Chem., 73 (1969) 962.
158 M. J. Day, G. Dixon-Lewis and K. Thompson, Proc. R. Soc. London, Ser. A, 330 (1972) 199.
159 M. J. Day, K. Thompson and G. Dixon-Lewis, 14th Int. Symp. Combust., Combustion Institute, Pittsburgh, 1973, p. 47.
160 G. Dixon-Lewis, Proc. R. Soc. London, Ser. A, 317 (1970) 235.
161 G. Dixon-Lewis, G. L. Isles and R. Walmsley, Proc. R. Soc. London, Ser. A, 331 (1973) 571.
162 M. A. A. Clyne and B. A. Thrush, Proc. R. Soc. London, Ser. A, 275 (1963) 559.
163 A. F. Dodonov, G. K. Lavrovskaya and V. L. Talrose, Kinet. Katal., 10 (1969) 701.
164 J. E. Bennett and D. R. Blackmore, 13th Int. Symp. Combust., Combustion Institute, Pittsburgh, 1971, p. 51.
165 A. A. Westenberg and N. de Haas, J. Chem. Phys., 50 (1969) 2512.
166 C. P. Fenimore and G. W. Jones, J. Phys. Chem., 62 (1958) 693.
167 C. P. Fenimore and G. W. Jones, J. Phys. Chem., 63 (1959) 1154.
168 K. H. Eberius, K. Hoyermann and H. Gg. Wagner, 13th Int. Symp. Combust., Combustion Institute, Pittsburgh, 1971, p. 713.
169 G. Dixon-Lewis, M. M. Sutton and A. Williams, Trans. Faraday Soc., 61 (1965) 255.
170 G. Dixon-Lewis, M. M. Sutton and A. Williams, J. Chem. Soc., (1965) 5724.
171 G. Dixon-Lewis, M. M. Sutton and A. Willaims, 10th Int. Symp. Combust., Combustion Institute, Pittsburgh, 1965, p. 495.
172 G. Dixon-Lewis, Proc. R. Soc. London, Ser. A, 330 (1972) 219.
173 G. Dixon-Lewis, F. A. Goldsworthy and J. B. Greenberg, Proc. R. Soc. London, Ser. A, 346 (1975) 261.
174 JANAF Thermochemical Tables, 2nd edn. Nat. Bur. Standards Publication NSRDS-NBS 37, Washington, D.C., 1971.
175 F. R. Del Greco and F. Kaufman, 9th Int. Symp. Combust., Academic Press, New York, 1963, p. 659.
176 W. P. Bishop and L. M. Dorfman, J. Chem. Phys., 52 (1970) 3210.
177 R. A. Svehla, Technical Report R-132, N.A.S.A., Washington, D. C., 1962.
178 D. L. Baulch and D. D. Drysdale, Combust. Flame, 23 (1974) 215.
179 W. E. Kaskan, Combust. Flame, 2 (1958) 286.
180 C. P. Fenimore and G. W. Jones, 10th Int. Symp. Combust., Combustion Institute, Pittsburgh, 1965, p. 489.
181 R. W. Getzinger and G. L. Schott, J. Chem. Phys., 43 (1965) 3237.
182 G. Dixon-Lewis, J. B. Greenberg and F. A. Goldsworthy, 15th Int. Symp. Combust., Combustion Institute, Pittsburgh, 1975, p. 717.

183 N. J. Friswell and M. M. Sutton, Chem. Phys. Lett., 15 (1972) 108.
184 S. N. Foner and R. L. Hudson, Adv. Chem. Ser., 36 (1962) 34.
185 T. T. Paukert and H. S. Johnston, J. Chem. Phys., 56 (1972) 2824.
186 E. A. Albers, Dissertation, Göttingen, 1969.
187 R. W. Getzinger, 11th Int. Symp. Combust., Combustion Institute, Pittsburgh, 1967, p. 117.
188 C. W. Hamilton and G. L. Schott, 11th Int. Symp. Combust., Combustion Institute, Pittsburgh, 1967, p. 635.
189 P. E. Rouse and R. Engleman, J. Quant. Spectrosc. Radiat. Transfer, 13 (1973) 1503.
190 D. M. Golden, F. P. del Greco and F. Kaufman, J. Chem. Phys., 39 (1963) 3034.
191 H. M. Smallwood, J. Am. Chem. Soc., 51 (1929) 1985.
192 I. Amdur, J. Am. Chem. Soc., 60 (1938) 2347.
193 A. A. Westenberg and N. de Haas, J. Chem. Phys., 43 (1965) 1550.
194 E. A. Albers, K. Hoyermann, H. Gg. Wagner and J. Wolfrum, 12th Int. Symp. Combust., Combustion Institute, Pittsburgh, 1969, p. 313.
195 M. A. A. Clyne and B. A. Thrush, Proc. R. Soc. London, Ser. A, 275 (1963) 544.
196 F. Dryer, D. Naegeli and I. Glassman, Combust. Flame, 17 (1971) 270.
197 I. W. M. Smith and R. Zellner, J. Chem. Soc., Faraday Trans. II, 70 (1974) 1045.
198 L. I. Avramenko and R. V. Lorentso, Zh. Fiz. Khim., 24 (1950) 207.
199 F. P. Del Greco and F. Kaufman, Discuss. Faraday Soc., 33 (1962) 128.
200 H. Wise, C. M. Ablow and K. M. Sancier, J. Chem. Phys., 41 (1964) 3569.
201 V. P. Balakhnin, Yu. M. Gershenzon, V. N. Kondratiev and A. B. Nalbandjan, Dokl. Akad. Nauk S.S.S.R., 170 (1966) 1117. English Translation, p. 659.
202 G. Dixon-Lewis, W. E. Wilson and A. A. Westenberg, J. Chem. Phys., 44 (1966) 2877.
203 W. G. Browne, R. P. Porter, J. D. Verlin and A. H. Clark, 12th Int. Symp. Combust., Combustion Institute, Pittsburgh, 1969, p. 1035.
204 R. G. Bennett and F. W. Dalby, J. Chem. Phys., 40 (1964) 1414.
205 N. R. Greiner, J. Chem. Phys., 46 (1967) 2795; 51 (1969) 5049.
206 E. L. Wong and F. E. Belles, NASA Rep. TN D-5707, N70-20629, 1970.
207 F. Stuhl and H. Niki, J. Chem. Phys., 57 (1972) 3671.
208 A. A. Westenberg and N. de Haas, J. Chem. Phys., 58 (1973) 4061.
209 C. P. Fenimore and G. W. Jones, J. Phys. Chem., 62 (1958) 1578.
210 A. Y. M. Ung and R. A. Back, Can. J. Chem., 42 (1964) 753.
211 R. R. Baldwin, R. W. Walker and S. J. Webster, Combust. Flame, 15 (1970) 167.
212 H. Kijewski and J. Troe, Int. J. Chem. Kinet., 3 (1971) 223.
213 D. L. Baulch, D. D. Drysdale, M. Din and D. J. Richardson, Proc. Eur. Symp. Combust., Academic Press, London, 1973, p. 30.
214 I. W. M. Smith and R. Zellner, J. Chem. Soc., Faraday Trans. II, 69 (1973) 1617.
215 D. L. Baulch and D. D. Drysdale, Combust. Flame, 23 (1974) 215.
216 V. N. Buneva, G. N. Kabasheva and V. N. Panfilov, Kinet. Katal., 10 (1969) 1221. English Translation, p. 1007.
217 A. A. Westenberg and N. de Haas, J. Chem. Phys., 47 (1967) 1393.
218 E. Hirsch and P. R. Ryason, J. Chem. Phys., 40 (1964) 2050.
219 G. L. Schott, Combust. Flame, 21 (1973) 357.
220 M. A. A. Clyne, 9th Int. Symp. Combust., Academic Press, New York, 1963, p. 211.
221 F. Kaufman, Ann. Geophys., 20 (1964) 106.
222 J. E. Breen and G. P. Glass, J. Chem. Phys., 52 (1970) 1082.
223 A. A. Westenberg, N. de Haas and J. M. Roscoe, J. Phys. Chem., 74 (1970) 3431.
224 E. L. Wong and A. E. Potter, J. Chem. Phys., 43 (1965) 3371.
225 K. Hoyermann, H. Gg. Wagner and J. Wolfrum, Ber. Bunsenges. Phys. Chem., 71 (1967) 599.
226 I. M. Campbell and B. A. Thrush, Trans. Faraday Soc., 64 (1968) 1265.

227 A. A. Westenberg and N. de Haas, J. Chem. Phys., 46 (1967) 490.
228 V. P. Balakhnin, V. I. Egorov and V. N. Kondratiev, Dokl. Akad. Nauk S.S.S.R., 193 (1970) 374.
229 G. L. Schott, R. W. Getzinger and W. A. Seitz, American Chemical Society, 164th National Meeting, New York, 1972. Phys. Chem. Abstract No. 113.
230 A. M. Dean and G. B. Kistiakowsky, J. Chem. Phys., 53 (1970) 830.
231 W. E. Wilson and J. T. O'Donovan, J. Chem. Phys., 47 (1967) 5455.
232 M. F. R. Mulcahy and R. H. Smith, J. Chem. Phys., 54 (1971) 5215.
233 A. A. Westenberg and N. de Haas, J. Chem. Phys., 58 (1973) 4066.
234 A. A. Westenberg and N. de Haas, J. Phys. Chem., 76 (1972) 1586.
235 G. K. Moortgat and E. R. Allen, Paper presented at 163rd Am. Chem. Soc. National Meeting, Boston, 1972.
236 T. Hikida, J. A. Eyre and L. M. Dorfman, J. Chem. Phys., 54 (1971) 3422.
237 M. J. Kurylo, J. Phys. Chem., 76 (1972) 3518.
238 W. Wong and D. D. Davis, Int. J. Chem. Kinet., 6 (1974) 401.
239 J. J. Ahumada, J. V. Michael and D. T. Osborne, J. Chem. Phys., 57 (1972) 3736.
240 D. D. Davis, W. Wong and R. Schiff, J. Phys. Chem., 78 (1974) 463.
241 D. E. Hoare and G. B. Peacock, Proc. R. Soc. London, Ser. A, 291 (1966) 85.
242 N. R. Greiner, J. Phys. Chem., 72 (1968) 406.
243 C. N. Hinshelwood, A. T. Williamson and J. H. Wolfenden, Proc. R. Soc. London, Ser. A, 147 (1934) 48.
244 J. W. Linnett and N. J. Selley, Z. Phys. Chem. N.F., 37 (1963) 402.
245 R. R. Baldwin, R. B. Moyes, B. N. Rossiter and R. W. Walker, Combust. Flame, 14 (1970) 181.
246 R. R. Baldwin, B. N. Rossiter and R. W. Walker, Trans. Faraday Soc., 66 (1970) 2004.
247 D. Appel and J. P. Appleton, 15th Int. Symp. Combust., Combustion Institute, Pittsburgh, 1975, p. 701.
248 N. R. Greiner, J. Chem. Phys., 48 (1968) 1413.
249 Z. G. Dzotsenidze, K. T. Aganesjan, G. A. Sachjan and A. B. Nalbandjan, Arm. Khim. Zh., 21 (1968) 68.
250 A. A. Westenberg and N. de Haas, J. Chem. Phys., 47 (1967) 4241.
251 A. A. Westenberg and W. E. Wilson, J. Chem. Phys., 45 (1966) 338.
252 V. V. Azatjan, R. R. Borodulin and E. I. Intezarova, Dokl. Akad. Nauk S.S.S.R., 213 (1974) 1053.
253 M. A. A. Clyne and B. A. Thrush, Trans. Faraday Soc., 57 (1961) 1305.
254 M. A. Clyne and B. A. Thrush, Discuss Faraday Soc., 33 (1962) 139.
255 R. Simonaitis, J. Phys. Chem., 67 (1963) 2227.
256 D. B. Hartley and B. A. Thrush, Proc. R. Soc. London, Ser. A, 297 (1967) 520.
257 D. L. Baulch, D. D. Drysdale, D. G. Horne and A. C. Lloyd, Evaluated Kinetic Data for High Temperature Reactions, Vol. 2. Butterworths, London, 1973.
258 E. M. Bulewicz and T. M. Sugden, Proc. R. Soc. London, Ser. A, 277 (1964) 143.
259 C. J. Halstead and D. R. Jenkins, Chem. Phys. Lett., 2 (1968) 281.
260 M. J. Day, G. Dixon-Lewis, M. M. Sutton and M. T. Walton, Combust. Flame, 10 (1966) 200.
261 P. G. Ashmore and B. P. Levitt, Trans. Faraday Soc., 52 (1956) 835; 53 (1957) 945; 54 (1958) 390.
262 W. A. Rosser, Jr. and H. Wise, J. Chem. Phys., 26 (1957) 571.
263 H. W. Thompson and C. N. Hinshelwood, Proc. R. Soc. London, Ser. A, 124 (1929) 219.
264 S. G. Foord and R. G. W. Norrish, Proc. R. Soc. London, Ser. A, 152 (1935) 196.
265 F. S. Dainton and R. G. W. Norrish, Proc. R. Soc. London, Ser. A, 177 (1941) 393.
266 F. S. Dainton and R. G. W. Norrish, Proc. R. Soc. London, Ser. A, 177 (1941) 411.

267 P. G. Ashmore and J. Chanmugam, Nature (London), 170 (1952) 1067.
268 C. J. Danby and C. N. Hinshelwood, J. Chem. Soc., (1940) 464.
269 A. T. Williamson and N. J. T. Pickles, Trans. Faraday Soc., 30 (1934) 926.
270 A. Davis, D. E. Hoare and A. D. Walsh, 13th Int. Symp. Combust., Combustion Institute, Pittsburgh, 1971, p. 63.
271 P. G. Ashmore and R. G. W. Norrish, Proc. R. Soc. London, Ser. A, 203 (1950) 454.
272 P. G. Ashmore, Trans. Faraday Soc., 51 (1955) 1090.
273 P. G. Ashmore and B. J. Tyler, J. Catal., 1 (1962) 39.
274 P. G. Ashmore and B. P. Levitt, Adv. Catal., 9 (1957) 367.
275 P. G. Ashmore and B. P. Levitt, 7th Int. Symp. Combust., Butterworths, London, 1959, p. 45.
276 P. G. Ashmore and B. J. Tyler, 9th Int. Symp. Combust., Combustion Institute, Pittsburgh, 1963, p. 201.
277 P. G. Ashmore and B. J. Tyler, Nature (London), 195 (1962) 279.
278 R. Simonaitis and J. Heicklen, J. Phys. Chem., 77 (1973) 1096; 78 (1974) 653.
279 W. A. Payne, L. J. Stief and D. D. Davis, J. Am. Chem. Soc., 95 (1973) 7614.
280 A. C. Lloyd, Int. J. Chem. Kinet., 6 (1974) 169.
281 W. Hack, K. Hoyermann and H. Gg. Wagner, Z. Naturforsch., Teil A, 29 (1974) 1236; Int. J. Chem. Kinet., 7 (Symp. 1) (1975) 329.
282 L. F. Phillips and H. I. Schiff, J. Chem. Phys., 37 (1962) 1233.
283 H. G. Wolfhard and W. G. Parker, 5th Int. Symp. Combust., Reinhold, New York, 1955, p. 718.
284 G. K. Adams, W. G. Parker and H. G. Wolfhard, Discuss. Faraday Soc., 14 (1953) 97.
285 G. K. Adams and G. W. Stocks, Rev. Inst. Fr. Pet. Ann. Combust. Liq., 13 (1958) 483.
286 W. G. Parker and H. G. Wolfhard, 4th Int. Symp. Combust., Williams and Wilkins, Baltimore, 1953, p. 420.
287 H. W. Melville, Proc. R. Soc. London, Ser. A, 142 (1933) 524.
288 H. W. Melville, Proc. R. Soc. London, Ser. A, 146 (1934) 737.
289 R. R. Baldwin, A. Gethin and R. W. Walker, J. Chem. Soc., Faraday Trans. I, 69 (1973) 352.
290 H. Henrici and S. H. Bauer, J. Chem. Phys., 50 (1969) 1333.
291 R. I. Soloukhin, 14th Int. Symp. Combust., Combustion Institute, Pittsburgh, 1973, p. 77.
292 E. A. Albers, K. Hoyermann, H. Schacke, K. J. Schmatjko, H. Gg. Wagner and J. Wolfrum, 15th Int. Symp. Combust., Combustion Institute, Pittsburgh, 1975, p. 765.
293 B. D. Fine and A. Evans, N.A.S.A. Technical Note D-1736, Washington, 1963.
294 J. E. Dove (comment), 14th Int. Symp. Combust., Combustion Institute, Pittsburgh, 1973, p. 82.
295 H. Navailles and M. Destriau, Bull. Soc. Chim. Fr., (1968) 2295.
296 M. Holliday and B. G. Reuben, Bull. Soc. Chim. Fr., (1969) 3087.
297 R. R. Baldwin, A. Gethin, J. Plaistowe and R. W. Walker, J. Chem. Soc., Faraday Trans. I, 71 (1975) 1265.
298 R. E. Tomalesky and J. E. Sturm, J. Chem. Phys., 55 (1971) 4299.
299 K. A. Wilde, Combust. Flame, 13 (1969) 173.
300 C. N. Hinshelwood and T. E. Green, J. Chem. Soc., (1926) 730.
301 C. N. Hinshelwood and J. W. Mitchell, J. Chem. Soc., (1936) 378.
302 W. M. Graven, J. Am. Chem. Soc., 79 (1957) 3697.
303 F. Kaufman and L. J. Decker, 8th Int. Symp. Combust., Williams and Wilkins Co., Baltimore, 1962, p. 133.
304 M. Koshi, H. Ando, M. Oya and T. Asaba, 15th Int. Symp. Combust., Combustion Institute, Pittsburgh, 1975, p. 809.

305 W. L. Flower, R. K. Hanson and C. H. Kruger, 15th Int. Symp. Combust., Combustion Institute, Pittsburgh, 1975, p. 823.
306 J. N. Bradley and P. Craggs, 15th Int. Symp. Combust., Combustion Institute, Pittsburgh, 1975, p. 833.
307 J. Duxbury and N. H. Pratt, 15th Int. Symp. Combust., Combustion Institute, Pittsburgh, 1975, p. 843.
308 A. J. Magnus, P. Chintapalli and M. Vanpée, Combust. Flame, 22 (1974) 71.
309 R. Zellner, K. Erler and D. Field, private communication.
310 C. Morley and I. W. M. Smith, J. Chem. Soc., Faraday Trans. II, 68 (1972) 1016.
311 R. Simonaitis and J. Heicklen, Int. J. Chem. Kinet., 4 (1972) 529.
312 A. A. Westenberg and N. de Haas, J. Chem. Phys., 57 (1972) 5375.
313 J. G. Anderson and F. Kaufman, Chem. Phys. Lett., 16 (1972) 375.
314 J. G. Anderson, J. J. Margitan and F. Kaufman, J. Chem. Phys., 60 (1974) 3310.
315 C. J. Howard and K. M. Evenson, J. Chem. Phys., 61 (1974) 1943.
316 R. Atkinson, D. A. Hansen and J. N. Pitts, Jr., J. Chem. Phys., 62 (1975) 3284.
317 G. W. Harris and R. P. Wayne, J. Chem. Soc., Faraday Trans. I, 71 (1975) 610.
318 D. D. Davis, J. T. Herron and R. E. Huie, J. Chem. Phys., 58 (1973) 530.
319 T. G. Slanger, B. J. Wood and G. Black, Int. J. Chem. Kinet., 5 (1973) 615.
320 P. P. Bemand, M. A. A. Clyne and R. T. Watson, J. Chem. Soc., Faraday Trans. II, 70 (1974) 564.
321 F. S. Klein and J. T. Herron, J. Chem. Phys., 41 (1964) 1285.
322 I. W. M. Smith, Trans. Faraday Soc., 64 (1968) 378.
323 A. A. Westenberg and N. de Haas, J. Chem. Phys., 50 (1969) 707.
324 H. A. Olschewski, J. Troe and H. Gg. Wagner, Ber. Bunsenges. Phys. Chem., 70 (1966) 450.
325 F. C. Kohout and F. W. Lampe, J. Am. Chem. Soc., 87 (1965) 5795.
326 E. Meyer, H. A. Olschewski, J. Troe and H. Gg. Wagner, 12th Int. Symp. Combust., Combustion Institute, Pittsburgh, 1969, p. 345.
327 J. Troe, Ber Bunsenges. Phys. Chem., 73 (1969) 946.
328 K. Glänzer and J. Troe, Ber. Bunsenges. Phys. Chem., 79 (1975) 465.
329 R. R. Baldwin, N. S. Corney and R. M. Precious, Nature (London), 169 (1952) 201.
330 A. Levy, 5th Int. Symp. Combust., Reinhold, New York, 1955, p. 495.
331 R. R. Baldwin, N. S. Corney and R. F. Simmons, 5th Int. Symp. Combust., Reinhold, New York, 1955, p. 502.
332 R. R. Baldwin, N. S. Corney, P. Doran, L. Mayor and R. W. Walker, 9th Int. Symp. Combust., Academic Press, New York and London, 1963, p. 184.
333 R. R. Baldwin and D. W. Cowe, Trans. Faraday Soc., 58 (1962) 1768.
334 R. R. Baldwin and R. F. Simmons, Nature (London) 175 (1955) 346.
335 R. R. Baldwin and R. F. Simmons, Trans. Faraday Soc., 51 (1955) 680.
336 R. R. Baldwin and R. F. Simmons, Trans. Faraday Soc., 53 (1957) 955, 964.
337 R. R. Baldwin, Trans. Faraday Soc., 60 (1964) 527.
338 R. R. Baldwin and R. W. Walker, Trans. Faraday Soc., 60 (1964) 1236.
339 R. R. Baldwin, N. S. Corney and R. W. Walker, Trans. Faraday Soc., 56 (1960) 802.
340 R. R. Baldwin, D. Booth and R. W. Walker, Trans. Faraday Soc., 58 (1962) 60.
341 R. R. Baldwin and R. W. Walker, Trans. Faraday Soc., 60 (1964) 1760.
342 R. R. Baker, R. R. Baldwin and R. W. Walker, 13th Int. Symp. Combust., Combustion Institute, Pittsburgh, 1971, p. 291.
343 R. R. Baldwin, C. J. Everett and R. W. Walker, Trans. Faraday Soc., 64 (1968) 2708.
344 R. R. Baldwin, C. J. Everett, D. E. Hopkins and R. W. Walker, Adv. Chem. Ser., 76 (1968) 124.
345 R. R. Baldwin, A. C. Norris and R. W. Walker, 11th Int. Symp. Combust., Combustion Institute, Pittsburgh, 1967, p. 889.

346 R. R. Baldwin, D. E. Hopkins, A. C. Norris and R. W. Walker, Combust. Flame, 15 (1970) 33.
347 R. R. Baldwin, D. E. Hopkins and R. W. Walker, Trans. Faraday Soc., 66 (1970) 189.
348 R. R. Baker, R. R. Baldwin and R. W. Walker, Trans. Faraday Soc., 66 (1970) 2812, 3016.
349 R. R. Baker, R. R. Baldwin, A. R. Fuller and R. W. Walker, J. Chem. Soc., Faraday Trans. I, 71 (1975) 736.
350 R. R. Baker, R. R. Baldwin and R. W. Walker, J. Chem. Soc., Faraday Trans. I, 71 (1975) 756.
351 R. R. Baldwin, A. R. Fuller, D. Longthorn and R. W. Walker, J. Chem. Soc., Faraday Trans. I, 68 (1972) 1362.
352 R. R. Baker, R. R. Baldwin and R. W. Walker, Combust. Flame, 14 (1970) 31.
353 R. R. Baker, R. R. Baldwin, C. J. Everett and R. W. Walker, Combust. Flame, 25 (1975) 285.
354 W. A. Bone and D. T. A. Towend, Flame and Combustion in Gases, Longmans Green, London, 1927.
355 G. Hadman, H. W. Thompson and C. N. Hinshelwood, Proc. R. Soc. London, Ser. A, 138 (1932) 297.
356 V. E. Cosslett and W. E. Garner, Trans. Faraday Soc., 27 (1931) 176.
357 D. E. Hoare and A. D. Walsh, Trans. Faraday Soc., 50 (1954) 37.
358 P. G. Dickens, J. E. Dove and J. W. Linnett, Trans. Faraday Soc., 60 (1964) 539.
359 D. R. Warren, Fuel, London, 33 (1954) 205.
360 V. V. Azatjan, V. V. Voevodsky and A. B. Nalbandjan, Kinet. Katal., 2 (1961) 340.
361 K. T. Oganesjan and A. B. Nalbandjan, Dokl. Akad. Nauk. S.S.S.R., 147 (1962) 361.
362 A. B. Nalbandjan, Dokl. Akad. Nauk S.S.S.R., 160 (1965) 162.
363 V. V. Azatjan, A. B. Nalbandjan and K. T. Oganesjan, Dokl. Akad. Nauk S.S.S.R., 157 (1964) 931.
364 A. B. Nalbandjan, Russ. Chem. Rev., 35 (1966) 244.
365 A. S. Gordon and R. H. Knipe, J. Phys. Chem., 59 (1955) 1160.
366 A. S. Gordon, J. Chem. Phys., 20 (1952) 340.
367 G. von Elbe, B. Lewis and W. Roth, 5th Int. Symp. Combust., Reinhold, New York, 1955, p. 610.
368 E. J. Buckler and R. G. W. Norrish, Proc. R. Soc. London, Ser. A, 167 (1938) 292.
369 E. J. Buckler and R. G. W. Norrish, Proc. R. Soc. London, Ser. A, 167 (1938) 318.
370 J. H. Burgoyne and H. Hirsch, Proc. R. Soc. London, Ser. A, 227 (1954) 73.
371 W. Jono, Rev. Phys. Chem. Jpn., 15 (1941) 17.
372 C. F. H. Tipper and R. K. Williams, Trans. Faraday Soc., 56 (1960) 1805.
373 F. Gaillard-Cusin and H. James, J. Chim. Phys. Phys. Chim. Biol., 63 (1966) 379.
374 G. Hadman, H. W. Thompson and C. N. Hinshelwood, Proc. R. Soc. London, Ser. A, 137 (1932) 87.
375 E. A. Th. Verdurmen, J Phys. Chem., 71 (1967) 681.
376 V. I. Tsvetkova, V. V. Voevodsky and N. B. Chirkov, J. Phys. Chim. U.S.S.R., 29 (1955) 380.
377 R. H. Knipe and A. S. Gordon, J. Chem. Phys., 27 (1957) 1418.
378 P. Harteck and S. Dondes, J. Chem. Phys., 27 (1957) 1419.
379 D. Garvin, J. Am. Chem. Soc., 76 (1954) 1523.
380 P. Harteck and S. Dondes, J. Chem. Phys., 26 (1957) 1734.
381 F. S. Dainton, Chain Reactions, Methuen, London, 1966.
382 E. J. Buckler and R. G. W. Norrish, Proc. R. Soc. London, Ser. A, 172 (1939) 1.

383 G. I. Kozlov, 7th Int. Symp. Combust., Butterworths, London, 1959, p. 142.
384 H. C. Hottel, G. C. Williams, N. M. Nerheim and G. R. Schneider, 10 Int. Symp. Combust., Combustion Institute, Pittsburgh, 1965, p. 111.
385 T. A. Brabbs and F. E. Belles, 11th Int. Symp. Combust., Combustion Institute, Pittsburgh, 1967, p. 125.
386 K. G. P. Sulzmann, B. F. Myers and E. R. Bartle, J. Chem. Phys., 42 (1965) 3969.
387 B. F. Myers, K. G. P. Sulzmann and E. R. Bartle, J. Chem. Phys., 43 (1965) 1220.
388 R. S. Brokaw, 11th Int. Symp. Combust., Combustion Institute, Pittsburgh, 1967, p. 1063.
389 E. S. Fishburne, K. R. Bilwakesh and R. Edse, Aerospace Res. Lab. Rep. ARL 67-0113, 1967.
390 P. Webster and A. D. Walsh, 10th Int. Symp. Combust., Combustion Institute, Pittsburgh, 1965, p. 463.
391 C. J. Halstead and D. R. Jenkins, Trans. Faraday Soc., 65 (1969) 3013.
392 R. A. Durie, G. M. Johnson and M. Y. Smith, Combust. Flame, 17 (1971) 197.
393 A. S. Kallend, Trans. Faraday Soc., 63 (1967) 2442.
394 J. Lebel and C. Ouellet, Can. J. Chem., 49 (1971) 447.
395 R. R. Baldwin, D. Jackson, A. Melvin and B. N. Rossiter, Int. J. Chem. Kinet., 4 (1972) 277.
396 A. G. Gaydon, Spectroscopy of Flames, Chapman and Hall, London, 1957.
397 G. A. Hornbeck and H. S. Hopfield, J. Chem. Phys., 17 (1949) 982.
398 A. R. Fairbairn and A. G. Gaydon, Trans. Faraday Soc., 50 (1954) 1256.
399 A. Fowler and A. G. Gaydon, Proc. R. Soc. London, Ser. A, 142 (1933) 3.
400 A. G. Gaydon, Proc. R. Soc. London, Ser. A, 176 (1940) 503.
401 A. D. Walsh, J. Chem. Soc., (1953) 2266.
402 R. N. Dixon, Proc. R. Soc. London, Ser. A, 275 (1963) 431.
403 A. G. Gaydon and F. Guedeney, Trans. Faraday Soc., 51 (1955) 894.
404 M. A. A. Clyne and B. A. Thrush, Proc. R. Soc. London, Ser. A, 269 (1962) 404.
405 W. E. Kaskan, Combust. Flame, 3 (1959) 39.
406 W. E. Kaskan, Combust. Flame, 3 (1959) 49.
407 G. Jahn, Der Zundvorgang in Gasgemischen, Oldenbourg, Berlin, 1934.
408 E. F. Fiock and C. H. Röder, N.A.C.A. Rep. No. 532 (1935); 553 (1936).
409 W. A. Strauss and R. Edse, 7th Int. Symp. Combust., Butterworths, London, 1959, p. 377.
410 L. A. Watermeier, J. Chem. Phys., 22 (1957) 1118.
411 R. Wires, L. A. Watermeier and R. A. Strehlow, J. Phys. Chem., 63 (1959) 989.
412 R. Friedman and J. A. Cyphers, J. Chem. Phys., 25 (1956) 448.
413 R. Friedman and J. A. Cyphers, J. Chem. Phys., 23 (1955) 1875.
414 C. P. Fenimore and G. W. Jones, J. Phys. Chem., 61 (1957) 651.
415 R. M. Fristrom, C. Grunfelder and S. Favin, J. Phys. Chem., 64 (1960) 1386; 65 (1961) 580.
416 A. A. Westenberg and R. M. Fristrom, J. Phys. Chem., 64 (1960) 1393; 65 (1961) 591.
417 T. Singh and R. F. Sawyer, 13th Int. Symp. Combust., Combustion Institute, Pittsburgh, 1971, p. 403.
418 R. Friedman and R. G. Nugent, 7th Int. Symp. Combust., Butterworths, London, 1959, p. 311.
419 G. Dixon-Lewis, P. Rhodes and R. J. Simpson, unpublished work.
420 W. Jost, H. Gg. Schecker and H. Gg. Wagner, Z. Phys. Chem. N. F., 45 (1965) 47.
421 J. P. Longwell and M. A. Weiss, Ind. Eng. Chem., 47 (1954) 1634.
422 G. C. Williams, H. C. Hottel and A. C. Morgan, 12th Int. Symp. Combust., Combustion Institute, Pittsburgh, 1969, p. 913.

423 G. K. Sobolev, 7th Int. Symp. Combust., Butterworths, London, 1959, p. 386.
424 M. A. Field, D. W. Gill, B. B. Morgan and P. G. W. Hawksley, Combustion of Pulverized Coal, B.C.U.R.A., Leatherhead, 1967.
425 L. I. Avramenko, Zhur. Fiz. Khim., 21 (1947) 1135.
426 J. T. Herron, J. Chem. Phys., 45 (1966) 1854.
427 R. P. Porter, A. H. Clarke, W. E. Kaskan and W. G. Browne, 11th Int. Symp. Combust., Combustion Institute, Pittsburgh, 1967, p. 907.
428 E. L. Wong, A. E. Potter and F. E. Belles, N.A.S.A. Rep. TN-4162, 1967.
429 J. Peeters and G. Mahnen, 14th Int. Symp. Combust., Combustion Institute, Pittsburgh, 1973, p. 133.
430 D. D. Davis, S. Fischer and R. Schiff, J. Chem. Phys., 61 (1974) 2213.
431 C. J. Howard and K. M. Evenson, J. Chem. Phys., 61 (1974) 1943.
432 T. P. J. Izod, G. B. Kistiakowsky and S. Matsuda, J. Chem. Phys., 55 (1971) 1406.
433 G. L. Tingey, J. Phys. Chem., 70 (1966) 1406.
434 V. F. Kochubei and F. B. Moin, Kinet. Katal., 10 (1969) 1203.
435 M. A. A. Clyne, B. A. Thrush and C. J. Halstead, Proc. R. Soc. London, Ser. A, 295 (1966) 355.
436 V. V. Azatjan, A. B. Nalbandjan and Ts'ui Meng-Yüan, Dokl. Akad. Nauk S.S.S.R., 147 (1962) 361.
437 H. A. Olschewski, J. Troe and H. Gg. Wagner, 11th Int. Symp. Combust., Combustion Institute, Pittsburgh, 1967, p. 155.
438 H. A. Olschewski, J. Troe and H. Gg. Wagner, Ber. Bunsenges. Phys. Chem., 70 (1966) 1060.
439 M. C. Lin and S. H. Bauer, J. Chem. Phys., 50 (1969) 3377.
440 V. N. Kondratiev and E. I. Intezarova, Kinet. Katal., 9 (1968) 477. English Trans., p. 395.
441 V. N. Kondratiev and E. I. Intezarova, Int. J. Chem. Kinet., 1 (1969) 105.
442 R. Simonaitis and J. Heicklen, J. Chem. Phys., 56 (1972) 2004.
443 D. L. Baulch, D. D. Drysdale and A. C. Lloyd, High Temperature Reaction Rate Data, No. 1, University of Leeds, 1968.
444 M. F. R. Mulcahy and D. J. Williams, Trans. Faraday Soc., 64 (1968) 59.
445 T. G. Slanger and G. Black, J. Chem. Phys., 53 (1970) 3722.
446 R. J. Donovan, D. Husain and L. J. Kirsch, Trans. Faraday Soc., 66 (1970) 2551.
447 V. V. Azatjan, N. E. Andreeva, E. I. Intezarova and V. N. Kondratiev, Kinet. Katal., 11 (1970) 290. English Trans., p. 244.
448 F. Stuhl and H. Niki, J. Chem. Phys., 55 (1971) 3943.
449 T. G. Slanger, B. J. Wood and G. Black, J. Chem. Phys., 57 (1972) 233.
450 W. B. DeMore, J. Phys. Chem., 76 (1972) 3527.
451 E. C. Y. Inn, J. Chem. Phys., 59 (1973) 5431.
452 E. C. Y. Inn, J. Chem. Phys., 61 (1974) 1589.
453 T. C. Clark, S. H. Garnett and G. B. Kistiakowsky, J. Chem. Phys., 52 (1970) 4692.
454 S. C. Barton and J. E. Dove, Can. J. Chem., 42 (1969) 521.
455 F. L. Dryer and I. Glassman, 14th Int. Symp. Combust., Combustion Institute, Pittsburgh, 1973, p. 987.
456 R. F. Heidner III, D. Husain and J. R. Wiesenfeld, J. Chem. Soc., Faraday Trans. II, 69 (1973) 927.
457 J. F. Noxon, J. Chem. Phys., 52 (1970) 1852.
458 I. D. Clark and J. F. Noxon, J. Chem. Phys., 57 (1972) 1033.
459 R. J. Donovan and D. Husain, Chem. Rev., 70 (1970) 489.
460 T. C. Clark, S. H. Garnett and G. B. Kistiakowsky, J. Chem. Phys., 51 (1969) 2885.
461 S. H. Garnett, G. B. Kistiakowsky and B. V. O'Grady, J. Chem. Phys., 51 (1969) 84.

462 W. T. Rawlins and W. C. Gardiner, Jr., J. Phys. Chem., 78 (1974) 497.
463 W. C. Gardiner, Jr., M. McFarland, K. Morinaga, T. Takeyama and B. F. Walker, J. Phys. Chem., 75 (1971) 1504.
464 L. J. Drummond, Aust. J. Chem., 21 (1968) 2631.
465 K. G. P. Sulzmann, L. Leibowitz and S. S. Penner, 13th Int. Symp. Combust., Combustion Institute, Pittsburgh, 1971, p. 137.
466 T. C. Clark, A. M. Dean and G. B. Kistiakowsky, J. Chem. Phys., 54 (1971) 1726.
467 H. Y. Wang, J. A. Eyre and L. M. Dorfman, J. Chem. Phys., 59 (1973) 5199.
468 V. V. Azatjan, N. E. Andreeva, E. I. Interzarova and L. A. Nersesjan, Arm. Khim. Zh., 26 (1973) 959.
469 R. Atkinson and R. J. Cvetanović, Can. J. Chem., 51 (1973) 370.
470 G. Herzberg, Electronic Spectra of Polyatomic Molecules, van Nostrand, Princeton, 1966.
471 J. W. C. Johns, S. H. Priddle and D. A. Ramsay, Discuss. Faraday Soc., 35 (1963) 90.
472 T. L. Cottrell, The Strengths of Chemical Bonds, Butterworths, London, 1958, p. 185.
473 D. B. Hartley, Chem. Commun., (1967) 1281.
474 M. A. Haney and J. L. Franklin, Trans. Faraday Soc., 65 (1969) 1794.
475 R. Atkinson, D. A. Hansen and J. N. Pitts, Jr., J. Chem. Phys., 63 (1975) 1703.
476 V. V. Azatjan, Dokl. Akad. Nauk S.S.S.R., 196 (1971) 617. English Trans., p. 51.
477 D. H. Volman and R. A. Gorse, J. Phys. Chem., 76 (1972) 3301.
478 D. D. Davis, W. A. Payne and L. J. Stief, Science, 179 (1973) 280.
479 A. Vardanjan, T. M. Dangjan, G. A. Sachjan and A. B. Nalbandjan, Dokl. Akad. Nauk S.S.S.R., 205 (1972) 619. English Trans., p. 632.
480 I. A. Vardanjan, G. A. Sachjan and A. B. Nalbandjan, Int. J. Chem. Kinet., 7 (1975) 23.
481 L. Farkas, F. Haber and P. Harteck, Naturwissenschaften, 18 (1930) 266.
482 R. H. Crist and O. C. Roehling, J. Am. Chem. Soc., 57 (1935) 2196.
483 R. H. Crist and G. M. Calhoun, J. Chem. Phys., 4 (1936) 696.
484 G. M. Calhoun and R. H. Crist, J. Chem. Phys., 5 (1937) 301.
485 F. B. Brown and R. H. Crist, J. Chem. Phys., 9 (1941) 840.
486 H. S. Johnston, W. A. Bonner and D. J. Wilson, J. Chem. Phys., 26 (1957) 1002.
487 J. H. Thomas and G. R. Woodman, Trans. Faraday Soc., 63 (1967) 2728.
488 A. Burcat and A. Lifshitz, J. Phys. Chem., 74 (1970) 263.
489 C. E. H. Bawn, Trans. Faraday Soc., 31 (1935) 461.
490 R. F. Strickland-Constable, Trans. Faraday Soc., 34 (1938) 1374.
491 D. G. Madley and R. F. Strickland-Constable, Trans. Faraday Soc., 49 (1953) 1312.
492 R. J. Emrich and D. B. Wheeler, Phys. Fluids, 1 (1958) 14.
493 D. Milks and R. A. Matula, 14th Int. Symp. Combust., Combustion Institute, Pittsburgh, 1973, p. 83.
494 R. I. Soloukhin, Dokl. Akad. Nauk S.S.S.R., 194 (1970) 143. English Trans., p. 678.
495 R. I. Soloukhin, 14th Int. Symp. Combust., Combustion Institute, Pittsburgh, 1973, p. 77.
496 I. S. Zaslonko, S. M. Kogarko, E. V. Mozzhukhin and V. N. Smirnov, Dokl. Akad. Nauk S.S.S.R., 212 (1973) 1138. English Trans., p. 846.
497 A. J. Arvia, P. J. Aymonino, C. H. Waldow and H. J. Schumacher, Angew. Chem., 72 (1960) 169.
498 J. M. Heras, A. J. Arvia, P. J. Aymonio and H. J. Schumacher, Z. Phys. Chem. N.F., 28 (1961) 250.
499 M. Wechsberg and G. H. Cady, J. Am. Chem. Soc., 91 (1969) 4432.
500 G. A. Kapralova, S. N. Buben and A. M. Chaikin, Kinet. Katal., 16 (1975) 591. English Trans., p. 508.

501 A. J. Arvia, P. J. Aymonino and H. J. Schumacher, Z. Phys. Chem. N.F., 51 (1966) 170.
502 P. J. Aymonio, Proc. Chem. Soc., London (1964) 341.
503 E. H. Appleman and M. A. Clyne, J. Chem. Soc., Faraday Trans. I, 71 (1975) 2072.
504 H. Henrici, M. C. Lin and S. H. Bauer, J. Chem. Phys., 52 (1970) 5834.
505 M. C. Lin and S. H. Bauer, J. Am. Chem. Soc., 91 (1969) 7737.
506 S. H. Bauer, P. Jeffers, A. Lifshitz and B. P. Yadava, 13th Int. Symp. Combust., Combustion Institute, Pittsburgh, 1971, p. 417.
507 T. P. J. Izod, G. B. Kistiakowsky and S. Matsuda, J. Chem. Phys., 56 (1972) 1083.
508 S. Matsuda, J. Chem. Phys., 57 (1972) 807.
509 S. Matsuda, J. Phys. Chem., 76 (1972) 2833.
510 P. G. Ashmore and R. G. W. Norrish, Nature (London), 167 (1951) 390.
511 J. W. Linnett, B. G. Reuben and T. F. Wheatley, Combust. Flame, 12 (1968) 325.
512 B. F. Gray, Chem. Soc. Specialist Periodical Repts., Reaction Kinetics, 1 (1975) 309.
513 J. E. Dove, D. Phil. Thesis, Oxford, 1959.
514 B. F. Gray, Trans. Faraday Soc., 66 (1970) 1118.
515 C. H. Yang, Combust. Flame, 23 (1974) 97.
516 C. H. Yang, Faraday Symp. Chem. Soc., 9 (1974) 114.
517 C. H. Yang and A. L. Berlad, J. Chem. Soc., Faraday Trans. I, 70 (1974) 1661.
518 C. H. Yang and B. F. Gray, J. Phys. Chem., 73 (1969) 3395.
519 K. K. Foo and C. H. Yang, Combust. Flame, 17 (1971) 223.
520 N. Minorsky, Non-linear Oscillations, van Nostrand, Princeton, New Jersey, 1962.
521 J. R. Bond, Faraday Symp. Chem. Soc., 9 (1974) 156 (discussion comment).
522 R. B. Klemm, W. A. Payne and L. J. Stief, Int. J. Chem. Kinet., 7 (Symp. 1) (1975) 61.
523 J. Vandooren, J. Peeters and P. J. van Tiggelen, 5th Int. Symp. Combust., Combustion Institute, Pittsburgh, 1975, p. 745.

Chapter 2

Hydrocarbons

R. T. POLLARD

1. Introduction

Modern society probably depends more upon the combustion of hydrocarbons than upon any other chemical reaction. The various applications of hydrocarbon combustion were recently described [1] in terms of a typical temperature—pressure ignition diagram for a hydrocarbon + oxygen (or air) mixture, as shown in Fig. 1.

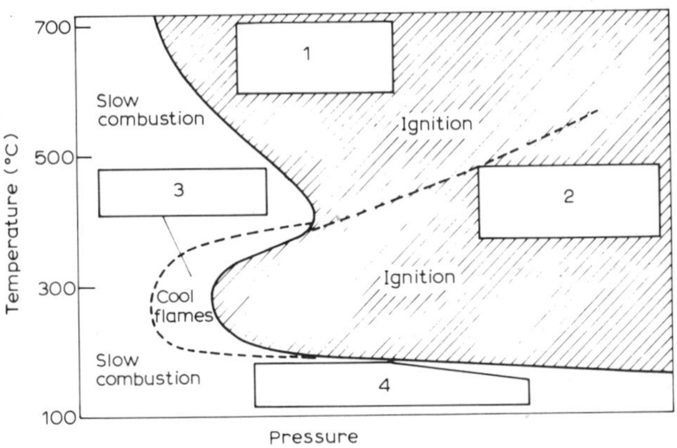

Fig. 1. Ignition diagram for a typical hydrocarbon and oxygen mixture: 1, conversion (by ignition) of chemical energy, e.g. turbo-jet engine; 2, conversion (by ignition) of chemical energy to heat and mechanical energy, e.g. internal combustion engine; 3, conversion of fuels to potentially useful chemicals, e.g. O-heterocycles; and 4, controlled conversion of fuels to useful chemicals, e.g. alcohols, peroxides, aldehydes, ketones, etc. (From ref. 1.)

Due to its importance with regard to the production of useful chemicals by selective oxidation and to its role in abnormal combustion phenomena such as "knock" and "run on" in the internal combustion engine, the spontaneous combustion of hydrocarbons in the temperature range 200—600 °C has been the most widely studied mode of oxidation. This Chapter, therefore, has a number of aims. First, the generally accepted theories for the mechanism of spontaneous combustion at the beginning of the 1960's will be briefly surveyed. A detailed discussion of current theories will then show how these and new theories have developed. Kinetic parameters will be given wherever they have been determined or estimated. Finally, the variation of the mechanism with the molecular weight and structure of the hydrocarbon will be discussed.

2. The prevalent theories on the mechanism of hydrocarbon oxidation in 1960

An excellent review of the development of research on the gas-phase oxidation of hydrocarbons from the end of the nineteenth century was published in 1960 by Shtern [2]. In common with many monographs it was not entirely unbiased, but nevertheless gave an up-to-date account of the current views on the mechanism of the gaseous oxidation of hydrocarbons at that time. Kinetically it is of course one of the best known degenerately branched-chain reactions, the theory of which has been developed in full by Semenov [3]. For the slow oxidation of hydrocarbons ($\geqslant C_2$) at pressures up to 380—760 torr and in the temperature range 200—600 °C the overall chemical mechanism was thought to involve two major routes, namely *strict oxidation*, which led to the formation of oxygen-containing products (aldehydes, alcohols, ketones, acids and carbon oxides) and *cracking*, which led to the formation of hydrogen and unsaturated and saturated hydrocarbons of a lower molecular weight than the initial fuel. The general mechanism of the oxidation route was summarized by Shtern as shown in Scheme 1.

The alkyl radical initially formed reacts readily with oxygen to give the corresponding alkylperoxy radical, which may abstract hydrogen from a fuel molecule to form the alkylhydroperoxide or alternatively decompose to yield an aldehyde and an alkoxy radical. Some workers thought that this decomposition was preceded by an isomerization of the alkylperoxy radical, the activation energy of which had been estimated by Semenov [3] to be ca. 20 kcal . $mole^{-1}$. Shtern was of the opinion that the major, if not the only, fate of the alkylperoxy radical was decomposition, but in contrast to other workers he believed that it must involve scission of a C—C bond and could not lead to the formation of a carbonyl compound and hydroxyl radical.

The fate of the alkoxy radicals is determined by their molecular weight. High molecular weight radicals decompose to formaldehyde and an alkyl radical of lower molecular weight than those initially formed from the fuel. Only methoxy radicals are considered sufficiently stable at these temperatures to react preferentially with the fuel to form the alcohol. Of the organic oxygenates formed, aldehydes are the main compounds which were thought to be further oxidized. The mechanism of their oxidation was not unequivocally established although reaction was believed to take place via the acyl radical as shown in Scheme 1.

Scheme 1

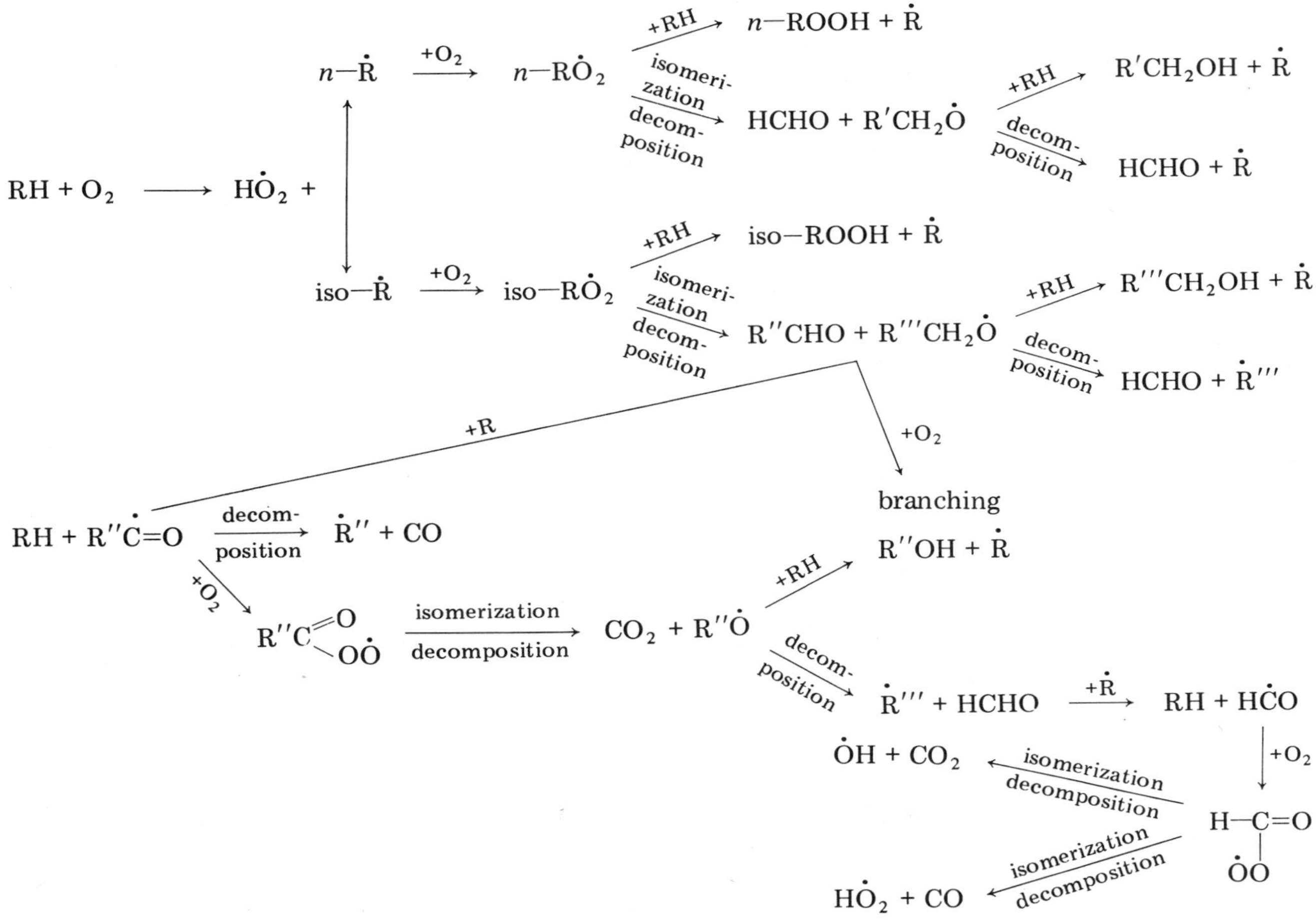

RH + O2 → HȮ2 +
n—Ṙ
iso—Ṙ
+O2
n—RȮ2
iso—RȮ2
+RH
n—ROOH + Ṙ
iso—ROOH + Ṙ
isomeri-
zation
decom-
position
HCHO + R′CH2Ȯ
R″CHO + R‴CH2Ȯ
R′CH2OH + Ṙ
HCHO + Ṙ
R‴CH2OH + Ṙ
HCHO + Ṙ‴
+R
+O2
branching
RH + R″Ċ=O
Ṙ″ + CO
R″C(=O)OȮ
isomerization
decomposition
CO2 + R″Ȯ
R″OH + Ṙ
Ṙ‴ + HCHO
+Ṙ
RH + HĊO
H—C(=O)—OȮ
ȮH + CO2
HȮ2 + CO

A detailed study of the oxidation of propane also led Knox and Norrish [4] to the conclusion that the main oxidation route was via the successive degradation of aldehydes. Their scheme differed from that shown above, however, since it did not include the intermediate formation of an alkylperoxy radical and hydroxyl radicals were thought to be the principal chain carriers rather than the alkyl, alkylperoxy and alkoxy radicals.

Considerable discussion had taken place during the period 1945–60 regarding the nature of the degenerate branching agent. On the one hand, Cullis and Hinshelwood [5], Walsh [6] and Nieman [7] had produced a considerable amount of evidence in favour of peroxides fulfilling this role from studies on alkanes of carbon number greater than five. On the other hand, Norrish [4, 8, 9], Shtern [10] and Knox [11, 12] had produced equally compelling evidence in favour of aldehydes acting as degenerate-branching agents mainly from studies on relatively low molecular weight alkanes (particularly propane). In his monograph, however, Shtern made the following theoretical assessment of the role of alkyl hydroperoxides in the oxidation of an equimolar mixture of propane and oxygen at 350 °C and an initial pressure of 282 torr, assuming the two processes

$$RO_2\cdot \xrightarrow{RH} ROOH + R\cdot \qquad (1')$$

$$RO_2\cdot \longrightarrow R'CHO + R''O\cdot \qquad (2')$$

Taking the difference in activation energies of these two reactions as 12.5 kcal . mole^{-1}, the ratio of the rate of decomposition of $RO_2^{\cdot}$ to its reaction with RH is ca. 20:1. Product analysis showed that the partial pressure of the decomposition products towards the end of the reaction was 40 torr, so that

$$\frac{\text{rate (2)}}{\text{rate (1)}} = \frac{[RO_2\cdot]_{\text{decomp.}}}{[RO_2\cdot]_{\text{perox.}}} = \frac{40}{[\text{peroxide}]} = 20$$

Hence the partial pressure of the peroxide was ca. 2 torr. Shtern then considered whether such small amounts of peroxide could lead to a rate of branching of the same magnitude as that given by the aldehyde, viz. branching due to either

$$ROOH \longrightarrow RO\cdot + \cdot OH \qquad (3')$$

or

$$RCHO + O_2 \longrightarrow R\dot{C}O + HO_2\cdot \qquad (4')$$

Using activation energies of 40 and 32 kcal . mole^{-1} for reactions (3′) and

(4′), respectively and the partial pressures of acetaldehyde (5 torr) and peroxide (0.1 torr) at the maximum rate, then

$$\frac{\text{rate (3)}}{\text{rate (4)}} = \frac{10^{13} \times e^{-\frac{40000}{2 \times 623}} \times 1.55 \times 10^{15}}{10^{-10} \times e^{-\frac{32000}{2 \times 623}} \times 2.2 \times 10^{18} \times 7.9 \times 10^{16}} = 1.3$$

He concluded, therefore, that such a small quantity of peroxide is capable of producing the same rate of branching as a quantity of aldehyde fifty times as great.

Shtern's theoretical consideration thus contrasted directly with his opinion based on experiment. In conclusion he wrote, "Undoubtedly, the solution of this contradiction is one of the outstanding problems for further research".

The mechanism of formation of the cracking products had not been resolved at the time of publication of Shtern's review [2]. Semenov [3], however, had pointed out that the direct decomposition of prop-1-yl and prop-2-yl radicals at 300 °C was most unlikely, due to the large activation energies involved (27—29 and 40 kcal . mole^{-1}, respectively). He therefore suggested the following alkylperoxy radical isomerization and decomposition reactions to explain the formation of propene and ethylene in propane oxidation, viz.

$$CH_3{-}\underset{\large O-O\cdot}{\underset{|}{CH}}{-}CH_3 \longrightarrow CH_3{-}\underset{\large O-OH}{\underset{|}{CH}}{-}\dot{C}H_2 \longrightarrow CH_3{-}CH{=}CH_2 + HO_2\cdot$$

$$\underset{\large O-O\cdot}{\underset{|}{CH_2}}{-}CH_2{-}CH_3 \longrightarrow \underset{\large O-OH}{\underset{|}{CH_2}}{-}CH_2{-}\dot{C}H_2 \longrightarrow CH_2{=}CH_2 + HCHO + \cdot OH$$

He considered that these isomerizations proceeded with an activation energy of not less than 27—29 kcal . mole^{-1} and that the isomerization was the rate-determining step. Experiment showed [13], however, that the difference in the activation energies was of the order of 19 kcal . mole^{-1} and not 7—9 kcal . mole^{-1} as required by Semenov's suggestion. This problem was to some extent overcome by the suggestion of Lewis and von Elbe [14] that the reaction

$$R\cdot + O_2 \longrightarrow \text{Alkene} + HO_2\cdot$$

occurred. Knox and Norrish [4] and Satterfield et al. [13, 15, 16] gave support to this mode of formation of the alkene. Satterfield and Wilson [16] showed that the ratio of cracking products/oxidation products was independent of the initial reactant concentrations at a given temperature and concluded that this gave experimental proof that the competing oxidation and cracking reactions are

$$\cdot C_3H_7 + O_2 \longrightarrow C_3H_7OO\cdot \longrightarrow \text{oxidation products}$$

$$\cdot C_3H_7 + O_2 \longrightarrow C_3H_6 + HO_2\cdot$$

Unfortunately, the same result would be expected from competition between the two unimolecular isomerizations and decompositions of the alkylperoxy radicals, viz.

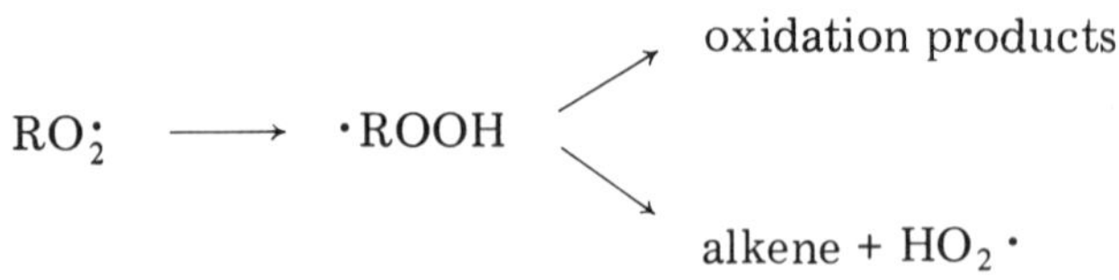

However, his work does show that the cracking involves reaction with oxygen.

Associated with the low temperature (200—450 °C) oxidation of hydrocarbons are the phenomena of the negative temperature coefficient, cool flames and their periodicity, and multiple-stage ignitions. Any mechanism must, therefore, be able to account for these phenomena in addition to explaining the modes of product formation.

An example of the occurrence of the negative temperature coefficient is shown in Fig. 2 for the slow oxidation of propane [17]. In general, the rate of oxidation decreases with temperature over a range of ca. 50 °C above initial temperatures of about 350 °C. Several explanations of this phenomenon had been put forward and although these have recently been

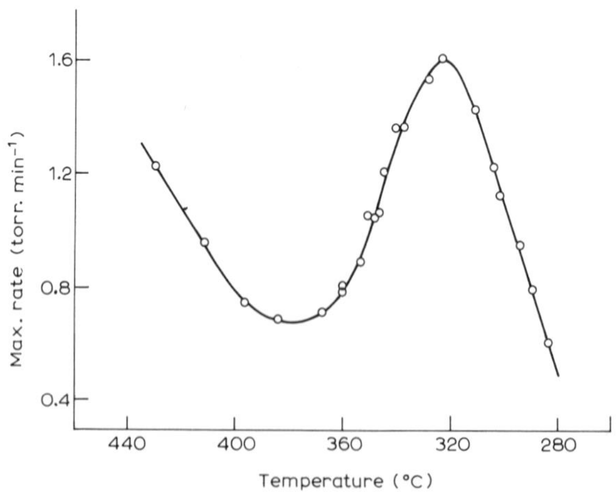

Fig. 2. The variation with temperature of the maximum rate of oxidation of propane. Initial pressure of propane = 60 torr; initial pressure of oxygen = 120 torr. (From ref. 17.)

reviewed by Déchaux [18], it will be useful to summarize them. Norrish [9, 19] suggested that above 350 °C reaction (5′)

$$R\cdot + O_2 \longrightarrow \text{alkene} + HO_2\cdot \qquad (5')$$

becomes more important that reaction (6′)

$$R\cdot + O_2 \longrightarrow \text{aldehyde} + \cdot OH \qquad (6')$$

because it has a higher activation energy and thus, since it does not directly form a branching agent, the rate of reaction falls. The peroxide theory is capable of explaining the effect in a similar manner, as Walsh [20] had previously pointed out, when $E_{7'} > E_{8'}$ for

$$RO_2\cdot \longrightarrow \text{non-branching products} \qquad (7')$$

$$RO_2\cdot + RH \longrightarrow ROOH + R\cdot \qquad (8')$$

Salooja [21] carried out extensive studies of this phenomenon and considered that it was due to the inhibiting effect of alkenes produced as initial products and that the decrease with temperature in the concentration of hydroxyl radicals was accompanied by a corresponding increase in the concentration of the less reactive hydroperoxy radicals. Norrish's mechanism concurs with this reasoning. Enikolopyan [22] put forward two reaction schemes in an attempt to explain the negative temperature coefficient. In the first he considered the reactions

$$R\cdot + O_2 \longrightarrow RO_2\cdot \qquad (9')$$

$$RO_2\cdot + R'CHO \longrightarrow RO\cdot + \cdot OH + R'\dot{C}O \qquad (10')$$

$$RO_2\cdot \longrightarrow R'CHO + R''O\cdot \qquad (11')$$

With increasing temperature, the unimolecular propagation reaction (11′) will predominate over the bimolecular branching reaction (10′). In addition, at these temperatures reaction (9′) will become rate-determining and thus the stationary concentration of alkylperoxy radicals will decrease. The net result of these two effects will be a sharp decrease in the rate of oxidation with increasing temperature [22]. In the second scheme, which was first proposed by Skirrow and co-workers [23, 24] to explain degenerate branching during the oxidation of propene, unimolecular decomposition reactions of the alkylperoxy, acyl and acylperoxy radicals were considered to predominate with increase in temperature over the bimolecular reactions (12′), (13′) and (14′)

$$RO_2\cdot + CH_3CHO \longrightarrow ROOH + CH_3\dot{C}O \qquad (12')$$

$$CH_3\dot{C}O + O_2 \longrightarrow CH_3CO_3\cdot \qquad (13')$$

$$CH_3CO_3\cdot + RH \longrightarrow CH_3CO_3H + R\cdot \qquad (14')$$

Thus, branching by the reaction

$$CH_3CO_3H \longrightarrow CH_3CO_2\cdot + \dot{O}H \qquad (15')$$

was decreased. In both schemes the subsequent increase in reaction rate was explained by the onset of further branching by the reaction

$$CH_3CHO + O_2 \longrightarrow CH_3\dot{C}O + HO_2\cdot \qquad (16')$$

Cool flames are observed at pressures lower than those necessary for two-stage ignition. These non-isothermal events occur at intervals of time during an otherwise almost isothermal reaction. The majority of workers consider cool-flame propagation to be the central part of reaction during which the bulk of the fuel is incompletely oxidized. Shtern, however, considers the cool flame to be a minor process which plays little part in the overall reaction, since the cool-flame oxidation and the slow oxidation are very similar in their chemical nature. Indeed, the pressure—time curves Shtern obtained for the cool-flame oxidation of propane have the same S-shape as those for the slow oxidation if the non-isothermal events are ignored, as can be seen in Fig. 3.

In any event, it was generally agreed that the propagation of a cool flame was associated with the accumulation of a critical concentration of a branching agent and hence a critical rate of branching. This being the case, the concentration of the branching agent would be expected to decrease rapidly during the propagation of the flame, as had been amply demonstrated for peroxides [25—27] and aldehydes [4, 9, 28].

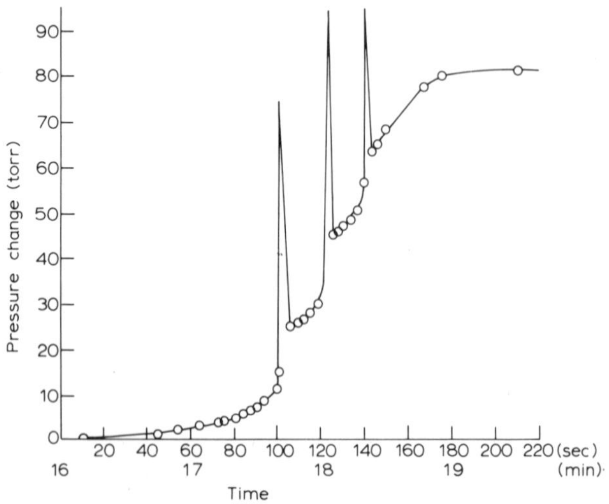

Fig. 3. The variation of pressure change with time during the cool-flame oxidation of propane. Initial temperature = 280 °C; initial pressure of propane = 210 torr; initial pressure of oxygen = 210 torr. (From ref. 2.)

Two explanations of the periodicity had been advanced. The "chain-thermal" theory [29] attributes the oscillations to the effect of the rise in temperature resulting from the propagation of the flame on the rates of formation and destruction of a single intermediate [4, 9, 25], viz.

$$\text{reactants} \xrightarrow[E_1]{\Delta H_1} \text{X} \xrightarrow[E_2]{\Delta H_2} \text{products}$$

Under suitable initial reaction conditions the intermediate can lead to multiple cool flames if $|\Delta H_2| > |\Delta H_1|$ and $E_2 > E_1$. Thus, as X accumulates the second reaction becomes more rapid and hence increases the temperature. Since $E_2 > E_1$, its rate is therefore accelerated relative to the first reaction and [X] falls. This in turn leads to a decrease in temperature and the first reaction is accelerated relative to the second leading to another increase in [X] and thus to a periodic thermokinetic phenomenon. The second theory is purely kinetic and depends on the production of critical concentrations of two different intermediate products which enter into branching reactions [30]. The reaction scheme may be represented as (where A and B are the reactant and final product, respectively, and X and Y are the intermediates)

$$\text{A} + \text{X} \longrightarrow \text{B} + 2\text{X} \quad \text{(a)}$$

$$\text{X} + \text{Y} \longrightarrow \text{B} + 2\text{Y} \quad \text{(b)}$$

$$\text{A} + \text{Y} \longrightarrow \text{B} \quad \text{(c)}$$

Hence

$$\text{d[X]}/\text{d}t = k_a[\text{A}][\text{X}] - k_b[\text{X}][\text{Y}]$$

$$\text{d[Y]}/\text{d}t = k_b[\text{X}][\text{Y}] - k_c[\text{A}][\text{Y}]$$

When [X] reaches the critical value $k_c[\text{A}]/k_b$, $\text{d[Y]}/\text{d}t$ becomes positive and [Y] increases at the expense of [X]. Similarly, when [Y] in turn reaches the critical value $k_a[\text{A}]/k_b$, $\text{d[X]}/\text{d}t$ becomes negative and [X] eventually falls below the value $k_c[\text{A}]/k_b$. [Y] will then begin to fall and when it becomes less than $k_a[\text{A}]/k_b$, [X] will increase again. Thus if the criteria for the odd-numbered cool flames is that $[\text{X}] > [\text{X}]_{crit}$ and the criteria for even-numbered cool flames is that $[\text{Y}] > [\text{Y}]_{crit}$, the periodicity is explained. This "two-product" theory has been discussed elsewhere [6, 14, 31, 32]. Not unexpectedly the identities of X and Y are thought to be hydroperoxides and aldehydes, respectively.

The phenomenon of two-stage ignition had not been extensively studied, but it had been suggested [14] that the temperature rise accompanying the passage of the cool flame is sufficient to cause rapid further oxidation in the high temperature region which leads to a thermal

References pp. 361—367

explosion. Thus, following the passage of the cool flame the reaction mixture contains a considerable concentration of aldehydes and the temperature of the mixture may be sufficiently high for branching associated with these compounds to be rapid, leading to further self-heating and explosion.

By 1960 then, much was known about the mechanism of hydrocarbon oxidation. The theory of degenerately-branched chain reactions had been fully developed (see Vol. 2, Chapter 2), the importance of aldehydes and peroxides as branching agents had been established and plausible explanations of all the low temperature combustion phenomena had been propounded. Even so, there was a lack of unequivocal quantitative data regarding the mechanism, particularly with respect to chain-propagation in the oxidation of high molecular weight hydrocarbons ($\geqslant C_5$). Also, very little reliable kinetic data had been obtained for the individual participating reactions. Fortunately, gas-chromatographic techniques were being extensively developed at this time and papers describing "Estimation of combustion products by gas-chromatography" were already appearing in the literature [33]. This new tool together with isotopic-tracer and spectrometric techniques allowed workers in the field of hydrocarbon combustion to enter the sixties with considerable hope of solving one of the most complicated chemical reactions ever encountered.

3. The low temperature mechanism

3.1 INITIATION

It is generally accepted that the initial attack on saturated hydrocarbons involves abstraction of a hydrogen atom to yield the alkyl radical and a hydroperoxy radical

$$C_nH_{2n+2} + O_2 \longrightarrow \cdot C_nH_{2n+1} + HO_2\cdot \qquad (1)$$

This mode of initiation was first suggested by Cullis and Hinshelwood [5] and substantiated theoretically by Semenov [34]. The activation energies of such abstraction reactions are ca. 40—55 kcal . mole^{-1} and reflect their high endothermicity. They are therefore slow and selective and as would be expected from the C—H bond strengths, tertiary C—H bonds are the most readily attacked and primary C—H bonds the least readily attacked.

It is extremely difficult to determine the nature of this reaction. Some workers are of the opinion that it takes place heterogeneously [35, 36] whilst others believe it is homogeneous [37].

3.2 PROPAGATION

The alkyl radicals initially generated will usually react exclusively with oxygen. Other reactions such as decomposition, disproportionation, isomerization, recombination and reaction with the fuel will only compete when the partial pressure of oxygen is low or the temperature high (ca. 450 °C). The nature of their reaction with oxygen, particularly in the temperature range 250—400 °C, has led to considerable experimentation and discussion.

Three requirements must be fulfilled before any mechanism can be accepted. Firstly, it must be capable of explaining the mode of formation of the reaction products; secondly, it must be acceptable from thermokinetic considerations and finally it must be capable of explaining phenomena such as the negative temperature coefficient and periodic cool flames. It was recently pointed out [38] that many mechanisms have been proposed in recent years which do not take the second consideration into account. Whilst this is a valid criticism, the reverse is also true, i.e. many mechanisms have been suggested which are based on inaccurate thermokinetic considerations and have not been confirmed experimentally. In any event, the system under consideration must be defined by experiment, which in this case requires extensive knowledge of the kinetics, the yields and nature of the products formed and their variation with the extent of reaction and the reaction conditions. Modern techniques have allowed the system to be reasonably well defined in these terms and this has led to two principal theories regarding chain-propagation.

3.2.1 Alkene theory

This theory was proposed by Knox [39] following a series of careful kinetic and analytical studies of the oxidation of ethane [40], propane [41], and isobutane [42] in the early stages of reaction (<1 % fuel consumption). For these lower alkanes, he found that at this stage of the reaction 75—80 % of the alkane consumed appeared in the products as the conjugate alkene both at low temperature (ca. 300 °C) and at high temperature (ca. 450 °C). He concluded, therefore, that the primary oxidation process for alkyl radicals is the same at both temperatures, viz.

$$\cdot C_nH_{2n+1} + O_2 \longrightarrow C_nH_{2n} + HO_2\cdot \qquad (2)$$

As the fuel consumption increased, the alkene concentration eventually reached a kinetically controlled stationary value [43]. Since the major product at all stages in the oxidation of isobutene is acetone [42] (Fig. 4), it was possible to follow the consumption of isobutene during the oxidation of isobutane by observing the formation of acetone. It can be seen from Fig. 5 that a marked increase in the yield of acetone in the later

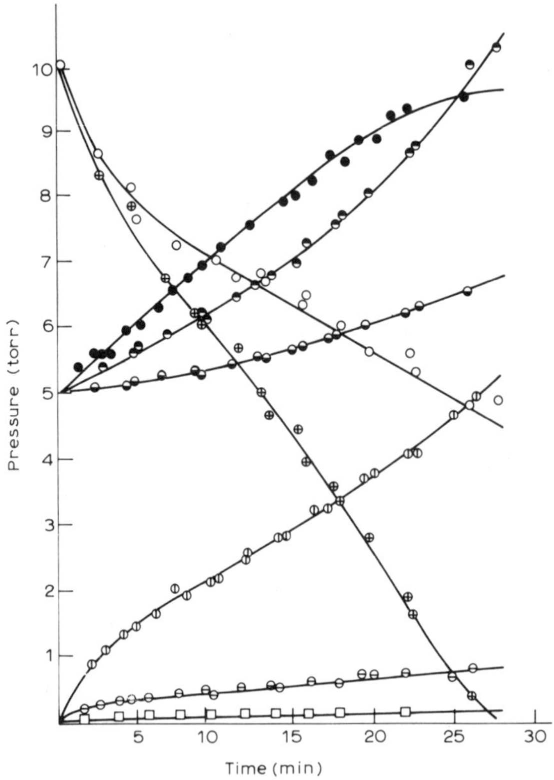

Fig. 4. The variation with time of product formation during the oxidation of isobutene. Initial temperature = 293 °C; initial pressure of isobutene = 100 torr; initial pressure of oxygen = 100 torr. ○, isobutene; ⊕, oxygen; ⦶, acetone; ⊖, isobutene oxide; □, isobutyraldehyde; ◒, carbon dioxide; ◓, carbon monoxide; ●, water. (From ref. 42.)

stages coincided with the stationary yield of isobutene in accord with the overall process [44]

$$C_4H_{10} \longrightarrow 80\ \%\ C_4H_8 \longrightarrow 80\ \%\ CH_3COCH_3$$

$$\searrow \qquad\qquad\qquad \searrow$$

$$20\ \%\ \text{minor products} \qquad 20\ \%\ \text{minor products}$$

Knox further concluded, therefore, that the conjugate alkenes must also play an important role in the intermediate and later stages of alkane oxidation.

The 20 % minor products were a complex mixture of oxygenates and propene, the composition of which varied with the nature of the reaction vessel surface as shown in Fig. 6. In contrast, however, the ratio of major/minor products was little affected by surface conditions. It was

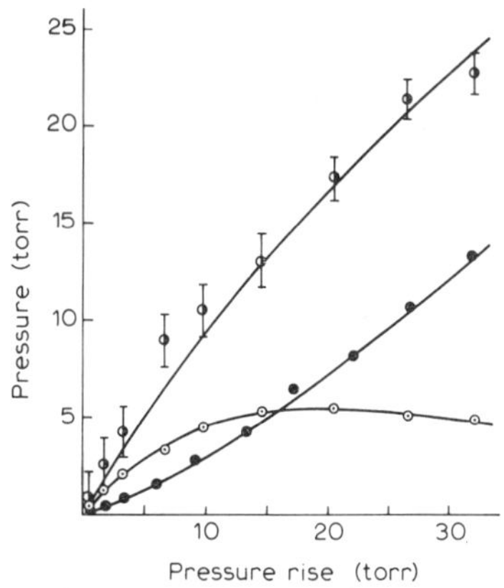

Fig. 5. Comparison of the yields of acetone and isobutene with consumption of isobutane. Curve for isobutane consumption calculated from mechanism, assuming 85 % conversion of isobutane to isobutene and 85 % conversion of isobutene to acetone. Initial temperature = 300°C; initial pressure of isobutane = 67 torr; initial pressure of oxygen = 133 torr. ◑, isobutane consumed; ●, acetone; ⊙, isobutene. (From ref. 44.)

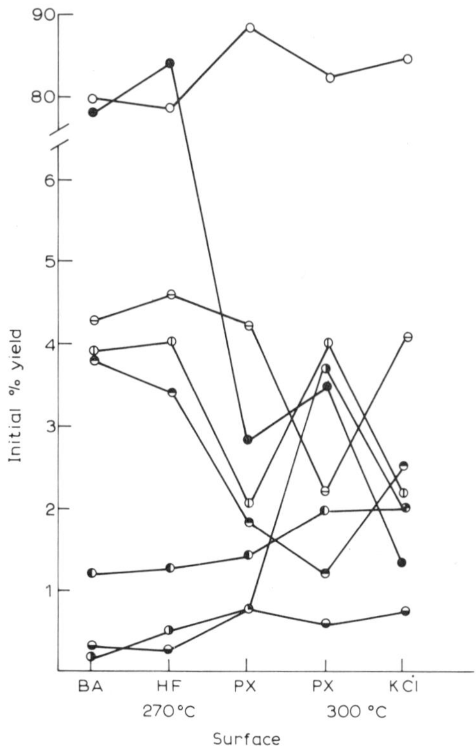

Fig. 6. The variation of initial percentage yields with surface at 270 and 300 °C. ○, isobutene; ●, propionaldehyde; ⊖, acetone; ⦶, propene; ◓, isobutyraldehyde; ◒, methacrolein; ◐, acetaldehyde; ◑, isobutene oxide. (From ref. 44.)

References pp. 361—367

concluded that two main homogeneous processes occur, the first being the formation of the alkene

$$(80\ \%)\ \cdot C_nH_{2n+1} + O_2 \longrightarrow C_nH_{2n} + HO_2\cdot \qquad (2)$$

and the second being the formation of alkylperoxy radicals

$$(20\ \%)\ \cdot C_nH_{2n+1} + O_2 \longrightarrow C_nH_{2n+1}OO\cdot \qquad (3)$$

The alkylperoxy radicals then form an unknown intermediate Z which is sufficiently stable to diffuse to the walls where it decomposes to yield the minor products [44]

$$C_nH_{2n+1}OO\cdot \longrightarrow Z \longrightarrow \text{minor products} \qquad (4)$$

Completion of the chain cycle posed a difficult problem, since it was clear from kinetic considerations [39] that reaction (5) would only compete with reaction (6), viz.

$$C_nH_{2n+2} + HO_2\cdot \longrightarrow \cdot C_nH_{2n+1} + H_2O_2 \qquad (5)$$

$$HO_2\cdot + HO_2\cdot \longrightarrow H_2O_2 + O_2 \qquad (6)$$

in the very early stages of reaction or if E_5 was less than ca. 10 kcal . mole^{-1}. Furthermore, competitive experiments in which the rates of formation of the corresponding alkenes were measured for ethane + propane at 25 % conversion [45] showed that attack on the alkane was unselective and similar experiments for isobutane + propane mixtures showed the selectivity ratio $k_{(\text{isobutane})}/k_{(\text{propane})}$ decreased from 2.8 ± 0.3 at <1 % conversion [46] to 1.3 ± 0.1 at 25 % conversion [47]. Thus, although reaction (5) might complete the chain cycle in the early stages, it was clear that the selectivity decreases as the reaction proceeds. In order to complete the mechanism, it was therefore necessary to include some means of converting $HO_2\cdot$ radicals to the more reactive ·OH radicals.

The reaction sequence (6) + (7)

$$HO_2\cdot + HO_2\cdot \longrightarrow H_2O_2 + O_2 \qquad (6)$$

$$H_2O_2 + M \longrightarrow 2\cdot OH + M \qquad (7)$$

was rejected since hydrogen peroxide decomposes heterogeneously below ca. 450 °C. In search of a low temperature conversion Knox proposed the sequence of reactions

$$HO_2\cdot + AB \longrightarrow HOOAB\cdot \qquad (8)$$

$$HOOAB\cdot + O_2 \longrightarrow HOOABOO\cdot \qquad (9)$$

$$HOOABOO\cdot + HO_2\cdot \longrightarrow HOOABOOH + O_2 \qquad (10)$$

$$HOOABOOH \longrightarrow HO\cdot + A{=}O + B{=}O + \cdot OH \qquad (11)$$

(where AB is the alkene)

The chain cycle may now be completed by reaction (12)

$$C_nH_{2n+2} + \cdot OH \longrightarrow \cdot C_nH_{2n+1} + H_2O \qquad (12)$$

This suggestion caused considerable discussion, some of which was due to a misunderstanding regarding the role of reaction (11). Semenov [48] pointed out that this scheme is for an unbranched-chain reaction and is not valid for a branched-chain reaction, since although reaction (11) yields two $\dot{O}H$ radicals these are converted into two $HO_2\cdot$ radicals by reactions (12) and (2) which then react further by reactions (8) and (10) to form one molecule of the dihydroperoxide. Semenov [48], in fact, went to great lengths to show that Knox's scheme was not viable from kinetic considerations, but apparently overlooked the fact that Knox considered chain branching to be due to the decomposition of alkyl monohydroperoxides and not reaction (11), which is part of his proposed chain propagation cycle. Semenov also objected to the inclusion of a radical—radical propagation reaction (reaction (10)), a proposal which was also unacceptable to Drysdale and Norrish [49].

With regard to Knox's experimental results, Semenov is of the opinion that they present evidence of a heterogeneous reaction such as

$$C_3H_8 + O_2 \xrightarrow{\text{wall}} C_3H_6 + H_2O_2 \qquad (13)$$

which occurs simultaneously with the homogeneous oxidation. He argues that in the early stages, when the rate of the homogeneous oxidation is very small, the heterogeneous reaction will predominate, but following the development of branching the reverse will be true. He therefore concluded that these results had no direct bearing on homogeneous oxidation.

Irvine and Knox [50] recently investigated Semenov's suggestion and found that, whilst there was evidence that the oxidation of isobutane is controlled in part heterogeneously prior to the onset of autocatalysis, the oxidation of ethane and propane are not.

Despite the kinetic objections to this scheme, which centred on the reaction sequence for the conversion of $HO_2\cdot$ to $\cdot OH$ radicals, considerable support has been furnished for the role of the conjugate alkene as the primary product. Brown and Tipper [51] studied the oxidation of propane and cyclohexane at initial temperatures in the range 240—325 °C. With propane the amount of propene present just prior to the cool flame was considerable (ca. 10 % of the propane introduced) and additions of

5 % V/V propene to equimolar mixtures of propane + oxygen reduced the cool-flame induction period by ca. 18 % at 300 °C. Like the previous work, these results showed the importance of the conjugate alkene in the autocatalytic oxidation of propane. However, at 247 °C, the yield of cyclohexene just prior to the cool flame was < 1 % of the total products. In contrast, at temperatures above 300 °C, it becomes the major product and is formed in roughly equal amounts with hydrogen peroxide prior to a stabilized cool flame [52]. Tipper concluded that above 300 °C reaction (2) occurs to an appreciable extent until well after the initial stage of oxidation since the differential yield of the alkene ($d[C_nH_{2n}]/d[C_nH_{2n+2}]$) was > 25 % over at least a quarter of the reaction.

Large yields of isobutene were also found during the induction period of the oxidation of isobutane, and Zeelenberg and Bickel [53] also concluded that its mode of formation was by reactions (5) and (2). In contrast to Knox, however, they suggested that the intermediate oxygenated products are formed from homogeneous isomerization and decomposition reactions of alkylperoxy radicals.

The importance of conjugate alkenes as primary products has inevitably led to neopentane being given much attention, since it does not have a conjugate alkene. Zeelenberg [54] studied the slow oxidation of this fuel at 260 °C and not surprisingly based his mechanism on reactions of alkylperoxy radicals. Fish [55] also studied the oxidation of neopentane, but over a much wider range of initial temperature, 275—425 °C, and pressure, 50—350 torr, and concluded that the Zeelenberg interpretation of the mechanism also applied to the cool-flame reaction. Whilst the formation of all the reaction products could be interpreted in this way the alkene theory could not be entirely discounted, since isobutene is the major primary product [55] and may therefore act as the conjugate alkene. However, the formation of large amounts of 3,3-dimethyloxetan cannot be explained in these terms [55] and therefore seriously questions the validity of the alkene theory at least when it is applied to neopentane.

Recently, Cullis and co-workers [56] studied the role of but-1-ene and but-2-ene during the oxidation of *n*-butane at 315 °C using ^{14}C-tracer techniques. Experiments in which [1-^{14}C]but-1-ene and [2-^{14}C]but-2-ene were added to reacting *n*-butane + oxygen mixtures showed that after 50 sec reaction at least 35 % of the initial alkane had been converted to the two conjugate alkenes and about 60 % of these had reacted further. At least 38 % of the 2-ethyloxiran originated from but-1-ene and at least 59 % of the *cis*-2,3-dimethyloxiran and 43 % of the *trans* compound originated from but-2-ene. Similarly, 8 % of the methyl ethyl ketone was produced from further reactions of but-1-ene and 16 % from but-2-ene. These results unequivocally demonstrate, therefore, the important role of the conjugate alkenes during the oxidation of this relatively low molecular weight alkane. Even so, it was not possible to prove that the mode of alkene formation was via reaction (2) and not via reactions (3) + (14α) + (15)

$$CH_3{-}CH_2{-}CH_2{-}\dot{C}H_2 + O_2 \longrightarrow CH_3{-}CH_2{-}CH{=}CH_2 + HO_2\cdot \quad (2)$$

$$CH_3{-}CH_2{-}CH_2{-}\dot{C}H_2 + O_2 \longrightarrow CH_3{-}CH_2{-}CH_2{-}\underset{\cdot O-O}{\underset{|}{CH_2}} \quad (3)$$

$$CH_3{-}CH_2{-}CH_2{-}\underset{\cdot O-O}{\underset{|}{CH_2}} \longrightarrow CH_3{-}CH_2{-}\dot{C}H{-}\underset{HO-O}{\underset{|}{CH_2}} \quad (14\alpha)$$

$$CH_3{-}CH_2{-}\dot{C}H{-}\underset{HO-O}{\underset{|}{CH_2}} \longrightarrow CH_3{-}CH_2{-}CH{=}CH_2 + HO_2\cdot \quad (15)$$

Cullis et al. considered, however, that but-1-ene is unlikely to arise via the latter route, since the predominant isomerization of the but-1-ylperoxy radical must involve 1:5 H-transfer, reaction (14β),

$$CH_3{-}CH_2{-}CH_2{-}\underset{\cdot O-O}{\underset{|}{CH_2}} \longrightarrow CH_3{-}\dot{C}H{-}CH_2{-}\underset{HO-O}{\underset{|}{CH_2}} \quad (14\beta)$$

because the strain energy of six-membered transition ring is only ca. 0.6 kcal . mole^{-1} compared with ca. 6.5 kcal . mole^{-1} of the five-membered transition ring required by reaction (14α).

Consideration of this problem from the thermokinetic standpoint led Benson [57] to a similar conclusion. His reasoning was as follows.* From the reaction scheme

$$>CH{-}\dot{C}H{-}CH_3 + O_2 \rightleftharpoons\ >CH{-}\underset{\cdot O-O}{\underset{|}{CH}}{-}CH_3 \quad (3)$$

$$>CH{-}\underset{\cdot O-O}{\underset{|}{CH}}{-}CH_3 \rightleftharpoons\ >\dot{C}{-}\underset{HO-O}{\underset{|}{CH}}{-}CH_3 \quad (14\alpha)$$

---

* Readers referring to Benson's work should note that he denotes the C-atoms adjacent to the substituted C-atom as β, γ, δ and the substituted C-atom as α, viz.

$$>CH{-}\overset{\alpha}{\underset{O-O\cdot}{\underset{|}{CH}}}{-}\overset{\beta}{CH_2}{-}\overset{\gamma}{CH_2}{-}\overset{\delta}{CH_3}$$

His reference to internal β abstraction is therefore, in fact, what is normally considered α abstraction, similarly γ is β and δ is γ.

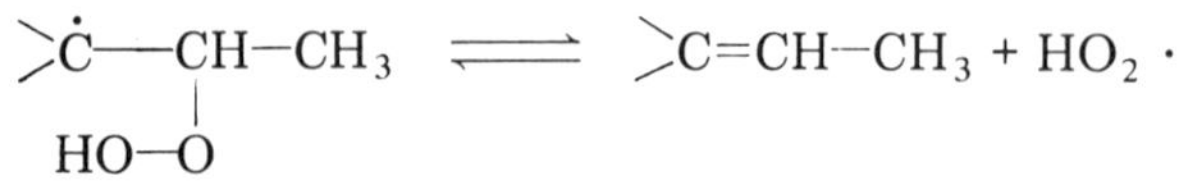

(15)

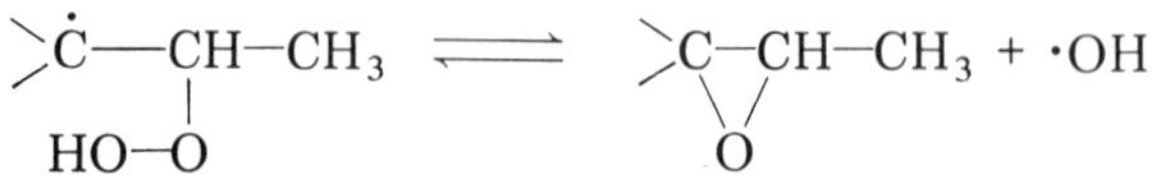

(16α)

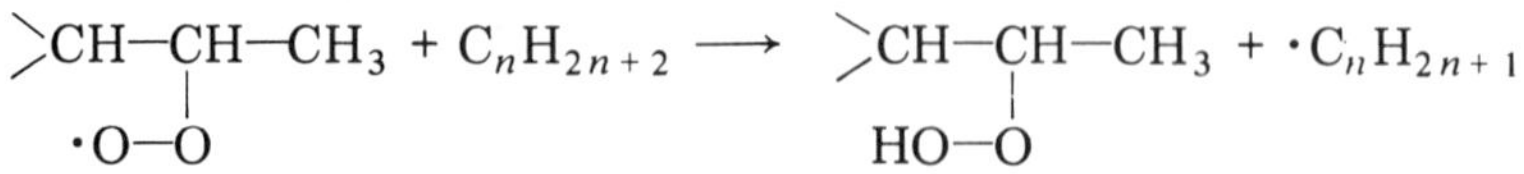

(17)

it can be seen that the maximum fraction of the alkylperoxy radicals converted to the alkene is given by the ratio

$$k_{14\alpha}/(k_{14\alpha} + k_{-3} + k_{17}[C_nH_{2n+2}])$$

$$= (1 + k_{-3}/k_{14\alpha} + k_{17}[C_nH_{2n+2}]/k_{14\alpha})^{-1}$$

$A_{14\alpha}$ = ca. $10^{11.5 \pm 0.3}$ sec^{-1}	(from transition state theory)
$E_{14\alpha}$ = 27 kcal . mole^{-1}	(sum of endothermicity + strain energy for 5-membered transition state ring + activation energy for normal H-abstraction = 4.5 + 6.3 + 16 $\simeq$ 27)
A_{-3} = ca. $10^{15.3}$ sec^{-1}	(from ΔS_3 = —32 cal . deg^{-1} . mole^{-1}, k_3 = $10^{9.6}$ l . mole^{-1} . sec^{-1}, $E_3 = 0$, $A_{-3} = \dfrac{10^{9.6}}{e^{-16}} = 10^{15.3}$)
E_{-3} = 27 kcal . mole^{-1}	(equal to the endothermicity)

Hence, the fraction converted $\lesssim k_{14\alpha}/k_{-3} = 10^{-3.8}$.

The competing path to alkene formation is the exothermic reaction (2)

$$\cdot C_nH_{2n+1} + O_2 \longrightarrow C_nH_{2n} + HO_2\cdot \qquad (2)$$

for which $A_2 \geqslant$ ca. $10^{8.7}$ l . mole^{-1} . sec^{-1}. E_2 is not known, but Benson believes it to be about 3 kcal . mole^{-1}. He points out, however, that even if it was as high as 6 kcal . mole^{-1} the rate of alkene production via reaction (2) at 300 °C would be ca. eight times faster than the maximum rate via the intramolecular rearrangement.

Benson's argument rests on the high activation energy (16—18 kcal . mole^{-1}) he uses for the bimolecular H-abstraction by $C_nH_{2n+1}OO^{\cdot}$. Unfortunately, there is no direct experimental evidence available to substantiate this value. In contrast, Fish [58], Heicklen [59] and Knox [38] have all estimated that the activation energy for 2°-H-abstraction by $C_nH_{2n+1}OO^\bullet$ is only ca. 11—13 kcal . mole^{-1}. Since the bond strengths $D[HOO{-}H] = 89 \pm 2$ [60], $D[C_nH_{2n+1}OO - H] = 89$ [39] and $D[H - Br] = 87$ are all similar, $HO_2^\bullet$, $C_nH_{2n+1}OO^\bullet$ and Br should all show similar selectivity in their initial attack on the alkane. The activation energy for $2^\bullet$-H-abstraction by Br [61] is 10.2 kcal . mole^{-1} and by $HO_2^\bullet$ [39, 62] it is in the range 6—13 kcal . mole^{-1}, thus a value of 11 kcal . mole^{-1} certainly seems more reasonable. Hence, as Benson points out, in this case $E_{14\alpha}$ would be ca. 5 kcal . mole^{-1} lower than he estimated and internal and external formation of the alkene would be competitive at 300 °C. As yet such thermokinetic arguments are rather tenuous and too much emphasis must not be placed upon them until more accurate kinetic information regarding individual propagation reactions is available.

More recently, Lucquin and co-workers [63, 64] have shown from studies of the oxidation of *n*-butane and isobutane that the alkene theory is in fact at variance with experiment. Thus, on this theory the negative temperature coefficient is seen as a direct consequence of the increasing instability with temperature of the hydroperoxyalkyl radical, viz.

$$C_nH_{2n} + HO_2^\bullet \rightleftharpoons {}^\bullet C_nH_{2n}OOH \qquad (8)$$

Whilst this is acceptable kinetically, it fails to explain the analytical observations. Thus, carbonyl compounds and consequently carbon oxides are necessarily formed in the branching reaction (11), but the yields of these compounds increase in the negative temperature region where the branching is suppressed. Furthermore, little further reaction of the intermediate conjugate alkene occurs, hence the carbonyl compounds must be formed by a different route from that proposed in the alkene theory under these conditions.

Fish and Wilson [65] studied the cool-flame oxidation of 2,3-dimethylbutane whose structure should be particularly conducive to conjugate alkene formation, since both 2,3-dimethylbutyl radicals have a tertiary hydrogen atom attached to a carbon adjacent to the carbon bearing the free electron. The results showed, however, that although conjugate alkene formation is important, abstractive attack by oxygen on the alkyl radical contributes less to the chain-propagation process than does its additive attack.

3.2.2 Alkylperoxy radical isomerization theory

The alkylperoxy radical isomerization theory was developed primarily as a result of studies of the oxidation of alkanes of carbon number greater than four during the later stages of the reaction, namely just prior to and at the cool flame.

References pp. 361—367

The main chain-propagating cycle in this theory may be summarized in general terms by the following reactions

$$\cdot C_nH_{2n+1} + O_2 \longrightarrow C_nH_{2n+1}OO\cdot \quad (3)$$

$$C_nH_{2n+1}OO\cdot \longrightarrow \cdot C_nH_{2n}OOH \quad (14)$$

$$\cdot C_nH_{2n}OOH \longrightarrow P + \cdot OH \quad (16)$$

$$C_nH_{2n+2} + \cdot OH \longrightarrow \cdot C_nH_{2n+1} + H_2O \quad (12)$$

In contrast to the alkene theory the predominant mode of oxidation of the alkyl radicals is by oxygen addition and the alkylperoxy radical so formed then undergoes homogeneous intramolecular rearrangement (reaction (14)). Decomposition of the rearranged radical (reaction (16)) usually leads to a hydroxyl radical and stable products which include O-heterocycles, carbonyl compounds and alcohols with rearranged carbon skeletons relative to the fuel and alkenes. The chain-cycle is then completed by unselective attack on the fuel by the hydroxyl radical (reaction (12)).

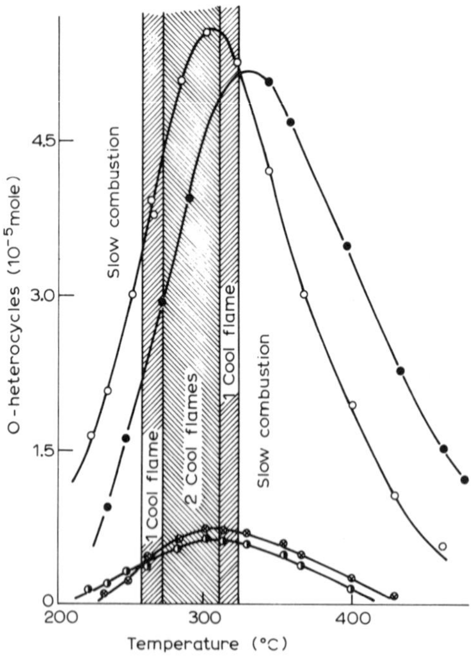

Fig. 7. The variation with temperature of the principal O-heterocycles formed under conditions of maximum rate during the oxidation of *n*-pentane. *n*-Pentane introduced = 39.9 x 10^{-5} mole; *n*-pentane:oxygen = 0.75. ○ 2,4-dimethyloxetan; ●, 2-methyl-3-ethyloxiran; ◑, 2-methyloxiran; ⊗, oxiran. (From ref. 70.)

Reaction (3) is a second-order reaction (except for methyl and ethyl radicals) and has an activation energy close to zero. The rate coefficient will therefore be approximately equal to the pre-exponential factor. Various estimations and experimental determinations of its value have been made, however, a value of $10^{9.3}$ l . mole^{-1} . sec.$^{-1}$ appears to be an acceptable mean value for most workers [38, 58].

The importance of alkylperoxy radicals as intermediates had long been realized (see Sect. 2) and their subsequent reaction to yield the alkylhydroperoxide or decomposition products such as aldehydes and alcohols had been reasonably successful in describing the mechanism of the autocatalytic oxidation of alkanes. However, even though O-heterocycles (which cannot be derived from intermediate aldehydes) had been found in the products of the oxidation of *n*-pentane as early as 1935 [66], the true extent of alkylperoxy radical isomerization reactions has been recognized only recently. Bailey and Norrish [67] first formulated the production of O-heterocycles in terms of alkylperoxy radical isomerization and subsequent cyclization in order to explain the formation of 2,5-dimethyltetrahydrofuran during the cool-flame oxidation of *n*-hexane. Their mechanism was a one-step process which involved direct elimination of •OH. However, it is now generally formulated as shown in reactions (14γ) and (16γ)

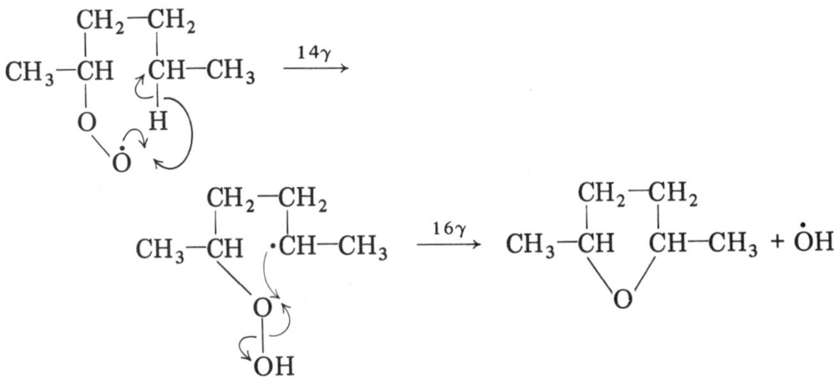

Since the work of Bailey and Norrish O-heterocycles have been found in the products of every alkane studied of carbon number ≥ 4, as shown in Table 1 and their yields are often considerable particularly under cool-flame conditions as shown in Table 2 and Fig. 7.

On the basis of the alkene theory the hydroperoxalkyl radical initially formed must necessarily be the α-hydroperoxyalkyl radical, e.g. for the oxidation of *n*-butane

$$CH_3{-}CH{=}CH{-}CH_3 + HO_2\cdot \longrightarrow CH_3{-}\underset{\displaystyle |\atop \displaystyle O{-}OH}{CH}{-}\dot{C}H{-}CH_3 \qquad (-15)$$

TABLE 1
O-heterocycles formed during the oxidation of hydrocarbons

Hydrocarbon	Oxidant	O-heterocycle	$F:O_2$ or air	Temp. (°C)	Pressure (torr)	Reactor	Nature of reaction	Ref.
Isobutane	Oxygen	2,2-Dimethyloxiran 2,Methyloxiran 3-Methyloxetan	9:1 to 1:4	260—360	100—400	Static	Slow	53
	Oxygen	2,2-Dimethyloxiran 2-Methyloxiran 3-Methyloxetan	1:2	310	240	Static	Cool flame	68
n-Butane	Oxygen	2-Methyloxetan 2-Ethyloxiran Tetrahydrofuran 2,3-Dimethyloxiran Oxiran	1:3.5	315	160	Static	Slow and cool flame	56
n-Pentane	Air	2-Methyltetrahydrofuran	1:80 to 3:80	450—500	760	Annular flow	Slow	69
	Oxygen	2,4-Dimethyloxetan 2-Methyl-3-ethyloxiran	3:4	250—450	20—200	Static	Slow and cool flame	70
	Oxygen	2-Methyltetrahydrofuran 2,4-Dimethyloxetan	1:1	257—280	150	Static	Slow and cool flame	71
	Oxygen + 99 % Argon	2-Methyltetrahydrofuran 2,4-Dimethyloxetan	8:1 to 1:1	600—850	1.5—5.0 atmos.	Shock-tube		72
Neopentane	Oxygen	3,3-Dimethyloxetan	3:1 to 2:3	260—290	100—400	Static	Slow	54
	Oxygen	3,3-Dimethyloxetan	1:2	310—340	170	Static	Cool flame	55
	Oxygen	3,3-Dimethyloxetan	1:1	280	200	Static	Slow	49
n-Hexane	Oxygen	2,5-Dimethyltetrahydrofuran 2-Methyltetrahydropyran	1:0.54 to 1:1.26	300	760	Flow	Cool flame	67
	Oxygen	2,5-Dimethyltetrahydrofuran 2-Methyl-4-ethyloxetan 2-Methyl-3-*n*-propyloxiran	1.25:1	200—450	20—200	Static	Cool flame	73
	Air	2-Ethyltetrahydrofuran 2,5-Dimethyltetrahydrofuran	1:48.5	—	760	Flow	Cool flame	74
	Oxygen	2-Ethyltetrahydrofuran 2,5-Dimethyltetrahydrofuran 2-Methyl-4-ethyloxetan 2-*n*-Propyloxetan 2-Methyl-3-*n*-propyloxiran	1:0.14 to 1:0.27	300—650	760	Flow	Slow	75, 76

2-Methylpentane	Oxygen	2,2-Dimethyltetrahydrofuran 2,4-Dimethyltetrahydrofuran 2,2,4-Trimethyloxetan 2,2-Dimethyl-3-ethyloxiran 2-Methyl-2-*n*-propyloxiran 2-Methyl-3-isopropyloxiran	1:2	230—310	40—220	Static	Slow and cool flame	77
	Oxygen/ nitrogen	As above plus 2-Methyltetrahydropyran	1:19:72	440—660	10—50 atmos.	Rapid compression machine	Slow and cool flame	78
	Oxygen/ nitrogen	As above	1:19:76	433	20 atmos.	Ricardo E6 engine	Cool flame	79
3-Methylpentane	Oxygen	2,3-Dimethyltetrahydrofuran 2,3,4-Trimethyloxetan 2-Methyl-3-ethyloxetan 2,2-Diethyloxiran 2-Ethyl-2,3-dimethyloxiran	1:2	295—405	80—150	Static	Cool flame	80
2,2-Dimethylbutane	Oxygen	2,3,3-Trimethyloxetan	1:0.7	420—450	760	Flow	Slow	81
2,3-Dimethylbutane	Oxygen	2,2,3-Trimethyloxetan 2,2,3,3-Tetramethyloxiran 2-Methyl-2-isopropyloxiran	1:2	318	325	Static	Cool flame	65
n-Heptane	Oxygen	2-Methyl-5-ethyltetrahydrofuran 2-*n*-Propyltetrahydrofuran 2-Methyl-4-*n*-propyloxetan 2,3-Diethyloxiran 2-*n*-Pentyloxiran	1:0.14 to 1:0.27	300—650	760	Flow	Slow	75, 76
	Oxygen	2,6-Dimethyltetrahydropyran 2-Methyl-5-ethyltetrahydrofuran 2-*n*-Propyltetrahydrofuran C_4—C_6 Tetrahydrofurans	1:1	235	150	Static	Cool flame	82
	Oxygen/ nitrogen	2-Methyl-5-ethyltetrahydrofuran 2-*n*-Propyltetrahydrofuran 2-Methyltetrahydrofuran 2-Methyl-4-*n*-propyloxetan 2-Methyl-3-*n*-butyloxiran 2-Ethyl-3-*n*-propyloxiran 2-*n*-Pentyloxiran	1:3.4:13.8	265—450	760	Flow	Cool flame	83
	Oxygen	2-Methyl-5-ethyltetrahydrofuran 2-*n*-Propyltetrahydrofuran 2-Methyltetrahydrofuran 2-Methyl-4-*n*-propyloxetan 2-Methyl-3-*n*-butyloxiran 2-Ethyl-3-*n*-propyloxiran 2-*n*-Pentyloxiran	1:9	>450	>35 atmos.	Fired engine	End gas	84

References pp. 361—367

TABLE 1—*continued*

3-Ethylpentane	Oxygen	2-Methyl-3-ethyltetrahydrofuran 2,4-Dimethyl-3-ethyloxetan 2,2-Diethyloxetan 2-Methyl-3,3-diethyloxiran	1:2	295—405	80—150	Static	Cool flame	85, 86
2,2,4-Trimethylpentane	Oxygen	2,2,4,4-Tetramethyltetrahydrofuran 2-Isobutyl-3-methyloxetan	2:1	450—475	760	Flow	Slow	81
	Air	As above	1:13.8	520—850	—	CFR engine	Pre-flame	87
	Air	As above plus 2-Isopropyl-3,3-dimethyloxetan 2,2-Dimethyl-3-isobutyloxiran	1:49.4	477—657	10—27 atmos.	Waukesha CFR engine	End gas	88, 89
	Air	As above	1:49.4	477—657	10—27 atmos.	Waukesha CFR engine	End gas and post-flame	90
Cyclohexane	Oxygen	1,2-Epoxycyclohexane 1,4-Epoxycyclohexane	1:0.5 to 1:2	260	50—100	Static	Slow	91
	Air	As above	1:0.5 to 1:2		760	Flow	Slow	91
o-Xylene	Air	*o*-Xylene oxide	1:105	455—525	760	Flow	Slow	92

TABLE 2
Percentage conversion[a] to conjugate O-heterocycles during the oxidation of different alkanes

Hydrocarbon	% Conversion to conjugate O-heterocycles[a]	Initial pressure (torr)	Initial temp. (°C)	Combustion regime	Hydrocarbon/ oxygen ratio
n-Butane	8.3	240	318	c.f.	2:7
n-Pentane	43.0	141	300	c.f.	3:4
n-Hexane	11.8	135	245	c.f.	5:4
2-Methylpentane	25.0	30	291	s.c.	1:2
3-Methylpentane	10.2	118	295	c.f.	1:2
n-Heptane	49.9	760	265—450	c.f.	1:3
3-Ethylpentane	19.8	115	295	c.f.	1:2

[a] Moles of product x 100 per mole of alkane consumed. c.f., cool flame; s.c., slow combustion. (From ref. 86.)

Subsequent cyclization will lead to 2,3-dimethyloxiran

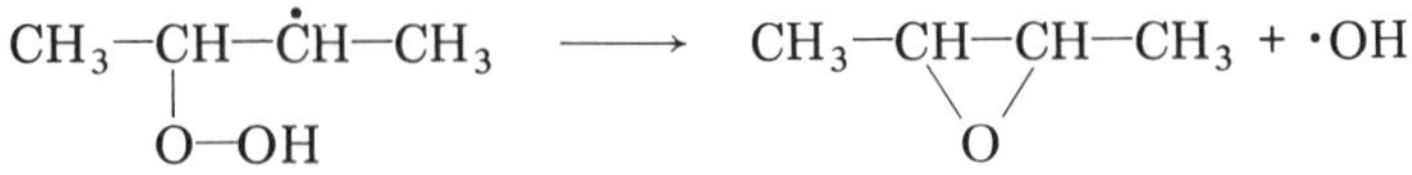
(16α)

2-Methyloxetan can only be formed if the α-hydroperoxyalkyl radical isomerizes to the but-2-ylperoxy radical, which then re-isomerizes to yield the β-hydroperoxyalkyl radical, reactions (−14α), (14β) and (16β),

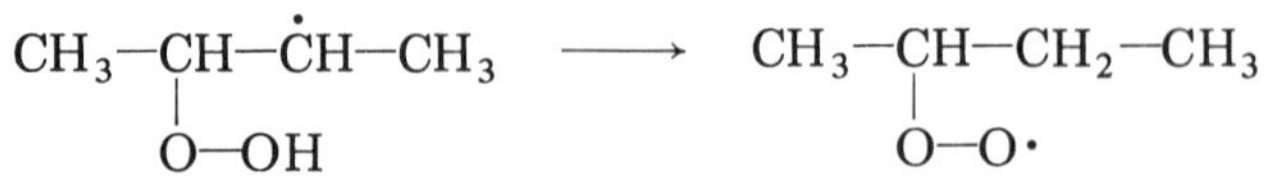
(−14α)

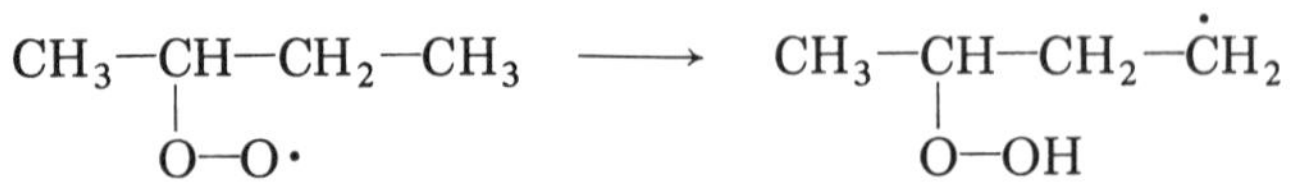
(14β)

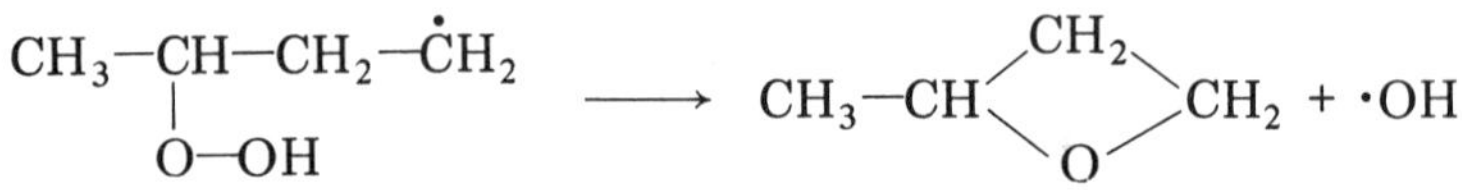
(16β)

The addition of [2-^{14}C]but-2-ene to reacting *n*-butane + oxygen mixtures at 315 °C [56] showed that the reverse isomerization reaction (− 14α) does not occur to any appreciable extent, since the 2-methyloxetan found in the products was inactive. It can be safely concluded, therefore, that the formation of derivatives of oxetans, furans and pyrans is diagnostic of alkylperoxy radical isomerization and subsequent decomposition, reactions (14) + (16). Table 1 thus presents a considerable volume of evidence for the wide occurrence of this chain-propagation step.

It is clear from the wide variety of intermediate products formed that the initial attack on the alkane is extremely unselective. Consideration of the mode of formation of the major products via alkylperoxy radical isomerization shows that ·OH is the radical predominantly formed in reaction (16) and spectroscopic studies have confirmed the presence of ·OH radicals in the oxidation of aldehydes [93] and methyl radicals [94]. Furthermore, Haskell and Read [95] have convincingly shown that the inhibition of the oxidation of 2-methylpentane by hydrogen is due to the participation of reactions (18) and (19)

$$H_2 + \cdot OH \longrightarrow H_2O + H \qquad (18)$$

$$H + O_2 + M \longrightarrow HO_2\cdot + M \qquad (19)$$

There is little doubt, therefore, that ·OH is the main chain carrier, particularly in the later stages of the reaction. The Arrhenius parameters of reaction (12)

$$C_nH_{2n+2} + \cdot OH \longrightarrow \cdot C_nH_{2n+1} + H_2O \qquad (12)$$

TABLE 3

Thermodynamic and kinetic parameters for alkylperoxy radical isomerization at 600 °K

$C_nH_{2n+1}OO\cdot \rightarrow \cdot C_nH_{2n}OOH$

Nature of C—H bond broken	Position of C atom from which H is transferred	Size of transition state ring (including H)	ΔH (kcal . mole^{-1})	$\log_{10} A$ (A in sec^{-1})	E (kcal . mole^{-1})	$\log_{10} k_{14}$	$\log_{10} K_{14}$	$\log_{10} k_{-14}$	Ref.
Primary	α, γ	5 or 7	8	11.5	20.5[a]	4.1	—2.9	7.0[d]	58
	β	6	8	11.5	14.6	6.2	—2.9	9.1	
	δ	8	8	11.5	24.0	2.8	—2.9	5.7	
Secondary	α, γ	5 or 7	4.5	11.5	17.0	5.4	—1.6	7.0	58
	β	6	4.5	11.5	11.1	7.5	—1.6	9.1	
	δ	8	4.5	11.5	20.5	4.1	—1.6	5.7	
Tertiary	α, γ	5 or 7	1.7	11.5	14.2	6.4	—0.6	7.0	58
	β	6	1.7	11.5	8.3	8.5	—0.6	9.1	
	δ	8	1.7	11.5	17.7	5.1	—0.6	5.7	
Secondary	α	5		12.7	30	18			103
Tertiary	α	5	4.5	11.5 ± 0.3	27[b]				57
	β	6		10.8					57
	β	6		11	15				104
	β	6		10.5	15				105
Unspecified	Unspecified	Unspecified		12.5	30—35[c]				106
Unspecified				12.0	22—27				38

[a] These activation energies were estimated as follows. The activation energy for H-abstraction by $C_nH_{2n+1}OO\cdot$ was first estimated using Benson's empirical formula for endothermic reactions ($E = \Delta H + 6$). This was then added to the corresponding strain energy for saturated cycloalkane rings (three-membered, 28; four-, 26; five-, 6.5; six-, 0.6; seven-, 6.5; eight, 10 kcal . mole^{-1}). Thus, for example E for β-3°-H-transfer = (1.7 + 6) + 0.6 = 8.3 kcal . mole^{-1}.

[b] This value of E is at least 2.8 kcal . mole^{-1} too high because Benson used the value of ΔH for abstraction from sec-C—H (4.5 kcal . mole^{-1}) instead of 1.7 kcal . mole^{-1} for *tert*-C—H. This value is also based on a high value of E(16–18 kcal . mole^{-1}) for abstraction from *tert*-C—H.

[c] These values of E are not very meaningful, since they embrace isomerizations involving H-abstraction from α, β and γ-secondary-C—H and from α, β, γ and δ-primary-C—H.

[d] Intramolecular isomerization is a reversible process in which $\Delta S = 0$. The equilibrium constants for each isomerization were calculated using the relationship $K = \exp(\Delta S/R) . \exp(-\Delta H/RT)$ and hence k_{-14} was obtained from k_{14} and K_{14}.

are difficult to determine accurately [96] but the activation energy is almost certainly < 4 kcal . mole^{-1} even for a primary C—H bond [97—99] (see Table 13, p. 316). At ambient temperature the selectivity of ·OH is similar to that of Cl atoms [100, 101]. Assuming that this is also the case at higher temperatures, the relative frequency of attack by ·OH on *p*-C—H, *s*-C—H and *t*-C—H will be 1:3:5 per H atom [77]. Recently, Greiner [102] found that this ratio is 2:3:5 per H atom at 525 °C, while the data of Baldwin and Walker [99] suggest that this ratio is of the order of 1:3:9 at 300 °C. In any event it is clear that attack by OH is unselective and in attempting to explain intermediate product formation, it is necessary therefore to consider each alkyl radical which may be formed from the alkane and hence the isomerization and subsequent decomposition of each of the corresponding alkylperoxy radicals.

The Arrhenius parameters of the isomerization reactions have not been directly measured experimentally, but their relative activation energies have been estimated. Table 3 shows the values of the activation energy and rate coefficient at 600 °K estimated by Fish [58] for isomerizations involving H-transfer from primary, secondary and tertiary atoms in the α, β, γ and δ positions to the peroxidized C-atom, viz.

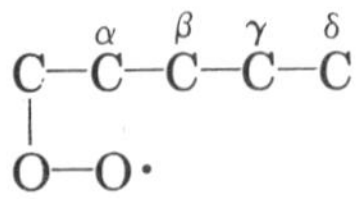

The pre-exponential factor for unimolecular reactions involving a cyclic transition state has been estimated by Benson [57, 104] to be $10^{11.5 \pm 0.3}$ sec^{-1}, and is taken to be the same for each of these isomerizations. Other workers have made experimental estimates of the activation energies of some of the isomerizations and these are included in Table 3 for comparison.

Schroder et al. [82] explained the formation of many of the products found during the oxidation of *n*-heptane in terms of the isomerization and subsequent decomposition of the various alkylperoxy radicals formed, while Zeelenberg similarly explained the formation of all the oxygenated intermediate products formed in the initial stages during the oxidation of isobutane [53], neopentane [54] and cyclohexane [91]. Perhaps the chief protagonist of the alkylperoxy radical isomerization theory, however, has been Fish, who has classified the modes of decomposition of hydroperoxyalkyl radicals on several occasions (see, for example, refs. 77 and 107) and has satisfactorily explained the mechanism of product formation during the cool-flame oxidation of *n*-hexane [73], 2-methylpentane [77], neopentane [55] and 2,3-dimethylbutane [65] in these terms.

(a) Modes of decomposition of hydroperoxyalkyl radicals to stable products

(i) Simple decomposition to O-heterocycles and ·OH

The simple decomposition of a hydroperoxyalkyl radical to an O-heterocycle with elimination of ·OH is an irreversible unimolecular process, e.g.

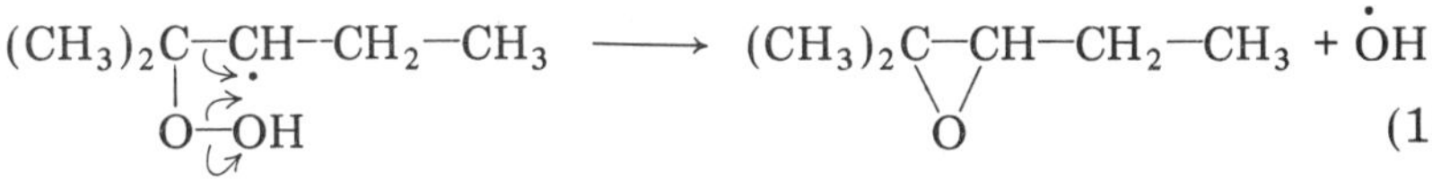

(16α)

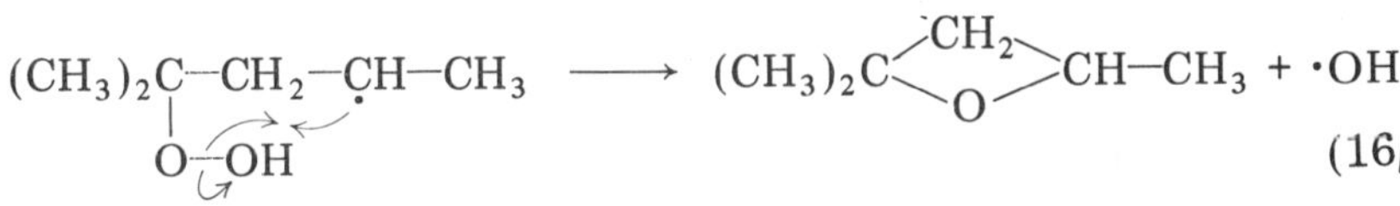

(16β)

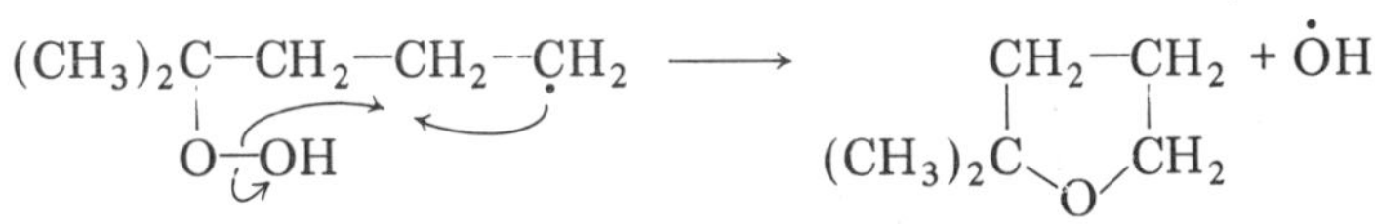

(16γ)

The Arrhenius parameters have not been determined experimentally for gas phase reactions, but the pre-exponential factor is believed [57] to be $10^{11.5 \pm 0.3}$ and Fish [58, 107] has estimated* the activation energies of these decompositions as shown in Table 4.

Knox [85, 106] believes that these values of the activation energies are too low. Consideration of the equal yields of 2-methylpropene and 3,3-dimethyloxetan obtained from the addition of small amounts of neopentane to slowly reacting mixtures of hydrogen and oxygen at 480 °C [108] allowed him to estimate $k_{16\beta} = 10^{4.0}$ sec^{-1} for

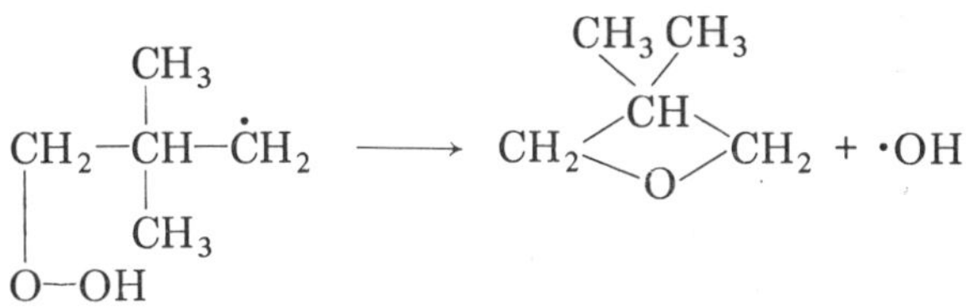

(16β)

---

* The only analogous reaction for which E is known is the thermoneutral cyclization of the 3-iodopropyl radical [104] for which E = 17 kcal . mole^{-1}. By analogy with the Semenov—Polanyi relationship [3] ($E = 11.5 + 0.25\Delta H$) the corresponding relationship for cyclization to 3-membered ring compounds would be $E = 17.0 + 0.25\Delta H$. 4-, 5- and 6-membered rings have lower ring strain energies and the value of E calculated from this relationship was therefore reduced by a quarter of the value of the difference from the strain energy of 3-membered rings.

TABLE 4

Thermodynamic and kinetic parameters for the decomposition of hydroperoxyalkyl radicals by cyclization at 600 °K (from ref. 107)

$\cdot C_nH_{2n}OOH \rightarrow$ O-heterocycle + $\cdot OH$

O-heterocycle	Ring size	ΔH (kcal . mole^{-1})	$17 + 0.25\Delta H$	Strain energy (kcal . mole^{-1})	E (kcal . mole^{-1})	$\log_{10} A$ (A in sec^{-1})	$\log_{10} k_{16}$
Oxiran	3	−13	14	28	14	11.5	6.4
Oxetan	4	−15	13	26	13	11.5	6.8
Tetrahydrofuran	5	−35	8	6.5	3	11.5	10.4
Tetrahydropyran	6	−40	7	0.6	0	11.5	11.5

Thus, for $A_{16\beta} = 10^{12}$—10^{13} sec^{-1}, $E_\beta = 28$—34 kcal . mole^{-1}. In contrast, Benson [36] estimated $k_{16\beta} = 10^{6.6 \pm 0.3}$ sec^{-1} from the same experimental results! For $A_{16\beta} = 10^{11.8}$ sec^{-1} this gave $E_{16\beta} = 18$ kcal . mole^{-1}. In earlier studies [104, 57] the same author suggested that $A_{16\beta} = 10^{11}$ sec^{-1} and $E_{16\beta} = 15$ kcal . mole^{-1}, and $A_{16\alpha} = 10^{11.5}$ sec^{-1} and $E_{16\alpha} = 10$—16 kcal . mole^{-1} (oxiran formation).

Recently, however, two pieces of direct experimental evidence have become available. First, in a low temperature (100 °C) liquid-phase investigation of the oxidation of 2,4-dimethylpentane Mill [105] found $A_{14\beta} = 10^{10.5}$ sec^{-1} and $E_{14\beta} = 15$ kcal . mole^{-1} for

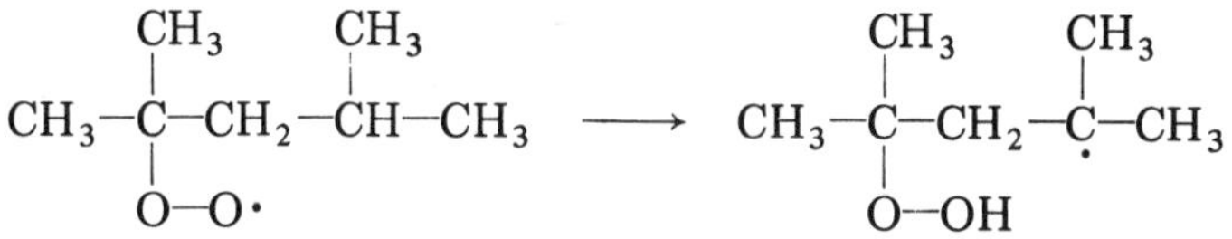

(14β)

and $A_{16\beta} = 10^{11.5}$ sec^{-1} and $E_{16\beta} = 14 \pm 2$ kcal . mole^{-1} for

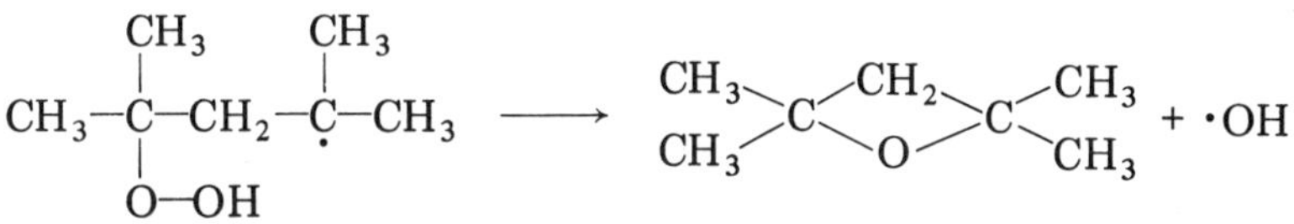

(16β)

Secondly, from studies of the yields of the conjugate oxiran and alkene found during the oxidation of 3-ethylpentane, see Sect. 3.2.2(a)(ii), Cullis and co-workers [86] have shown that if the values of the Arrhenius parameters for reaction (15)

$$\alpha\text{-}\dot{C}_nH_{2n+1}OOH \longrightarrow C_nH_{2n} + HO_2\cdot \qquad (15)$$

suggested by Knox [39] and Benson [104] ($A = 10^{13}$—$10^{13.5}$ sec^{-1}. $E = 18$—20 kcal . mole^{-1}) are accepted, then at 600 °K $k_{16\alpha} = 10^{6.2}$—$10^{6.5}$ sec^{-1} for the formation of 2,2-diethyl-3-methyloxiran. For $A_{16\alpha} = 10^{11.5}$ sec^{-1}, this gives $E_{16} = 13.8$—14.6 kcal . mole^{-1}. Both these experimentally determined values are in good agreement with those estimated by Fish [58, 107].

(ii) β-Scission of C—O to give conjugate alkenes and $HO_2\cdot$

The two modes of alkene formation were discussed in Sect. 3.2.1, viz.

$$\dot{C}_nH_{2n+1} + O_2 \longrightarrow C_nH_{2n} + HO_2\cdot \qquad (2)$$

or

$$\dot{C}_nH_{2n+1} + O_2 \longrightarrow C_nH_{2n+1}OO\cdot \qquad (3)$$

$$C_nH_{2n+1}OO\cdot \longrightarrow \alpha\text{-}\dot{C}_nH_{2n}OOH \qquad (14\alpha)$$

$$\alpha\text{-}\dot{C}_nH_{2n}OOH \longrightarrow C_nH_{2n} + HO_2\cdot \qquad (15)$$

Experimentally it is difficult to distinguish between these modes of formation. However, it has been shown [70] that in the cool-flame oxidation of n-pentane the yields of C_5 O-heterocycles and C_5 alkenes vary with time in a similar manner, suggesting that they are formed from a common precursor, e.g.

$$CH_3{-}CH(OOH){-}\dot{C}H{-}CH_2{-}CH_3 \longrightarrow CH_3{-}\underset{\diagdown O \diagup}{CH{-}CH}{-}CH_2{-}CH_3 + \dot{O}H \qquad (16\alpha)$$

$$CH_3{-}CH(OOH){-}\dot{C}H{-}CH_2{-}CH_3 \longrightarrow CH_3{-}CH{=}CH{-}CH_2{-}CH_3 + HO_2\cdot \qquad (15)$$

A recent investigation of the oxidation of 3-ethylpentane [86] and of 3-methylpentane [80] over a wide range of initial conditions has confirmed this conclusion. Thus, over the temperature range 295—405 °C, the conjugate alkenes and oxirans are primary products and the ratio of their yields is constant, except in the later stages of reaction where their own further reaction is extensive. For these higher molecular weight (C_5—C_7)

TABLE 5

Kinetic parameters for the β-scission decomposition of α-hydroperoxyalkyl radicals

$\alpha\text{-}\cdot C_nH_{2n}OOH \rightarrow C_nH_{2n} + HO_2\cdot$

$\log_{10} A$ (A in sec^{-1})	E (kcal . $mole^{-1}$)	Ref.
13.5	20	39[a]
13.0	18	104[a]
13.4	20	57
13.7	25.5	58

[a] Estimations [39, 104] may be made from considerations of the equilibrium constant K_{-15} and the rate coefficient for the reverse reaction k_{-15}. The entropy change in reaction (—15) is —32 cal . deg^{-1} . $mole^{-1}$ and the enthalpy change will equal the dissociation energy $D[\dot{C}_nH_{2n}{-}OOH]$ which has been estimated [39] to be 14 kcal . $mole^{-1}$. Thus, $K_{-15} = 10^{-5.2}$ exp $(-14{,}000/RT)$ l . $mole^{-1}$. The Arrhenius parameters for reaction (—15) may be estimated by analogy with similar radical addition reactions. Taking values of $A_{-15} = 10^{8.3}$ l . $mole^{-1}$. sec^{-1} and E_{-15} = 6 kcal . $mole^{-1}$ gives $k_{-15} = 10^{8.3}$ exp $(-6{,}000/RT)$, whence $k_{15} = 10^{13.5}$ exp $(-20{,}000/RT)$ [39]. Using slightly different values for the assumed parameter, Benson [104] obtained $k_{15} = 10^{13}$ exp $(-18{,}000/RT)$ for the decomposition of alkylperoxyalkyl radicals.

alkanes then the formation of alkenes via reactions (3) + (14α) + (15) appears to be important.

Once again there are no quantitative data available for the Arrhenius parameters for reaction (15), but several estimations have been made as shown in Table 5.

(iii) O—O Homolysis and alkyl group shift to give a rearranged carbonyl compound and ·OH

It has already been demonstrated (Sect. 3.2.1) that ^{14}C-tracer techniques provide an extremely powerful tool for diagnosing reaction mechanisms. They have also been successfully used to show that this mode of decomposition may also occur particularly during the passage of a cool flame. Thus, for example, the formation of butanone from isobutane occurs via reaction (20)

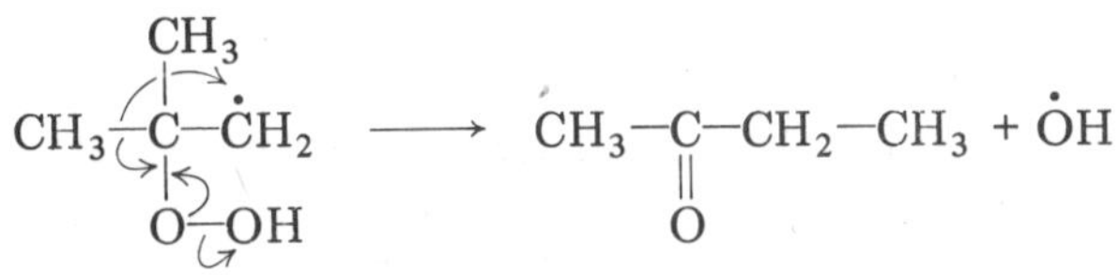
(20)

i.e. the *tert*-butylperoxy radical undergoes isomerization involving 1:4 H-transfer followed by O—O homolysis and 1:2 CH_3-shift [109]. It had been suggested [110] that butanone was formed from the further reaction of acetone

$$CH_3COCH_3 \xrightarrow{-H} CH_3CO\dot{C}H_2 \xrightarrow{+CH_3} CH_3COCH_2CH_3$$

but the addition of [1,3-^{14}C] acetone to reacting mixtures of isobutane + oxygen and comparison of the specific activity of the acetone after combustion with that of the butanone formed showed that in the cool-flame oxidation 98 % of the butanone formed did not involve acetone. Even during slow combustion 75 % of the butanone was formed from reactions not involving acetone. In view of the large yields of butanone formed during cool-flame oxidation it was concluded that reaction (20) was of considerable importance under these conditions. Moreover, this mode of isomerization and decomposition of *tert*-butylperoxy radicals was shown to be 20 times faster than intermolecular H-abstraction followed by decomposition of the *tert*-butylhydroperoxide to acetone. ^{14}C-tracer studies on the decomposition of *tert*-butylhydroperoxide confirmed the unimportance of the latter route [111].

Cullis and co-workers [80, 85, 86] have shown that group shifts involving ethyl radicals may also take place. Thus the presence of small yields (ca. 2 %) of 4-methylhexan-3-one and heptan-3-one in the products of the cool-flame oxidation of 3-ethylpentane show that the 3-ethyl-3-

hydroperoxypent-2-yl and the 3-ethyl-3-hydroperoxypent-1-yl radicals may decompose as follows

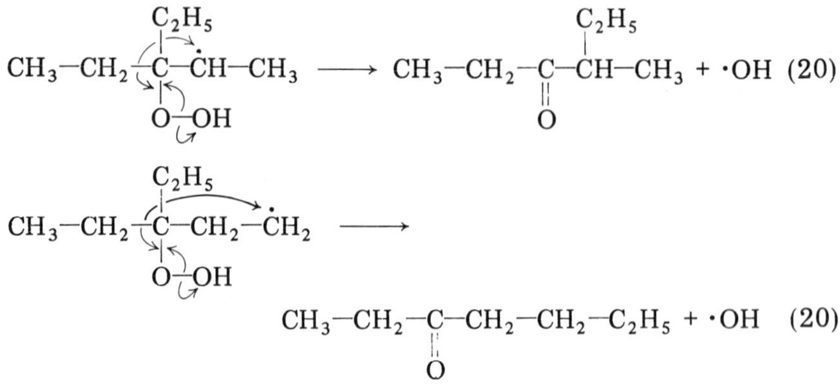

These compounds are also formed prior to the cool flame, so reaction (20) does not take place solely during the transient temperature rise accompanying the passage of the flame.

Similar reactions involving O—O homolysis but accompanied by H-shifts may yield carbonyl compounds with the same carbon skeleton as the fuel. Studies of the oxidation of alkanes in which H-, methyl- or ethyl-group shifts have been recorded are summarized in Table 6.

Estimation of the enthalpy change for this reaction from bond strength considerations allows the activation energy to be calculated by analogy with the thermoneutral cyclization of the iodopropyl radical

TABLE 6
Products resulting from group shifts during alkylperoxy radical isomerization

Hydrocarbon	Product	Shift	Ref.
n-Butane	Butanone	H	3
Isobutane	Butanone	CH_3	109
n-Pentane	Pentan-2-one	H	70
Neopentane	3-Methylbutanal	CH_3	55
2-Methylpentane	3-Methylpentan-2-one	CH_3	77
	4-Methylpentan-2-one	H, CH_3	
	Hexan-2-one	CH_3	
	Hexan-3-one	CH_3	
	2-Methylpentan-3-one	H	
	2-Methylpentanal	H	
3-Methylpentane	2-Methylpentan-2-one	H	80
	2-Methylpentan-3-one	CH_3	
	Hexan-3-one	CH_3	
	Hexan-2-one	C_2H_5	
2,3-Dimethylbutane	2,2-Dimethylbutan-3-one	CH_3	65
3-Ethylpentane.	3-Ethylpentan-2-one	H	86
	4-Methylhexan-3-one	C_2H_5	
	Heptan-3-one	C_2H_5	

[104] for which $E = 17$ kcal . mole^{-1} (see footnote, p. 277). Thus, since $D[C_nH_{2n+1}O{-}OH]$ [36] = 42 kcal . mole^{-1}, $D[C_2{-}C_3]$ [112] = 77 kcal . mole^{-1}, and the heat release associated with π-bond formation is 75 kcal . mole^{-1} for $>$C=O [3], ΔH, for 4-methylhexan-3-one, the formation of which involves a 1:2 ethyl group shift, is given by $\Delta H = 77 + 42 - 77 - 75 = -33$ kcal . mole^{-1}, so that $E = 17 - 8.25 \simeq 9$ kcal . mole^{-1}. Similarly, for heptan-3-one, the formation of which involves a 1:3 ethyl group shift, $E = 7.5$ kcal . mole^{-1}. From consideration of the yields of these ketones and those of 2,2-diethyl-3-methyloxiran and 2,2-diethyloxetan, Cullis and co-workers [86] have suggested that A_{20} = ca. 10^9 sec^{-1}.

(iv) O—O Homolysis and β-scission of a C—C bond to give an alkene, either a carbonyl compound or an O-heterocycle and ·OH

β-Hydroperoxyalkyl radicals may decompose by scission of the C—C bond in the β-position to the free electron and O—O homolysis, viz

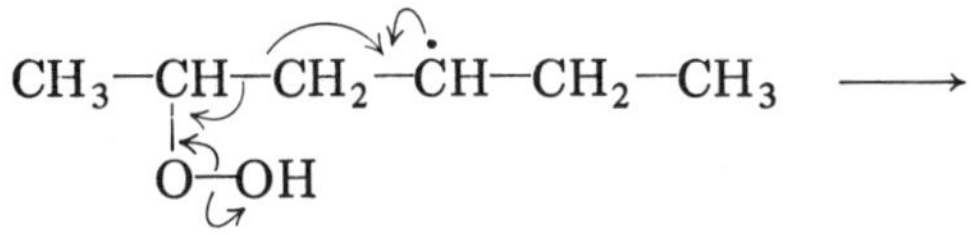

$CH_3{-}CHO + CH_2{=}CH{-}CH_2{-}CH_3 + \cdot OH$ (21)

The production of carbonyl compounds and lower alkenes in pairs may be accounted for qualitatively in this way [77]. ^{14}C-tracer studies [113, 114] leave little doubt that these reactions occur to an appreciable extent. Even so they have received little attention from the thermokinetic point of view. The likely order of magnitude of k_{21} may be assessed, however, from consideration of the strengths of the bonds broken and of those formed. Thus, for the decomposition of the 2-hydroperoxy-3-ethylpent-4-yl radical

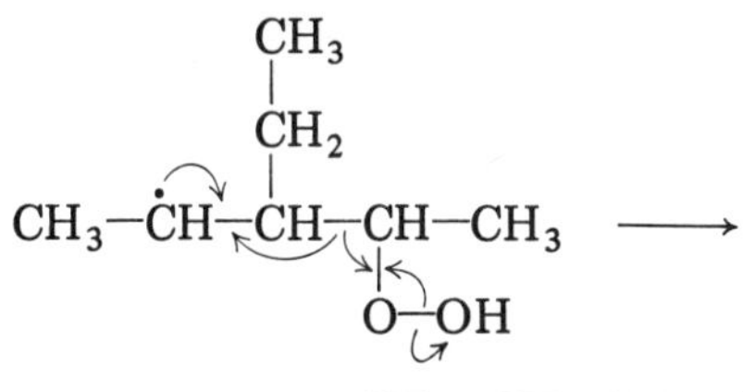

$CH_3{-}CH{=}CH{-}CH_2{-}CH_3 + CH_3{-}CHO + \cdot OH$ (21)

$\Delta H = 77(C_2{-}C_3) + 42(C_nH_{2n}O{-}OH) - 57(>C{=}C<) - 75(>C{=}O) = -13$ kcal . mole^{-1}. (The heat release associated with π-bond formation is 57 kcal . mole^{-1} for $>$C=C$<$ [3].)

The activation energy may now be estimated from the Semenov—Polanyi equation and is found to be 8.25 kcal . mole^{-1}. For a pre-exponential factor of 10^{12} sec^{-1} this gives a value of 10^9 sec^{-1} for k_{21} at 600 °K. Fish [107] has suggested $E_{21} \simeq 12$ kcal . mole^{-1} and $k_{21} = 10^8$ sec^{-1} at 550 °K.

α-Scission reactions are energetically unfavourable and are therefore rare in comparison. Although it is tempting to describe the formation of the large quantities of acetone from *n*-pentane by this route, viz.

$$\mathrm{CH_3{-}\underset{\displaystyle O{-}OH}{\overset{\displaystyle H}{C}}{-}CH_2{-}CH{-}\overset{\displaystyle H}{CH_2}} \longrightarrow \mathrm{CH_3{-}\underset{\displaystyle O}{\overset{}{\underset{\|}{C}}}{-}CH_3 + CH_2{=}CH_2 + {\cdot}OH} \qquad (21\alpha)$$

the low yields of ethylene preclude its acceptability [71]. Clearly, scission decompositions of hydroperoxyalkyl radicals are not entirely responsible for the production of lower carbonyl compounds and decomposition of dihydroperoxy compounds and alkoxy radicals must be taken into account.

(*v*) *O—O Homolysis and β-Scission of C—C bond accompanied by a group shift to give rearranged scission products and ·OH*

^{14}C-tracer techniques have also shown that this is a viable mode of decomposition. Cullis et al. [113] studied the oxidation of ^{14}C-labelled 2-methylpentane and showed that the four C atoms in butanone had the arrangement $\mathrm{\genfrac{}{}{0pt}{}{C}{C}{>}C{-}C}$ in the fuel (Table 7).

TABLE 7

The relative activities of ketones formed from ^{14}C-labelled 2-methylpentanes (From ref. 113.)

Hydrocarbon	Activity (relative to 2-methylpentane = 100)	
	Acetone	Butanone
$\mathrm{\genfrac{}{}{0pt}{}{C}{C}{>}C{-}C{-}C{-}{}^{14}C}$	1	2
$\mathrm{\genfrac{}{}{0pt}{}{C}{C}{>}C{-}C{-}{}^{14}C{-}C}$	3	3
$\mathrm{\genfrac{}{}{0pt}{}{C}{C}{>}C{-}{}^{14}C{-}C{-}C}$	3	102
$\mathrm{\genfrac{}{}{0pt}{}{C}{C}{>}{}^{14}C{-}C{-}C{-}C}$	106	97

Since butanone was formed in the early stages of the reaction and in high yield (ca. 13 %) it is unlikely that it is formed either by further reaction of acetone or by a radical—radical reaction. A β-scission decomposition accompanied by a group shift appears to be the only acceptable explanation, viz.

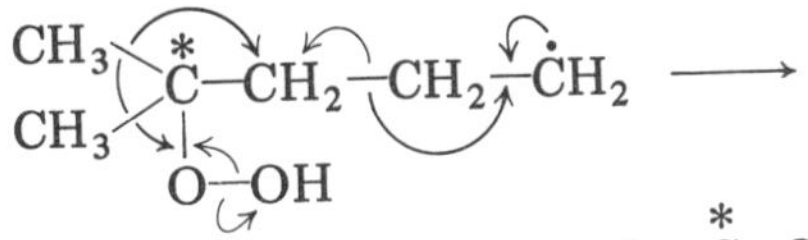

$$CH_3-\overset{*}{\underset{\underset{O}{\|}}{C}}-CH_2-CH_3 + CH_2{=}CH_2 + {\cdot}OH \quad (22)$$

The activation energy of this reaction will be a little larger than the corresponding β-scission without group transfer since the strain energy of the transitional 3-membered ring must be taken into account. By analogy with the cyclization of the iodopropyl radical, see Sect. 3.2.2(a)(i), the activation energy may be expressed as $E = 17 + 0.25\Delta H$. For the example given

$$\Delta H = 42(C_nH_{2n}O{-}OH) + 80(C_2{-}C_2) + 80(C_1{-}C_3) - 84(C_1{-}C_2)$$
$$- 57(\rangle C{=}C\langle) - 75(\rangle C{=}O) = -14 \text{ kcal . mole}^{-1}$$

Hence $E = 13.5$ kcal . mole^{-1}. For a pre-exponential factor of $10^{11.5}$ sec^{-1} this yields a value of $k_{22} = 10^{6.6}$ sec^{-1} at 600 °K.

(vi) O—O Homolysis and scission of two other bonds to give an unsaturated carbonyl compound, an alkyl radical and water

There is no direct experimental evidence for this complex decomposition and it may well occur by several steps [107]. However, substantial yields of unsaturated carbonyl compounds are formed particularly at high pressures [78] under initial reaction conditions where cool flames propagate. For example, the cool-flame oxidation of 2-methylpentane at 525 °C and 19.7 atm in a rapid compression machine [78] yields no less than 14 unsaturated carbonyl compounds viz: acrolein, methacrolein, but-1-en-3-one, pent-2-enal, pent-1-en-3-one, pent-1-en-4-one, *trans*-pent-2-en-4-one, 2-methylbut-1-en-3-one, 2-methylpent-1-en-3-one, 4-methylpent-1-en-3-one, 2-methylpent-1-en-4-one, 2-methylpent-2-en-4-one, 2-methylpent-2-enal and 4-methylpent-2-enal. Spectroscopic studies of the pre-flame reactions [78] have shown that the unsaturated ketones account for ca. 90 % of the absorption which occurs at 2600 A. At lower initial temperatures and pressures acrolein and crotonaldehyde are formed from *n*-pentane [69, 70] and *n*-heptane [82], and acrolein is also formed from isobutane [68].

In view of the known extensive isomerization and subsequent decomposition of alkylperoxy radicals which occurs under the same reaction conditions it is not unreasonable to suppose that these unsaturated carbonyl compounds are also the result of hydroperoxyalkyl decomposition even though it requires the scission of three bonds. Fish *et al.* [78] have formally explained the formation of the above compounds from

2-methylpentane in detail elsewhere; only a simple example is, therefore, given herein, viz.

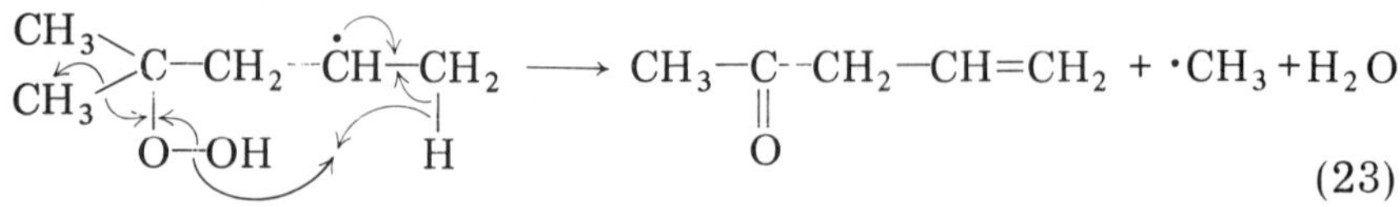

(23)

If this is a two-stage process, O—O homolysis occurring initially, then the C—C and C—H bonds which undergo scission are both in the β-position to atoms with unpaired electrons. Fish [107] feels, therefore, that this decomposition could be simply another example of a complex β-scission process.

(b) Reverse isomerization of hydroperoxyalkyl radicals

Reaction (14) is of course a reversible reaction and hence the reverse isomerization of the hydroperoxyalkyl radical will compete with its decomposition and further oxidation, viz.

$$\cdot C_nH_{2n}OOH \xrightarrow{+O_2} \cdot O_2C_nH_{2n}OOH \qquad (24)$$

$$\cdot C_nH_{2n}OOH \rightleftharpoons C_nH_{2n+1}OO\cdot \qquad (-14)$$

$$\cdot C_nH_{2n}OOH \longrightarrow P + \cdot OH \qquad (16)$$

The estimated values of k_{-14} for α, β, γ and δ-H transfer are given in Table 3 (p. 275) from which it can be seen that the reverse reaction is faster than the forward reaction. However, it can be seen by comparison with Table 4 (p. 278) that the reverse isomerization of γ and δ-hydroperoxyalkyl radicals is much slower than their decomposition to tetrahydrofurans and tetrahydropyrans, respectively. Reactions (—14γ) and (—14δ) will therefore be of little significance. On the other hand, the reverse isomerization of α- and β-hydroperoxyalkyl radicals would be expected faster than either their decomposition to oxirans and oxetans, respectively or than their decomposition to the conjugate alkene and β-scission products respectively. Reactions (—14*a*) and (—14β) would thus be expected to be important. As discussed in Sect. 3.2.2, however, ^{14}C-tracer studies on the oxidation of *n*-butane at 315 °C showed that reaction (—14α) did not participate to any measurable extent [56]. Hence, despite careful thermokinetic considerations the estimated rate coefficients appear to be in error. The reason for this might be in estimation of the equilibrium constant K_{14}. It was assumed that the entropy change for intramolecular H-transfer was close to zero [104] and so k_{14} was taken as being equal to $\exp(-\Delta H/RT)$. If this assumption is not correct, then the values of K_{14} will be too low and hence k_{-14} will be too high.

In any event, only a small error in the estimation of $k_{-14\alpha}$ or $k_{16\alpha}$ will lead to an erroneous conclusion. This is a good example, therefore, of the need for thermokinetic considerations to be backed by quantitative experimental results when considering the mechanism of formation of intermediate products.

(*c*) *Oxidation of hydroperoxyalkyl radicals*

Further oxidation of hydroperoxyalkyl radicals by addition of oxygen to give hydroperoxyalkylperoxy radicals, viz.

$$\cdot C_nH_{2n}OOH + O_2 \longrightarrow \cdot O_2C_nH_{2n}OOH \qquad (24)$$

will compete with the decomposition reactions of hydroperoxyalkyl radicals particularly at high partial pressures of oxygen. This reaction is usually followed by intermolecular abstraction to yield dihydroperoxide

$$\cdot O_2C_nH_{2n}OOH + C_nH_{2n+2}(\text{or AH}) \longrightarrow HOOC_nH_{2n}OOH + \cdot C_nH_{2n+1} \qquad (25)$$

the decomposition of which leads to carbonyl scission products or dicarbonyl compounds.

Organic peroxides are difficult to determine analytically by conventional gas-chromatographic techniques. The development of paper-chromatographic techniques [115], however, have allowed the identification of small amounts of organic peroxides in the products of the oxidation of *n*-heptane, 2,2,3-trimethylbutane, *n*-butane, propane and cyclohexane [52, 116–119]. Dihydroperoxides from propane and 2,2,3-trimethylbutane were not found, while dihydroperoxybutane was detected over a very limited temperature range only (335–345 °C). In contrast, dihydroperoxyheptane is formed over a wide range of initial reaction conditions. These results are in accord with the relative ease of isomerization of the corresponding alkylperoxy radicals.

Studies of the oxidation of *n*-heptane in the presence of hydrogen bromide [120] have yielded direct evidence for participation of these reactions and similar studies with *n*-pentane as the fuel have yielded indirect evidence [121]. Thus, in the absence of HBr no monohydroperoxides from *n*-heptane were detected [119], whilst in its presence they were easily isolated from the combustion products. Similarly, in the absence of HBr, acetone is the major product formed during the early stages of the oxidation of *n*-pentane, but in the presence of HBr the conjugate ketones are the major products [121]. These results were interpreted as follows. In the absence of HBr the propagation proceeds via reactions (3) + (14) + (24) + (25) to give the dihydroperoxide which may decompose to give acetone when *n*-pentane is the fuel [71]. However, in the presence of HBr the propagation proceeds via reactions (3) + (26)

$$\cdot C_nH_{2n+1} + O_2 \longrightarrow C_nH_{2n+1}OO\cdot \qquad (3)$$

$$C_nH_{2n+1}OO\cdot + HBr \longrightarrow C_nH_{2n+1}OOH + Br \qquad (26)$$

since $D[\mathrm{H{-}Br}]$ is relatively small (87 kcal . mole^{-1} compared with 94.5 kcal . mole^{-1} for a secondary C—H bond), and reaction (26) therefore competes effectively with reaction (24). Thus, monohydroperoxides will be formed and their further reaction with Br atoms will lead to the formation of the conjugate ketones [121], viz.

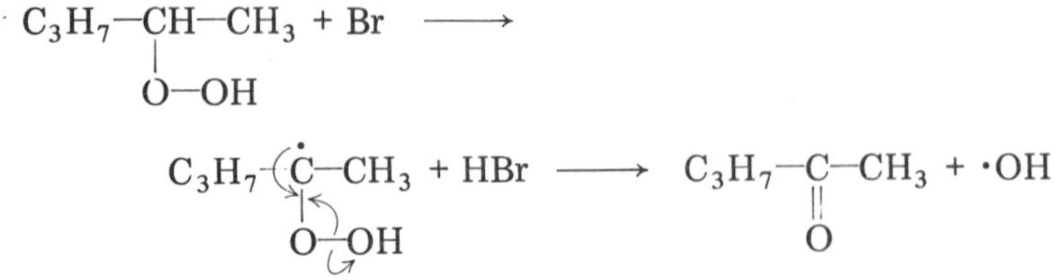

(27)

α-Dihydroperoxides may decompose by homolysis of both O—O bonds accompanied by β-scission of the C—C bond of the peroxidized carbon atoms

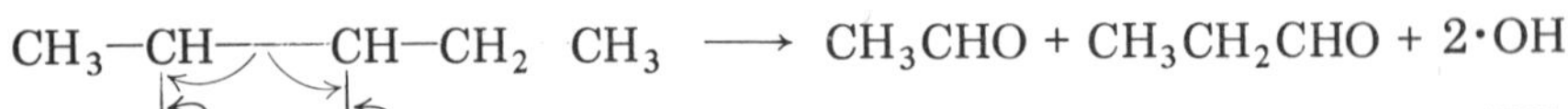

(28)

β-Dihydroperoxides are likely to decompose to dicarbonyl compounds, since they do not have a common C—C bond in the β-position to the O—O bonds

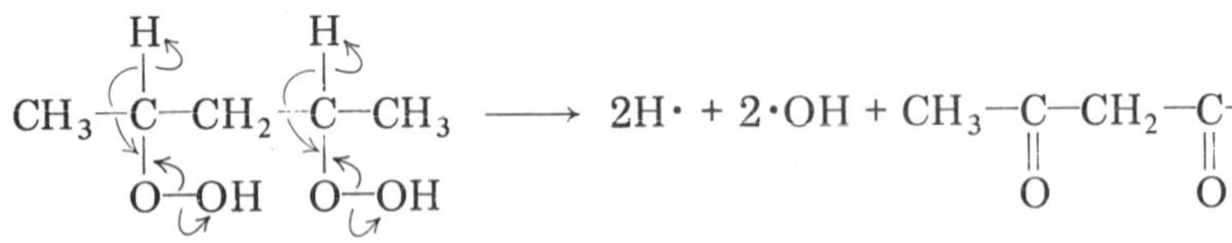

(29)

The spectroscopic studies [78] of the pre-flame reactions occurring during the oxidation of 2-methylpentane have shown that the strong adsorption at 2600 A is also partly due to the β-dicarbonyl compounds, which are subsequently consumed during the passage of the cool flame.

The formation of large yields of acetone from n-pentane, however, has led to the suggestion that decomposition analogous to that of the α-dihydroperoxides is important

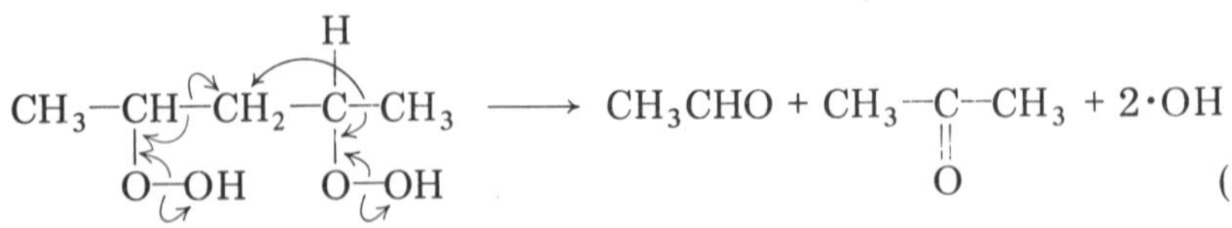

(30)

Ray and Waddington [122] have recently suggested that the formation of isobutyraldehyde from neopentane also occurs via this mechanism.

It was pointed out in Sect. 3.2.2(a)(iv) that on the basis of β-scission decompositions of hydroperoxyalkyl radicals equal yields of lower carbonyl compounds and their corresponding lower alkene would be expected, but that this is not found in experiment where the yield of the carbonyl compounds is always greater. Decomposition reactions such as reactions (29) and (30) afford a possible explanation for the disparity.

Reaction (24) is analogous to reaction (3) and would also be expected to proceed with zero activation energy. The larger $\cdot C_nH_{2n}OOH$ radical will, however, reduce the value of the pre-exponential factor and it has therefore been suggested [58] that a reasonable value for k_{24} would be $10^{8.8}$ l . mole^{-1} . sec^{-1} at 550 °K.

(*d*) *Semi-quantitative test of the alkylperoxy radical isomerization theory*

In the preceding sections it has been seen that the alkylperoxy radical isomerization theory is successful in describing qualitatively the mode of formation of the majority of the intermediate products found during the slow and cool-flame oxidation of hydrocarbons, particularly when the carbon number is greater than 3. Furthermore, from thermokinetic considerations it has been possible to assign reasonable Arrhenius parameters to the individual reaction steps. Clearly a good test of the theory is to see if it can predict the relative yields of the intermediate products which are found experimentally. To do this for all the products (e.g. 65 from 2-methylpentane) is a prohibitive task, since reliable Arrhenius parameters are not available for the more complex propagation steps. However, a semi-quantitative test can be made if one makes some reasonable assumptions and considers only those products formed by the simple propagation steps.

Three such tests have been made [77, 78, 85] and good correlations between the theoretical and experimental relative yields were obtained for the O-heterocycles formed from 2-methylpentane [77] and from 3-ethylpentane [85] at sub-atmospheric pressures. The most exhaustive test was carried out for the oxidation of 2-methylpentane [78] at 525 °C and 19.7 atm and considered the relative yields of O-heterocycles, conjugate alkenes and β-scission products. The reaction scheme considered is shown below.

A pseudo-steady-state treatment was developed for this scheme in which it was assumed that the chains are long, that the rates of formation of the products are controlled by the rates of propagation of the chains, that the formation of alkyl radicals occurs only via reaction (12) and that the sum of the rates of further reactions of hydroperoxyalkylperoxy radicals is large compared with their rate of formation. In order to simplify the mathematical treatment, it was also necessary to assume that reactions (2) and (−4, 1) are slow and that the more complex decomposition reactions of hydroperoxyalkyl radicals can be ignored.

The rate of formation of a product via hydroperoxyalkyl radical decomposition is then given by

$$\mathrm{d[product]}/\mathrm{d}t = k_{5x}[\cdot C_6H_{12}OOH_{(x)}]$$

$$\text{or } k_{4,x}[\cdot C_6H_{12}OOH_{(x)}]$$

Scheme 2

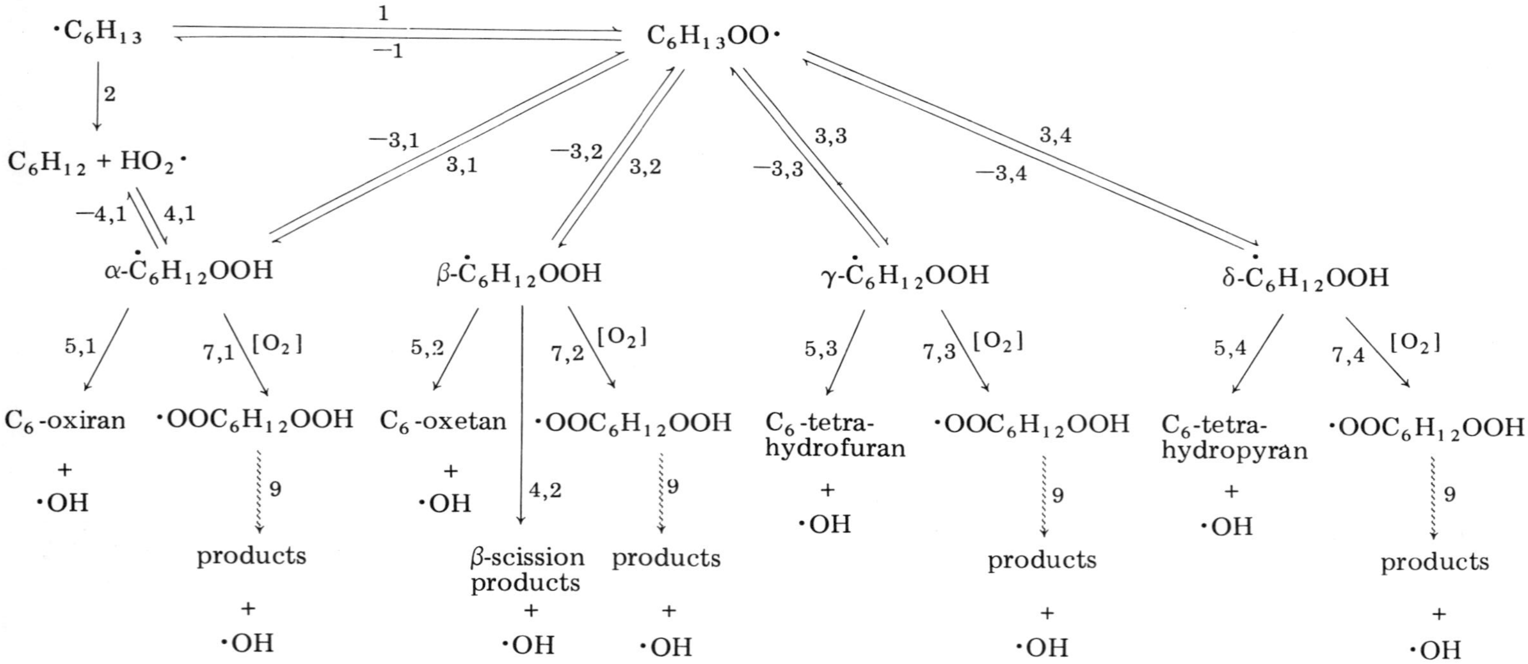

$\cdot C_6H_{12}OOH \xrightarrow{6}$ products other than *O*-heterocycles and β-scission products + $\cdot OH$

$\cdot OH + C_6H_{14} \xrightarrow{8} C_6H_{13}\cdot + H_2O$

depending on whether the product is an O-heterocycle or an alkene or β-scission product. The steady state concentrations of the corresponding hydroperoxyalkyl radicals $[\cdot C_6H_{12}OOH_{(x)}]$ and of $[C_6H_{13}OO\cdot]$ are given by

$$[\cdot C_6H_{12}OOH_{(x)}] = \frac{k_{3,x}[C_6H_{13}OO\cdot]}{k_{-3,x} + k_{5,x} + k_{7,x}[O_2] + k_{4,x}}$$

and

$$[C_6H_{13}OO\cdot] = k_1[\cdot C_6H_{13}][O_2] \times$$

$$\left\{ k_{-1} + \sum_{x=1}^{x=n} k_{3,x} - \sum_{x=1}^{x=n} \frac{k_{3,x}k_{-3,x}}{(k_{-3,x} + k_{5,x} + k_{7,x}[O_2] + k_{4,x})} \right\}^{-1}$$

Hence

$$[\cdot C_6H_{12}OOH_{(x)}] = k_1 k_{3-,x}[\cdot C_6H_{13}][O_2] \times$$

$$\left\{ [k_{-3,x} + k_{5,x} + k_{7,x}[O_2] + k_{4,x}] \times \right.$$

$$\left. \left[k_{-1} + \sum_{n=1}^{n=x} \frac{k_{3,x}(k_{5,x} + k_{7,x}[O_2] + k_{4,x})}{(k_{-3,x} + k_{5,x} + k_{7,x}[O_2] + k_{4,x})} \right] \right\}^{-1}$$

where n is the number of hydroperoxyalkyl radicals that can be obtained from a given alkylperoxy radical (in this case, n = 4 or 5). The relative frequency of attack at the C atoms in 2-methylpentane will be (see p. 276)

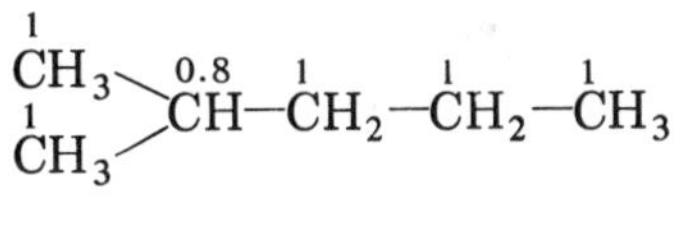

and so the relative concentrations of the various 2-methylpentyl radicals may be calculated. Each O-heterocycle (except 3-n-propyloxetan) and each alkene can be formed from two alkylperoxy radicals and addition of the rates of formation of each compound allows its theoretical yield to be calculated relative to that of 2,2-dimethyltetrahydrofuran and to be compared with experiment as shown in Table 8.

Direct comparison of the theoretical and experimental yields is complicated by the fact that further reactions of the products have not been taken into account. Even so, some excellent agreements are obtained and in most cases where there is disparity it is clear that this is due to competing propagating steps not embraced by the scheme [78]. The alkylperoxy radical isomerization theory provides a good account, then, of the mechanism of chain-propagation, at least semi-quantitatively.

TABLE 8

The yields of C_6 O-heterocycles, C_6 alkenes and β-scission products from the oxidation of 2-methylpentane at 525 °C and 19.7 atm (From ref. 78.)
(Arbitrary units, relative to yield of 2,2-dimethyltetrahydrofuran = 1250)
($RH:O_2:N_2$ = 1:19:72)

Product	Experimental relative yield	Theoretical yield
2,2-Dimethyltetrahydrofuran	1250	1250
2,4-Dimethyltetrahydrofuran	870	400
2,2,4-Trimethyloxetan	160	370
2-Ethyl-3-methyloxetan		210
2-Isopropyloxetan		110
3-*n*-proplyoxetan		20
2,2-Dimethyl-3-ethyloxiran	60	140
2-Methyl-2-*n*-propyloxiran	ca. 10	120
2-Isopropyl-3-methyloxiran	6	30
2-Isobutyloxiran		10
3-Methyltetrahydropyran	150	10
2-Methylpent-1-ene	930	350
2-Methylpent-2-ene	420	280
4-Methylpent-1-ene	150	30
4-Methylpent-2-ene		50
Pent-2-ene	570	1240
3-Methylbut-1-ene	280	620
Pent-1-ene	130	120
Isobutene	1900	1560
Propene	3870	860
Ethylene	2310	90
Acetone	4320	780
Acetaldehyde	5820	1560
Propionaldehyde	1730	90
Isobutyraldehyde	280	90

3.3 BRANCHING

Prior to 1960 the mechanism of branching during hydrocarbon oxidation received more attention than the mechanism of propagation. Although the reverse has been true in recent years, some important advances have been made which have to a large extent resolved the old arguments outlined in Sect. 2 regarding the nature of the degenerate branching agent, viz. peroxides, aldehydes or peracids.

A typical temperature—pressure ignition diagram is shown in Fig. 8 for mixtures of 3-ethylpentane and oxygen in the molar ratio 1:2. The low temperature region of the diagram which is associated with the propagation of multiple cool flames shows "fine structure", i.e. there are several minima and maxima in the value of the initial pressure required for

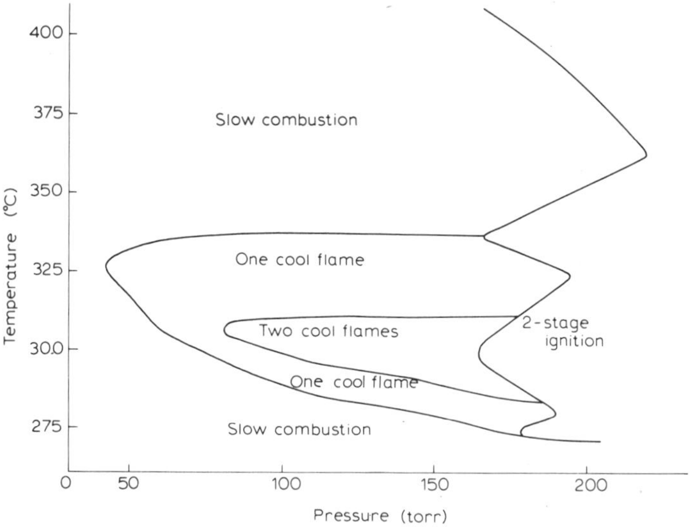

Fig. 8. The temperature—pressure ignition diagram for mixtures of 3-ethylpentane and oxygen in the molar ratio 1:2. Cylindrical pyrex reaction vessel, volume 280 cm^3. (From ref. 85.)

two-stage ignition as the initial temperature is increased. These give rise to the "lobes" which are labelled L_0, L_1, L_2 and L_3 in order of increasing temperature (3-ethylpentane + oxygen mixtures do not exhibit the L_3 lobe). The variation of initial pressure with temperature would not be expected to exhibit such turning points unless there are repeated fluctuations in the value of the net branching factor. Such variations may be explained by the onset and decay of various branching reactions or of the reactions leading to the formation of the branching agent. From detailed studies of the "morphology" (fine structure) of the ignition diagrams of a series of hydrocarbons (notably methane, propane, butane, *n*-pentane and neopentane) in the presence and absence of additives and of the variation of the induction periods with initial alkane and oxygen concentration, Antonik and Lucquin [123] have been able to explain occurrence of the lobes in these terms.

Thus, the addition of acetaldehyde (5 %) to *n*-pentane + oxygen mixtures causes the L_1 and L_2 lobes to be moved to lower temperatures, while the addition of HBr (1 %) to neopentane + oxygen mixtures not only moves the L_2 lobe to lower temperatures but also causes the appearance of the L_1 lobe and periodicity, both of which are absent in the binary system. Periodicity is thus associated with the L_1 lobe and not with the L_2 lobe. Secondly, two types of behaviour of the variation of the induction period with reactant concentration are observed. The induction period decreases with increase in hydrocarbon concentration in the L_1

lobe, but increases with increase in hydrocarbon concentration in the L_2 lobe. Furthermore, in the neopentane + oxygen system, which only exhibits the L_2 lobe, the induction period always increases with hydrocarbon concentration, but upon addition of only 0.2 % HBr it decreases with hydrocarbon concentration in the region of the new L_1 lobe. It may be concluded, therefore, that the L_1 mechanism is favoured by high hydrocarbon concentration, whilst the L_2 mechanism is favoured by high oxygen concentration. Finally, two types of slow combustion/cool-flame boundary are evident. With *n*-pentane + oxygen at low temperatures (ca. 320 °C), i.e. L_1, a sudden jump is noticed between the light emission of the slow reaction and that of the cool-flame reaction at the boundary. However, at high temperatures (ca. 380 °C), i.e. L_2, there is no sudden appearance of the cool flame and the light intensity and transient pressure rise increase continuously across the boundary. This is also the case with methane + oxygen at ca. 485 °C, i.e. L_3. Thus the L_1 and probably L_0 mechanisms give rise to distinct boundaries, whilst the L_2 and L_3 mechanisms do not.

These results allowed Antonik and Lucquin to interpret the L_0, L_1, L_2, and L_3 mechanisms as follows.

L_0 mechanism

This mechanism which occurs at the lowest temperatures at which cool flames propagate is thought to be a process of pure peroxidation, and only occurs for hydrocarbons with tertiary C—H bonds, viz.

$$C_nH_{2n+1}OO\cdot + C_nH_{2n+2} \longrightarrow C_nH_{2n+1}OOH + \cdot C_nH_{2n+1} \qquad (17)$$

Fish [58] has estimated that the value of k_{17} at 550 °K is 4.8 l . mole^{-1} . sec^{-1} using the Arrhenius parameters A_{17} = $10^{7.9}$ l . mole^{-1} . sec^{-1} (assuming free rotation in the transition state) and E_{17} = 7.7 kcal . mole^{-1} (from $E = 6 + \Delta H$ for endothermic reactions [104]). Using the same authors value for the activation energy for 1:6 intramolecular transfer of secondary hydrogen atoms, $E \simeq$ 17 kcal . mole^{-1} and a pre-exponential factor of $10^{11.5}$ sec^{-1} it can be estimated that reaction (17) will be about 100 times slower than the isomerization reaction for an initial alkane pressure of 200 torr. The rate of production of the branching agent is therefore slow in comparison with the rate of other product formation. Similar estimations for the rate of hydrogen abstraction from secondary carbon atoms show that this will be about 1000 times slower than isomerization at this temperature.

Several values for the Arrhenius parameters of the branching reaction (31) may be found in the literature as shown in Table 9.

$$C_nH_{2n+1}OOH \longrightarrow C_nH_{2n+1}O\cdot + \cdot OH \qquad (31)$$

TABLE 9
Arrhenius parameters for the decomposition of hydroperoxides
$C_nH_{2n+1}OOH \rightarrow C_nH_{2n+1}O\cdot + \cdot OH$

C_nH_{2n+1}	$\log_{10} A$ (A in sec^{-1})	E (kcal . $mole^{-1}$)	Ref.
CH_3	12.63	38.5	124
C_2H_5		37.7	125
iso-C_3H_7		40.7	125
tert-C_4H_9		37.8 ± 0.7	125
C_nH_{2n+1}	15	43	57[a]
C_nH_{2n+1}	16	44	126[a]
H	15.4 (14.9)	48 (47)	127 (99)

[a] Estimated.

L_1 mechanism

The nature of the cool flames which propagate in the L_1 lobe are similar to those of the L_0 lobe and this mechanism is also thought to be caused by primary peroxidation, but in this case it is catalysed by hydrogen donors. It was seen in Sect. 3.2.2(a)(iv) that aldehydes are formed in the primary chain-propagation cycle and hence alkylhydroperoxides may be formed by the reaction sequence (14) + (21) + (32), (where A*H* is an aldehyde)

$$C_nH_{2n+1}OO\cdot \longrightarrow \cdot C_nH_{2n}OOH \quad (14)$$

$$\cdot C_nH_{2n}OOH \longrightarrow AH + \text{alkene} + \cdot OH \quad (21)$$

$$C_nH_{2n+1}OO\cdot + AH \longrightarrow C_nH_{2n+1}OOH + A\cdot \quad (32)$$

The acyl radical (A·) will also be formed by ·OH attack on the aldehyde, reaction (33), and secondary peroxidation of this radical will lead to the formation of the peracid, viz.

$$AH + \cdot OH \longrightarrow A\cdot + H_2O \quad (33)$$

$$A\cdot + O_2 \longrightarrow AO_2\cdot \quad (34)$$

$$AO_2\cdot + AH \longrightarrow AO_2H + A\cdot \quad (35)$$

Hence, a secondary mode of branching is also possible in this mechanism.

The strength of the labile C—H bond in aldehydes is 86.5 kcal . $mole^{-1}$ (cf. secondary C—H = 94.5 kcal . $mole^{-1}$). Since D[ROO—H] = 90 kcal . $mole^{-1}$ reaction (32) will be exothermic by about

3.5 kcal . mole^{-1} and will thus facilitate easier formation of the alkylhydroperoxide. The catalytic action of acetaldehyde and hydrogen bromide (D[H—Br] = 87 kcal . mole^{-1}) are thus explained.

The oxidation of methane exhibits a special case of this mechanism, viz.

$$CH_3OO\cdot \longrightarrow HCHO + \cdot OH \tag{21}$$

$$CH_3OO\cdot + HCHO \longrightarrow CH_3OOH + \cdot CHO \tag{32}$$

but only in the slow reaction.

L_2 mechanism

This is probably caused by secondary peroxidation of aldehydes

$$\cdot C_nH_{2n}OOH \longrightarrow AH + \text{alkene} + \cdot OH \tag{21}$$

$$AH + O_2 \longrightarrow A\cdot + HO_2\cdot \tag{36}$$

$$A\cdot + O_2 \longrightarrow AO_2\cdot \tag{34}$$

$$AO_2\cdot + AH \longrightarrow AO_2H + A\cdot \tag{35}$$

The essential difference in the L_1 and L_2 mechanisms is therefore the different fate of the aldehyde produced by alkylperoxy radical isomerization and subsequent decomposition.

$$AH + C_nH_{2n+1}OO\cdot \longrightarrow C_nH_{2n+1}OOH + A\cdot \tag{32}$$

$$AH + O_2 \longrightarrow A\cdot + HO_2\cdot \tag{36}$$

Since reaction (32) is a radical—molecule reaction and reaction (36) is a molecule—molecule reaction, E_{L_1} is less than E_{L_2} (E_{32} = ca. > 7 and E_{36} = 14 kcal . mole^{-1} for acetaldehyde [128]. Furthermore, at the higher temperature associated with the L_2 lobe the concentration of alkylperoxy radicals will be lower due to the increased rate of isomerization, hence it is to be expected that the L_2 mechanism will become more important with increase in temperature. The influence of oxygen concentration is understandable and if the aldehyde forms more readily than it reacts the lack of a distinct boundary between slow combustion and cool flame also becomes apparent [123].

L_3 mechanism

This mechanism is similar to the L_2 mechanism, the only difference being that the aldehyde is formaldehyde. The activation energy of reaction (36), when AH is formaldehyde [129], is 32 kcal . mole^{-1} and (36) will thus be important only at higher temperatures. The methane + oxygen system would therefore be expected to exhibit this lobe as is

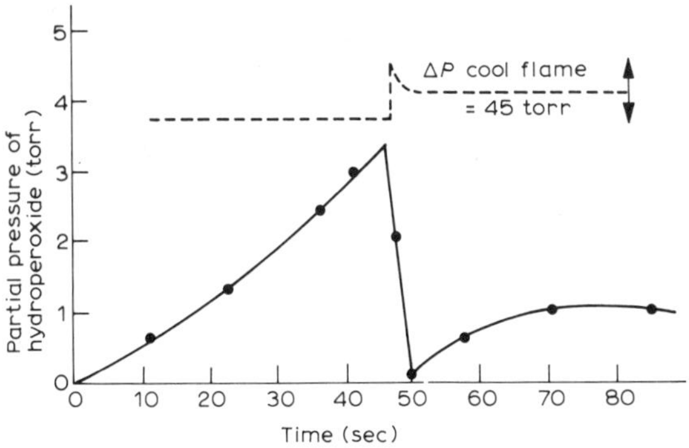

Fig. 9. The variation with time of the yield of heptyl hydroperoxide during the cool-frame oxidation of *n*-heptane. Initial temperature = 242 °C; initial pressure of *n*-heptane = 50 torr; initial pressure of oxygen = 50 torr. (From ref. 130.)

found experimentally. However, more recently, Baldwin et al. [159] have found that $E_{36} \sim 40$ kcal . mole^{-1} for both these aldehydes and their work suggests that the L_3 lobe may be due to the homogeneous decomposition of hydrogen peroxide. Nevertheless, Antonik and Lucquin have reconciled to a large extent the aldehyde and peroxide theories.

Burgess and Laughlin [130] reported the most convincing evidence to date on the role of alkyl hydroperoxides during cool-flame oxidation. The variation with time of the concentration of organic peroxides (calculated as $C_nH_{2n+1}OOH$) found in the products of reacting mixtures of *n*-heptane + oxygen at 242 °C was measured before and after the propagation of a cool flame (Fig. 9). The partial pressure of the peroxides builds up to a maximum of 3.4 torr immediately before the cool flame

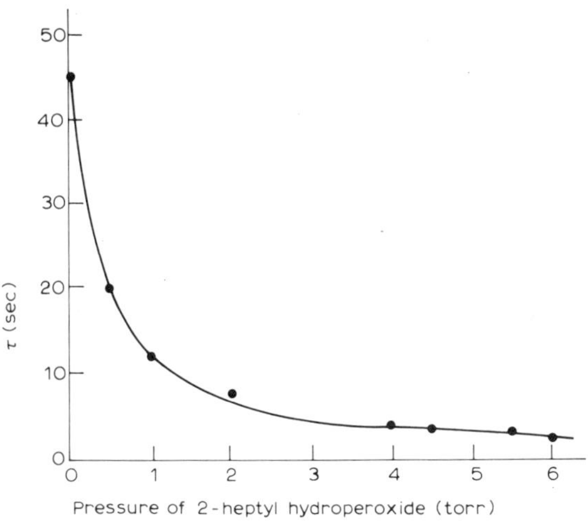

Fig. 10. The variation of τ with the concentration of added 2-heptyl hydroperoxide. (From ref. 130.)

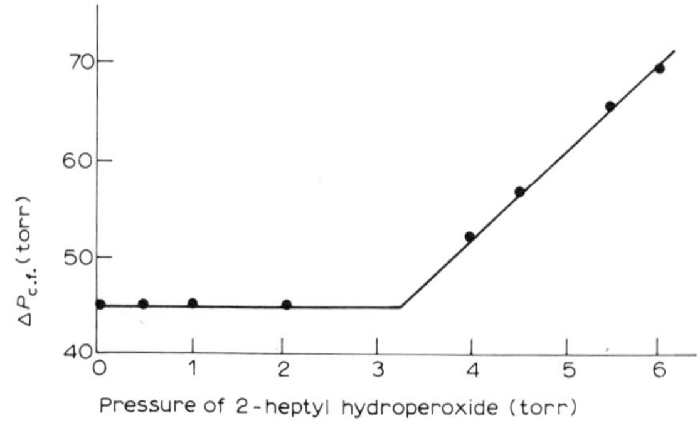

Fig. 11. The variation of $\Delta P_{c.f.}$ with the concentration of added 2-heptyl hydroperoxide. (From ref. 130.)

and they are completely destroyed during its passage. In a second experiment, small amounts of 2-heptyl hydroperoxide were added to the reacting mixture and its effect on the induction period (τ) and pressure rise accompanying the passage of the cool flame ($\Delta P_{c.f.}$) noted. It can be seen from Figs. 10 and 11 that τ is drastically reduced in the presence of very small amounts of the peroxide (<2 % v/v) whilst $\Delta P_{c.f.}$ is unaffected. However, additions greater than ca. 3.2 % have little further effect on τ, but in contrast $\Delta P_{c.f.}$ increases rapidly. These results strongly suggest that a critical concentration of heptyl hydroperoxides is required before a cool flame will propagate at this temperature which corresponds to the L_1 lobe for n-heptane + oxygen mixtures. The addition of neopentyl hydroperoxide to reacting mixtures of neopentane and oxygen gave similar results [55].

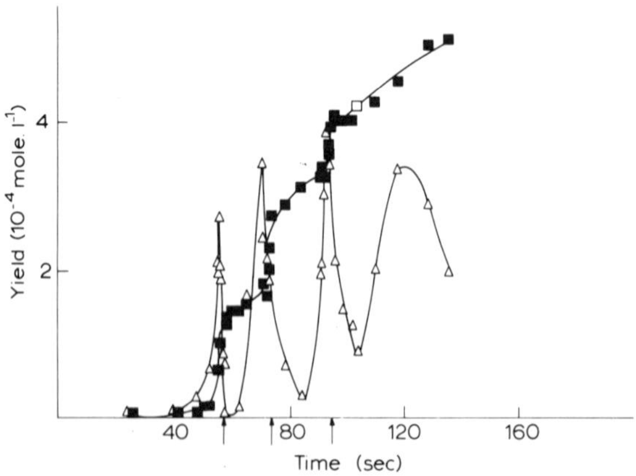

Fig. 12. The variation with time of peroxide yield during the cool-flame oxidation of isobutane. Initial temperature = 310 °C; isobutane introduced = 1.44×10^{-3} mole . l^{-1} ; isobutane: oxygen = 1:2; volume of reaction vessel = 500 cm^3. △, *tert*-butyl hydroperoxide × 10; ■, hydrogen peroxide. (From ref. 132).

Further support for the attainment of a critical concentration of hydroperoxide prior to the passage of a cool flame at temperatures corresponding to the L_0 and L_1 lobes has been obtained by Taylor [131], and more recently by Pollard and co-workers [68, 132], who determined the maximum concentrations of *tert*-butyl hydroperoxide found during the cool-flame oxidation of isobutane. Again, the concentration of hydroperoxide increased prior to the cool flame and it was almost entirely consumed during its passage (Fig. 12). Also, in common with other hydrocarbon + oxygen systems, (e.g. refs. 55, 65, 78, 133) the induction period to the first cool flame (τ_1) was related to the initial reactant pressure (p_0) by the expression

$$\tau_1 = kp_0^{-n} + c \tag{3.1}$$

where k, n and c are constants [134]. In practical terms, the value of c at a given temperature is the shortest possible induction period to the first cool flame and as such would be expected to bear a close relationship to the rate of branching. Its variation with the maximum concentration of *tert*-butyl hydroperoxide $[\mathrm{TBH}]_{max}$ was therefore examined.

The value of c at a given temperature was taken as the value which gave a maximum value for the correlation coefficient of the linear relationship

$$\log(\tau_1 - c) = \log k - n \log p_0 \tag{3.2}$$

The variation of c with temperature obtained in this way is shown in Fig. 13, from which it can be seen that below ca. 320 °C

$$c = 10^{-33.4}\, e^{4.7 \times 10^4/T} \tag{3.3}$$

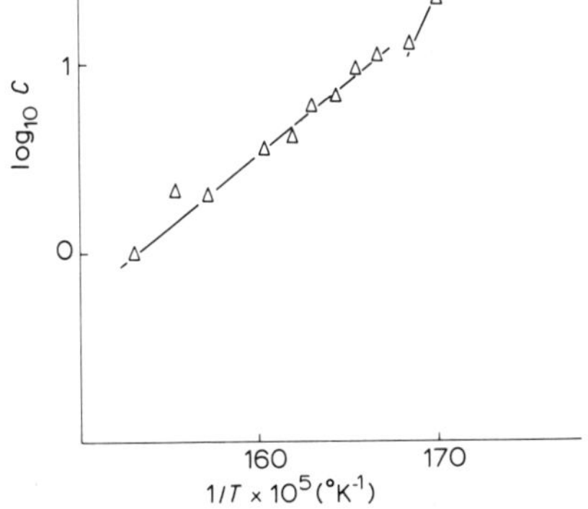

Fig. 13. The variation of C with termperature.

References pp. 361—367

TABLE 10
The maximum concentration of *tert*-butyl hydroperoxide at the first cool flame

Temp. (°C)	Initial pressure (torr)	$[TBH]_{max}$ (10^{-5} mole . l^{-1})	$[H_2O_2]$ (10^{-5} mole . l^{-1})
305	300	2.18	8.22
	250	2.22	7.92
	Average	2.20	
310	340	1.98	8.66
	315	2.02	8.02
	270	1.92	7.22
	240	1.86	6.98
	210	2.04	6.28
	Average	1.96	
315	290	1.20	4.28
	200	1.16	4.16
	Average	1.18	
320	308	1.44	5.72
	270	1.24	4.04
	256	1.08	3.90
	200	0.94	3.70

$[TBH]_{max}$ is the maximum concentration of *t*-butyl hydroperoxide detected.
$[H_2O_2]$ is the concentration of hydrogen peroxide at the same time.

$[TBH]_{max}$ was measured for a wide range of initial reactant conditions and as can be seen from Table 10 its value is independent of initial pressure for initial temperatures below ca. 320 °C. This result allowed the variation of c with $[TBH]_{max}$ to be determined for the whole of the low-temperature cool-flame region (below 320 °C) as shown in Fig. 14, whence

$$\frac{1}{c} = 0.0925 - 3.74 \times 10^3 [TBH]_{max} \tag{3.4}$$

By combining eqns. (3.3) and (3.4) it is therefore possible to relate $[TBH]_{max}$ to the absolute temperature, viz.

$$3.74 \times 10^3 [TBH]_{max} = 0.0925 - 10^{33.4} e^{-4.7 \times 10^4/T} \tag{3.5}$$

If *tert*-butyl hydroperoxide is the major branching agent below 320 °C, $[TBH]_{max}$ should be related to the critical concentration required for cool-flame propagation even though it may not be an exact measure of $[TBH]_{crit.}$. This indeed appears to be so, since values of $[TBH]_{max}$ observed at temperatures too low for cool-flame propagation are always lower than that predicted by eqn. (3.5) whilst they are in excellent

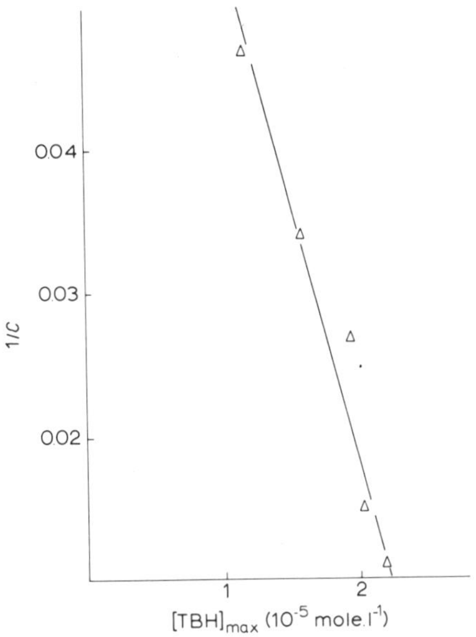

Fig. 14. The relationship between *C* and the maximum concentration of *tert*-butyl hydroperoxide.

agreement at the slow combustion/cool-flame boundary, the critical value of $[TBH]_{max}$ being 2.44×10^{-5} mole . l^{-1}.

It can be seen from Fig. 13 that a sharp break occurs in the variation of *c* with temperature at 320 °C. Above this temperature the relationship (3.5) no longer holds, the experimental value of $[TBH]_{max}$ always exceeds that predicted. It is believed that this is due to a change in the mode of formation of *tert*-butyl hydroperoxide and the onset of further branching reactions. At these temperatures, considerable amounts of acetaldehyde and propionaldehyde are formed and hence the L_1 mechanism is predominant. The yields of these compounds increase with initial pressure [68], and hence the yield of *tert*-butyl hydroperoxide would also be expected to increase if the L_1 mechanism is operative. This is indeed found to be so. At more elevated temperatures (ca. 350 °C) the concentration of *tert*-butyl hydroperoxide does not oscillate with cool-flame propagation. This behaviour is now exhibited by the aldehydes, the direct oxidation of which appears to be ultimately responsible for branching (L_2 mechanism), possibly in the form of peracetic acid which is thought to be an important agent of degenerate branching above ca. 320 °C, particularly for oxygen rich mixtures (H.C.:O_2 = 1:4) [63]. Indirect evidence of this has recently been furnished by Dechaux *et al.* [63, 135] from determinations of the relative concentrations of carbon dioxide and carbon monoxide found during the oxidation of *n*-butane. Thus, the acyl radical formed in reactions (32) and (33) may either react further with oxygen or decompose, viz.

$$CH_3CHO + C_nH_{2n+1}OO\cdot \longrightarrow CH_3\dot{C}O + C_nH_{2n+1}OOH \quad (32)$$

$$CH_3CHO + \dot{O}H \longrightarrow CH_3\dot{C}O + H_2O \quad (33)$$

$$CH_3\dot{C}O + O_2 \longrightarrow CH_3CO_3\cdot \quad (34)$$

$$CH_3\dot{C}O + M \longrightarrow \cdot CH_3 + CO + M \quad (37)$$

The activation energy of reaction (37) [136] is ~10.8 kcal . mole^{-1} while that of reaction (34) will be close to zero, hence reaction (37) will be favoured at higher temperatures. The acylperoxy radical formed in reaction (34) may also decompose or it may abstract hydrogen to form peracetic acid, viz.

$$CH_3\dot{C}O_3 \longrightarrow CH_3\dot{C}O + O_2 \quad (-34)$$

$$CH_3CO_3\cdot \longrightarrow CH_3O\cdot + CO_2 \quad (38)$$

$$CH_3CO_3\cdot + CH_3CHO \longrightarrow CH_3CO_3H + CH_3\dot{C}O \quad (39)$$

E_{-34} [99] = ΔH_{34} = −22.4 kcal . mole^{-1}, E_{38} [137] = 17 kcal . mole^{-1} and E_{39} [138] = 7 kcal . mole^{-1}, so reaction (38) will also be favoured at higher temperatures. The onset of the region of negative temperature coefficient and the increase in the CO/CO_2 ratio above ca. 350 °C found by Déchaux et al. is therefore explained in terms of these competing reactions. Clearly, the rate of formation of the branching agent decreases with temperature and the formation of CO by reaction (37) will prevail over the formation of CO_2 via decomposition of the peracid or via reactions (34) + (38). The precise temperature at which the onset of the region of negative temperature coefficient occurs and the yields of carbon oxides would be expected to increase with initial oxygen concentration as found experimentally [63, 135].

The *n*-heptane, isobutane and *n*-butane studies give strong support therefore to Antonik and Lucquin's theory. It should be noted, however, that attempts by gas—liquid chromatography to identify 2-methylbut-2-yl hydroperoxide in the products of the oxidation of 2-methylbutane at 320—460 °C failed [139] and that Hoare and Lill [140] have recently put forward evidence which suggests that in systems in which acetaldehyde is formed methyl hydroperoxide becomes more important than peracetic acid as the temperature approaches 350 °C, the methyl hydroperoxide being formed as a result of the methyl radicals produced in reaction (37).

The role of dihydroperoxides as branching agents will also be important for lean mixtures. The decomposition of these compounds has already been discussed in detail in Sect. 3.2.2(c) and raises an important question. Thus, it has been suggested [71] that the large yields of acetone formed during the oxidation of n-pentane arise via the formation and subsequent decomposition of 2,4-dihydroperoxypentane, viz.

$$CH_3{-}\underset{\underset{O{-}OH}{|}}{CH}{-}CH_2{-}\overset{H}{\overset{|}{\underset{\underset{O{-}OH}{|}}{C}}}{-}CH_3 \longrightarrow CH_3CHO + CH_3{-}\underset{\underset{O}{\|}}{C}{-}CH_3 + 2{\cdot}OH \qquad (30)$$

Clearly, the formation of the major product via a branching reaction is not viable in terms of the Semenov theory for degenerately branched-chain reactions, unless the chain length is very short or the branching agent very unstable (in which case the reaction would not be a good example of degenerate branching). On the other hand, acetone is unlikely to arise via the monohydroperoxyalkyl radical as explained in Sect. 3.2.2(c). This anomaly is one of the major remaining problems to be resolved regarding the mechanistic interpretation of product formation from the simpler alkanes.

Finally, branching reactions involving free radicals as reactants may also play a part. Thus, for example, in the oxidation of n-heptane the hydroperoxyalkylperoxy radicals formed may decompose sufficiently rapidly, e.g.

$$CH_3{-}\underset{\underset{O{-}OH}{|}}{CH}{-}CH_2{-}\underset{\underset{O{-}O\cdot}{|}}{CH}C_3H_7 \longrightarrow 2{\cdot}OH + CH_3{-}\underset{\underset{O}{\|}}{C}{-}CH_3 + {\cdot}C_3H_7 + CO \qquad (40)$$

to compete with the corresponding dihydroperoxide formation.

3.4 TERMINATION

Chain termination may be either heterogeneous or homogeneous in nature. At low pressure the chain carriers and branching agents may diffuse to the walls of the reactor where they are destroyed either by capture or conversion to inactive products. These heterogeneous termination reactions are either diffusion-controlled in which case the rate coefficient is given by the expression

$$k = \frac{\pi D}{r^2}$$

(D = diffusion coefficient, r = radius of reactor) or they are efficiency-controlled in which case

$$k = \frac{(\bar{c}^2)^{1/2}}{4r}\,\epsilon$$

(where ϵ is the efficiency of the wall in destroying the species involved) [141].

At higher pressures and radical concentrations such as those present during cool-flame propagation homogeneous termination by dis-

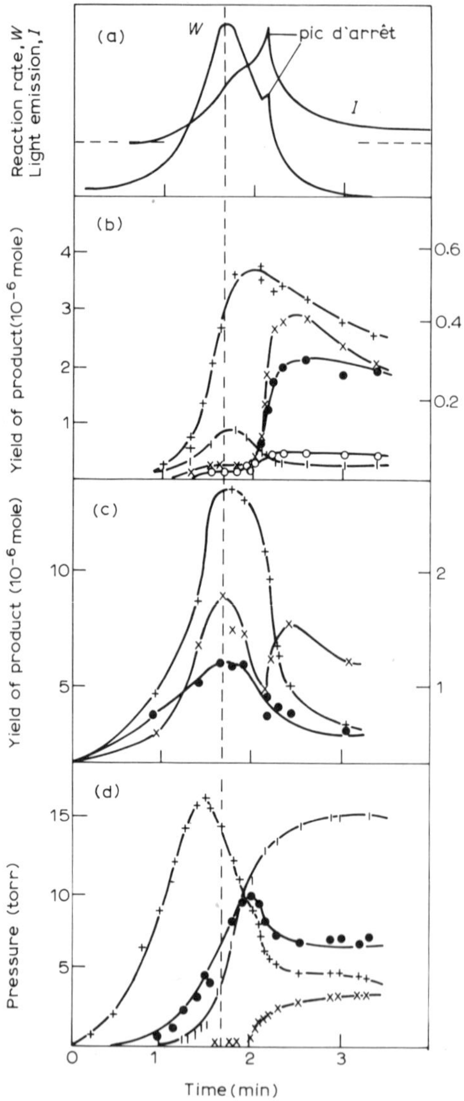

Fig. 15. The influence of the "pic d'arrêt" on product formation during the oxidation of propane. Initial temperature = 430 °C; initial pressure of propane = 90 torr; initial pressure of oxygen = 210 torr; volume of reaction vessel = 30 cm^3. (b) Left ordinate: +, methyl alcohol. Right ordinate: ×, isopropyl alcohol; ●, ethyl alcohol; ○, *n*-propyl alcohol; |, allyl alcohol. (c) Left ordinate: +, hydrogen peroxide; ●, formaldehyde. Right ordinate: ×, total aldehydes. (d) +, propene; |, methane; ●, ethylene; × ethane. (From ref. 147.)

proportionation reactions of radicals which have relatively long lifetimes such as $C_nH_{2n+1}OO\cdot$ and $HO_2\cdot$ will predominate.

Investigations of the mode of termination by studying the effect of surface on the overall kinetics of the gas-phase oxidation of hydrocarbons cannot lead to unequivocal conclusions, because under certain conditions the major propagation steps may also be heterogeneous (see Sect. 5.2). However, Lucquin and co-workers have thrown considerable light on some of these reactions.

In the final stages of the reaction, under favourable initial reaction conditions, a sudden temporary acceleration of the reaction is observed. This phenomenon, known as the "pic d'arrêt", was first observed by Lucquin in the low temperature slow oxidation of *n*-pentane [142] and subsequently in the high temperature oxidation of other hydrocarbons, e.g. refs. 143, 144. The pic d'arrêt manifests itself as a sudden increase in the intensity (I) of the emission of light and as a peak on the recording of the derivative of the pressure change (W) against time as shown in Fig. 15 [145].

Systematic studies of the slow reaction in terms of initial pressure, temperature and mixture composition at both low [146] and high temperature [143, 147] showed that the extent of reaction at the pic d'arrêt is proportional to the initial oxygen concentration, but independent of the initial hydrocarbon concentration and that this phenomenon is, in fact, directly connected with the consumption of the remaining oxygen in the reacting mixture. Figure 16 shows a typical pressure—temperature ignition diagram for propane + oxygen mixtures (1:1) in

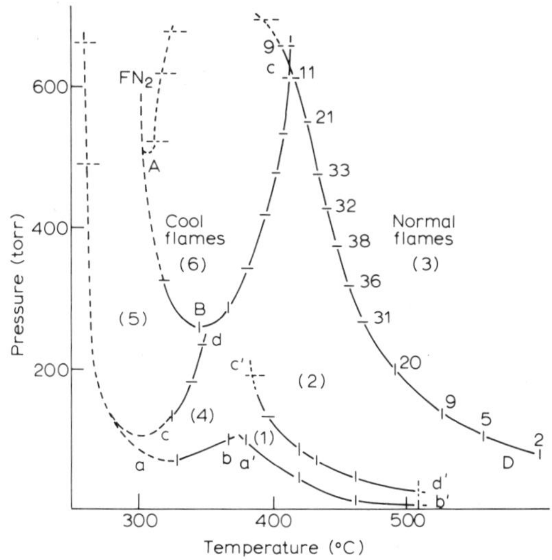

Fig. 16. The pressure—temperature ignition diagram for propane—oxygen mixtures in the molar ratio 1:1. Cylindrical silica reaction vessel, volume = 30 cm^3. (1), (4) slow reaction; (2), (5) slow reaction with pic d'arrêt; (3) normal flames; (6) cool flames. (From ref. 147.)

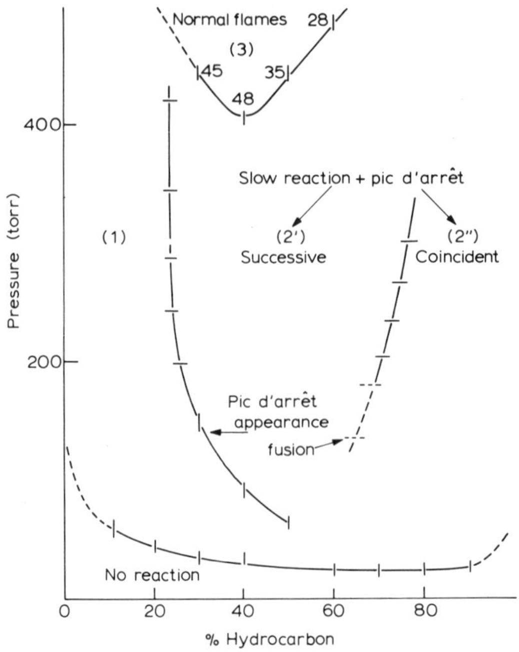

Fig. 17. The pressure—composition ignition diagram for propane—oxygen mixtures at 429 °C. (From ref. 147.)

which the regions where the pic d'arrêt may be observed have been mapped out. At 429 °C, regions 1, 2 and 3 are observed and the corresponding pressure—composition ignition diagram at this temperature is shown in Fig. 17. Studies of the variation of product formation with time for total initial pressures of 300 torr at different mixture compositions thus allowed viable comparisons to be made in the presence and absence of the pic d'arrêt (Figs. 15 and 18). The formation of products in the early stages is the same in both cases, but when the pic d'arrêt appears notable differences occur. The yields of acetaldehyde, ethanol, propanol, isopropanol and ethane increase and the maximum rate of methane and ethane formation occur not at the maximum reaction rate but during the pic d'arrêt. In contrast, the yields of hydrogen peroxide, formaldehyde and alkene decrease.

These observations may be explained in terms of radical recombination and disproportionation reactions. The essential reactions considered [147, 148] are (41), (42), (43) and (44)

$$\cdot C_nH_{2n+1} + \cdot C_nH_{2n+1} \longrightarrow C_{2n}H_{4n+2} \tag{41}$$

$$C_nH_{2n+1}OO\cdot + C_nH_{2n+1}OO\cdot \longrightarrow 2C_nH_{2n+1}O\cdot + O_2 \tag{42}$$

$$\cdot C_nH_{2n+1} + C_nH_{2n+1}OO\cdot \longrightarrow C_nH_{2n+1}OOC_nH_{2n+1} \tag{43}$$

$$C_nH_{2n+1}OOC_nH_{2n+1} \longrightarrow 2C_nH_{2n+1}O\cdot \tag{44}$$

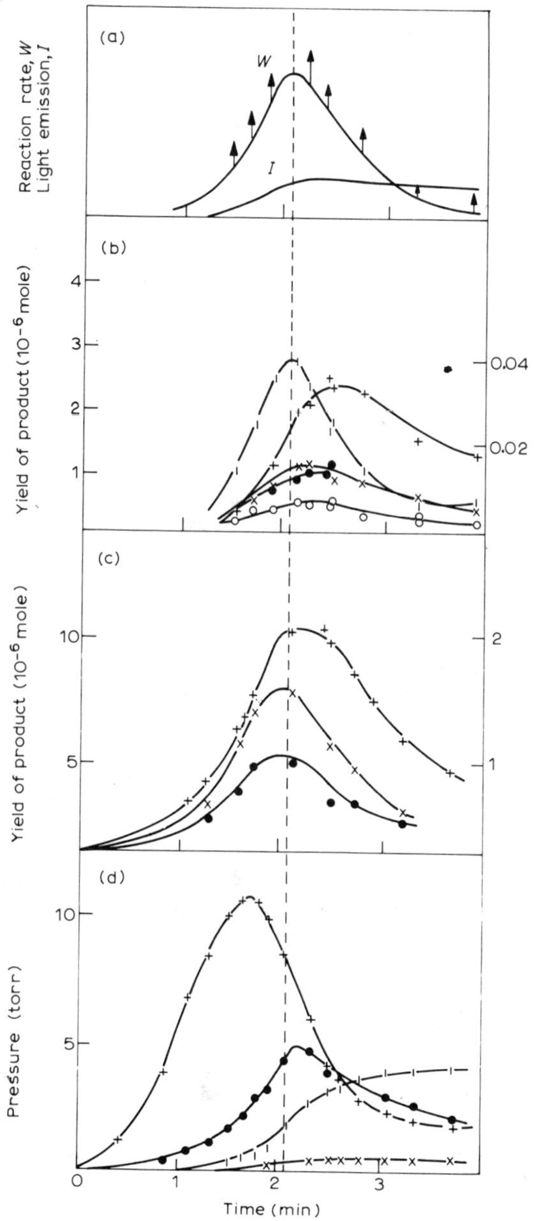

Fig. 18. Product formation during the oxidation of propane in the absence of pic d'arrêt. Initial temperature = 430 °C; initial pressure of propane = 60 torr; initial pressure of oxygen = 240 torr. (b) Left ordinate: +, methyl alcohol. Right ordinate: ×, isopropyl alcohol; ●, ethyl alcohol; ○, *n*-propyl alcohol; |, allyl alcohol (×0.5). (c) Left ordinate: +, hydrogen peroxide; ●, formaldehyde. Right ordinate: ×, total aldehydes. (d) +, propene; |, methane; ●, ethylene; ×, ethane. (From ref. 147.)

Using values [59] of $k_{42} = 10^{9.4}$ l . mole^{-1} . sec^{-1} and $k_{43} = 10^{8.8}$ l . mole^{-1} . sec^{-1}, the ratio of the rates of reactions (42) and (43) will be

$$R_{42}/R_{43} = 10^{0.6}[C_nH_{2n+1}OO\cdot]/[\cdot C_nH_{2n+1}]$$

Assuming reaction (3) is in equilibrium, viz.

$$\cdot C_nH_{2n+1} + O_2 \rightleftharpoons C_nH_{2n+1}OO\cdot \qquad (3)$$

then $R_{42}/R_{43} = 10^{0.6}[O_2]K_{eq}$. It is possible, therefore, to calculate R_{42}/R_{43} for different oxygen concentrations by taking [104] $K_{eq} = 10^{3.61}$ l . mole^{-1}, viz.

P_{O_2} (torr)	150	15	1.5	0.15
R_{42}/R_{43}	55	5.5	0.5	0.05

Measurements of the oxygen concentration at the pic d'arrêt show that it is sufficiently low for reaction (43) to predominate [147]. The onset of the final stage of the reaction is thus explained by the conversion of alkylperoxy radicals to alkoxy radicals via reactions (43) and (44). The production of the dialkylperoxide and its subsequent branching leads to the autocatalytic increase in rate and the increase in the intensity of the emitted light. The oxygen concentration, which is already small, rapidly falls to zero and thereby quenches the reaction. Reaction (43) is then replaced by the terminating reaction (41) which accounts for the formation of ethane [147]. The alkoxy radicals formed at the pic d'arrêt react as shown in the scheme below, thus accounting for the observed increases in product formation. Under conditions where the oxygen concentration is higher, reaction (45)

$$R_2CHO_2\cdot + R_2CHO_2\cdot \longrightarrow R_2C{=}O + R_2CHOH + O_2 \qquad (45)$$

is also an important terminating reaction for primary and secondary alkylperoxy radicals [148]. The absence of acetone in the products found at the pic d'arrêt during the oxidation of propane gives strong support, therefore, to the fact that reaction (43) predominates under these conditions.

The light emission is caused by excited formaldehyde, which is believed to be formed in highly exothermic termination reactions such as reactions (46), (47) and (48), since its concentration does not increase as would be

Scheme 3. Mechanism of pic d'arrêt

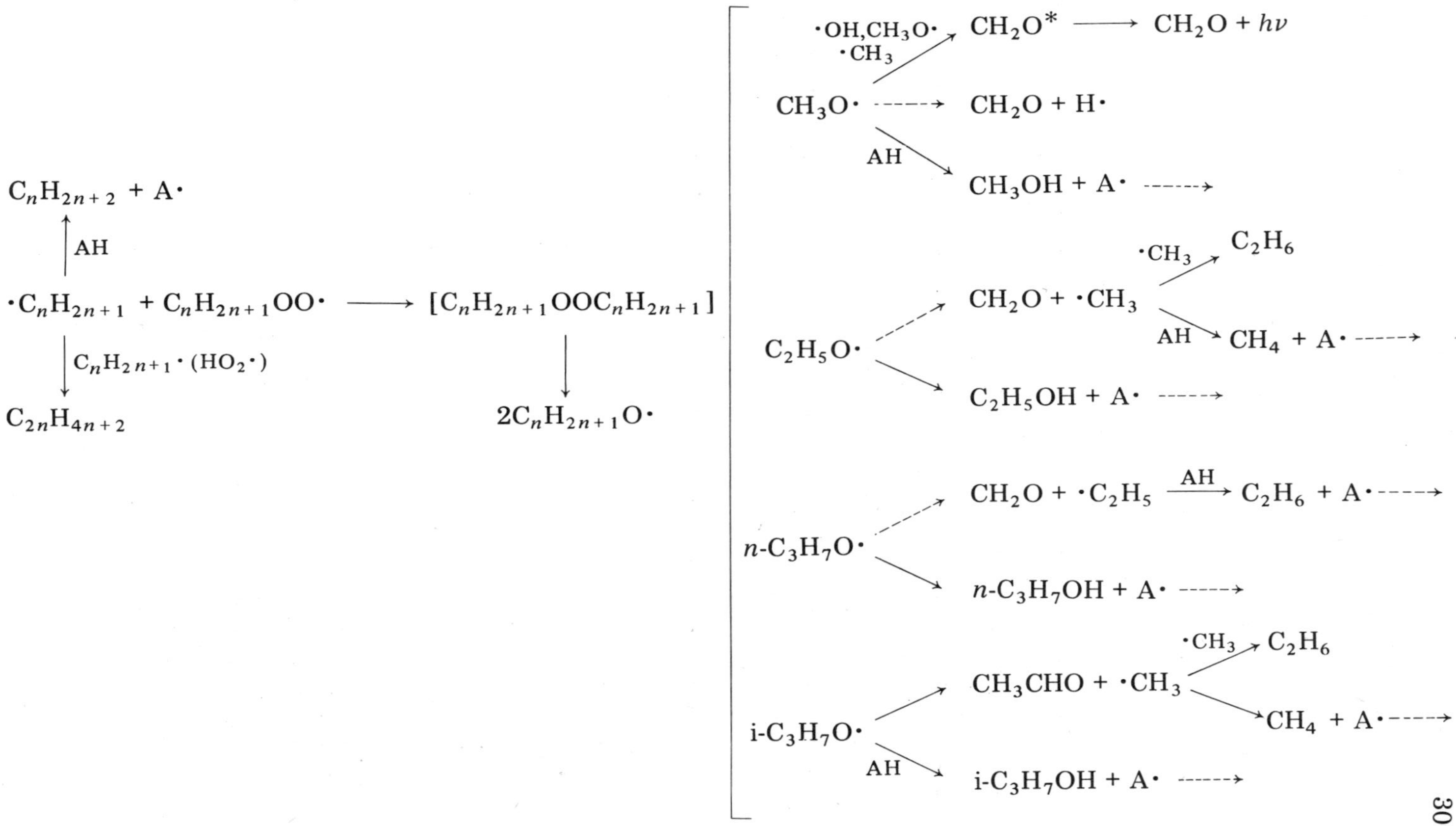

expected if it were formed by propagation reactions such as reaction (49) [147].

$$CH_3O\cdot + CH_3O\cdot \longrightarrow CH_2O + CH_3OH \quad (46)$$

$$CH_3O\cdot + \cdot OH \longrightarrow CH_2O + H_2O \quad (47)$$

$$CH_3O\cdot + \cdot CH_3 \longrightarrow CH_2O + CH_4 \quad (48)$$

$$C_3H_7O\cdot \longrightarrow CH_2O + \cdot C_2H_5 \quad (49)$$

The mechanism of the pic d'arrêt has an important analogy with "oxygen cut-off" observed in liquid-phase reactions [148]. At the lower temperature of these reactions the dialkylperoxide is more stable and reaction (43) is not followed by reaction (44), so that it is a bona fide terminating reaction. Autocatalysis at the final stage of the reaction does not occur therefore and the pic d'arrêt is not observed. There is, however, a rapid decrease in the luminescence which is due to reaction (42) ceasing when all the oxygen is exhausted.

Absolute rate coefficients for these reactions at high temperatures (600 °K) have not been determined, but Heicklen [59] has estimated

TABLE 11
Rate coefficients for reactions of methylperoxy radicals with other radicals (From ref. 59.)

Reaction	$\log_{10} k$ (k in l. mole^{-1} . sec^{-1})
$2CH_3O_2\cdot \rightarrow 2CH_3O\cdot + O_2$	~10.2
$CH_3O_2\cdot + HO_2\cdot \rightarrow CH_3OOH + O_2$	~9.5
$CH_3O_2\cdot + CH_3O\cdot \rightarrow CH_3OOH + CH_2O$	~9.5
$CH_3O_2\cdot + \cdot OH \rightarrow CH_3OH + O_2$	<7
$CH_3O_2\cdot + \cdot CH_3 \rightarrow CH_3OOCH_3$	~8.8
$CH_3O_2\cdot + \cdot CH_3 \rightarrow CH_3OOH + C_nH_{2n}$	~9.5

values for these and other radical—radical reactions by comparison with analogous reactions of alkoxy radicals (Table 11). Electron spin resonance and rotating sector studies, however, have allowed several workers [149, 150] to determine Arrhenius parameters for termination reactions of secondary and tertiary alkylperoxy radicals at low temperature (—30 to +40 °C) in the liquid phase (Table 12). Bennett et al. [150] believe that the predominant reactions to which these values refer are reaction (45) for secondary and reaction (50) for tertiary radicals, viz.

$$2R_2CHO_2\cdot \longrightarrow R_2C{=}O + R_2CHOH + O_2 \quad (45)$$

$$2R_3CO_2\cdot \longrightarrow R_3COOCR_3 + O_2 \quad (50)$$

TABLE 12

Kinetic parameters for the second-order termination reactions of some alkylperoxy radicals (From refs. 149, 150)

Peroxy radical	$\log_{10} k$ (k in l . mole^{-1} . sec^{-1})	Temp. (°C)	Solvent	E (kcal . mole^{-1})	$\log_{10} A$ (A in l . mole^{-1} . sec^{-1})
n-Butyl	7.60	30	α-Methylstyrene		
sec-Butyl	6.18	30	Tetralin		
tert-Butyl	3.45	30	CF_2Cl_2	8.7	9.7
tert-Butyl	4.72	−24	3-Methylpentane	8.4	12.1
Neopentyl	>8.60	−20	Neopentane		
sec-Heptyl	6.38	30	Heptane	1.9	7.7
2-Methylpent-2-yl	4.18	30	2-Methylpentane	9.3	11.1
2,2,3-Trimethylbut-3-yl	0.0077	30	2,2,3-Trimethylbutane	7.5	9.2

References pp. 361—367

Kinetic, isotopic and product studies of autoxidation suggest that these reactions and reaction (42) proceed via a tetroxide intermediate (RO_4R) which may be a transition state or a molecule of finite lifetime [151]. Russell [152] has suggested that the termination reactions of primary and secondary alkylperoxy radicals in fact involve the formation of a cyclic transition state, viz.

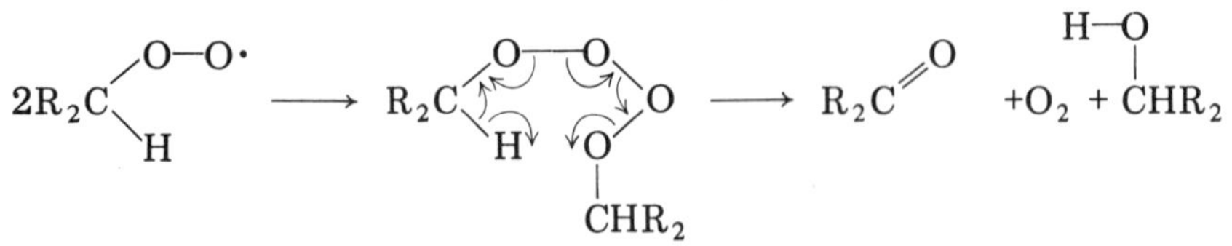

Bennett et al. [151] have been able to measure the equilibrium constants for reaction (51) at −120 °C for several tertiary alkylperoxy radicals, but whether tetroxides are also formed as intermediates at high temperature remains a problem for conjecture.

$$2R_3CO_2\cdot \rightleftharpoons R_3CO_4CR_3 \qquad (51)$$

4. The high temperature mechanism

Above ca. 400—450 °C abstraction of a hydrogen atom from alkyl radicals by oxygen to yield the conjugate alkene and hydroperoxy radical

$$\cdot C_nH_{2n+1} + O_2 \rightleftharpoons C_nH_{2n} + HO_2\cdot \qquad (2)$$

becomes important.* The primary chain cycle is then completed by reaction (5)

$$C_nH_{2n+2} + HO_2\cdot \rightleftharpoons \cdot C_nH_{2n+1} + H_2O_2 \qquad (5)$$

At these temperatures, the hydrogen peroxide so formed suffers appreciable homogeneous decomposition to give to ·OH radicals [153]

$$H_2O_2 + M \longrightarrow 2\cdot OH + M \qquad (7)$$

$$(A_7 = 7.4 \times 10^{14}\ \text{l . mole}^{-1}\ \text{. sec}^{-1}, E_7 = 47\ \text{kcal . mole}^{-1})\ [99]$$

and can thus act as a source of degenerate chain-branching. This leads to a further propagation step

$$C_nH_{2n+2} + \cdot OH \longrightarrow \cdot C_nH_{2n+1} + H_2O \qquad (12)$$

In addition, in the temperature range 450—650 °C pyrolysis will begin to compete effectively with the oxidation of alkyl radicals [38].

---

* It will be seen in Sect. 5 that the temperature at which this occurs depends on the molecular weight and structure of the hydrocarbon.

One of the few recent investigations of alkane oxidation in this temperature range was carried out by Sampson [154, 155] using a high speed flow reactor. For ethane + oxygen mixtures in the ratios 2.5:1 to 10:1 at 1.1 atm and temperatures between 600 and 632 °C large yields of ethylene are obtained [154]. Thus, for example, at 623 °C and 2 % ethane conversion the yield of ethylene is ca. 71 %. The yield of hydrogen peroxide is considerably smaller (ca. 6.7 %), however, and Sampson concluded that the chain cycle (2) + (5) is unimportant except in the very early stages and that it is rapidly superseded by the cycle (2) + (6) + (7) + (12)

$$C_nH_{2n+1}\cdot + O_2 \longrightarrow C_nH_{2n} + HO_2\cdot \tag{2}$$

$$HO_2\cdot + HO_2\cdot \longrightarrow H_2O_2 + O_2 \tag{6}$$

$$H_2O_2 + M \longrightarrow 2\cdot OH + M \tag{7}$$

$$C_nH_{2n+2} + \cdot OH \longrightarrow \cdot C_nH_{2n+1} + H_2O \tag{12}$$

Pyrolysis of ethyl radicals was found to be relatively unimportant although the reverse is true for larger alkyl radicals such as *n*-propyl, isobutyl and *tert*-butyl. Indeed, Sampson estimated that almost half the isobutane consumed gives radicals whose fate is pyrolysis [155]. On the other hand, radical—radical reactions are important and many of the minor products are believed to be formed from further reactions of ethoxy radicals formed in reaction (52)

$$\cdot C_2H_5 + HO_2\cdot \longrightarrow C_2H_5OOH \longrightarrow C_2H_5O\cdot + \cdot OH \tag{52}$$

For many years conjugate alkene formation had been accepted as being diagnostic of the high temperature mechanism and it was suggested that the existence of the upper temperature limit of cool flame might be due to the change with temperature of the predominant mode of reaction of alkyl radicals with oxygen [77]. It is now clear that this is not necessarily the case (Sect. 3.3) particularly for high molecular weight alkanes. On the other hand, however, there is indirect evidence which suggests that the mechanism of one-stage ignition observed at high temperatures may be associated with branching by decomposition of hydrogen peroxide [62, 156].

It is extremely difficult from direct studies of hydrocarbon oxidation to unravel the contributions made to the overall mechanism at these temperatures by the additive, abstractive and pyrolysis routes. Baldwin, Walker and their co-workers have recently developed a new technique, however, which has overcome this problem to some extent and has also yielded quantitative Arrhenius parameters for many elementary reactions [108, 157—161]. This technique utilizes the fact that the kinetics of slowly reacting hydrogen and oxygen mixtures in aged boric acid coated

References pp. 361—367

vessels at ca. 500 °C are extremely reproducible. Furthermore, a detailed mechanism has been established for this reaction and the velocity constants of the individual steps have been evaluated. The reaction can therefore be used as a reliable and controllable source of the H, O, ·OH and HO_2· radicals. By adding small amounts (⩽1 %) of hydrocarbon and related compounds to slowly reacting mixtures of $H_2 + O_2$ quantitative information can be therefore obtained on their reactions with H, O, ·OH and HO_2· and on the subsequent reactions of the free radicals produced.

This technique has many advantages over direct studies of hydrocarbon oxidation. Thus, in direct studies the nature and concentration of the radicals are controlled by the hydrocarbon and its oxidation products and these will change as the reaction proceeds. In the addition studies, however, the concentration of H, O, ·OH and HO_2· radicals is controlled by the $H_2 + O_2$ system and so the hydrocarbon and its products are effectively in a constant radical environment. Secondly, a wide range of additive concentration (0.01—1.0 %), hydrogen concentration (7—86 %) and oxygen concentration (7—72 %) can be investigated. Thirdly, by plotting the product yields against the pressure change due to the $H_2 + O_2$ reaction it is possible to distinguish between primary, secondary and tertiary products.

The basic mechanism of the $H_2 + O_2$ reaction used by Baldwin and Walker is

$$OH + H_2 \longrightarrow H_2O + H \quad (53)$$

$$H + O_2 \longrightarrow OH + O \quad (54)$$

$$O + H_2 \longrightarrow OH + H \quad (55)$$

$$H + O_2 + M \longrightarrow HO_2 + M \quad (56)$$

$$H_2O_2 + M \longrightarrow 2OH + M \quad (57)$$

$$H + HO_2 \longrightarrow 2OH \quad (58)$$

$$2HO_2 \longrightarrow H_2O_2 + O_2 \quad (59)$$

$$HO_2 + H_2 \longrightarrow H_2O_2 + H \quad (60)$$

$$H + H_2O_2 \longrightarrow H_2O + OH \quad (61)$$

$$H + H_2O_2 \longrightarrow H_2 + HO_2 \quad (62)$$

$$\cdot OH + H_2O_2 \longrightarrow HO_2 + H_2O \quad (63)$$

Computer programs for solving the kinetics due to this mechanism may be modified to include reactions (5), (12), (64) and (65) in which the hydrocarbon is removed, viz.

$$HO_2\cdot + C_nH_{2n+2} \longrightarrow \cdot C_nH_{2n+1} + H_2O_2 \quad (5)$$

$$\cdot OH + C_nH_{2n+2} \longrightarrow \cdot C_nH_{2n+1} + H_2O \quad (12)$$

$$H + C_nH_{2n+2} \longrightarrow \cdot C_nH_{2n+1} + H_2 \quad (64)$$

$$O + C_nH_{2n+2} \longrightarrow \cdot C_nH_{2n+1} + \cdot OH \quad (65)$$

Although computer treatment is ultimately the most convenient way to interpret the results accurately, a preliminary analytical treatment is possible if reactions (58) and (62), which are of minor importance, and reactions (60) and (63) are neglected. Reactions (5) and (65) are also of minor importance [158], so combining reactions (12) and (64) in turn with the simplified H_2/O_2 scheme expressions for the relative rates of removal of hydrogen and hydrocarbon may be obtained. Assuming only the hydrocarbon concentration varies, on integration these give the following approximate expressions for 20 % hydrocarbon conversion [158, 161]

$$\Delta H_2 = 0.223k_{53}[H_2]/k_{12}$$

$$\Delta H_2 = 0.223k_{56}[O_2][M]/k_{64}$$

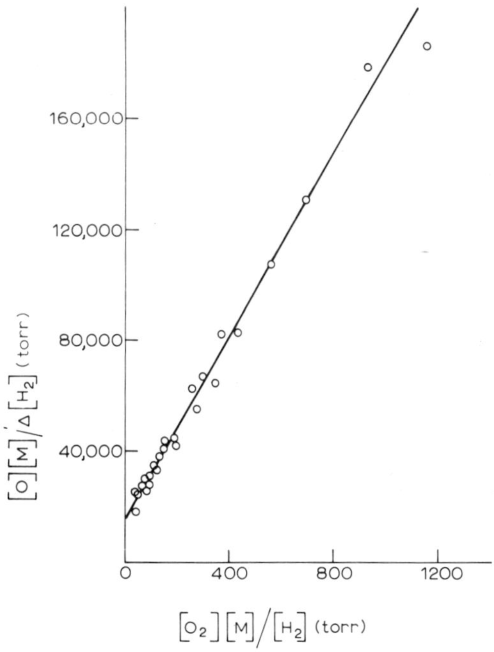

Fig. 19. Variation of $[O_2][M]/\Delta(H_2)$ with $[O_2][M]/[H_2]$ for *n*-butane at 480 °C (From ref. 158.)

For initial attack by H and ·OH the hydrocarbon consuming terms are additive and with rearrangement these expressions give the following linear relationship

$$\frac{[O_2][M]}{\Delta H_2} = \frac{k_{12}[O_2][M]}{0.223k_{53}[H_2]} + \frac{k_{64}K}{0.223k_{54}}$$

where $K = k_{54}/k_{56} = 13.0$ torr ($M = H_2$) at 480 °C. A plot of $[O_2][M]/\Delta H_2$ against $[O_2][M]/[H_2]$ is shown in Fig. 19 for n-butane; the gradient and intercept of which allow the ratios of k_{12}/k_{53} and k_{64}/k_{54} to be determined at 480 °C. (Allowance was made for the pressure change caused by the formation of H_2O_2 and by the oxidation of the hydrocarbon.)

The results obtained for a range of hydrocarbons were treated similarly and values of k_{12}/k_{53} at 480 °C found to be

CH_4	1.1	n-C_4H_{10}	34.0
C_2H_6	10.5	i-C_4H_{10}	31.0
C_3H_8	21.5	$(CH_3)_4C$	16.0

(These are preliminary values, accurate to about 10 % [157].)

Examination of these results has allowed the relative importance of OH attack at primary, secondary and tertiary C—H positions to be determined. Thus, from the value for ethane, the value of $k_{12}{}^{p}/k_{53}$ (where p denotes attack at one CH_3 group) may be taken as 5.25. Assuming additivity, the value for $k_{12}{}^{s}/k_{53}$ for one secondary CH_2 group is 11.0 from C_3H_8 and 11.7 from n-C_4H_{10}, and the value for $k_{12}{}^{t}/k_{53}$ for one tertiary CH group is 15.3 from i-C_4H_{10}. Combining these results with known values of k_{53} at lower temperatures Baldwin and Walker have

TABLE 13
Arrhenius parameters for OH and H attack at individual C—H bonds
$C_nH_{2n+2} + \cdot OH \text{ (or H)} \rightarrow \cdot C_nH_{2n+1} + H_2O \text{ (or } H_2)$

Hydrocarbon	Bond	E (kcal . mole^{-1})		$\log_{10} A$ per C—H bond (A in l . mole^{-1} . sec^{-1})		Ref.
		OH	H	OH	H	
CH_4	—	5.00	11.90	9.80	10.49	158
C_2H_6	Primary	3.52	9.70	10.16	10.51	
C_3H_8	Secondary	2.55	8.45	10.34	10.71	
n-C_4H_{10}	Secondary	—	8.45	—	10.74	
iso-C_4H_{10}	Tertiary	2.39	7.06	10.72	10.94	
						98
	Primary	1.64		8.91		
	Secondary	0.85		9.15		
	Tertiary	−0.19		9.10		

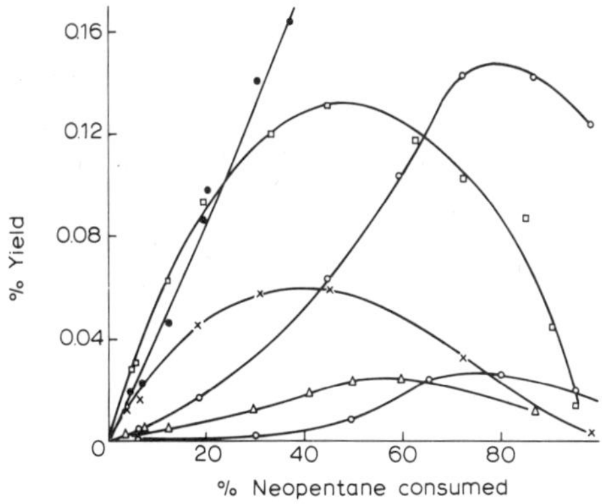

Fig. 20. The variation of product formation with neopentane consumed. Initial temperature = 480 °C; initial pressure of hydrogen = 140 torr; initial pressure of oxygen = 70 torr; total pressure = 500 torr. ●, methane; ○, acetone; □, but-2-ene; ×, 3,3-dimethyloxetan; △, propene; ○, ethylene. (From ref. 108).

calculated the Arrhenius parameters for initial attack at individual C—H bonds as shown in Table 13 where they are compared with the values of Greiner [98].

Detailed analysis of the reaction products at various stages of the reaction has allowed the fate of radicals produced from ethane [161], propane [162], *n*- and isobutane [162], neopentane [108], ethylene [108] and propene [163] to be examined.

With ethane at 500 °C, 96 ± 2 % of the ethyl radicals react with oxygen to give ethylene, the other primary products being ethylene oxide (2 ~ 4 %) and acetaldehyde (ca. 0.5 %). Carbon monoxide, formaldehyde and methane are formed as secondary products. With propane, *n*-butane and isobutane the initial yields of the conjugate alkenes are 80, 65 and 80 %, respectively, the remaining products resulting almost equally from radical—pyrolysis and radical—oxygen reactions. For neopentane, there is, of course, no conjugate alkene and the primary products are isobutene, 3,3-dimethyloxetan, methane and formaldehyde [108] as shown in Fig. 20. Later results [164] which show that the ratios of acetone/3,3-dimethyloxetan and of (acetone + 3,3-dimethyloxetan)/isobutene are both directly proportional to the oxygen concentration over a ten-fold range suggest that acetone is also a primary product and that it is formed via further oxidation of the hydroperoxyneopentyl radicals (c.f. refs. 108, 49, 54, 55 and 122)

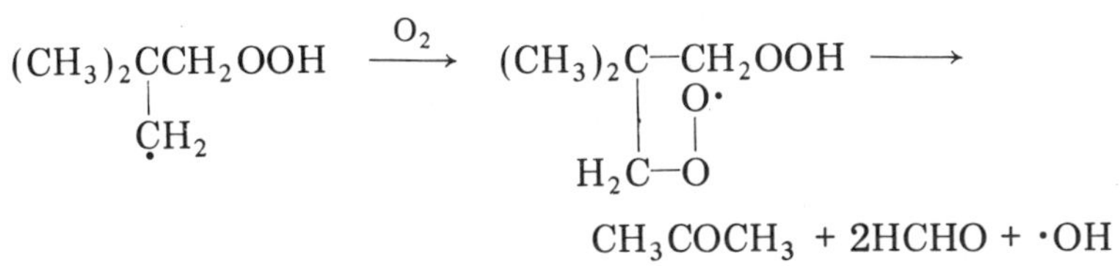

(66)

The yield of formaldehyde is about twice that of acetone and therefore gives support for this mechanism. As Baldwin and Walker [99] recently pointed out, this is one of the few cases where kinetic evidence supports the proposed mechanism of product formation via hydroperoxyalkylperoxy radicals (see Sect. 3.2.2(c)).

Measurement of the rates of formation of the primary products allows the ratios of several important rate coefficients to be determined. Thus, for example, for propane the following reactions are in competition, viz.

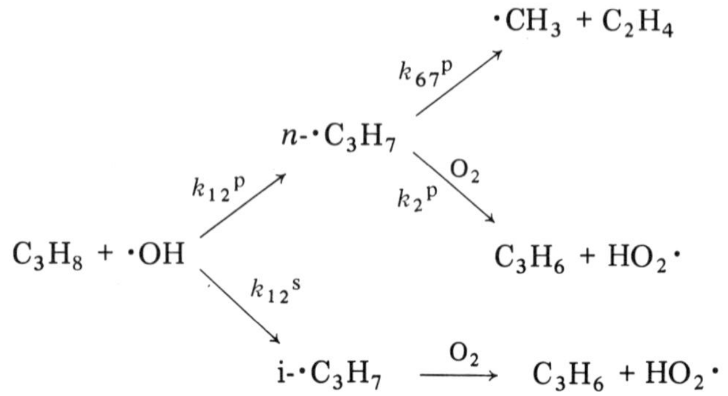

The relative rate of formation of propene and ethylene is given by [158]

$$\frac{d[C_3H_6]}{d[C_2H_4]} = \frac{k_{12}^{s}}{k_{12}^{p}} + \frac{k_2^{p}}{k_{67}^{p}} + \left(1 + \frac{k_{12}^{s}}{k_{12}^{p}}\right)[O_2]$$

A plot of $d[C_3H_6]/d[C_2H_4]$ against $[O_2]$ allows k_{12}^{s}/k_{12}^{p} and k_2^{p}/k_{67}^{p} to be determined. Values for k_2/k_{67} have been obtained similarly

TABLE 14
Rate coefficients for reactions of alkyl radicals with oxygen (From ref. 158.)

$\cdot C_nH_{2n+1}$	Product	$\log_{10} k_2$ (k in l. mole^{-1}. sec^{-1})	$\log_{10} k_{68}$	Temp. (°C)
C_2H_5	Ethylene	7.91		440
C_2H_5	Ethylene	8.00		623
n-C_3H_7	Propene	7.58		480
i-C_4H_9	Isobutene	7.36		480
n-C_4H_9	But-1-ene	8.45		480
s-C_4H_9	*cis*-But-2-ene	8.08		480
s-C_4H_9	*trans*-But-2-ene	8.32		480
C_2H_5	Acetaldehyde		5.53	500
C_2H_5	Ethylene oxide		6.42	500
n-C_3H_7	Propionaldehyde		5.04	480
n-C_3H_7	Propene oxide		6.49	480
s-C_4H_9	Butanone		6.36	480
i-C_4H_9	Isobutyraldehyde		6.36	480

for each hydrocarbon (except neopentane) and using the values of k_{67} given by Kerr and Lloyd [165], values of k_2 have been interpolated as shown in Table 14. Combining Slater and Calvert's [166] estimated value of $k_2 = 2.6 \times 10^5$ l . mole^{-1} . sec^{-1} for the isobutyl radical at 40 °C with the above value at 480 °C gives $A_2 = 10^{8.7}$ l . mole^{-1} . sec^{-1} and E = 4.7 kcal . mole^{-1}.

This technique also allowed values of k_2/k_{68} for

$$\cdot C_nH_{2n+1} + O_2 \longrightarrow C_nH_{2n} + HO_2\cdot \qquad (2)$$

$$\cdot C_nH_{2n+1} + O_2 \longrightarrow C_nH_{2n}O + \cdot OH \qquad (68)$$

to be estimated and hence k_{68} can be interpolated (Table 14). However, as Baldwin and co-workers [158] point out any simple interpretation of the values of k_2 and k_{68} may not be valid, since the overall reactions (2) and (68) may be complex and occur via the alkylperoxy radical. A rough calculation for the oxidation of ethyl radicals suggests that oxidation via ethylperoxy radicals does indeed contribute to the overall reaction, viz.

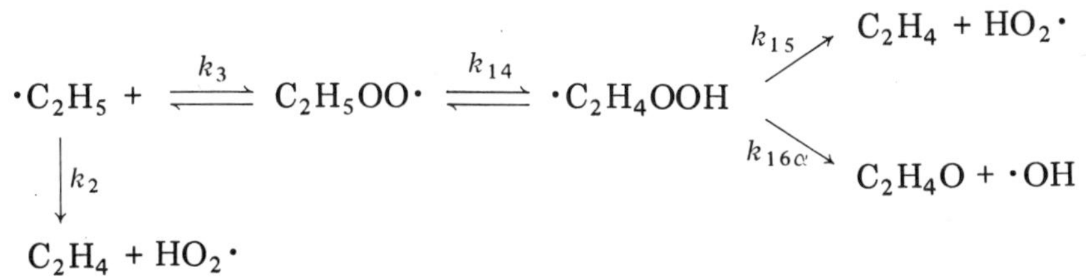

Using the Arrhenius parameters given in Sect. 3.2.2 for reactions (15) and (16α),

$$\text{rate (15)/rate (16}\alpha) = 10^{13.5}\, e^{\frac{-20{,}000}{2 \times 770}} \Big/ 10^{11.5}\, e^{\frac{-14{,}000}{2 \times 770}} \simeq 3.$$

At 500 °C, $[C_2H_4]/[C_2H_4O] \simeq 39$ at 5 % ethane conversion [161], hence since ethylene oxide is a primary product, 36 ethyl radicals are consumed via reaction (2) for every 4 consumed via reaction (3). This data suggests, therefore, that the route via the ethylperoxy radical contributes to the overall oxidation by ca. 10 %.

TABLE 15
Rate coefficients for the initial attack on aldehydes at 440 °C (From ref. 157.)

Aldehyde	k_{36} (l . mole^{-1} . sec^{-1})	$k_{69}/k_{59}^{1/2}$ (l . mole^{-1} . sec^{-1})$^{1/2}$
HCHO	1.33×10^{-2}	32
C_2H_5CHO	6.23×10^{-2}	43
C_3H_7CHO	—	57

The addition of aldehydes to slowly reacting mixtures of $H_2 + O_2$ provides information of the reactions of $HO_2\cdot$ radicals and of the reactions of alkyl radicals [157, 159, 167]. Assuming the basic reaction scheme (here for propionaldehyde)

$$C_2H_5CHO + O_2 \longrightarrow C_2H_5\dot{C}O + HO_2\cdot \quad (36)$$

$$C_2H_5\dot{C}O + M \longrightarrow \cdot C_2H_5 + CO + M \quad (37)$$

$$\cdot C_2H_5 + O_2 \longrightarrow C_2H_4 + HO_2\cdot \quad (2)$$

$$HO_2\cdot + C_2H_5CHO \longrightarrow C_2H_5\dot{C}O + H_2O_2 \quad (69)$$

$$HO_2\cdot + HO_2\cdot \longrightarrow H_2O_2 + O_2 \quad (59)$$

$$H_2O_2 + M \longrightarrow 2\cdot OH + M \quad (57)$$

$$\cdot OH + C_2H_5CHO \longrightarrow C_2H_5\dot{C}O + H_2O \quad (33)$$

Baldwin and Walker have been able to measure the values of k_{36} and $k_{69}/k_{59}{}^{1/2}$ as shown in Table 15.

In the oxidation of propionaldehyde detectable amounts of ethane are formed by reaction (70)

$$\cdot C_2H_5 + C_2H_5CHO \longrightarrow C_2H_6 + C_2H_5\dot{C}O \quad (70)$$

which competes with reaction (2)

$$\cdot C_2H_5 + O_2 \longrightarrow C_2H_4 + HO_2\cdot \quad (2)$$

hence,

$$\frac{d[C_2H_4]}{d[C_2H_6]} = \frac{k_2[O_2]}{k_{70}[AH]}$$

At 440 °C, the plot of $[C_2H_4]/[C_2H_6]$ against $[O_2]/[AH]$ is indeed linear and passes through the origin. From the gradient $k_2/k_{70} = 41$ at 440 °C which leads to a value of $k_2 = 8.2 \times 10^7$ l . mole^{-1} . sec^{-1} which is in good agreement with the estimate of 1.0×10^8 l . mole^{-1} . sec^{-1} made by Sampson [154].

Estimations of other important rate coefficients can be made similarly, provided the mechanism is simple and the routes of product formation are known with some degree of certainty. Baldwin and Walker's technique provides a powerful tool, therefore, not only for measuring rate coefficients of the elementary steps, but also for elucidating the reaction mechanism.

5. The variation of mechanism with the molecular weight and structure of the hydrocarbon

During the last ten years much discussion concerning the nature of the principal chain-propagating steps in the reactions leading to the passage of cool flames during the oxidation of hydrocarbons has been focussed on the seemingly contradictory results obtained when low and high molecular weight alkanes are used as fuels. This led to the evolution of two apparently opposing theories for the mechanism of chain propagation, namely the alkene theory and the alkylperoxy radical isomerization theory, which were discussed in detail in Sect. 3. It will be remembered that the former theory was put forward following studies of the oxidation of low molecular weight alkanes (C_2—C_4) in the early stages of the reaction, usually at 1 % conversion long before a cool flame propagates, whilst the latter was put forward principally following studies during the later stages of the reaction, usually just prior to or at the passage of a cool flame. Comparative studies of the oxidation of *n*-butane [168] and 2-methylpentane [169] both at 1 % conversion and at the cool flame have shown that the nature of the primary products is a function of the fuel rather than the stage of the reaction, so it appears, therefore, that the predominating chain-propagation steps vary with the molecular weight and structure of the fuel.

The link between these mechanisms and hence the nub of the whole problem is the alkylperoxy radical as can be seen from the scheme

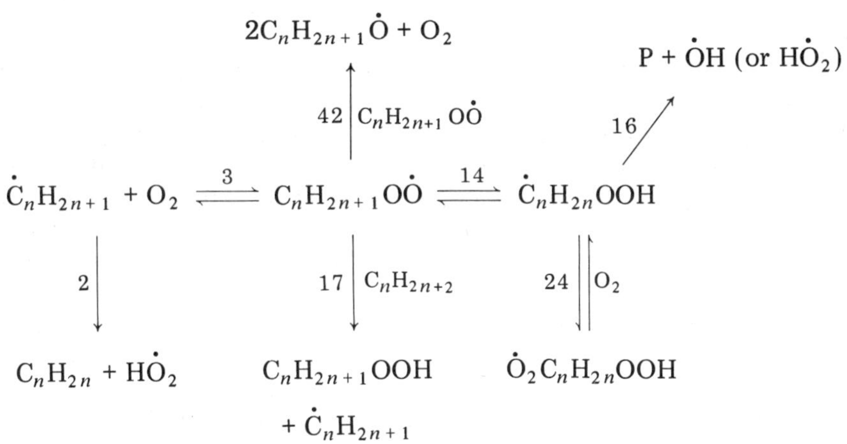

Whilst this scheme is not intended to be fully definitive it does illustrate that the predominant reaction path will depend on the relative values of k_2, K_3, k_{17}, k_{42}, K_{14}, K_{24}, and k_{16} and on the equilibrium value of $[C_nH_{2n+1}O\dot{O}]/[\dot{C}_nH_{2n+1}]$. For most alkanes, except methane, reactions (2), (3), (17) and (42) are all possible, but reaction (14) will be highly

TABLE 16

Comparison of the ease of formation of β-hydroperoxyalkyl radicals for different parent alkanes at 600 °K

$C_nH_{2n+1}OO\cdot \rightarrow \cdot C_nH_{2n}OOH$

C_nH_{2n+2}	Initial attack	$C_nH_{2n+1}OO\cdot$	Nature of C—H broken in 1:5 H-transfer	$\log_{10} k_1$ (k in sec)
Ethane	Primary	Ethylperoxy	—	—
Propane	Primary	Prop-1-ylperoxy	Primary	6.2
	Secondary	Prop-2-ylperoxy	—	—
n-Butane	Primary	But-1-ylperoxy	Secondary	7.5
	Secondary	But-2-ylperoxy	Primary	6.2
n-Pentane	Primary	Pent-1-ylperoxy	Secondary	7.5
	Secondary	Pent-2-ylperoxy	Secondary	7.5
	Secondary	Pent-3-ylperoxy	Primary	6.2

dependent on the structural constraints placed upon the ability of the particular alkylperoxy radical to isomerize.

It can be seen from Table 3 that the most facile isomerizations are those which involve 1:5 H-transfer (i.e. the formation of a 6-membered transition state ring) to form a β-hydroperoxyalkyl radical, e.g.

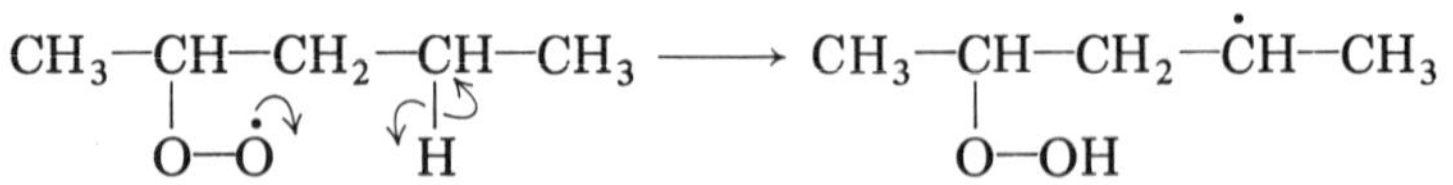

(14β)

The ease of formation of β-hydroperoxyalkyl radicals from the alkane increases with molecular weight as shown in Table 16. Thus, for example, isomerization involving 1:5 H-transfer is impossible for ethylperoxy and prop-2-ylperoxy radicals, while isomerization of the pent-2-ylperoxy radicals leads to the lowest molecular weight hydroperoxyalkyl radical which can be formed by initial attack at a secondary C—H bond followed by isomerization involving 1:5 H-transfer from another secondary C—H bond.

1:5 H-transfer is always ca. 10^2 faster than 1:4 H-transfer at 600 °K (see Table 3), so it will predominate when the molecular structure of the fuel permits. Simple estimation of the relative concentrations of the hydroperoxyalkyl radicals derived from propane, *n*-butane and *n*-pentane illustrates this. Thus, if the relative frequency of attack by $\dot{O}H$ at primary, secondary and tertiary C—H bond is taken as 2:3:5 [102], then the relative concentrations of propyl, butyl and pentyl radicals may be obtained. The equilibrium constant for reaction (3)

$$\dot{C}_nH_{2n+1} + O_2 \rightleftharpoons C_nH_{2n+1}O\dot{O} \qquad (3)$$

TABLE 17
Relative concentrations of hydroperoxyalkyl radicals in the early stages of reaction at 600 °K

C_nH_{2n+2}	Nature of $\cdot C_nH_{2n}OOH$	Relative concentrations of $\cdot C_nH_{2n}OOH$	$\frac{\alpha}{\beta+\gamma}$
Propane	α	13.2	
	β	1	13.2
	γ	0	
n-Butane	α	1.94	
	β	1	1.8
	γ	0.08	
n-Pentane	α	1.46	
	β	1	0.96
	γ	0.53	

is independent of the structure of the alkyl radical and so, as a first approximation, the relative concentrations of the hydroperoxyalkyl radicals are given by $K_{14}[\dot{C}_nH_{2n+1}]$ (cf. Sect. 3.2.2(d)), K_{14} being the equilibrium constant for

$$C_nH_{2n+1}O\dot{O} \rightleftharpoons \dot{C}_nH_{2n}OOH \qquad (14)$$

Summation of the concentrations of like hydroperoxyalkyl radicals calculated in this way shows that the relative concentrations of α-hydroperoxyalkyl radicals decreases rapidly with increase in molecular weight (Table 17).

Since cyclization of γ-hydroperoxyalkyl radicals and β-scission of β-hydroperoxyalkyl radicals are ca. 10^{2-3} times faster than β-scission of cyclization of α-hydroperoxyalkyl radicals, product formation via β- and γ-hydroperoxyalkyl radicals will increase rapidly.

Alkane fuels can be divided roughly into two classes, therefore, namely the low molecular weight alkanes ($<C_5$) whose alkylperoxy radicals are restricted in their ability to isomerize and the high molecular weight alkanes ($>C_5$) whose alkylperoxy radicals are unrestricted in their ability to isomerize.

5.1 ALKANES OF CARBON NUMBER $\leqslant$ 5

The kinetics and intermediate products observed during the oxidation of low molecular weight alkanes at low temperatures (ca. $<$ 350 °C) are very sensitive to the initial reactant pressure [68, 170] and to the surface of the reactor in the early stages [42, 106, 123, 134, 171]. Neither phenomenon has been unequivocally elucidated, although some plausible explanations have been made in terms of radical—radical propagation reactions [103] and heterogeneous reactions of alkylperoxy radicals [106, 171].

5.1.1 *Heterogeneity*

Kinnear and Knox [106, 172] studied the oxidation of *n*-pentane at low conversions at 290 °C and found the acetone, a major product, was particularly sensitive to the nature of the reactor surface (Fig. 21) and that the ratio of the yields of the other products to acetone varies linearly with pentane concentration (Fig. 22). Addition of inert gas showed that this ratio is also directly proportional to the diffusion time for the pentylperoxy radicals (Fig. 23).

From these results they concluded that the pentylperoxy radicals have three fates, viz.

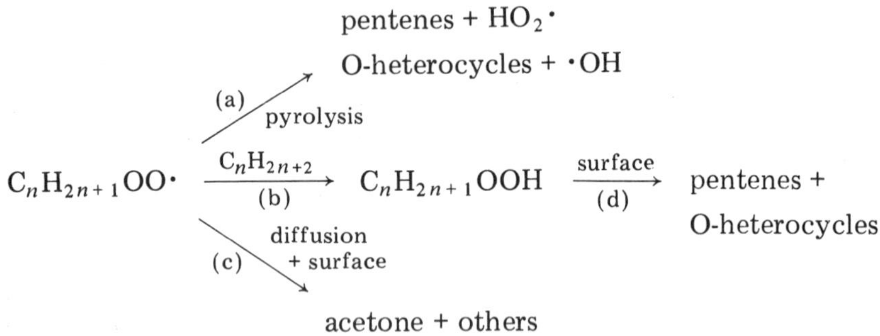

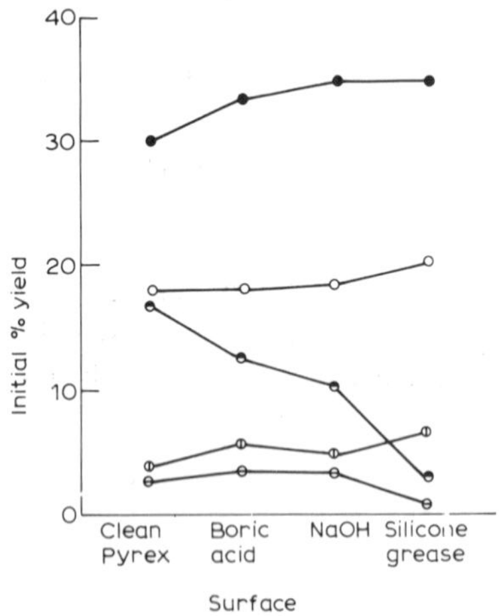

Fig. 21. The variation with surface of the initial percentage yield of major products from the oxidation of *n*-pentane. Initial temperature = 290 °C; initial pressure of *n*-pentane = 25 torr; initial pressure of oxygen = 12.5 torr; total pressure = 82 torr; volume of reaction vessel = 500 cm^3. ●, pent-2-ene; ○, 2-methyltetrahydrofuran; ◓, acetone; ⦶, pent-1-ene; ⊖, butanone. (From ref. 106.)

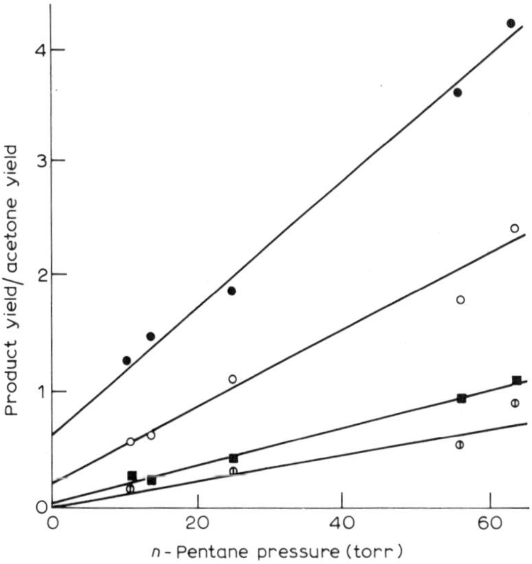

Fig. 22. The variation with *n*-pentane pressure of the ratio of product yield/acetone yield at 290 °C. ●, pent-2-ene; ○, 2-methyltetrahydrofuran; ■, 2,4-dimethyloxetan; ⦶, pent-1-ene. (From ref. 106.)

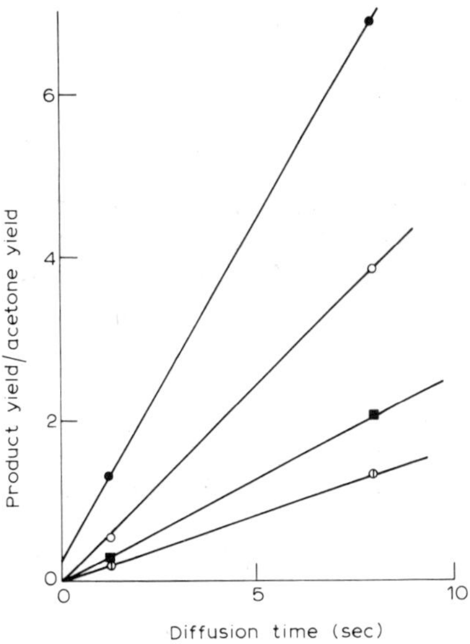

Fig. 23. The variation with diffusion time of the ratio of product yield/acetone yield at a constant *n*-pentane pressure of 10 torr. ●, pent-2-ene; ○, 2-methyltetrahydrofuran; ■, 2,4-dimethyloxetan; ⦶, pent-1-ene. (From ref. 106.)

Figure 22 shows that

$$[\text{product}]/[\text{acetone}] = m + n[C_5H_{10}]$$

where $m = k_a/k_c$ and $n = k_b/k_c$. Consideration of the diffusion time for pentylperoxy radicals and the values of the intercepts in Fig. 22 allowed estimations of E_c and E_a to be made as follows. For a typical reaction mixture at an initial pressure of 82 torr the diffusion time was estimated to be 1.2 sec, which gives $k_c \simeq 0.8$ sec^{-1}. At initial pentane pressures of 25 torr the ratio of the products for reaction (b) to acetone is ca. 3, hence $k_b[C_nH_{2n+2}]/k_c \simeq 3$ and thus $k_b = 10^{3.5}$ l . mole^{-1} . sec^{-1} . Assuming $A_b = 10^{8.0}$ l . mole^{-1} . sec^{-1}, $E_b = 12$ kcal . mole^{-1} which is in good agreement with the previously estimated activation for H-abstraction by alkylperoxy radicals (see Sect. 3.2.1). From the intercepts in Fig. 22,

$$k_{a(\text{pent-2-ene})}/k_c = 0.6$$

$$k_{a(\text{2-methyltetrahydrofuran})}/k_c = 0.2$$

whence

$$k_{a(\text{pent-2-ene})} = 0.5 \text{ sec}^{-1}$$

$$k_{a(\text{2-methyltetrahydrofuran})} = 0.15 \text{ sec}^{-1}$$

For $A_a = 10^{12.5}$ sec^{-1}, $E_{a(\text{pent-2-ene})} = 33$ kcal . mole^{-1} and $E_{a(\text{2-methyltetrahydrofuran})} = 34.5$ kcal . mole . mole^{-1}. Again, these activation energies are in good agreement with those obtained by Baldwin et al. [108, 161] for alkylperoxy radical decompositions, although they are higher than those proposed by Fish [58].

Two criticisms of this mechanism can be made. First, these activation energies are "overall" activation energies for a two-step process for the decomposition of different alkylperoxy radicals [106]; see opposite page. For the formation of 2-methyltetrahydrofuran both steps will involve cyclization and will have pre-exponential factors [104] of ca. $10^{11.5}$ sec^{-1}, whereas the formation of pent-2-ene involves only one such step and a second step for which [39] $A = 10^{13.5}$ sec^{-1}. Since the strain energy involved in the isomerizations of each of the alkylperoxy radicals is the same (ca. 6.5 kcal. mole^{-1}) the activation energies of this step will only differ by the difference in primary and secondary C—H strengths (ca. 3.5 kcal . mole^{-1}). It is difficult, therefore, to see how the "overall" activation energies for the formation of pent-2-ene and 2-methyltetrahydrofuran can be approximately equal.

Secondly, the effect of pressure is difficult to interpret, since in practice, the oxygen pressure was increased for a corresponding decrease in *n*-pentane pressure and vice versa. However, comparison of the initial yield of acetone at different total initial pressures [71, 171], but for the

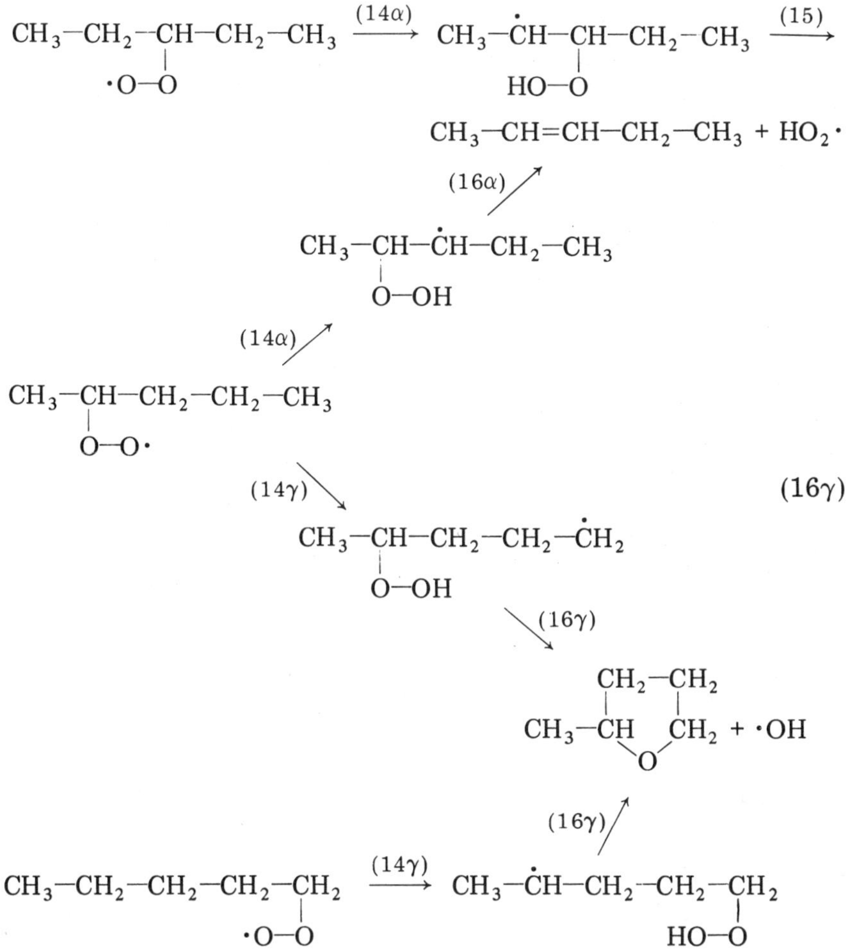

same molar ratio, shows that it increases with initial pressure (Table 18). Clearly, the reverse would be expected if acetone is formed heterogeneously.

In contrast, a recent study of the oxidation of isobutane [171] suggests that the principal heterogeneous reaction involves the formation of isobutene. Thus, a decrease in the S/V ratio of the reactor from 0.99 to 0.63 cm^{-1} led to a decrease ca. 10—17.5 % in the yield of isobutene over the pressure range 250—350 torr at 310 °C as shown in Fig. 24. Conversely, the yield of acetone increased by ca. 5 %, which again suggests that the minor oxygenates are not formed heterogeneously (compare refs. 42 and 44, p. 259). Pollard and co-workers [171] also interpreted these results in terms of the heterogeneous reaction of the alkylperoxy radicals. They pointed out, however, that such an interpretation was only valid if

TABLE 18

The variation with pressure of the percentage conversion to initial products during the slow oxidation of *n*-pentane at ~250 °C (From refs. 71 and 171.)

Temp. (°C)			250			250	257
n-Pentane: oxygen			3:4			2:1	2:1
Pyrex reaction vessel			Untreated			Untreated	HF treated
Pressure (torr)	70	110	142	168	197	90	150
Ethylene	2.5	2.3	2.1	1.7	2.0	2	
Propene	0.4	0.3	0.2	0.2	0.2		
Pent-2-ene	10.8	8.9	7.0	6.0	5.1	15	14
Acetaldehyde	26.4	24.5	25.9	28.4	24.5	4	5
Propionaldehyde	9.4	10.3	11.1	12.7	14.8	3	3
Acetone	23.3	28.0	31.8	30.9	32.7	29	40
Butanone	10.5	11.1	10.5	10.9	11.8	10	9
C_5 Ketones	1.1	0.9	1.3	1.4	1.7	2	
Methyl vinyl ketone	1.6	1.2	0.8	0.9	1.2		
Pent-2-en-4-one	1.3	1.1	0.9	0.7	0.6		
2-Methyltetrahydrofuran	4.0	2.9	2.0	1.9	1.7	6	6
2-Methyl-3-ethylorixan	1.0	0.7	0.8	0.6	0.6		
2-Methyloxiran	2.2	2.2	2.0	2.5	1.2	3	3
Ethanol	0.8	0.5	0.7	0.5	1.0		
Pentan-3-ol	0.4	0.4	0.4	0.4	0.7		
Pentan-2-ol	0.2	0.1	0.1	0.1	0.2		

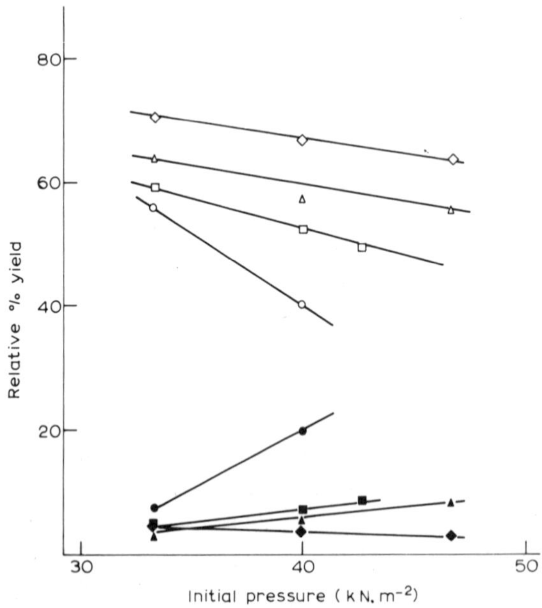

Fig. 24. The effect of stoichiometry and S/V ratio on the variation of the yields of isobutene and acetone with initial pressure at 1 % conversion during the oxidation of isobutane at 310 °C. Open symbols, isobutene; filled symbols, acetone; ◇, fuel/O_2 = 1:2 S/V = 0.99 cm^{-1}; △, fuel/O_2 = 4:1, S/V = 0.63 $cm.^{-1}$; □, fuel/O_2 = 1:2, S/V= 0.63 cm^{-1}; ○, fuel/O_2 = 1:4, S/V = 0.63 cm^{-1}.

the Arrhenius parameters for the isomerization of the *tert*-butylperoxy radicals

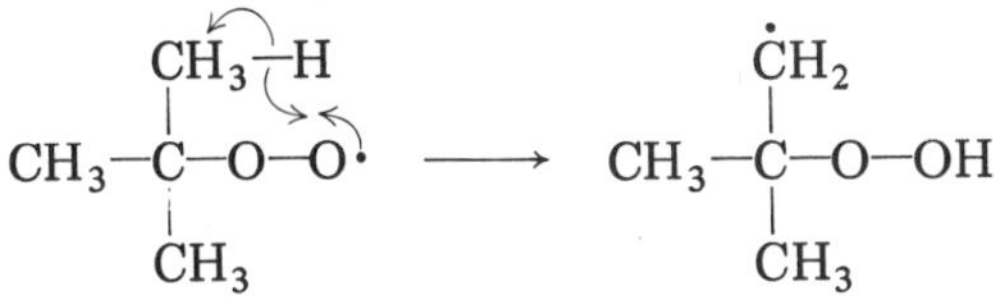

(14α)

are of the order of $A_{14\alpha} = 10^{10.5}$ sec^{-1} and $E_{14\alpha} = 28.5$ kcal . mole^{-1} (cf. the values suggested by Fish, see Table 3, $A_{14\alpha} = 10^{11.5}$ sec^{-1} and $E_{14\alpha} = 20.5$ kcal . mole^{-1}). Using these values for (14α) and $k_{wall} = 0.5$ sec^{-1}, $K_3 = 10^{5.4}$ l . mole^{-1} and $k_2 = 10^{5.6}$ l . mole^{-1} . sec^{-1}, at 600 °K, and so

$$\frac{d[\text{isobutene}]_{het}}{d[\text{isobutene}]_{homo}} = \frac{k_{wall}[C_nH_{2n+1}O\dot{O}]}{k_2[\dot{C}_nH_{2n+1}][O_2] + K_{14\alpha}[C_nH_{2n+1}O\dot{O}]}$$

$$= \frac{k_{wall}}{k_2/K_3 + k_{14}} = \frac{1}{5}$$

Hence, although the heterogeneous formation of isobutene may be significant at low conversion, the rate of its homogeneous formation is about five times as fast.

More recent studies by Irvine and Knox [50] on the competitive oxidation of isobutane with ethane and propane at 300 °C have also led them to conclude that at low rates of reaction of isobutane a heterogeneous component leading to isobutene does indeed occur in parallel, but independently of the homogeneous reaction under most experimental conditions used in slow oxidation studies. They have suggested, however, in agreement with Semenov, that the reaction responsible probably involves the direct reaction of oxygen with isobutane adsorbed on the surface of the reactor (see p. 263), viz.

$$C_4H_{10} + O_2 \xrightarrow{\text{wall}} \text{i-}C_4H_8 + H_2O_2$$

Baldwin and Walker [99] have pointed out that, from kinetic considerations, surface reactions of alkylperoxy radicals cannot play a significant role except at very low overall rates of reaction and conclude that it is more likely that surface destruction of relatively stable intermediates such as the alkyl hydroperoxides or hydrogen peroxide are the main cause of surface effects in hydrocarbon oxidation. Luckett and Pollard [68, 134] have provided evidence, which suggests that the surface destruction of *tert*-butylhydroperoxide is indeed important during the oxidation of isobutane below ca. 320 °C. Since isobutene and acetone are known products of the decomposition of *tert*-butylhydroperoxide, it is clear that many of the foregoing results can be explained in these terms, but if this is the predominant heterogeneous reaction the yield of acetone would be expected to increase with increase in S/V ratio, particularly in the early stages of the reaction where radical concentrations are low and hence radical—radical reactions are thought to be relatively unimportant, whereas experiment shows a small decrease (Fig. 24). Hence, there are either other important heterogeneous reactions or radical—radical reactions leading to the formation of t-butoxy radicals, and hence acetone, which are more important under these conditions than has hitherto been realized.

5.1.2 *Radical—radical reactions*

Marked decreases in the yields of the conjugate alkenes formed during the oxidation of low molecular weight alkanes at sub-atmospheric pressures have been observed during both the early [171] and last [135, 173] stages of the reaction as shown in Fig. 25 for isobutane and Table 19 for n-butane, whilst at 10—20 atm and 350 °C the yield of the conjugate alkene is virtually zero [170]. These results were interpreted by Pollard and co-workers [171] and Lucquin and co-workers [173] to be a consequence of the heterogeneous formation of alkene compounds via the alkylperoxy radicals at low pressure. Whilst this is possible, a more plausible explanation is in terms of radical—radical reactions, particularly in view of the fact that the parallel homogeneous formation of the alkene appears to be faster. Thus, Baldwin and Walker [99, 103], Barnard and

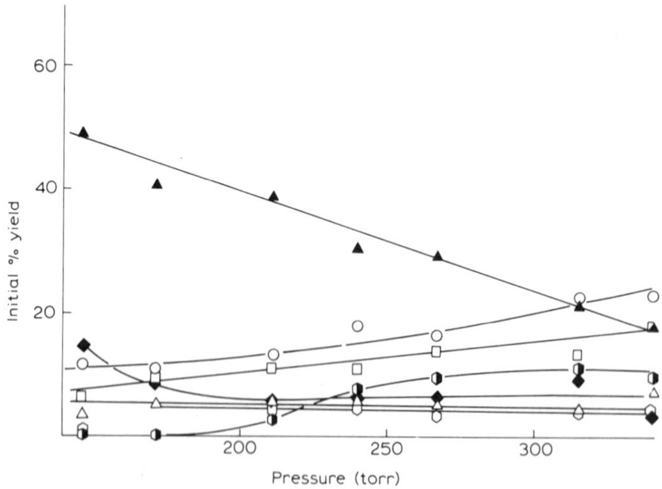

Fig. 25. The variation with initial pressure of the initial percentage yield of products from the oxidation of isobutane at 310 °C. Isobutane: oxygen = 1:2; volume of reaction vessel = 500 cm^3. ▲, isobutene; □, acetaldehyde; ○, propionaldehyde; △, propene; ◑, *tert*-butyl hydroperoxide; ◆, isobutene oxide; ○, acetone.

TABLE 19
The variation with initial pressure of the percentage conversion to ethylene and but-1-ene measured at the completion of the oxidation of *n*-butane at 290 °C (From ref. 173.)

Pressure (torr)	Ethylene (35% *n*-butane)	But-1-ene (35% *n*-butane)
52.5	5.92	3.29
75.0	4.72	2.34
101.0	3.69	1.83
133.5	2.70	1.23

Handscombe [174] and Mill et al. [175] have all recently suggested that alkylperoxy radical disproportionation reactions are important propagation steps at least during the oxidation of low molecular weight alkanes, and Quinn and co-workers [176] were unable to simulate cool flames during the oxidation of propane using a model with propylperoxy radical propagation and propyl hydroperoxide as the branching agent unless it was assumed that the chains were propagated at least in part by the reaction

$$C_nH_{2n+1}O\dot{O} + C_nH_{2n+1}O\dot{O} \longrightarrow 2C_nH_{2n+1}\dot{O} + O_2 \tag{42}$$

Unfortunately, the Arrhenius parameters for this reaction have not been determined. Heicklen [59], however, believes that reactions (42) and (6)

$$HO_2\cdot + HO_2\cdot \longrightarrow H_2O_2 + O_2 \tag{6}$$

have similar rate coefficients at room temperature, viz. $k_{42} = k_6 = 10^{9.5\pm0.3}$ l . mole^{-1} . sec^{-1}, but Knox [38] has argued that k_{42} will be less than k_6 and lies in the range 10^4—10^8 l . mole^{-1} . sec^{-1} at 300 °K. Using a value of $k_{42} = 10^9$ l . mole^{-1} . sec^{-1} for *tert*-butylperoxy radicals, however, Baldwin and Walker [103] have shown that propagation via reaction (42) gives a consistent and quantitative explanation of the widely differing rates of formation and yields of acetaldehyde found by Cullis and co-workers [56] and by Euker and Leinroth [177] during the oxidation of *n*-butane at low and high pressures.

Clearly, if $[C_nH_{2n+1}O\dot{O}]$ reaches a value of ca. 10^{-8} mole . l^{-1} then reaction (42) will become an important propagation reaction for alkanes whose alkylperoxy radicals are restricted in their ability to isomerize, if k_{42} is a large as 10^9 l . mole^{-1} . sec^{-1} and $k_{14\alpha}$ is only ca. $10^{0.5}$ sec^{-1} (for $[C_nH_{2n+2}]$ = 50 torr, $k_{17}[C_nH_{2n+2}] \simeq 1$ sec^{-1} at 600 °K [38]).

$$\dot{C}_nH_{2n+1} + O_2 \underset{}{\overset{3}{\rightleftharpoons}} C_nH_{2n+1}O\dot{O} \overset{14}{\rightleftharpoons} \alpha\text{-}\dot{C}_nH_{2n}OOH$$

$$C_nH_{2n+1}O\dot{O} \xrightarrow[42]{C_nH_{2n+1}O\dot{O}} 2C_nH_{2n+1}\dot{O} + O_2$$

$$C_nH_{2n+1}O\dot{O} \xrightarrow[17]{C_nH_{2n+2}} C_nH_{2n+1}OOH + \dot{C}_nH_{2n+1}$$

Baldwin and Walker [103] suggest that the decrease in the yield of conjugate alkenes with increase in pressure is probably due to the concomitant increase in $[C_nH_{2n+1}O\dot{O}]$ and hence stress the importance of alkylperoxy radical disproportionation reactions.

The available evidence suggests, therefore, that alkylperoxy radical disproportionation reactions are important for low molecular weight alkanes, but reliable values of the Arrhenius parameters for reactions (42) and (14α) are urgently needed to confirm this. However, since $k_{14\beta}$ is greater than $k_{14\alpha}$ by ca. 10^2 this will not be the case for alkanes which can form secondary or tertiary alkylperoxy radicals capable of undergoing extensive isomerization involving 1:5 or 1:6 H-transfer from further secondary or tertiary carbon atoms.

5.2 ALKANES OF CARBON NUMBER > 5

It was seen in Sect. 3.2.2 that, for high molecular weight alkanes, alkylperoxy radical isomerization and subsequent decomposition of the hydroperoxyalkyl radicals so formed is the major chain-propagating step throughout the cool-flame region. Indeed, during the oxidation of 3-ethylpentane [85, 86], 3-methylpentane [80] and 2-methylpentane [178] its importance is maintained at quite high temperatures (ca. 400 °C) at sub-atmospheric pressures. This product distribution is continuous across the slow combustion/cool-flame boundary and is little affected by carbon deposits resulting from two-stage ignition or very large increases in the initial pressure [78, 79, 84] which suggests that these reactions are predominantly homogeneous.

Examination of the alkylperoxy radicals which may be formed from 3-ethylpentane, 3-methylpentane and 2-methylpentane shows that, in each case, some may undergo isomerization reactions with relatively low activation energies [58], examples of which are shown in Table 20.

TABLE 20
Estimated activation energies for isomerization of some alkylperoxy radicals derived from 3-ethylpentane, 3-methylpentane and 2-methylpentane [58]

C_nH_{2n+2}	$C_nH_{2n+1}OO\cdot$	$\cdot C_nH_{2n}OOH$	E (kcal . mole^{-1})
3-Ethylpentane	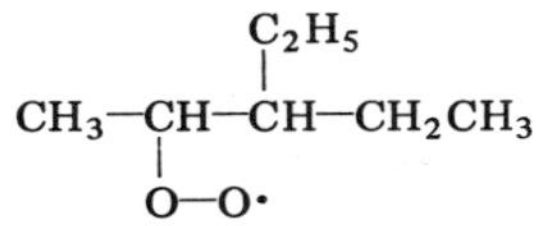	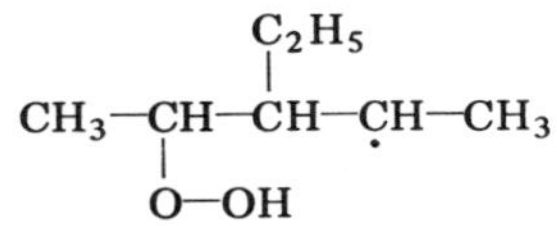	11.1
	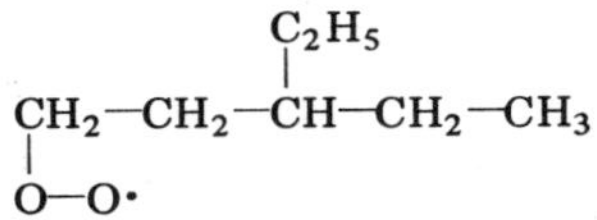	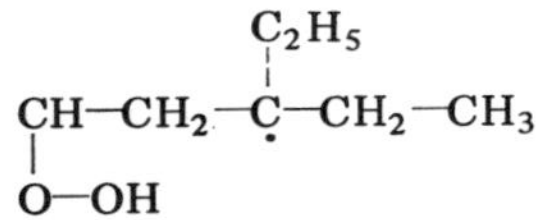	8.3
3-Methylpentane	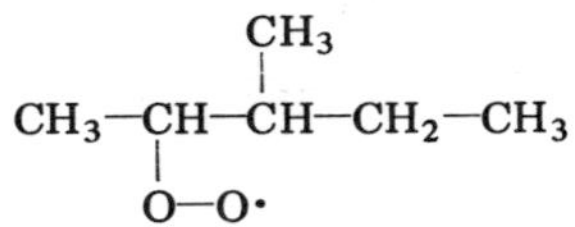	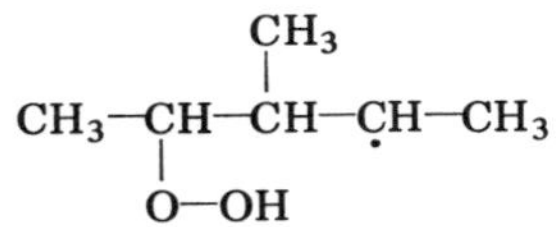	11.1
	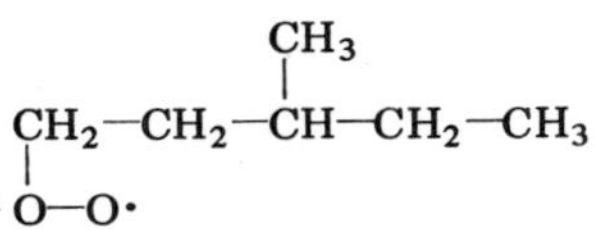	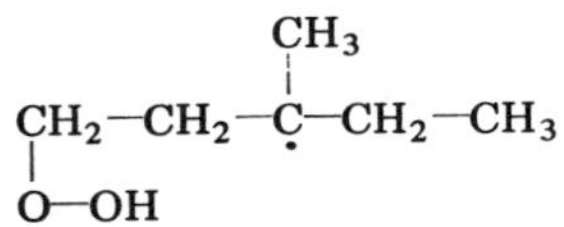	8.3
2-Methylpentane	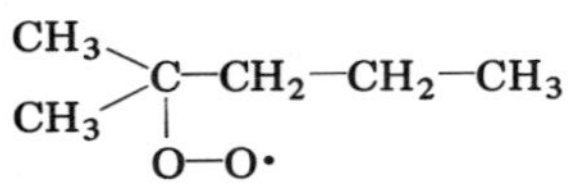	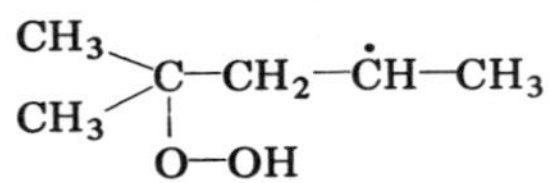	11.1
	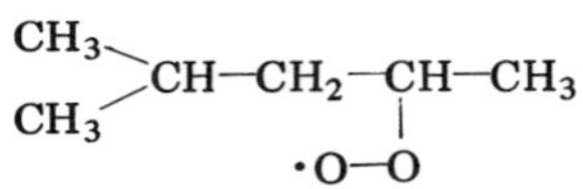	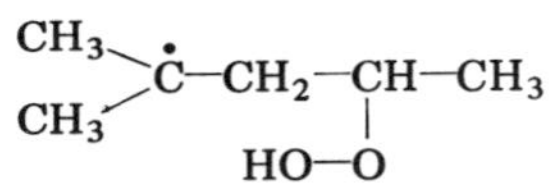	8.3

TABLE 21
The variation of product distribution with the degree of branching of the carbon skeleton of the alkane

C_nH_{2n+2}	$\cdot C_nH_{2n}OOH$	O-heterocycle	% Conversion[a]	β-Scission product	% Conversion[a]
n-Heptane [83]	$CH_3-CH(O{-}OH)-CH_2-CH_2-\dot{C}H-CH_2-CH_3$			But-1-ene	<0.3
		2-Methyl-5-ethyltetrahydrofuran	29.3		
	$CH_3-\dot{C}H-CH_2-CH_2-CH(O{-}OH)-CH_2-CH_3$			Butanone	<2.0
	$CH_3-CH(O{-}OH)-CH_2-\dot{C}H-CH_2-CH_2-CH_3$	2-Methyl-4-*n*-propyloxetan	10.2	Pent-1-ene	1.6
	$CH_3-\dot{C}H-CH_2-CH(O{-}OH)-CH_2-CH_2-CH_3$			*n*-Butyraldehyde	<2.0
	$CH_2(O{-}OH)-\dot{C}H-CH_2-CH_2-CH_2-CH_2-CH_3$	2-*n*-Pentyloxiran	1.6	Hept-1-ene	10.4
	$CH_3-CH(O{-}OH)-\dot{C}H-CH_2-CH_2-CH_2-CH_3$	2-Methyl-3-*n*-butyloxiran	1.1	Hept-2-ene	5.2

3-Ethylpentane [86]	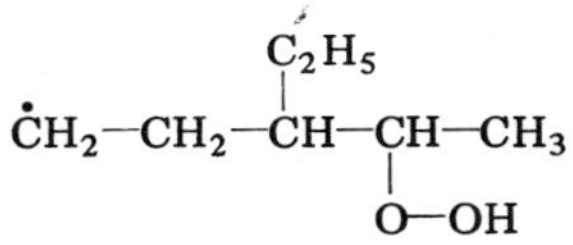			2-Methyl-3-ethyloxiran	0.5
		2-Methyl-3-ethyltetrahydrofuran	10.7		
	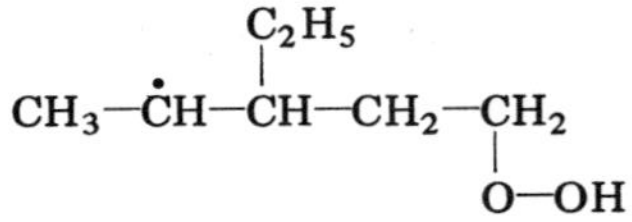			Pent-2-ene	12.9
	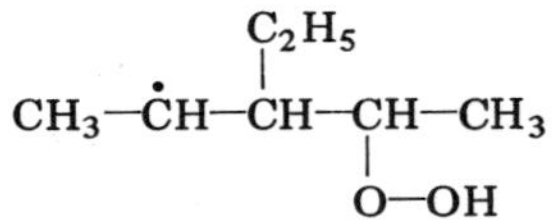	2,4-Dimethyl-3-ethyloxetan	3.7	Pent-2-ene	12.9
	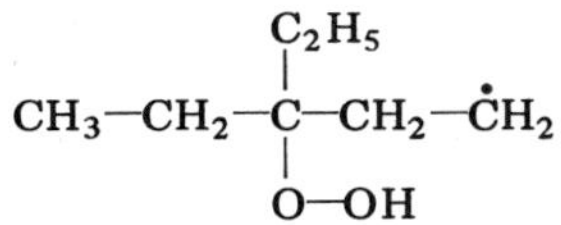			Pentan-3-one	6.4
		2,2-Diethyloxetan	1.9		
	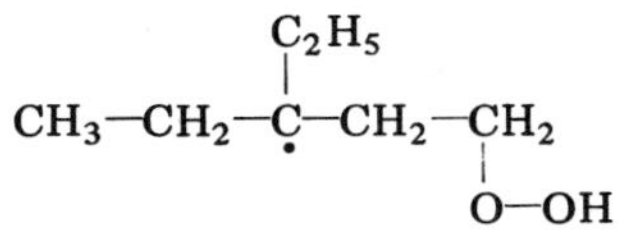			2-Ethylbut-1-ene	2.5
	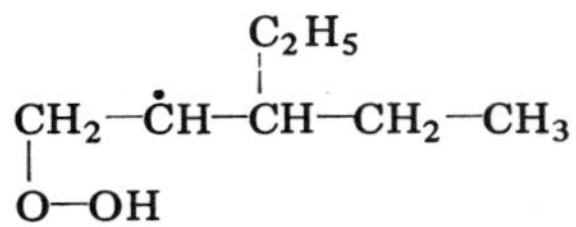	2-Isoamyloxiran	0	3-Ethylpent-1-ene	1.4

TABLE 21—*continued*

C_nH_{2n+2}	$\cdot C_nH_{2n}OOH$	O-heterocycle	% Conversion[a]	β-Scission product	% Conversion[a]
	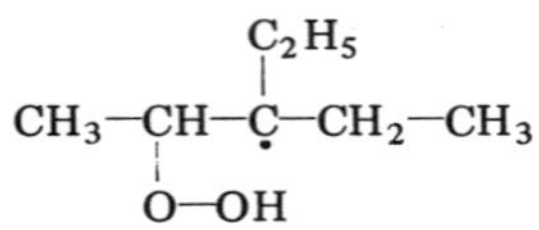	2-Methyl-3,3-diethyloxiran	3.5	3-Ethylpent-2-ene	5.3
2,3-Dimethylbutane [65]	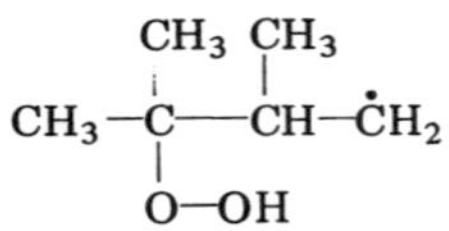	2,2,3-Trimethyloxetan	Major	Acetone	Major
	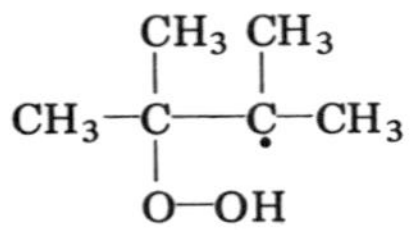	2,2,3,3-Tetramethyloxiran	Minor	2,3-Dimethylbut-2-ene 2,2-Dimethylbutan-3-one	Major Major
	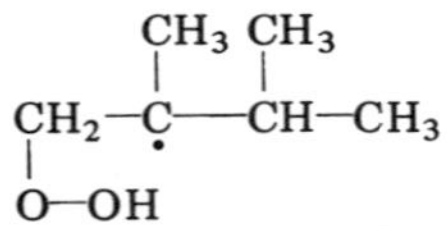	2-Methyl-2-isopropyloxiran	Minor	2,3-Dimethylbut-1-ene	Major

[a] Moles of product × 100 per mole of alkane consumed.

Thus, in the case of 3-ethylpentane initial attack at a secondary C—H bond may always be followed by oxygen addition and 1:5 H-transfer involving another secondary C—H bond. Furthermore, since the initial attack is unselective during cool-flame oxidation a considerable proportion of primary alkylperoxy radicals will be formed from this alkane and these may all undergo the relatively easy isomerization involving 1:5-hydrogen transfer from a tertiary C—H bond

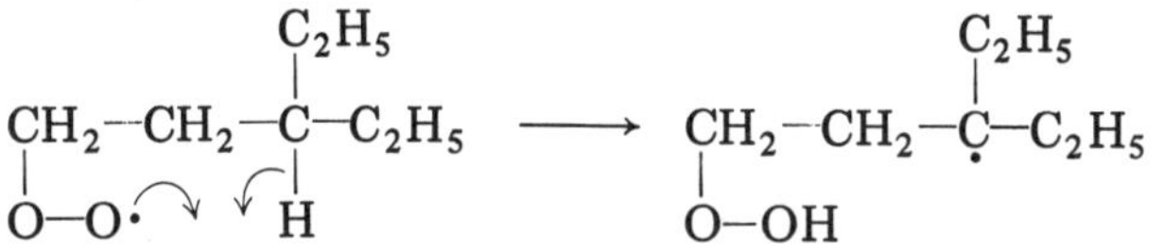

(14β)

The isomerization of alkylperoxy radicals derived from these hydrocarbons may therefore be considered to be unrestricted.

Although the nature of chain propagation under conditions where cool flames propagate is essentially the same for high molecular weight alkanes ($>C_5$), the distribution of the intermediate products depends upon the degree of branching of the carbon skeleton of the alkane as shown in Table 21.

The major product from the oxidation of *n*-heptane [83, 84] is the conjugate 0-heterocycle 2-methyl-5-ethyltetrahydrofuran. The predominant chain cycle therefore involves initial attack at a secondary C—H, followed by addition of oxygen, 1:6-hydrogen transfer from another secondary C—H and decomposition of the γ-hydroperoxyalkyl radical by simple cyclization and loss of $\dot{O}H$, e.g.

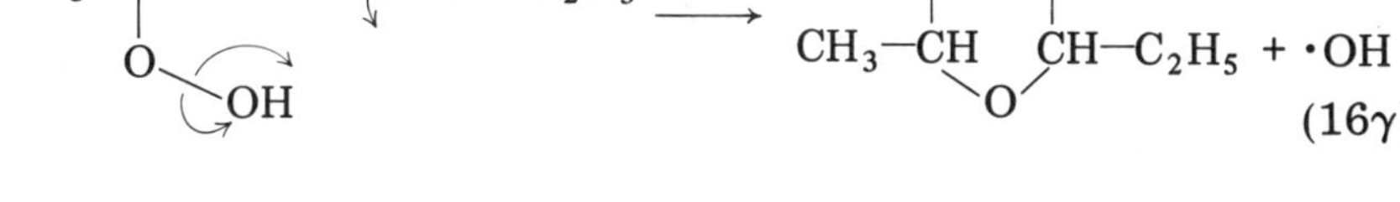

(14γ)

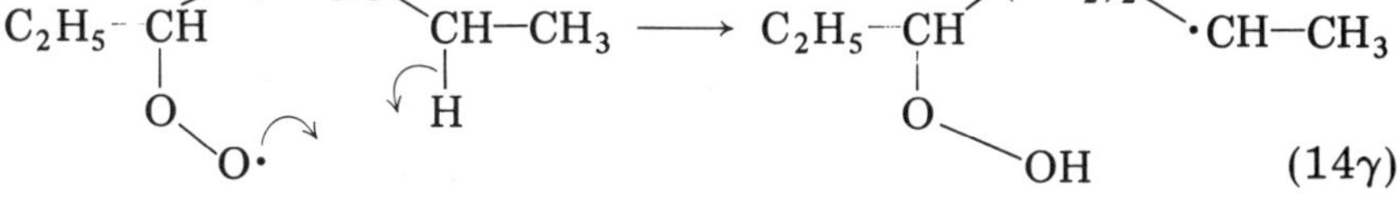

(16γ)

or

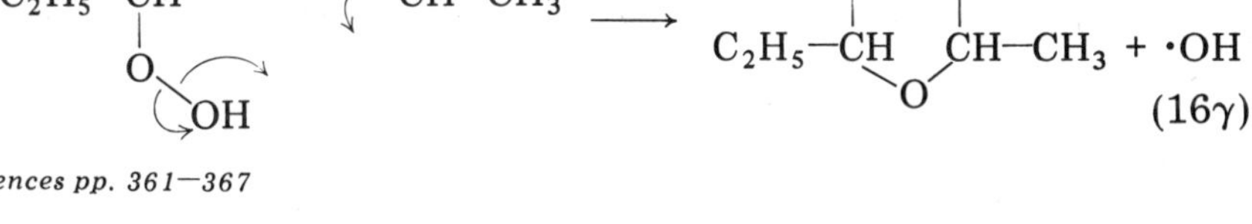

(14γ)

(16γ)

The yields of the corresponding β-scission products are much smaller and clearly β-scission decomposition of γ-hydroperoxyalkyl radicals, reaction (21γ) below, does not compete effectively with their decomposition by cyclization to tetrahydrofurans.

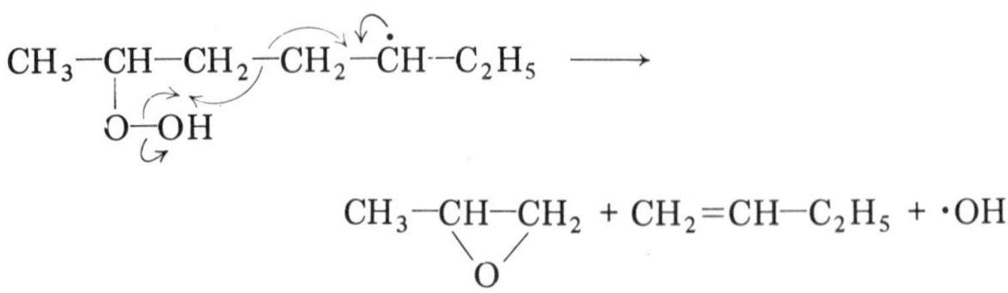

(21γ)

In contrast, compounds arising from β-scission of C—C bonds are the major products formed during the oxidation of 3-ethylpentane, their yields being ca. 3 times as large as those of the corresponding oxetans. β-Scission decomposition of β-hydroperoxyalkyl radicals competes effectively, therefore, with their decomposition by cyclization to oxetans.

The molecular structure of 2,3-dimethylbutane favours the formation of a-hydroperoxyalkyl radicals and the yields of the conjugate alkenes formed from this alkane are correspondingly larger than those found in the cool-flame oxidation of n-heptane and 3-ethylpentane. Decomposition of α-hydroperoxyalkyl radicals by a β-scission reaction usually leads to the conjugate alkene, while cyclization leads to the conjugate oxiran.* However, for this highly branched alkane, decomposition of α-hydroperoxyalkyl radicals involving a methyl group shift (by β-scission) also appears to be important, 2,2-dimethylbutan-3-one being a major product, viz.

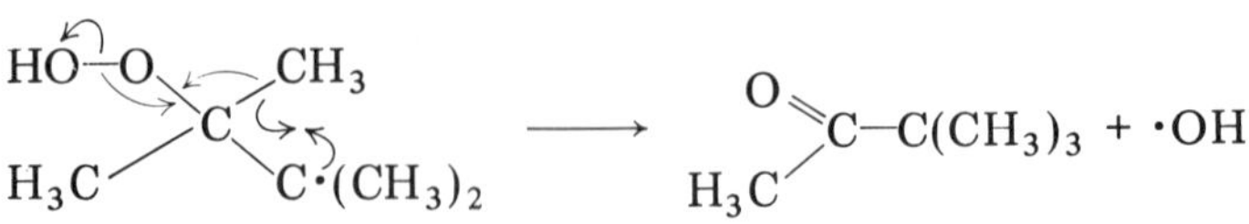

(20)

Alternatively, the ketone may be formed from 2,2,3,3-tetramethyloxiran and its yield may therefore be a measure of the ease of isomerization of the oxiran.

It is clear from these results that the relative rates of decomposition by cyclization and β-scission depend on the nature of the hydroperoxyalkyl

* Comparison of the rates of these decompositions is difficult since oxirans may isomerize and the alkene may, in principle, be formed also by the abstractive route.

radical, viz.

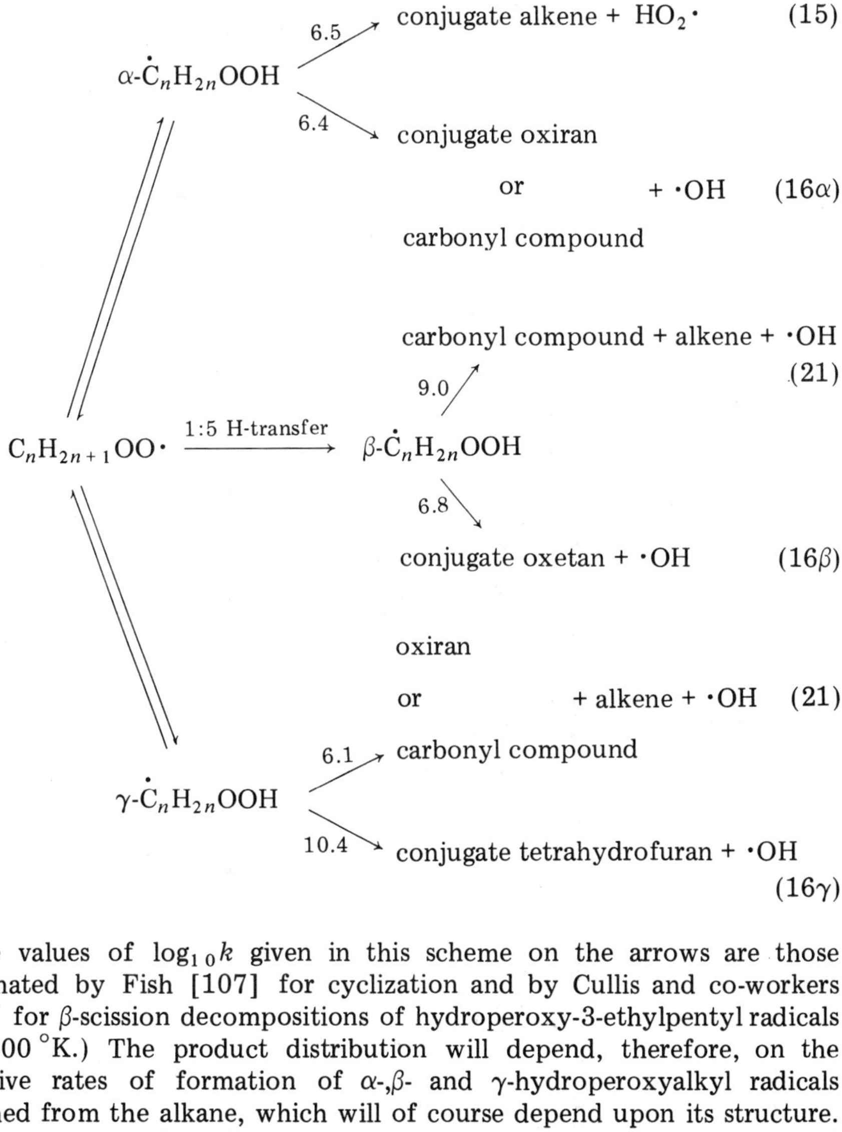

(The values of $\log_{10}k$ given in this scheme on the arrows are those estimated by Fish [107] for cyclization and by Cullis and co-workers [86] for β-scission decompositions of hydroperoxy-3-ethylpentyl radicals at 600 °K.) The product distribution will depend, therefore, on the relative rates of formation of α-,β- and γ-hydroperoxyalkyl radicals formed from the alkane, which will of course depend upon its structure.

Estimation of the relative concentrations of these radicals (see p. 276) formed from n-heptane, 3-ethylpentane and 2,3-dimethylbutane shows that the experimental findings may be anticipated from theoretical considerations. Thus, Table 22 shows that the importance of γ-hydroperoxyalkyl radicals for these alkanes is in the order n-heptane $>$ 3-ethylpentane $\gg$ 2,3-dimethylbutane, while that of α-hydroperoxyalkyl

TABLE 22
Relative concentrations of hydroperoxyalkyl radicals and rates of product formation at 600 °K

C_nH_{2n+2}	Research Octane Number	Nature of $\cdot C_nH_{2n}OOH$	Relative concentrations of $\cdot C_nH_{2n}OOH$	Relative rates of product formation	
				Cyclization	β-Scission
n-Heptane	0	α	1.65	1.6×10^{-4}	2.0×10^{-4}
		β	1.33	3.2×10^{-4}	5.0×10^{-2}
		γ	1	1	5.0×10^{-5}
3-Ethylpentane	65	α	5.85	6.3×10^{-4}	7.9×10^{-4}
		β	5.76	1.6×10^{-3}	2.5×10^{-1}
		γ	1	1	5.0×10^{-5}
2,3-Dimethylbutane	92	α	140	1.6×10^{-2}	2.0×10^{-2}
		β	99	2.5×10^{-2}	4.0
		γ	1	1	5.0×10^{-5}

radicals is in the reverse order. Estimation of the relative rates of product formation correctly predicts the major intermediate products from *n*-heptane and 2,3-dimethylbutane and the relatively higher yields of conjugate alkene from the latter fuel. It underestimates, however, the relative yields of β-scission products from 3-ethylpentane. This is also the case when the semi-quantitative test is applied to 2-methylpentane, Sect. 3.2.2(d). Clearly, a more detailed model and more accurate rate coefficient data are required for a quantitative test of this theory. In particular, no allowance has been made for the effect of structure on the rate coefficients for the decomposition reactions and it would appear that $k_{16\gamma}$ decreases with increase in branching of the carbon skeleton of the parent alkane.

The foregoing discussion has shown, however, that the molecular structure of the parent alkane profoundly affects the distribution of the intermediate products of its cool-flame oxidation and clearly, there is a strong correlation between the distribution, the degree of branching of the carbon skeleton, the rate of formation of the hydroperoxyalkyl radicals and the Research Octane Number of the alkane.

5.3 TRANSITION FROM LOW TO HIGH TEMPERATURE MECHANISM

The transition from the low to high temperature mechanism is essentially due to a change in the relative rates of reactions (3), (−3), (2) and (14)

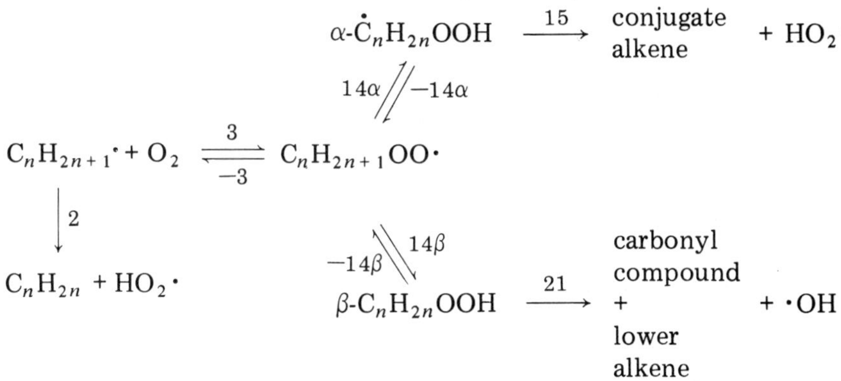

Since $k_{15} \gg k_{14\alpha}$, $k_{15} \simeq k_{-14\alpha}$, $k_{21} \gg k_{14\beta}$ and $k_{21} \simeq k_{-14\beta}$ (see Sect. 3.2.2), a general expression for the ratio of the rates of product formation by the low and high temperature mechanisms is given by

$$\frac{\text{rate(low } T)}{\text{rate(high } T)} = \frac{(k_{14\alpha} + k_{14\beta})[\text{C}_n\text{H}_{2n+1}\text{OO}\cdot]}{k_2[\cdot\text{C}_n\text{H}_{2n+1}][\text{O}_2]}$$

At a steady state,

$$\frac{[C_nH_{2n+1}OO\cdot]}{[\cdot C_nH_{2n+1}]} = \frac{k_3[O_2]}{k_{14\alpha} + k_{14\beta} + k_{-3}}$$

Hence

$$\frac{\text{rate(low } T)}{\text{rate(high } T)} = \frac{k_3(k_{14\alpha} + k_{14\beta})}{k_2(k_{14\alpha} + k_{14\beta} + k_{-3})}$$

Since the rate of isomerization of alkylperoxy radicals depends upon their molecular weight and structure, it can be seen that the temperature at which the transition occurs will be dependent upon the molecular weight and structure of the hydrocarbon. Thus, for example, in the case of propane 1:5 hydrogen transfer is impossible for the prop-2-ylperoxy radical and 1:4 hydrogen transfer involves the cleavage of a primary C—H bond. The expression therefore reduces to

$$\frac{\text{rate(low } T)}{\text{rate(high } T)} = \frac{k_3 k_{14\alpha^1}}{k_2(k_{14\alpha^1} + k_{-3})}$$

At a given temperature, this ratio is always higher for *n*-pentane than for propane (Table 23). It can be seen, therefore, that once again a change in mechanism depends on the restrictions imposed by the molecular weight and structure of the hydrocarbon on the ability of its alkylperoxy radicals to isomerize. In this respect neopentane is again an interesting example,

TABLE 23

The relative rates of the high and low temperature mechanisms for propane and *n*-pentane

Temp. (°K)	Rate (low T)/rate (high T)	
	Propane	*n*-Pentane
600	10	25
700	3.2	16
800	1.6	10

Arrhenius parameters used in the above estimation

Reaction	$\log_{10} A$ (A in sec^{-1} or $l.mole^{-1}.sec^{-1}$)	E ($kcal.mole^{-1}$)
2	9.5	5.0
3	9.1	0
−3	14.3	29.0
$14\alpha^1$	11.5	20.5
$14\alpha^2$	11.5	17.0
$14\beta^2$	11.5	11.1

since it cannot yield a conjugate alkene. The "low temperature" mechanism would therefore be expected to play an important role even at high temperatures as has been recently shown [108, 157].

The values of the ratio shown in Table 23 are higher than might previously have been anticipated from high temperature analytical studies. This may be due to the inaccuracies in the Arrhenius parameters, many of which have been estimated and to the simplicity of the scheme. It may also be due to the fact that the conjugate alkenes are formed via alkylperoxy radical isomerization even above ca. 700 °K to a larger extent than has hitherto been realized, as several recent results have suggested [63, 80, 86, 135]. Even so, this simplified scheme illustrates the dependence of the predominant reaction mechanism on the molecular weight and structure of the hydrocarbon and certainly the rapid decrease in conjugate alkene yield from ethane to *n*-butane found by Baldwin and co-workers [158] is reflected in Table 23.

6. Mathematical models

The detailed knowledge of the chemistry and phenomenological behaviour observed during the oxidation of hydrocarbons has inevitably led to attempts to build mathematical models which can describe these systems. With present day high-speed computers and programming techniques the integration of sets of conservation equations is no longer a prohibitive problem and several models have recently been described.

Gray and Yang [179--182] showed that many experimental observations can be explained by treating cool flames as thermokinetic oscillations in a radical chain reaction which is linearly branched and terminated. By postulating two linear chain-terminating reactions one of which has a larger and the other a smaller activation energy than the branching reaction, they were able to explain the negative temperature coefficient for the slow oxidation and to show that a lobe on the cool-flame/two-stage ignition boundary is to be expected. The boundaries of the cool-flame region were located by identifying the conditions for which oscillatory solutions exist for the set of simultaneous differential equations which describe the conservation of mass and energy in the reaction system. Unfortunately, the boundaries to the oscillatory solutions cannot be uniquely identified with cool-flame limits observed experimentally [183]. Gray and Yang's model also neglects fuel consumption, which may be considerable at the first cool flame (ca. 35—40 % [77, 86]), and this precludes the possibility of explaining both the number of cool flames observed experimentally and the variation in their amplitude. Indeed, the model bears only a small resemblance to the chemistry of hydrocarbon oxidation and could certainly not throw light on the variation of mechanism with the molecular structure of the fuel for example.

References pp. 361—367

Lucquin et al. [184—188], have tested several models which allow for fuel consumption and include degenerate branching. Their models are therefore more realistic and give good accounts of the effect of promoters and inhibitors. As yet, however, they have not identified the specific chemical reactions in the models, but they are attempting to use them to describe the observed kinetics and morphology of propane—oxygen mixtures.

The best attempt to build a realistic model has been that of Halstead et al. [183, 189] for the oxidation, in the presence of excess argon, of acetaldehyde, which is known to be an important intermediate in the oxidation of alkanes (except methane and ethane). The chemical model they used is

$$CH_3CHO + O_2 \longrightarrow CH_3\dot{C}O + HO_2\cdot \quad (a)$$

$$CH_3CHO + HO_2\cdot \longrightarrow CH_3\dot{C}O + H_2O_2 \quad (b)$$

$$CH_3CHO + \cdot OH \longrightarrow CH_3\dot{C}O + H_2O \quad (c)$$

$$CH_3\dot{C}O + O_2 \longrightarrow CH_3CO\cdot O\dot{O} \quad (d)$$

$$CH_3CO\cdot O\dot{O} + CH_3CHO \longrightarrow CH_3CO\cdot OOH + CH_3\dot{C}O \quad (e)$$

$$CH_3\dot{C}O + M \longrightarrow \cdot CH_3 + CO + M \quad (f)$$

$$\cdot CH_3 + O_2 + M \longrightarrow CH_2O + \cdot OH + M \quad (g)$$

$$CH_3CO\cdot OOH \longrightarrow \cdot CH_3 + \cdot OH + CO_2 \quad (h)$$

$$\cdot CH_3 + \cdot CH_3 \longrightarrow C_2H_6 \quad (i)$$

$$CH_3CO\cdot O\dot{O} + \cdot CH_3 \longrightarrow CH_3CO\cdot OOCH_3 \quad (j)$$

$$CH_3CO\cdot O\dot{O} + CH_3CO\cdot O\dot{O} \longrightarrow CH_3CO\cdot OO\cdot COCH_3 + O_2 \quad (k)$$

$$CH_3CO\cdot OOCH_3 \longrightarrow CH_3O\cdot + \cdot CH_3 + CO_2 \quad (l)$$

$$CH_3O\cdot + CH_3CHO \longrightarrow CH_3OH + CH_3\dot{C}O \quad (m)$$

$$CH_3CO\cdot OO\cdot COCH_3 \longrightarrow \cdot CH_3 + \cdot CH_3 + 2CO_2 \quad (n)$$

During the induction period, the agent of degenerate branching, peracetic acid, is formed by the low activation energy sequence reactions (d) + (e). During the cool flame the temperature rise facilitates the high activation energy reaction (f) which then competes effectively with reaction (d). The increase in the rate of reaction (f) relative to that of reaction (d) and the rapid increase in the rate of branching with temperature results in the concentration of peracetic acid falling rapidly to a very low value. Consequently the radical concentration and reaction rate fall and the temperature relaxes. The self-quenching of the cool flame is thus attributed to a "thermal switch" between reactions (d) and (f).

The thermal model is that of Semenov. Space-averaged variables were used in order to minimize the mathematical difficulties and hence the model does not give an account of the spatial propagation of cool flames. The energy conservation equation used is

$$\frac{dT}{dt} = \frac{1}{C} \sum_i q_i v_i - \frac{\alpha S}{CV} (T - T_0)$$

where T = an instantaneous, spatially averaged gas temperature, T_0 = initial or bath temperature, C = heat capacity per unit volume, S/V = surface : volume ratio of the vessel, α = a heat transfer coefficient, q_i = molar heat of the ith reaction and v_i = rate of the ith reaction. The function $CV/\alpha S$ can be identified with the thermal relaxation time, t_0.

The chemical model may be expressed as a set of eight conservation equations governing the consumption of acetaldehyde and oxygen, and the rates of change of the concentrations of the radicals CH_3, CH_3CO and $CH_3CO_3\cdot$, the branching agent $CH_3CO{\cdot}OOH$ and the unstable products $CH_3CO{\cdot}OOCH_3$ and $CH_3CO{\cdot}OO{\cdot}COCH_3$. These chemical equations were coupled to the thermal equation to form a chain-thermal model comprising nine non-linear first-order differential equations. The solutions

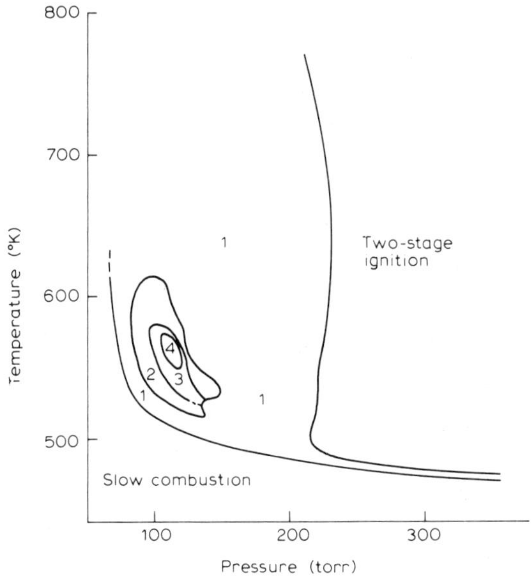

Fig. 26. The temperature—pressure ignition diagram obtained from the model for acetaldehyde + oxygen + argon mixtures in the molar ratio 1:1:1. (From ref. 183.)

TABLE 24
The parameters used in the computations for the acetaldehyde—oxygen model [183]

Reaction	q (kcal . mole^{-1})	$\log_{10} A$ (A in sec^{-1}, l . mole^{-1} . sec^{-1} or l^2 . mole^{-2} . sec^{-1})	E (kcal . mole^{-1})
1. $CH_3CHO + O_2 \rightarrow CH_3\dot{C}O + HO_2\cdot$	$q_1 + q_2 = 38.8$	9.48^a	40.9^a
2. $HO_2\cdot + CH_3CHO \rightarrow CH_3\dot{C}O + H_2O_2$			
3. $CH_3CO{\cdot}OOH \rightarrow {\cdot}CH_3 + CO_2 + {\cdot}OH$	$q_3 + q_5 = -7.9$	14.30^a	40.2^a
4. ${\cdot}CH_3 + O_2 + M \rightarrow CH_2O + {\cdot}OH + M$	$q_4 + q_5 = 84.2$	9.73^a	0^a
5. ${\cdot}OH + CH_3CHO \rightarrow H_2O + CH_3\dot{C}O$			
6. $CH_3\dot{C}O + O_2 \rightarrow CH_3CO{\cdot}OO\cdot$	68.9^a	6.85^a	0^a
7. $CH_3CO{\cdot}O\dot{O} + CH_3CHO \rightarrow CH_3CO{\cdot}OOH + CH_3\dot{C}O$	12.0^a	10.00^a	10.1^a
8. $CH_3\dot{C}O + M \rightarrow {\cdot}CH_3 + CO + M$	−12.0	11.97^a	10.8
9. ${\cdot}CH_3 + {\cdot}CH_3 \rightarrow C_2H_6$	86.1	10.30	0
10. ${\cdot}CH_3 + CH_3CO{\cdot}OO\cdot \rightarrow CH_3CO{\cdot}OOCH_3$	60.1	10.60^a	0^a
11. $CH_3CO{\cdot}OO\cdot + CH_3CO{\cdot}OO\cdot \rightarrow CH_3CO{\cdot}OO{\cdot}COCH_3 + O_2$	34.9^a	10.30^a	0^a
12. $CH_3CO{\cdot}OOCH_3 \rightarrow {\cdot}CH_3 + CH_3O\cdot + CO_2$	$q_{12} + q_{13} = -15.1^a$	10.00^a	34.9^a
13. $CH_3O\cdot + CH_3CHO \rightarrow CH_3OH + CH_3\dot{C}O$			
14. $CH_3CO{\cdot}OO{\cdot}COCH_3 \rightarrow {\cdot}CH_3 + {\cdot}CH_3 + 2CO_2$	29.9^a	10.00^a	34.9^a

$t_0 = 8.3 \times 10^4$ [M] sec. $C = 6.77$ [M] cal. °C^{-1} . cm^{-3} where [M] is the total gas concentration, mole . cm^{-3}.
^{a} = assumed value. Exothermicities were estimated.

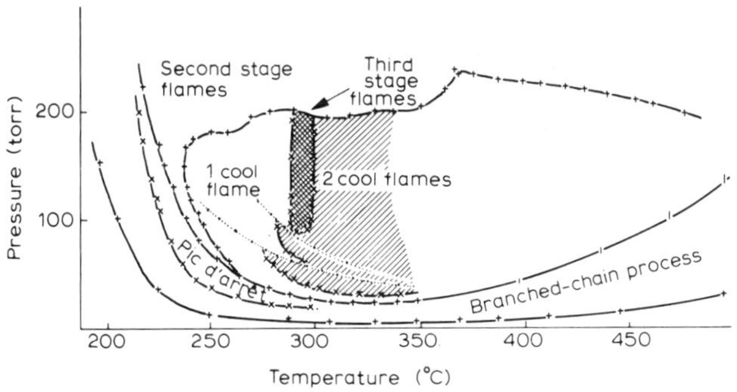

Fig. 27. The experimental pressure—temperature ignition diagram for acetaldehyde + oxygen mixtures in the molar ratio 1:1. (From ref. 191.)

to the model must be obtained over a wide range of initial conditions and these differential equations will therefore suffer from so-called "stiffness" which has hitherto prevented their solution. However, Prothero [190] has recently developed a technique specifically designed to solve such sets of stiff equations and this has allowed these workers to obtain accurate solutions of a chain-thermal model of this type for the first time.

Values of the coefficients for the set of differential equations were chosen to give cool flames at realistic initial pressures and temperatures. Further restrictions on the choice of coefficients were imposed by requiring that the fuel conversion should not exceed 25 % at the maximum of the temperature pulse, that the induction period should be between 15 and 20 sec, and that the thermal relaxation time should be 0.25 sec. To achieve this the rate coefficients of reactions (d), (f), (h) and (g) were varied about reasonable estimates of their likely values. The parameters chosen for the model are given in Table 24. The computer was used in a conversational mode to map out an ignition diagram (Fig. 26) which compares favourably with that found experimentally [191] (Fig. 27).

Figure 28 shows the propagation of four successive cool flames and illustrates the mechanism of self-quenching. As the gas temperature starts to increase rapidly the formation of peracetic acid is halted and the existing concentration of it is consumed. The overall reaction rate consequently falls to zero and the gas temperature relaxes to that of the bath. A further point of interest is that the amplitudes of successive cool flames varies in an irregular fashion as has been found experimentally. The model also showed that the peracetic acid concentration did not increase during the second stage of two-stage ignition and gave a good account of the sharp transition from slow oxidation to cool-flame behaviour, the dependence on initial conditions of the maximum temperature rise during the cool flame and the negative temperature coefficient.

A simplified model was used for a study by the analytical methods employed by Gray and Yang [179—182]. Three dimensionless differential

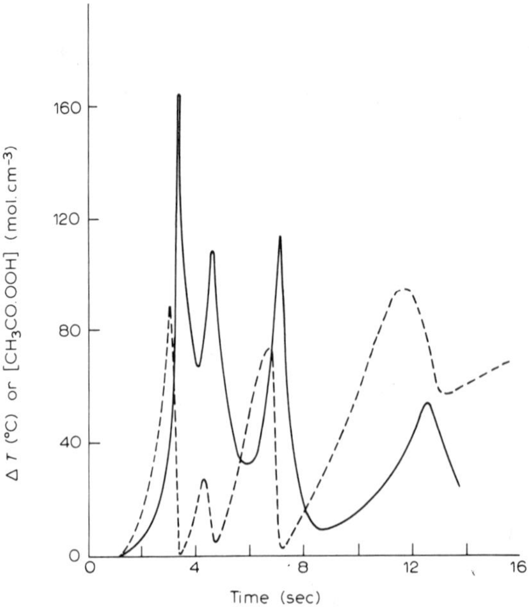

Fig. 28. The simulation of four consecutive cool flames at an initial temperature of 561 °K and 112.5 torr. ——, ΔT; ----, [$CH_3CO{\cdot}OOH$]. (From ref. 183.)

equations were set up describing the rates of change of reactant concentrations and temperature. The solutions trace out paths in three-dimensional space of the dependent variables $\theta = C(T - T_0)/q[O_2]_0$, $\beta =$

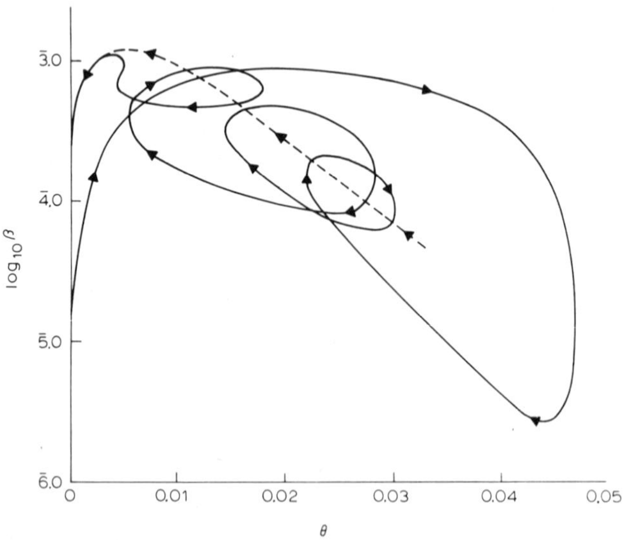

Fig. 29. The trajectory of a multiple-cool-flame solution in the (θ, β) plane for an initial temperature and pressure of 561 °K and 112.5 torr. ——, solution of simplified model; ---, locus of the pseudo-stationary point, S_1. (From ref. 183.)

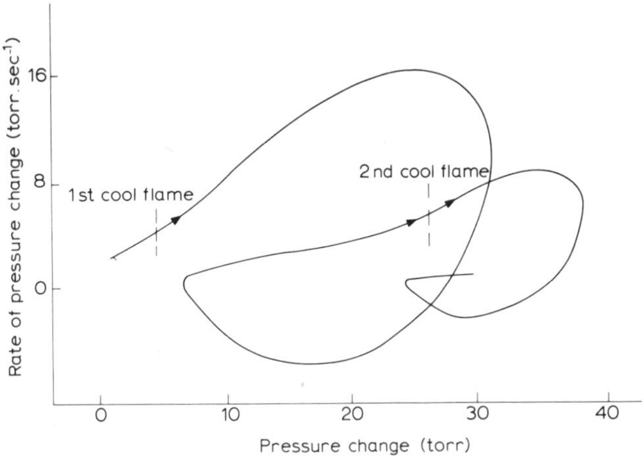

Fig. 30. Phase diagram for the propagation of two cool flames during the oxidation of isobutane at 315 °C and 220 torr. Isobutane:oxygen = 1:2; volume of reaction vessel = 500 cm^3. (From ref. 193.)

$[CH_3CO{\cdot}OOH]/[O_2]_0$ and $\epsilon = [CH_3CHO]/[O_2]_0$. The path of a multiple cool flame in this phase-space would be a spiral-type curve, which may be illustrated by projecting the path onto a (θ, β) plane (Fig. 29). Similar phase diagrams are obtained experimentally for the variation of rate with pressure change for example, as shown in Fig. 30 [192, 193].

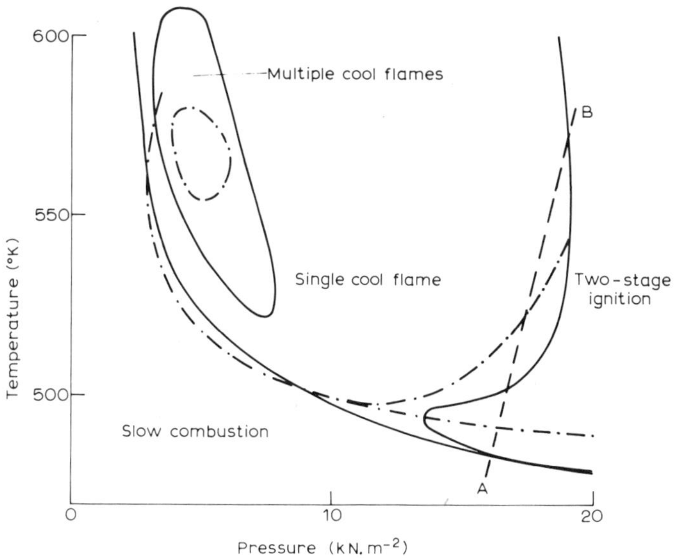

Fig. 31. The temperature—pressure ignition diagrams for equimolar mixtures of acetaldehyde + oxygen + argon determined by the modified model and experimentally. ——, Complete diagram; — · — ·, experimental. (From ref. 194.)

The form of the solutions to the simplified model were analysed by examining the existence and types of the pseudo-stationary points of the equations for $d\theta/d\tau = d\beta/d\tau = 0$ and values of ϵ in the range 0—1 ($\tau = t/t_0$). Figure 29 shows the oscillation of a multiple-cool-flame solution about the locus of such a pseudo-stationary point, S_1. The initial oscillation is damped while S_1 is a stable focus. The changing of S_1 into a unstable focus surrounded by a stable limit cycle leads to an amplification of the oscillation which approaches the amplitude of the limit cycle. When S_1 reverts to a stable focus, and then a stable node, the solution approaches the locus of the pseudo-stationary point. In this way an insight may be gained into the oscillatory behaviour of multiple cool flames.

Halstead et al. [194] subsequently determined the ignition diagram for equimolar mixtures of acetaldehyde, oxygen and argon experimentally and then modified their model to fit the experimental results. The most significant improvement is the inclusion of radical branching by hydrogen atoms and degenerate branching by hydrogen peroxide, which become effective at high transient temperatures and thus carry the cool flame over into ignition. In this way, the new model is capable of describing the low temperature ignition peninsula and hence is more realistic as can be seen in Fig. 31.

This work has recently been extended to the oxidation of propane [176, 195]. The addition of the following reactions to their first acetaldehyde model [183] allowed the simulation of multiple cool flames for propane—oxygen mixtures in the molar ratio 2:1 and gave a reasonable

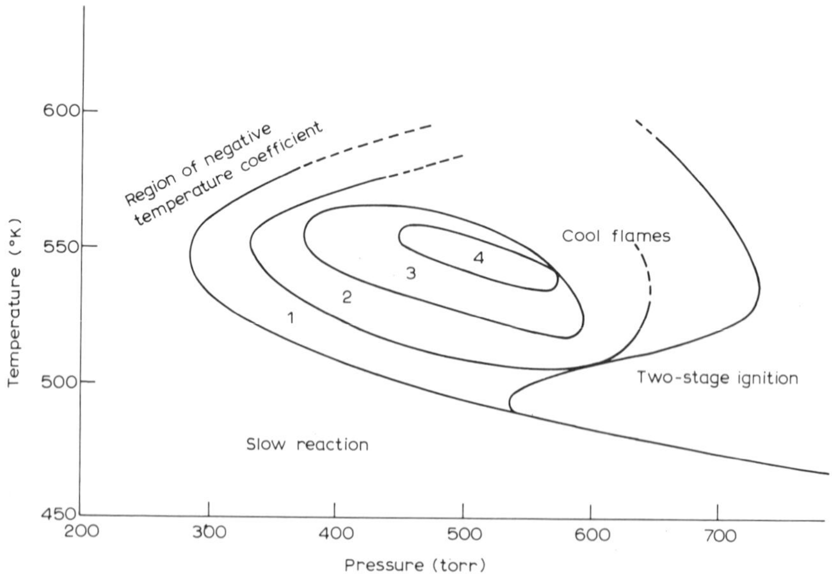

Fig. 32. The temperature—pressure ignition diagram obtained from the model for propane + oxygen mixtures in the molar ratio 2:1. (From ref. 195.)

ignition diagram as shown in Fig. 32, although the cool flame region is ca. 55 °C lower than that observed experimentally [195].

$$C_3H_8 + \dot{O}H \longrightarrow \dot{C}_3H_7 + H_2O$$

$$C_3H_8 + H\dot{O}_2 \longrightarrow \dot{C}_3H_7 + H_2O_2$$

$$\dot{C}_3H_7 + O_2 \longrightarrow \text{Products} + \dot{O}H \text{ (or } H\dot{O}_2\text{)}$$

$$H_2O_2 + M \longrightarrow 2\dot{O}H + M$$

$$2\dot{C}_3H_7 \longrightarrow \text{Termination}$$

In contrast, however, the following simple model [176] for propane oxidation, in which propyl hydroperoxide is the agent of degenerate branching, viz.

$$C_3H_7OOH \longrightarrow C_3H_7\dot{O} + \dot{O}H$$

$$C_3H_7\dot{O} + C_3H_8 \longrightarrow C_3H_7OH + \dot{C}_3H_7$$

$$C_3H_8 + \dot{O}H \longrightarrow \dot{C}_3H_7 + H_2O$$

$$\dot{C}_3H_7 + O_2 \longrightarrow C_3H_7\dot{O}_2$$

$$C_3H_7\dot{O}_2 + C_3H_8 \longrightarrow C_3H_7OOH + \dot{C}_3H_7$$

$$C_3H_7\dot{O}_2 \longrightarrow \text{Product} + \dot{O}H$$

$$C_3H_7\dot{O}_2 + C_3H_7\dot{O}_2 \longrightarrow \text{Termination}$$

failed. Satisfactory simulation can only be obtained if the chain terminating radical propagates the chain by a low activation energy reaction. Thus, cool-flame simulation was obtained if the chains were terminated by propyl radical recombination, but since $[C_3H_7\dot{O}_2]$ is expected to be much larger than $[\dot{C}_3H_7]$ under these conditions such termination appears unjustifiable. This led Halstead and co-workers [176] to suggest that the chains may be propagated by alkylperoxy radical disproportionation reactions at least in part.

Both Lucquin et al. [184, 185] and Halstead et al. [183, 189] stress that the phenomenological complexity observed during hydrocarbon oxidation may be explained by relatively simple, but realistic, models, even though the overall chemistry is known to be complex. In this respect, it is interesting to note that Enikolopyan [22] explained the negative temperature coefficient observed during hydrocarbon oxidation in terms of the same "thermal-switch" between reactions (d) and (f) in 1958. Even

References pp. 361—367

so, it is acknowledged that the complex morphology of the ignition diagrams of higher alkanes reflects a change in branching reactions and their precursors. Since these depend to a large extent on the ability of the alkylperoxy radicals to isomerize it is clear that a more complex model will have to be developed to explain the change in morphology and mechanism with alkane structure in general.

Before such a complex model can be achieved, however, more accurate information regarding the Arrhenius parameters of the many elementary reaction steps are required. The work of Baldwin and Walker (see Sect. 4) will certainly help in this respect, but it is clear that far more effort in this direction is needed. Likewise, more accurate information regarding the phenomenological behaviour of hydrocarbon oxidation is also required. At present much of the accumulated experimental data has been obtained from reactions carried out in unstirred static reactors. This leads to a complex spacial temperature profile, which during the slow oxidation of propane, for example, is intermediate between that predicted for either purely conductive or strongly convective heat transfer [196]. The use of stirred reactors overcomes this problem since the resulting temperature and concentration distributions become nearly uniform in space [196]. Again, experimental techniques such as this which simplify the system being studied will allow more viable comparisons to be made between theoretical and experimental results.

7. Appendix

In addition to the ignition diagram for 3-ethylpentane + oxygen given on p. 293 some recent diagrams for some C_3—C_8 hydrocarbons are given in Figs. 33—50 for reference. With regard to these diagrams, it should be noted that the upper temperature boundaries of multiple cool-flame

(Text continues on p. 361.)

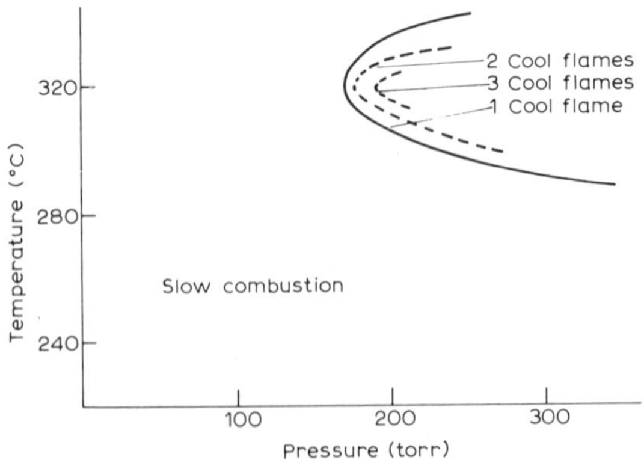

Fig. 33. Propane—oxygen. Molar ratio 1:1; cylindrical pyrex reaction vessel, volume 150 cm^3. (From ref. 119.)

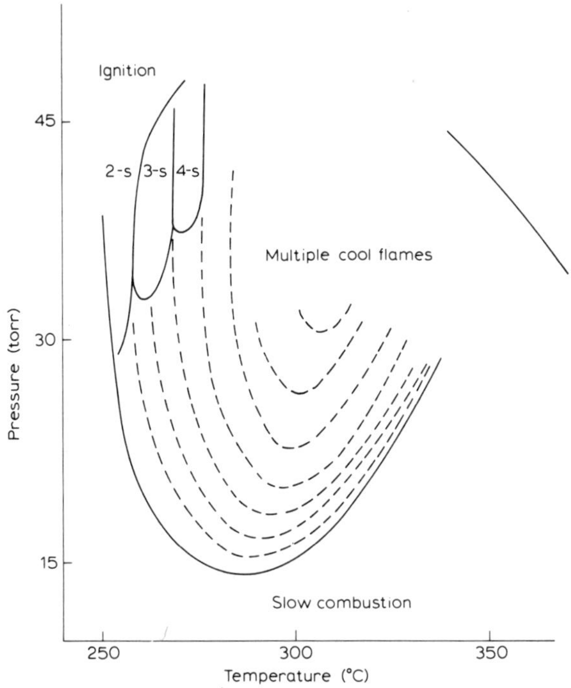

Fig. 34. Propane—oxygen. Molar ratio 1:1; stirred spherical pyrex reaction vessel, volume 1000 cm^3. (From ref. 196.)

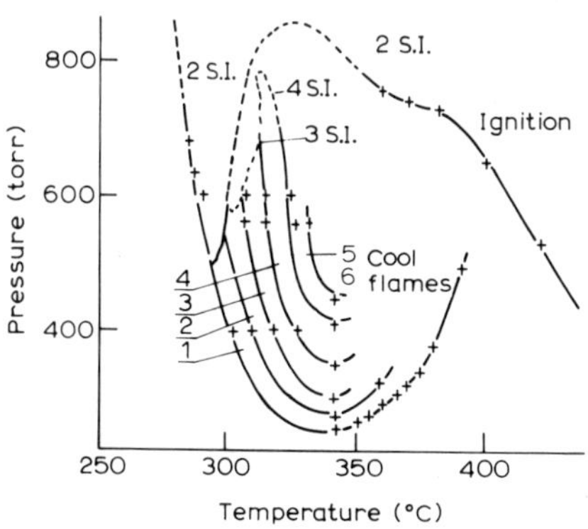

Fig. 35. Propane—oxygen. Molar ratio 1:1; cyclindrical pyrex reaction vessel, volume 33 cm^3. (From ref. 197.)

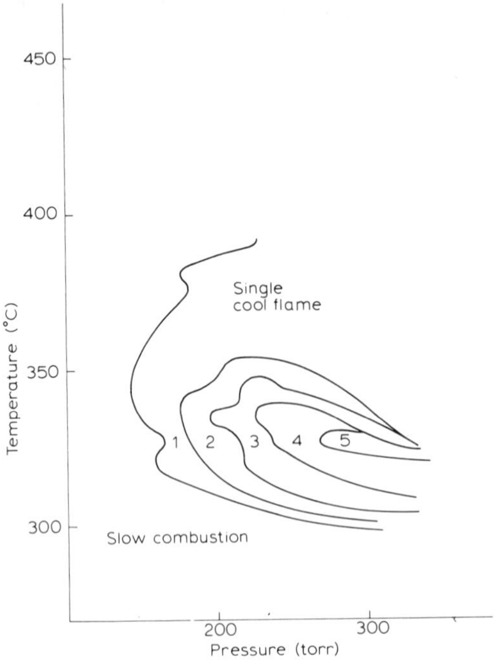

Fig. 36. Isobutane—oxygen. Molar ratio 1:2; spherical pyrex reaction vessel, volume 500 cm^3. (From ref. 134.)

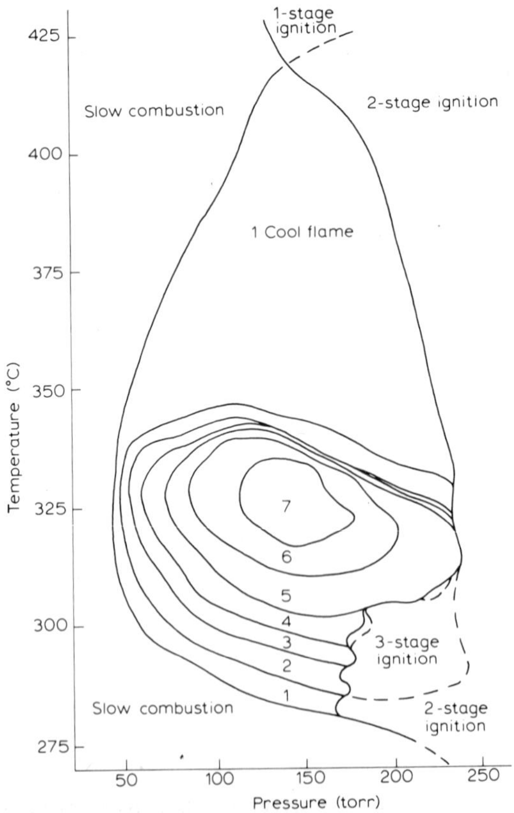

Fig. 37. *n*-Butane—oxygen. Molar ratio 1:2; spherical pyrex reaction vessel, volume 500 cm^3. (From ref. 132.)

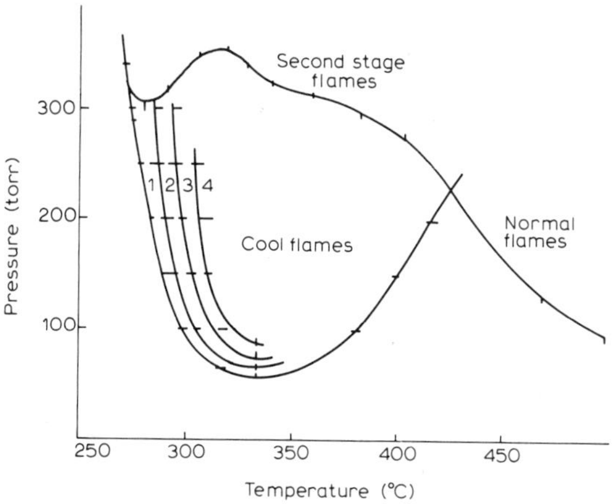

Fig. 38. *n*-Butane—oxygen. Molar ratio 1:1; cylindrical pyrex reaction vessel, volume 33 cm^3. (From ref. 135.)

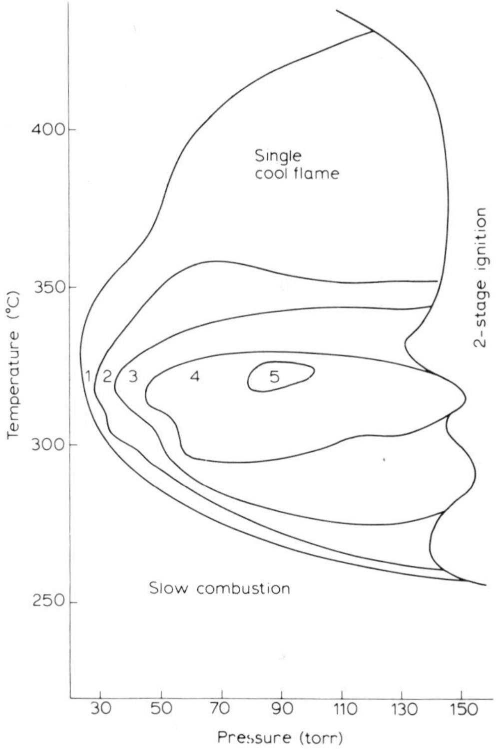

Fig. 39. *n*-Pentane—oxygen. Molar ratio 3:4; spherical pyrex reaction vessel, volume 500 cm^3. (From ref. 198.)

References pp. 361–367

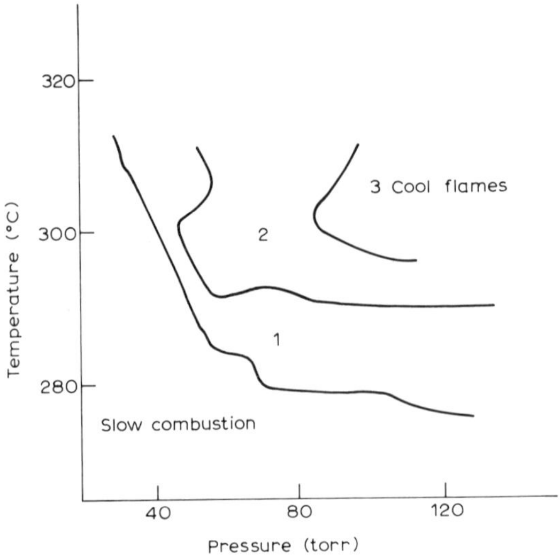

Fig. 40. *n*-Pentane—oxygen. Molar ratio 1:2; cylindrical pyrex reaction vessel, 48 mm i.d., 190 mm length. (From ref. 199.)

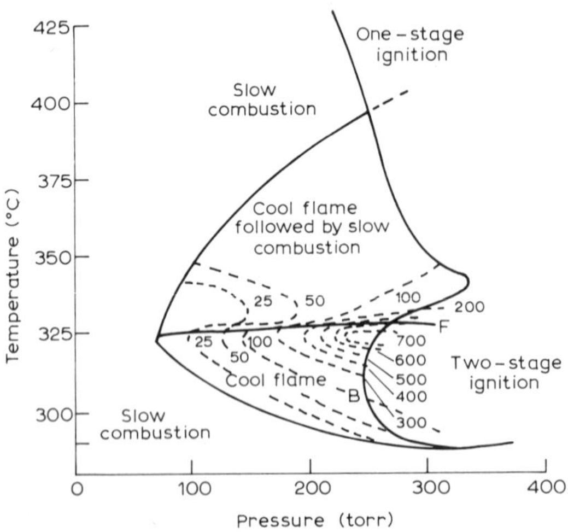

Fig. 41. Neopentane—oxygen. Molar ratio 1:2; spherical pyrex reaction vessel, volume 450 cm^3. The dotted curves are contours of equal rate of pressure rise at the cool flame. (From ref. 55.)

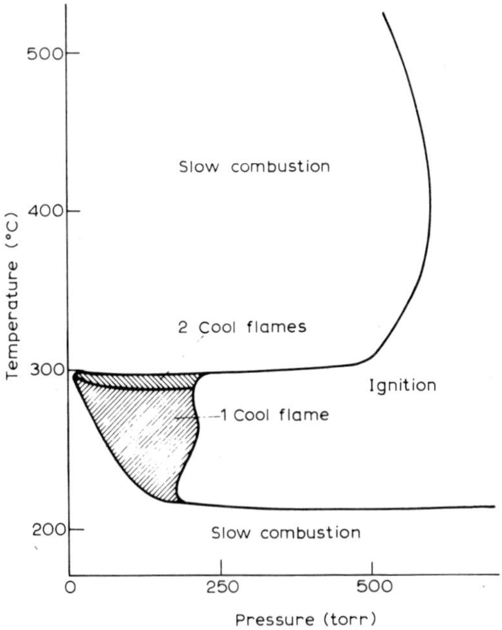

Fig. 42. *n*-Hexane—oxygen. Molar ratio 1.25:1; cylindrical pyrex reaction vessel, volume 200 cm^3. (From ref. 73.)

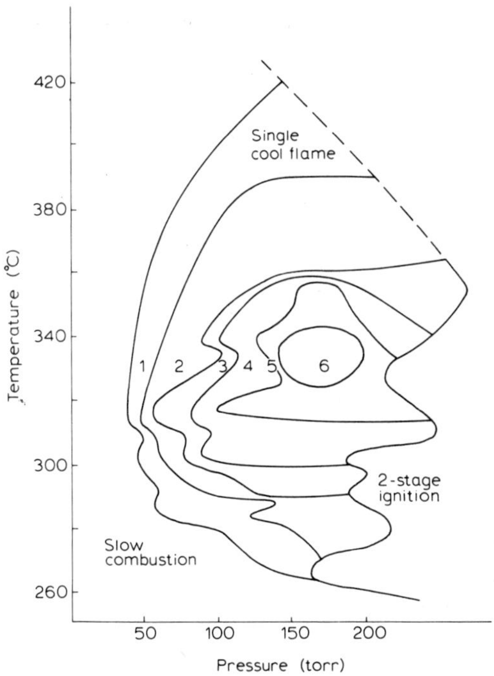

Fig. 43. 2-Methylpentane—oxygen. Molar ratio 1:2; spherical pyrex reaction vessel, volume 500 cm^3. (From ref. 178.)

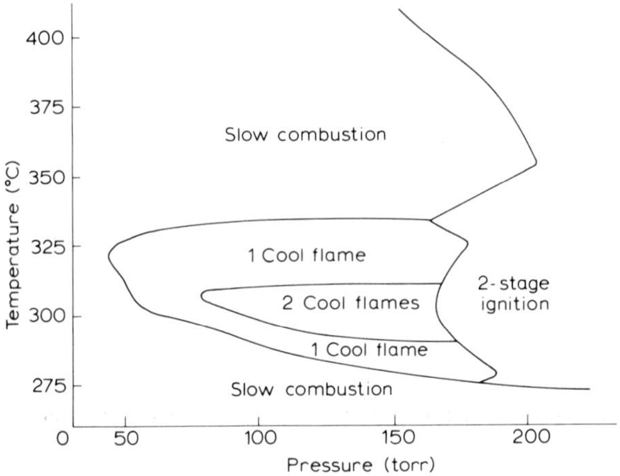

Fig. 44. 3-Methylpentane—oxygen. Molar ratio 1:2; cylindrical pyrex reaction vessel, volume 280 cm^3. (From ref. 80.)

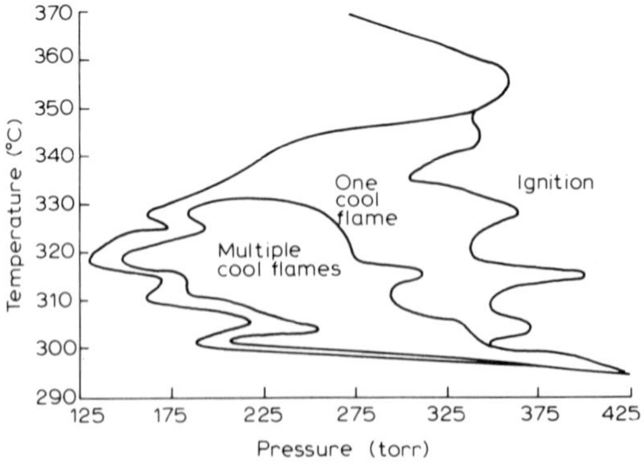

Fig. 45. 2,3-Dimethylbutane—oxygen. Molar ratio 1:2; spherical pyrex reaction vessel, volume 450 cm^3. (From ref. 65.)

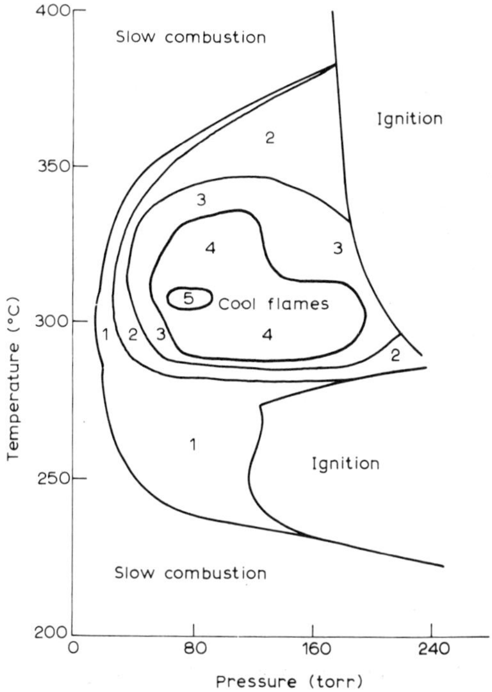

Fig. 46. *n*-Heptane—oxygen. Molar ratio 1:2; cylindrical silica reaction vessel, volume 320 cm^3. (From ref. 200.)

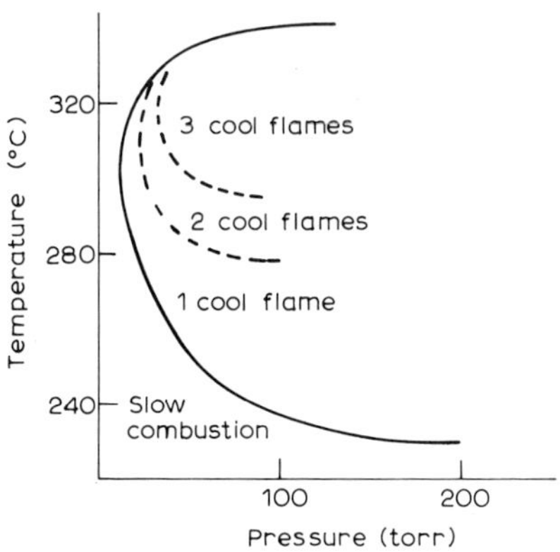

Fig. 47. *n*-Heptane—oxygen. Molar ratio 1:1; cylindrical pyrex reaction vessel, volume 150 cm^3. (From ref. 119.)

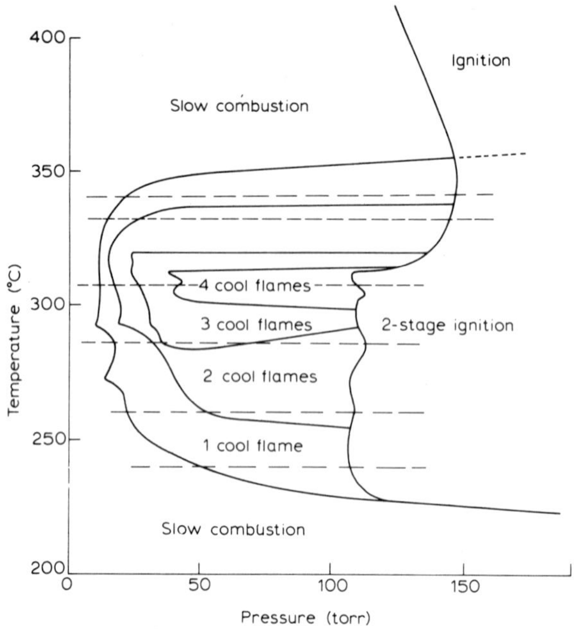

Fig. 48. *n*-Heptane—oxygen. Molar ratio 1:1; cylindrical pyrex reaction vessel, volume 330 cm^3. (From ref. 201.)

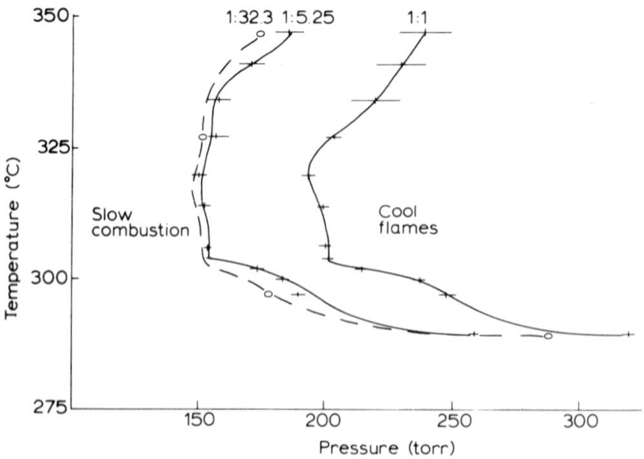

Fig. 49. 2,2,4-Trimethylpentane—oxygen. Molar ratios 1:32.3, 1:5.25 and 1:1; cylindrical pyrex reaction vessel, volume 500 cm^3. (From ref. 202.)

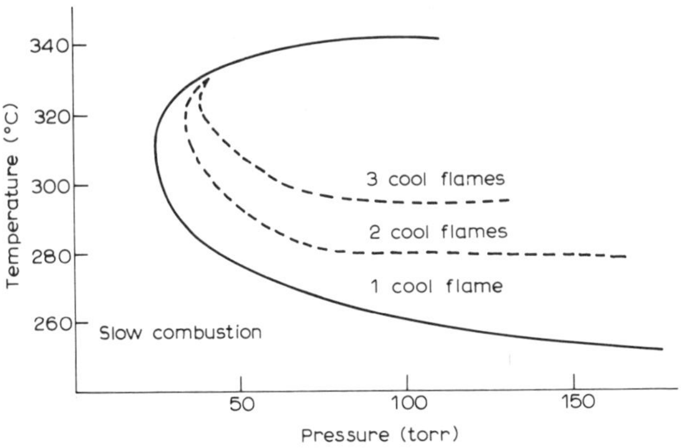

Fig. 50. Cyclohexane—oxygen. Molar ratio 1:1; cylindrical pyrex reaction vessel, volume 150 cm^3. (From ref. 203.)

propagation are difficult to determine and should not be regarded as being fully definitive. Also, for the 3-ethylpentane, *n*-hexane and 3-methylpentane—oxygen systems it is probable that a further extensive single cool flame envelope exists at higher temperatures than those indicated as has recently been found for *n*-pentane + oxygen (cf. Fig. 39 and ref. 70) and for 2-methylpentane + oxygen (cf. Fig. 43 and ref. 77). The cool flames propagating in this region are characterized by the rapid "slow combustion" which follows them and can only be observed by a fast-response recording system.

Acknowledgement

Figure 49 is reproduced by permission of the Ministry of Defence.

REFERENCES

1 C. F. Cullis, Chem. Br., 3 (1967) 370.
2 V. Ya Shtern, The Gas Phase Oxidation of Hydrocarbons, Pergamon, London, 1964.
3 N. N. Semenov, Some Problems of Chemical Kinetics and Reactivity, Pergamon, London, 1958.
4 J. H. Knox and R. G. W. Norrish, Proc. R. Soc. London, Ser. A, 221 (1954) 151.
5 C. F. Cullis and Sir C. N. Hinshelwood, Discuss. Faraday Soc., 2 (1947) 117.
6 A. D. Walsh, Trans. Faraday Soc., 42 (1964) 269; 43 (1947) 297; 43 (1947) 305.
7 M. B. Neiman, Usp. Khim., 7 (1938) 341.
8 R. G. W. Norrish, Cinetique et Mecanisme de Reactions d'Inflammation et de Combustion en Phase Gazeuse, Société des Editions Technique, Paris, 1948.
9 R. G. W. Norrish, Discuss. Faraday Soc., 10 (1951) 269.
10 V. Ya Shtern, The Gas Phase Oxidation of Hydrocarbons, Pergamon, London, 1958, p. 310.

11 W. E. Falconer and J. H. Knox, Proc. R. Soc. London, 250 (1959) 493.
12 J. H. Knox, Trans. Faraday Soc., 55 (1959) 1362.
13 C. N. Satterfield and R. C. Reid, J. Phys. Chem., 59 (1955) 283.
14 B. Lewis and G. von Elbe, Combustion, Flames and Explosions in Gases, Academic Press, New York, 1951.
15 C. N. Satterfield and R. C. Reid, Fifth Symposium (International) on Combustion, Reinhold, New York, 1955, p. 511.
16 C. N. Satterfield and R. E. Wilson, Ind. Eng. Chem., 46 (1954) 1001.
17 M. Seakins and Sir C. N. Hinshelwood, Proc. R. Soc. London, Ser. A, 276 (1963) 324.
18 J. C. Déchaux, Oxid. Combust. Rev., 6 (1973) 75.
19 J. H. Knox and R. G. W. Norrish, Trans. Faraday, Soc., 50 (1954) 928.
20 A. D. Walsh, Trans. Faraday Soc., 43 (1947) 297.
21 K. C. Salooja, Nature (London), 185 (1960) 32.
22 N. S. Enikolopyan, Dokl. Adad. Nauk SSSR, 119 (1958) 520.
23 C. E. H. Bawn and G. Skirrow, Fifth Symposium (International) on Combustion, Reinhold, New York 1955, p. 521.
24 J. D. Mullen and G. Skirrow, Proc. R. Soc. London, Ser. A, 244 (1958) 312.
25 J. Bardwell and Sir C. N. Hinshelwood, Proc. R. Soc. London, Ser. A, 205 (1951) 375.
26 J. Bardwell, Fifth Symposium (International) on Combustion, Reinhold, New York, 1955, p. 1529.
27 W. M. MacNevin, P. F. Urone, M. L. B. Omietanski and M. L. Dunton, Fifth Symposium (International) on Combustion, Reinhold, New York, 1955, p. 402.
28 Z. G. Szabo and D. Gal, Acta. Chem. Hung., 16 (1958) 29.
29 R. N. Pease, J. Am. Chem. Soc., 62 (1940) 2234.
30 D. A. Frank—Kamenetskii, Diffusion and Heat Exchange in Chemical Kinetics, Princeton University Press, Princeton, 1955.
31 A. D. Walsh, Trans. Faraday Soc., 42 (1946) 264.
32 G. H. N. Chamberlain and A. D. Walsh, Third Symposium (International) on Combustion, Williams and Wilkins, Baltimore, 1949, p. 375.
33 C. F. Cullis, A. Fish, F. R. F. Hardy and E. A. Warwicker, Chem. Ind. (1961) 1158.
34 N. N. Semenov, Some Problems of Chemical Kinetics and Reactivity, Vol. 2, Pergamon, London, 1958, p. 128.
35 G. J. Minkoff and C. F. H. Tipper, Chemistry of Combustion Reactions, Butterworths, London, 1962.
36 S. W. Benson, Mechanisms of Pyrolysis, Oxidation and Burning of Organic Compounds, NBS Spec. Publ. 357, U.S. Department of Commerce, 1972, p. 121.
37 A. G. Szabo, Thirteenth Symposium (International) on Combustion, The Combustion Institute, Pittsburgh, 1971, p. 216.
38 J. H. Knox, Adv. Chem. Ser., 76 (1968) 1.
39 J. H. Knox, Combust. Flame, 9 (1965) 297.
40 J. H. Knox and C. H. J. Wells, Trans. Faraday Soc., 59 (1963) 2786, 2801.
41 J. H. Knox, Trans. Faraday Soc., 55 (1959) 1362; 56 (1960) 1225.
42 J. Hay, J. H. Knox and J. M. C. Turner, Tenth Symposium (International) on Combustion, The Combustion Institute, Pittsburgh, 1965, p. 331.
43 J. M. C. Turner, Thesis, Edinburgh, 1964.
44 J. H. Knox, in P. G. Ashmore, F. S. Dainton and T. M. Sugden (Eds.), Photochemistry and Reaction Kinetics, Cambridge University Press, Cambridge, 1967, p. 250.
45 J. H. Knox, R. F. Smith and A. F. Trotman-Dickenson, Trans. Faraday Soc., 54 (1958) 1509.
46 J. H. Knox and J. M. C. Turner, J. Chem. Soc., (1960) 3210.

47 W. E. Falconer, J. H. Knox and A. F. Trotman-Dickenson, J. Chem. Soc. (1961) 782.
48 N. N. Semenov, in P. G. Ashmore, F. S. Dainton and T. M. Sugden (Eds.), Photochemistry and Reaction Kinetics, Cambridge University Press, Cambridge, 1967, p. 229.
49 D. D. Drysdale and R. G. W. Norrish, Proc. R. Soc. London, Ser. A, 308 (1969) 305.
50 G. W. Irvine and J. H. Knox, Symposium on the Mechanisms of Hydrocarbon Reactions, Siofok, Hungary, 1973.
51 J. Brown and C. F. H. Tipper, Combust. Flame, 12 (1968) 79.
52 J. F. Griffiths, G. Skirrow and C. F. H. Tipper, Combust. Flame, 12 (1968) 444.
53 A. P. Zeelenberg and A. F. Bickel, J. Chem. Soc. (1961) 4014.
54 A. P. Zeelenberg, Rec. Trav. Chim. Pays-Bas, 81 (1962) 720.
55 A. Fish, Combust. Flame, 13 (1969) 23.
56 T. Berry, C. F. Cullis and D. L. Trimm, Proc. R. Soc. London, Ser. A, 316 (1970) 377.
57 S. W. Benson, Adv. Chem. Ser., 76 (1968) 143.
58 A. Fish, Adv. Chem. Soc., 76 (1968) 69.
59 J. Heicklen, Adv. Chem. Ser., 76 (1968) 23.
60 S. N. Foner and R. L. Hudson, J. Chem. Phys., 36 (1962) 2681.
61 G. C. Fettis and F. F. Trotman-Dickenson, J. Am. Chem. Soc., 81 (1959) 5260.
62 C. F. Cullis, A. Fish and J. F. Gibson, Proc. R. Soc. London, Ser. A, 284 (1965) 108.
63 J. C. Déchaux, J. L. Flamant and M. Lucquin, Combust. Flame, 17 (1971) 205.
64 J. P. Sawerysyn., M. Van de Steene and M. Lucquin, C. R. Acad. Sci., Ser. C, 272 (1971) 264.
65 A. Fish and J. P. Wilson, Thirteenth Symposium (International) on Combustion, The Combustion Institute, Pittsburgh, 1971, p. 221.
66 A. R. Ubbelohde, Proc. R. Soc. London, Ser. A. 152 (1935) 354, 378.
67 H. C. Bailey and R. G. W. Norrish, Proc. R. Soc. London, Ser. A, 212 (1952) 311.
68 G. A. Luckett and R. T. Pollard, unpublished results.
69 Y. H. Chung and S. Sandler, Combust. Flame 6 (1962) 295.
70 C. F. Cullis, M. Saeed and D. L. Trimm, Proc. R. Soc. London, Ser. A, 300 (1967) 455.
71 R. Hughes and R. F. Simmons, Twelfth Symposium (International) on Combustion, The Combustion Institute, Pittsburgh, 1969, p. 449.
72 J. A. Barnard, private communication.
73 C. F. Cullis, A. Fish, M. Saeed and D. L. Trimm, Proc. R. Soc. London, Ser. A, 289 (1966) 402.
74 G. Kyryacos, H. R. Menapace and C. E. Boord, Anal. Chem., 31 (1959) 222.
75 J. H. Jones and M. R. Fenske, Ind. Eng. Chem., 51 (1959) 262.
76 J. H. Jones, H. D. Allendorf, D. W. Hutton and M. R. Fenske, J. Chem. Eng. Data, 6 (1961) 620.
77 A. Fish, Proc. R. Soc. London, Ser. A, 298 (1967) 204.
78 A. Fish. W. W. Haskell and I. A. Read, Proc. R. Soc. London, Ser. A, 313 (1969) 261.
79 W. S. Affleck and A. Fish. Eleventh Symposium (International) on Combustion, The Combustion Institute, Pittsburgh, 1967, p. 1003.
80 P. Barat, C. F. Cullis and R. T. Pollard, Proc. R. Soc. London, Ser. A, 329 (1972) 433.
81 F. F. Rust and D. O. Collamer, J. Am. Chem. Soc., 76 (1954) 1055.
82 E. Schröder, G. Ohlmann and E. Leibnitz, Z. Phys. Chem. (Leipzig), 225 (1964) 175.
83 D. M. Whitehead, Discussion on Low Temperature Ignition of Organic Substances in The Gas Phase, Donnan Laboratories, Liverpool, 1967.

84 A. R. Burgess, C. J. Luck, D. H. Desty, D. M. Whitehead and G. Pratley, Fourteenth Symposium (International) on Combustion, The Combustion Institute, Pittsburgh, 1973, p. 501.
85 P. Barat, C. F. Cullis and R. T. Pollard, Thirteenth Symposium (International) on Combustion, The Combustion Institute, Pittsburgh, 1971, p. 171.
86 P. Barat, C. F. Cullis and R. T. Pollard, Proc. R. Soc. London, Ser. A, 325 (1971) 469.
87 J. B. Maynard, C. E. Legate and L. B. Graiff, Combust. Flame, 11 (1967) 115.
88 M. Alpestein and R. L. Bradow, Combust. Flame, 11 (1967) 26.
89 M. Alpestein and R. L. Bradow, Soc. Automotive Engineer Mid-Year Meeting, Paper 660410, Detroit, 1966.
90 M. Alpestein and R. L. Bradow, Soc. Automotive Engineer Fuels and Lubricants Meeting, Paper 660781, Houston, 1966.
91 A. P. Zeelenberg and H. W. de Bruijin, Combust. Flame, 9 (1965) 281.
92 J. Loftus and C. N. Satterfield, J. Phys. Chem., 69 (1965) 909.
93 J. F. McKellar and R. G. W. Norrish, Proc. R. Soc. London, Ser. A, 254 (1960) 147.
94 J. F. McKellar and R. G. W. Norrish, Proc. R. Soc. London, Ser. A, 263 (1961) 51
95 W. W. Haskell and I. A. Read, Symposium on Gas Kinetics, University of Szeged, Hungarian Chemical Society, 1969, p. 245.
96 N. R. Greiner, J. Chem. Phys., 46 (1967) 3389.
97 T. Berces and A. F. Trotman-Dickenson, J. Chem. Soc. (1961) 4281.
98 N. R. Greiner, J. Chem. Phys., 53 (1970) 1070.
99 R. R. Baldwin and R. W. Walker, Fourteenth Symposium (International) on Combustion, The Combustion Institute, Pittsburgh, 1973, p. 241.
100 J. H. Knox and R. L. Nelson, Trans. Faraday, Soc., 55 (1959) 937.
101 H. O. Pritchard, J. B. Pyke and A. F. Trotman-Dickenson, J. Am. Chem. Soc., 77 (1955) 2629.
102 N. R. Greiner, Paper presented at 156th Meeting of the American Chemical Society, Atlantic City, 1968.
103 R. R. Baldwin and R. W. Walker, Combust. Flame, 21 (1973) 55.
104 S. W. Benson, J. Am. Chem. Soc., 87 (1965) 972.
105 T. Mill, Thirteenth Symposium (International) on Combustion, The Combustion Institute, Pittsburgh, 1971, p. 229.
106 G. G. Kinnear and J. H. Knox, Thirteenth Symposium (International) on Combustion, The Combustion Institute, Pittsburgh, 1971, p. 209.
107 A. Fish, in D. Swern (Ed.), Organic Peroxides, Interscience, New York, 1970, p. 141.
108 R. R. Baldwin, C. J. Everett, D. E. Hopkins and R. W. Walker, Adv. Chem. Ser., 76 (1968) 124.
109 C. F. Cullis, A. Fish and D. L. Trimm, Proc. R. Soc. London, Ser. A, 273 (1963) 427.
110 A. N. Bose, Trans. Faraday. Soc., 55 (1959) 778.
111 C. F. Cullis, J. A. Garcia-Dominguez, D. Kiraly and D. L. Trimm, Proc. R. Soc. London, Ser. A, 291 (1966) 235.
112 S. W. Benson, Thermochemical Kinetics, Wiley, New York, 1968.
113 C. F. Cullis, F. R. F. Hardy and D. W. Turner, Proc. R. Soc. London, Ser. A, 251 (1959) 265.
114 A. Fish and J. P. Wilson, unpublished results.
115 J. Cartlidge and C. F. H. Tipper, Anal. Chim. Acta, 22 (1960) 106.
116 J. Cartlidge and C. F. H. Tipper, Proc. R. Soc. London, Ser. A, 261 (1961) 388.
117 J. Cartlidge and C. F. H. Tipper,Proc. Chem. Soc., London (1959) 190; (1960) 219.

118 J. Cartlidge and C. F. H. Tipper, Combust. Flame, 5 (1961) 87.
119 B. H. Bonner and C. F. H. Tipper, Combust. Flame, 9 (1965) 387.
120 A. Hardacre, G. Skirrow and C. F. H. Tipper, Combust. Flame, 7 (1963) 100.
121 A. W. Bastow and C. F. Cullis, Symposium on the Mechanisms of Hydrocarbon Reactions, Siofok, Hungary, 1973.
122 D. J. M. Ray and D. J. Waddington, Symposium on the Mechanism of Hydrocarbon Reactions, Siofok, Hungary 1973.
123 S. Antonik and M. Lucquin, Bull. Soc. Chem. Fr., 10 (1968) 4043.
124 J. A. Barnard, Adv. Chem. Ser., 76 (1968) 98.
125 A. D. Kirk and J. H. Knox, Trans. Faraday Soc., 56 (1960) 1296.
126 S. W. Benson and R. Shaw, in D. Swern (Ed.), Organic Peroxides, Interscience, New York, 1970, p. 106.
127 D. E. Hoare, J. B. Protheroe and A. D. Walsh, Trans. Faraday Soc., 55 (1959) 548.
128 C. A. Mc. Dowall and J. A. Thomas, J. Chem. Soc. (1949) 2208, 2217; (1950) 1462.
129 L. V. Karmilova, N. S. Enikolopyan, A. B. Nalbandyan and N. N. Semenov, Z. Fiz. Chim., 34 (1960) 562.
130 A. R. Burgess and R. G. W. Laughlin, Chem. Commun. (1967) 769.
131 C. W. Taylor, Can. J. Chem. 36 (1958) 1213.
132 A. J. Brown, N. Burt, G. A. Luckett and R. T. Pollard, Symposium on the Mechanisms of Hydrocarbon Reactions, Siofok, Hungary, 1973.
133 A. Fish, Proc. R. Soc. London, Ser. A, 293 (1966) 378.
134 G. A. Luckett and R. T. Pollard, Combust. Flame, 21 (1973) 265.
135 J. C. Déchaux, Thèse, La Faculté, Des Sciences de L'Université de Lill , 1971.
136 S. W. Benson, The Foundation of Chemical Kinetics, McGraw-Hill, New York, 1960.
137 G. R. MacMillan and J. G. Calvert, Oxid. Combust. 1 (1965) 121.
138 J. F. Griffiths and G. Skirrow, Oxid. Combust. Rev., 3 (1968) 47.
139 C. F. Cullis and E. Fersht, Combust. Flame, 7 (1963) 353.
140 D. E. Hoare and D. E. Lill, J. Chem. Soc., Faraday Trans. I, 69 (1973) 603.
141 N. N. Semenov, Acta. Physicochim. URSS 18 (1943) 93.
142 M. Lucquin, J. Chim. Phys., 55 (1958) 827.
143 L. R. Sochet and M. Lucquin, J. Chim. Phys., 65 (1965) 796.
144 L. R. Sochet, J. Egret and M. Lucquin, J. Chim. Phys., 63 (1966) 1555.
145 L. R. Sochet and M. Lucquin, J. Chim. Phys., 65 (1968) 977.
146 M. Lefebvre and M. Lucquin, J. Chim. Phys. 62 (1965) 775, 784.
147 L. R. Sochet, J. P. Sawerysyn and M. Lucquin, Adv. Chem. Ser., 76 (1968) 111.
148 L. R. Sochet and M. Lucquin, Combust. Flame, 13 (1969) 319.
149 J. A. Howard and K. U. Ingold, J. Am. Chem. Soc., 90 (1968) 1058.
150 J. E. Bennett, D. M. Brown and B. Mile, Trans. Faraday Soc., 66 (1970) 386.
151 J. E. Bennett, D. M. Brown and B. Mile, Trans. Faraday Soc., 66 (1970) 397.
152 G. A. Russell, J. Am. Chem. Soc., 79 (1957) 3871.
153 R. R. Baldwin and D. Brattan, Eighth Symposium (International) on Combustion, Williams and Wilkins, Baltimore, 1962, p. 110.
154 R. J. Sampson, J. Chem. Soc. (1963) 5095.
155 R. J. Sampson, Discussion on Oxidation in Organic Chemistry, Manchester, 1964.
156 C. F. Cullis, A. Fish and J. F. Gibson, Proc. R. Soc. London, Ser. A, 292 (1966) 575.
157 R. R. Baldwin and R. W. Walker, Discussion on Low Temperature Oxidation in The Gas Phase, Donnan Laboratories, Liverpool, 1969.
158 R. R. Baker, R. R. Baldwin and R. W. Walker, Thirteenth Symposium (International) on Combustion, The Combustion Institute, Pittsburgh, 1971, p. 291.

159 R. R. Baldwin, D. H. Langford, M. J. Matchan, R. W. Walker and D. A. Yorke, Thirteenth Symposium (International) on Combustion, The Combustion Institute, Pittsburgh, 1971, p. 251.
160 R. R. Baldwin, A. C. Norris, and R. W. Walker, Eleventh Symposium (International) on Combustion, The Combustion Institute, Pittsburgh, 1967, p. 889.
161 R. R. Baldwin, D. E. Hopkins and R. W. Walker, Trans. Faraday Soc., 66 (1970) 189.
162 R. R. Baker, R. R. Baldwin and R. W. Walker, Trans. Faraday Soc., 66 (1970) 2821.
163 R. R. Baldwin, B. Tunnicliffe and R. W. Walker, unpublished results.
164 R. R. Baker, R. R. Baldwin and R. W. Walker, Combust. Flame, 14 (1970) 31.
165 J. A. Kerr and A. C. Lloyd, Q. R. Chem. Soc., 22 (1968) 549.
166 D. H. Slater and J. G. Calvert Adv. Chem. Ser., 76 (1968) 58.
167 R. R. Baldwin, M. J. Matchan and R. W. Walker, Combust. Flame, 15 (1970) 109.
168 A. J. Brown and R. T. Pollard, unpublished results.
169 S. F. Rehman and R. T. Pollard, unpublished results.
170 H. D. Medley and S. D. Cooley, Adv. Pet. Chem. Refin., 3 (1960) 309.
171 J. G. Atherton, A. J. Brown, G. A. Luckett and R. T. Pollard, Fourteenth Symposium (International) on Combustion, The Combustion Institute, 1973, p. 513.
172 C. G. Kinnear and J. H. Knox, Symposium on Gas Kinetics, University of Szeged, Hungarian Chemical Society, 1969, p. 356.
173 J. C. Déchaux, F. Langrand, G. Hermant and M. Lucquin, Bull. Soc. Chim. Fr., 10 (1968) 403.
174 J. A. Barnard and R. D. Handscombe, European Symposium on Combustion, Sheffield, England, 1973.
175 T. Mill, F. Mayo, H. Richardson, K. Irwin and D. L. Allara, J. Am. Chem. Soc., 94 (1972) 6802.
176 F. Baronnet, M. P. Halstead, A. Prothero and C. P. Quinn, European Symposium on Combustion, Sheffield, England, 1973.
177 C. A. Euker and J. P. Leinroth, Combust. Flame, 15 (1970) 275.
178 R. T. Miles, R. T. Pollard and J. P. Wilson, unpublished results.
179 B. F. Gray, Trans. Faraday Soc., 65 (1968) 1603.
180 B. F. Gray and C. H. Yang, Trans Faraday Soc., 65 (1968) 1614.
181 B. F. Gray and C. H. Yang, Trans. Faraday Soc., 65 (1968) 2133.
182 B. F. Gray and C. H. Yang, J. Phys. Chem., 73 (1969) 3395.
183 M. P. Halstead, A. Prothero and C. P. Quinn, Proc. R. Soc. London, Ser. A, 322 (1971) 377.
184 M. Lucquin, J. Montastier, F. Langrand and A. Perche, J. Chim. Phys., 66 (1969) 1389.
185 M. Lucquin, J. Montastier, F. Langrand, A. Perez and A. Perche, J. Chim. Phys., 66 (1969) 1714.
186 A. Perche, A. Perez and M. Lucquin, Combust. Flame, 15 (1970) 89.
187 A. Perche, A. Perez and M. Lucquin, Combust. Flame, 17 (1971) 179.
188 A. Perche, A. Perez and M. Lucquin, J. Chim. Phys. 69 (1972) 389.
189 M. P. Halstead, A. Prothero and C. P. Quinn, Chem. Commun. (1970) 1150.
190 A. Prothero, unpublished results.
191 J. Chamboux and M. Lucquin, J. Chim. Phys., 59 (1962) 979.
192 C. P. Quinn and J. P. Wilson, unpublished results.
193 J. G. Atherton and R. T. Pollard, unpublished results.
194 M. P. Halstead, A. Prothero and C. P. Quinn, Combust. Flame, 20 (1973) 211.
195 F. Baronnet, M. P. Halstead, A. Prothero and C. P. Quinn, C. R. Acad. Sci., Ser. C, 275 (1972) 17.

196 J. F. Griffiths, B. F. Gray and P. Gray, Thirteenth Symposium (International) on Combustion, The Combustion Institute, Pittsburgh, 1971, p. 239.
197 A. Antonik and M. Lucquin, Combust. Flame, 19 (1972) 311.
198 A. W. Bastow and C. F. Cullis, Proc. R. Soc. London, Ser. A, 338 (1974) 327.
199 R. Hughes and R. F. Simmons, Combust. Flame, 14 (1970) 103.
200 C. F. Cullis, A. Fish and J. F. Gibson, Proc. R. Soc. London, Ser. A, 311 (1969) 253.
201 A. R. Burgess and R. G. W. Laughlin, Combust. Flame, 19 (1972) 315.
202 J. A. Barnard and B. Harwood, private communication.
203 B. H. Bonner and C. F. H. Tipper, Combust. Flame, 9 (1965) 317.

Chapter 3

The gas phase combustion of aldehydes

D. J. DIXON and G. SKIRROW

1. Introduction

The main aims of this chapter are to examine the gas phase oxidation of the lower aldehydes in various temperature regions, to attempt to recognize the principal elementary steps comprising the overall mechanism for particular systems and, where possible, to give the kinetic parameters for these steps.

Aldehydes are often intermediates in the oxidation of other fuels [1—4, 29], and the ease with which they themselves oxidize and give rise to peroxidic materials or active radicals means that their role in these systems is likely to be important. For example, formaldehyde is produced during the oxidation of most hydrocarbons, and is known to behave as a branching intermediate during the high temperature combustion of methane [1—6]. However, in certain systems, and particularly at lower temperatures, formaldehyde may behave as a retarder [7—9, 57]. Acetaldehyde is an intermediate in the oxidation of propene [10] and other olefins [11, 12], and its addition to these systems reduces the induction period or enhances the maximum rate. Many other examples are known both of the occurrence of aldehydes amongst the combustion products and of the ability of aldehydes to influence the oxidation of systems in which they occur [1, 13—19].

The facility of aldehydes for undergoing oxidation at a measurable rate at low temperatures (often below 100 °C) means that they are convenient subjects for study in their own right since the relative simplicity of the stoichiometry of their oxidation under these conditions, viz.

$$RCHO + O_2 \longrightarrow RCO_3H$$

coupled with the comparative absence of complicating side or secondary reactions, enables many conclusions to be drawn with greater confidence than is possible for those oxidations which proceed readily only at higher temperatures. As the temperature is raised aldehyde oxidation becomes at first more complex [20] because of the growing importance of intermediate decomposition and the onset of processes characteristic of the high temperature oxidation (Sect. 4). At a sufficiently high temperature, (>400 °C) many of these decompositions, particularly those of acyl radicals, are virtually complete. As a result, peracid branching does not occur and the oxidation is again relatively simple [21, 22]. The simplicity of the oxidation at the lower and higher temperatures means that these systems should be useful subjects with which to study additive action. Formaldehyde differs from its higher homologues in that it reacts with oxygen only slowly at temperatures below about 250 °C, and most of the published studies of formaldehyde [23—41] combustion have been made at temperatures in the order of 400 °C.

A review dealing principally with the oxidations of acetaldehyde and propionaldehyde below about 200 °C and with the oxidation of formaldehyde has recently been given [42]. In the present account, those areas dealt with in the earlier review are treated somewhat briefly except where the original conclusions appear to be in need of modification or where additions are necessary. The scope of this chapter is greater in that it extends the temperature range considered and examines what published material is available concerning the oxidation of aldehydes containing more than three carbon atoms. Some aspects of cool flame phenomena are examined, although no attempt is made to develop a comprehensive theoretical interpretation.

2. Some general aspects of aldehyde combustion

A few comments on some general aspects of aldehyde oxidation should be made, and the reader is also referred to Vol. 1 of this series for discussion of experimental techniques.

For most static low temperature oxidation studies of aldehydes other than formaldehyde it is customary to use acid-washed pyrex vessels. With this surface condition and at temperatures below about 150 °C the oxidation proceeds with a pressure decrease corresponding approximately to the stoichiometry

$$RCHO + O_2 \longrightarrow RCO_3H$$

up to >50 % reaction. At lower temperatures this stoichiometry persists to even later stages. For example, at 70 °C there is a 99.5 % stoichiometry based on pressure change and oxygen consumption almost throughout the reaction [43]. There is a period of acceleration to the maximum rate of pressure decrease, although at low oxygen/fuel ratios, this autocatalysis is evident only when sensitive continuous monitoring equipment is used. Near the maximum rate the reaction is mainly homogeneous, and reproducibility is surprisingly good, even between different groups of workers [42], provided the surface/volume ratio is not too high. However, there is evidence that the reaction is at least partially heterogeneous, even with acid-washed pyrex surfaces [9, 42, 43], and certain vessel coatings (notable KCl) considerably alter the reaction and introduce pronounced heterogeneous characteristics [44—46]. Care is necessary in ensuring reproducibility of the early stages of the reaction, and it is important that the products should be removed from the vessel by a route which avoids contaminating the inlet lines with peracids since small amounts of these profoundly modify the initial stages of reaction. Care should also be taken to avoid the access of mercury or mercury vapour to the vessel since this also modifies the reaction. It might be noted that fouling of a mercury diffusion pump quickly occurs if products from the low temperature oxidation are allowed to come into contact with the pump fluid.

The simple stoichiometry observed below about 150 °C means that for many preliminary purposes the pressure decrease provides a satisfactory method of following the reaction. The need to exclude Hg vapour makes the use of diaphragm gauges such as the glass spoon gauge, or preferably the continuously recording pressure transducer gauge desirable. For more comprehensive investigations some sort of analytical investigation is necessary. Continuous monitoring by mass spectrometry has been used [47], although more commonly, for the slow oxidation, the reaction is interrupted for analysis by allowing the contents of the vessel to expand through a graded series of cold traps so as to obtain a preliminary separation. The products in each trap and the non-condensable permanent gases are examined, where possible, by gas chromatography [20, 48]. Products difficult to examine in this way because of their reactivity or their unsatisfactory behaviour on the column (peroxides, formaldehyde) are best examined by conventional chemical methods. Difficulties are often encountered in relating the composition of the product mixture found in the cold traps to that present in the vessel at the moment of interruption since it is not easy to separate, for example, unreacted aldehyde which has co-condensed with peracid [20, 43].

The high temperature oxidations of acetaldehyde and propionaldehyde have been studied mainly using boric acid coated vessels [21, 22]. This surface is inert towards HO_2 and H_2O_2 destruction and leads to good reproducibility. KCl treated surfaces appear to be destructive towards HO_2 and H_2O_2 formed during the combustion of formaldehyde [23, 37, 38], but do not seem to have been used during investigations of the high temperature oxidation of the higher aldehydes.★ However, it seems reasonable to expect that, on the basis of the known differences in behaviour of the formaldehyde oxidation in salt coated and boric acid coated vessels, the characteristics of the oxidation of the higher aldehydes above 400 °C would also be different in these two types of vessel. This would be particularly so for propionaldehyde which, in a boric acid coated vessel, oxidizes by a mechanism in which hydrogen peroxide branching occurs.

The change in temperature accompanying rapidly developing reactions such as cool flames has been used by Griffiths et al. [49] to monitor

★ Recent work by Baldwin and co-workers on the oxidation of propionaldehyde using KCl coated vessels is considered in Sect. 4.6.2.

events in hydrocarbon oxidations, and the method can no doubt also be applied to aldehyde cool-flame studies. In this method the rapid temperature rise is recorded by means of a storage oscilloscope coupled to a fine-wire thermocouple rendered inert towards the reacting system by means of a coating of silica. Care is necessary in the interpretation of observations made in this way, and in particular, proper attention should be given to ensuring that the response times of the thermocouple and the ancillary equipment are sufficiently short [49, 50]. It should be realized that convection may become important when appreciable temperature changes occur in static systems. A stirred system, as described for example by Griffiths et al. [49, 50], eliminates the influence of inhomogeneous temperature distributions on non-isothermal behaviour.

The emission of radiant energy accompanying reaction can be used to recognize features of the oxidation which would not normally be detected by more conventional methods. In certain regions of the oxidation the light emission is sufficiently high not to require specially sensitive apparatus; Chamboux and Lucquin [51, 52] have been able to show by means of sensitive photomultipliers that light emission is not restricted to the cool-flame region.

Flow systems [33—36, 53, 54] are not often used in aldehyde oxidation studies since it is not always easy to obtain kinetic data from them and there are problems associated with the purification of the relatively large amounts of reactant needed. However, they have been used successfully, particularly by Russian workers [33—36], to establish many features of formaldehyde oxidation. Under certain circumstances, flow systems offer special advantages. In particular, it may be possible to transform what may be a small separation in time between the cool flame and the second stage into a separation in distance of several centimeters along the axis of a flow reactor and thereby considerably facilitate analysis by means of a probe into a mass spectrometer [55—57] (see Sect. 5.2).

3. Low temperature aldehyde oxidation

This section describes the oxidation in the temperature range below that at which appreciable peracid, peracyl or acyl radical decomposition occurs. This is not an entirely satisfactory basis for classification since some peracid decomposition must occur if autocatalysis — a characteristic feature of aldehyde oxidation — is to take place. Furthermore, the proportion of RCO radicals generated which decompose depends not only on the temperature but also on the oxygen pressure. Nevertheless, provided that the oxygen pressure is sufficiently high and the temperature below about 150 °C, peracid formation is almost quantitative, at least over the first 50 % or so of reaction.

Formaldehyde has no slow oxidation regime below about 250 °C. The reasons for this become apparent when the mechanism of oxidation of other aliphatic aldehydes is understood. This problem is further discussed in Sect. 3.5.1.

3.1 ACETALDEHYDE

The main experimental features may be summarized as follows [42, 60]. In acid-washed pyrex vessels, below 150 °C acetaldehyde oxidizes autocatalytically and with a pressure decrease over most of the reaction, peracid being produced in approximate accordance with

$$CH_3CHO + O_2 \longrightarrow CH_3CO_3H$$

The initial period of acceleration is not very evident if the initial O_2/fuel ratio is high. For ratios greater than unity the maximum rate of pressure decrease (ρ_{max}) is attained somewhere between one-quarter and one-third of the total reaction. Most investigators have found ρ_{max} to be oxygen independent (except for O_2/fuel ratios very much less than unity) and approximately second order with respect to the initial aldehyde pressure (Table 1). The overall activation energy (based on the maximum rate of pressure change) between 90 and 150 °C is about 15 kcal . mole^{-1} (Table 2).

When the initial O_2/fuel ratio is much greater than unity the rate of oxidation, after passing through a maximum, decreases to zero only

TABLE 1

Orders of reaction for the low temperature oxidation of aldehydes ($\rho_{max} = k[RCHO]^n[O_2]^m$)

Ref.	Temp. (°C)	n	m
Acetaldehyde			
125	120	1.7	0
45	120	2 (approx.)	0
126	119	1.87	0
19	119	1.8—2.0	0
Propionaldehyde			
127	150	1.8—2.1	0
19	118—148	2.0	0
20	155	1.89	0
n-Butyraldehyde			
84	124	2 (approx.)	Mainly oxygen dependent[a]
iso-Butyraldehyde			
84	124	1.5 (approx.)	Mainly oxygen dependent[a]

[a] See Fig. 6.

TABLE 2
Overall activation energies for aldehyde oxidation

Acetaldehyde		Propionaldehyde	
Ref.	E_a (kcal . mole^{-1})	Ref.	E_a (kcal . mole^{-1})
125	10	127	15.4
126	14.0—15.6	19	14.5
19	14.5 ± 2	20	16.5
128	8.7	20	20[a]

[a] Determined from indirect measurement of the rate of oxygen consumption.

gradually. When this ratio is close to unity the decrease in rate following the maximum is rather more abrupt; for initial ratios less than about 0.25 the "expected" maximum rate may not be attained because total consumption of the oxygen occurs during the period of acceleration.

The oxidation is only slightly influenced by inert gases, although hydrogen is reported to cause acceleration in flow systems [53, 54]. Until recently, because (*a*) variation of the surface/volume ratio appeared to have little effect on ρ_{max}, and (*b*) the results of different workers for

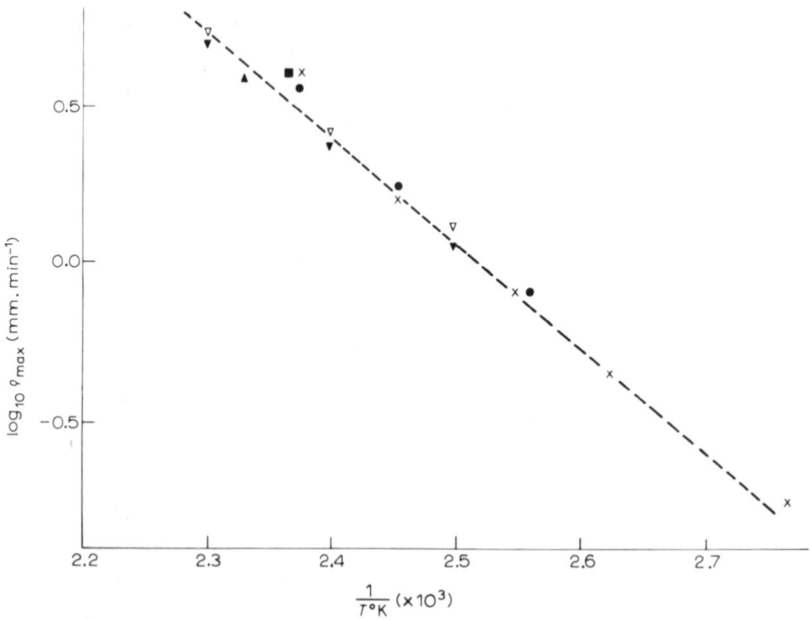

Fig. 1. Arrhenius plot of the maximum rate of oxidation of acetaldehyde and propionaldehyde [42]. Aldehyde pressure, 100 torr; excess oxygen. × and ▽, CH_3CHO; ●, ■, ▲ and ▼, C_2H_5CHO.

ρ_{max} (for both acetaldehyde and propionaldehyde oxidations) agreed well [42] (Fig. 1), it seemed likely that the main propagation and termination steps were largely homogeneous, and that the enhancement of both initial and maximum rates noted when the surface/volume ratio was increased was a consequence of the rate of surface initiation becoming comparable with that of branching for higher surface areas. However, recent data [43] indicate that the branching process is heterogeneous (p. 379). It might be noted that certain surfaces (e.g. KCl [44—46] do modify the reaction considerably, possibly as a consequence of heterogeneous peracid decomposition. A typical pressure—time plot for KCl coated surfaces is shown in Fig. 2.

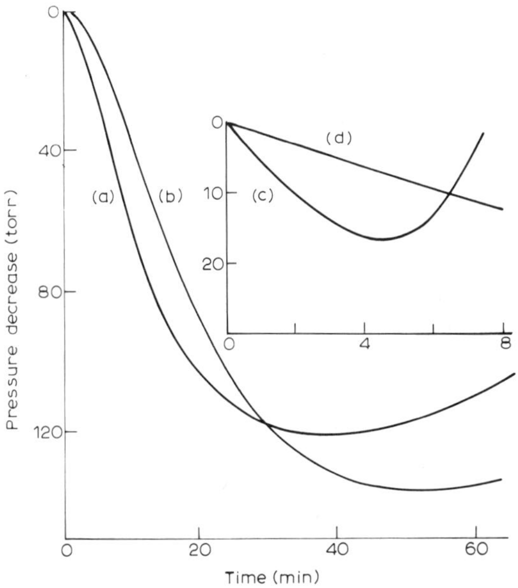

Fig. 2. Pressure—time plots for acid-washed and KCl coated vessels [44]. (a) Acetaldehyde; (b) propionaldehyde: aldehyde and oxygen pressures, 208 torr; temperature, 143 °C; uncoated vessel. Inset: acetaldehyde and oxygen pressures, 100 torr; temperature, 126 °C; (c) KCl coated and (d) uncoated vessels.

Additives normally regarded as sources of radicals (peracids [19], diperoxides [58]) and also HBr [59], result in an enhanced rate of oxidation. The reaction is also accelerated by UV light [62]. Many additives (notably HCHO [7, 9], alcohols [46, 63, 64], and amines [46, 65—69], but also ethane [70] and higher hydrocarbons [53], olefins [64, 71], NH_3 [65] and NO_2 [61, 137]) cause inhibition or retardation. The possible role of some of these retarders is considered in Sect. 3.5.1.

The main features of the low temperature oxidation in static systems can be explained in terms of the scheme [15—19, 42, 72]

$$\text{initiation} \longrightarrow RCO \tag{1}$$

$$RCO + O_2 \longrightarrow RCO_3 \tag{2}$$

$$RCO_3 + RCHO \longrightarrow RCO_3H + RCO \tag{3}$$

$$RCO_3 + RCO_3 \longrightarrow \text{termination} \tag{4a}$$

$$RCO_3 + RCO \longrightarrow \text{termination} \tag{4b}$$

$$RCO + RCO \longrightarrow \text{termination} \tag{4c}$$

$$RCO_3H \longrightarrow \text{branching (RCO)} \tag{5a}$$

$$2RCO_3H \longrightarrow \text{branching (RCO)} \tag{5b}$$

$$RCO_3H + RCHO \longrightarrow \text{branching (RCO)} \tag{5c}$$

$$RCO + M \longrightarrow R + CO + M \tag{6a}$$

$$RCO \longrightarrow R + CO \tag{6b}$$

For reasons outlinded below, (4b) and (4c) contribute insignificantly to termination at all except the lowest O_2/fuel ratios and termination can generally be assumed to be solely by (4a). Branching is by (5c) and, under the conditions considered here, (6) can be disregarded (see Sect. 3.1.1). Initiation (which is almost certainly heterogeneous at low temperatures) is tentatively assumed to be by

$$RCHO + O_2 \longrightarrow RCO + HO_2 \tag{1a}$$

The low reactivity of the HO_2 radical means that at these temperatures

$$RCHO + HO_2 \longrightarrow RCO + H_2O_2 \tag{1b}$$

will not be important. The combination of (1a), (2), (3), (4a) and (5c) leads to the instantaneous rate expression

$$-d[RCHO]/dt = k_{1a}[RCHO][O_2] + k_{5c}[RCHO][RCO_3H] + k_3[RCHO][RCO_3] \tag{I}$$

$$= k_{1a}[RCHO][O_2] + k_{5c}[RCHO][RCO_3H] + k_3/(2k_{4a})^{1/2}\{k_{1a}[RCHO][O_2] + k_{5c}[RCHO][RCO_3H]\}^{1/2}[RCHO] \tag{II}$$

where the unsubscripted terms refer to the instantaneous concentrations or pressures.

At the maximum rate $k_{5c}[RCHO][RCO_3H] \gg k_{1a}[RCHO][O_2]$, and provided that the aldehyde consumed is approximately equal to the instantaneous concentration of peracid, the maximum rate can be shown [19, 42] to be given by

$$\rho_{max} = \frac{3\sqrt{3}k_3}{16}\left[\frac{k_{5c}}{2k_{4a}}\right]^{1/2} [RCHO]_0^2[O_2]_0^0 \qquad \text{(III)}$$

where $[RCHO]_0$ is the initial concentration of aldehyde and $[O_2]_0$ that of oxygen. Expression (III) is in reasonable agreement with experiment.

The reasons for adopting this scheme will be outlined only briefly but the reader is referred to the earlier review for a more detailed account.

3.1.1 *Propagation*

Griffiths and Skirrow [42] have discussed various estimates of the rate for (2) and concluded that it was around 10^8 l . mole^{-1} . sec^{-1}. The most recent value, based on the kinetics of the final stage of the oxidation at 60—80 °C (with large excess of aldehyde) [43], is lower, $(1.2 \pm 0.2) \times 10^7$ l . mole^{-1} . sec^{-1}. However, it is clear that k_2 is high enough to ensure that RCO radicals produced directly or indirectly in the branching step will react by (2) rather than by (4b) or (4c) except at very low oxygen pressures. Thus reaction (3) for which the rate coefficient is [42, 62, 73]

$$k_3 = 1.9 \times 10^9 \exp(-7{,}200/RT) \text{ l . mole}^{-1}\text{ . sec}^{-1}$$

will normally be the rate controlling propagation step. Steps (2) and (3) are the simplest explanations of propagation consistent with the almost quantitative yields of peracid.

The limiting high pressure RCO decomposition (6b) is unimportant at temperatures below about 15 °C as the rate coefficient (for R = CH_3)

$$k_{6b} = 2 \times 10^{13} \exp(-22{,}000/RT) \text{ sec}^{-1} \quad [105(a)]$$
$$= 3 \times 10^{13} \exp(-17{,}200/RT) \text{ sec}^{-1} \quad [105(b)]$$

is such that the competitive reaction with oxygen (2) will predominate.

3.1.2 *Branching*

Although branching by (5b) may be important in some liquid phase oxidations [74], there is no evidence for its occurrence in this system and it will not be further considered. Second-order branching by (5c) is chosen in preference to the first-order decomposition (5a) largely on the basis of the work of Combe et al. [15—19] who showed

(*i*) the order with respect to aldehyde is consistent with branching by (5c) but not with branching by (5a). First-order branching (5a) would require ρ_{max} to be proportional to $(RCHO)_0^{3/2}$;

(*ii*) for the related processes — the peracid induced pyrolysis of acetaldehyde — the rate is more consistent with initiation by (5c) than by (5a). (See also ref. 138.)

Analysis of the pressure—time curves for the oxidation shows that when the initial oxygen concentration is in excess of that of the aldehyde, if branching were to occur by (5c), the maximum rate should occur at about 25 % reaction; branching by (5a) would result in the occurrence of ρ_{max} at about 33 % reaction. In practice, the position of ρ_{max} is not sufficiently well defined to enable this test to be used.

Although the simplest branching step which is consistent with the observed kinetics is first-order in both peracid and aldehyde, Combe et al. [19] suggested that the overall branching process may be more complicated than (5c) implies. It was considered that interaction of aldehyde and peracid may lead to the formation of an addition compound similar to that proposed for the liquid phase oxidation of acetaldehyde [75—79], and that this compound could either regenerate the aldehyde and peracid, or, alternatively, decompose to give radicals, viz.

$$RCHO + RCO_3H \longrightarrow \text{compound} \qquad (A)$$

$$\text{compound} \longrightarrow RCHO + RCO_3H \qquad (B)$$

$$\text{compound} \longrightarrow \text{radicals} \qquad (C)$$

For a stationary state concentration of the compound

$$k_{5c} = \frac{k_A k_C}{k_B + k_C}$$

If $k_C \ll k_B$, then

$$k_{5c} = k_C K_e$$

where K_e ($= k_A/k_B$) is the equilbrium quotient for compound formation.

Supporting evidence [19] for this compound formation was reported in that when an excess of aldehyde reacted to completion with a small amount of oxygen at temperatures sufficiently low to minimize complications of interpretation arising from excessive peracid decomposition, the overall pressure decrease on exhaustion of the oxygen appeared to be greater than the initial oxygen pressure (Table 3). On the basis of these results, obtained at 51 and 66 °C, compound formation was indicated, the enthalpy of formation being about —5.5 kcal . mole^{-1}. However, recent

TABLE 3
Compound formation during low temperature acetaldehyde oxidation [19]

Temp. (°C)	Initial pressure (torr)		Pressure decrease (torr)	$K_e \times 10^3$
	O_2	CH_3CHO		
66	14	227	17	1.3
	14	118	16	1.45
	14	298	18	1.4
	57	227	67	1.25
51	14	227	16.5	1.9
	20	285	27	2.08
	21	190	28	1.9

K_e given by $P_{compound}/P_{RCO_3H}P_{RCHO}$ and calculated by assuming that $P_{compound} = P - P_{initial\ O_2}$ and that $P_{RCO_3H} = P_{initial\ O_2} - P_{compound}$.

TABLE 4
Pressure decrease during the low temperature oxidation of acetaldehyde (62.5 °C)[43]

Initial pressure (torr)		Pressure decrease (torr)
O_2	CH_3CHO	
18.8	300	18.7
37.6	300	38.0
79.4	300	129.5
9.2	107	9.15

work [43] has failed to substantiate these findings (Table 4 and Fig. 3), and suggests that this compound, if formed at all, exists at much lower concentrations than the previous workers results would indicate. However, it should be noted that under circumstances such that the yield of peracid approaches its saturated vapour pressure, the simple pressure stoichiometry no longer applies. The condensation of liquid products accompanying reaction results in a characteristic form of the pressure—time curve shown in Fig. 3.

For expression (III) the overall activation energy for $\rho_{max}(E_a)$ is $E_3 + 1/2\ (E_{5c} - E_{4a})$. Taking E_a, E_3 and E_{4a} to be 15, 7.0 and 0 kcal . mole^{-1}, respectively, E_{5c} for the range 100—140 °C is calculated to be 16 kcal . mole^{-1}. This value is considerably lower than that normally associated with peroxide or hydroperoxide decomposition [42], although it must be remembered that the branching process is second-order, probably heterogeneous and that it is possible that there may be transient formation of a compound in which the O—O bond strength is less than that observed in peracids. However, it must be admitted that this is one of the least satisfactory features of the present mechanistic interpretation of

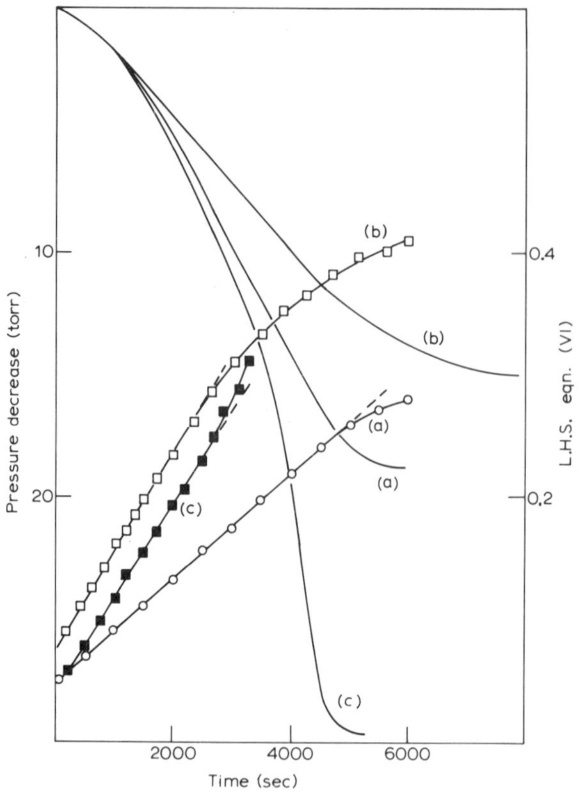

Fig. 3. Typical pressure—time curves and plots according to eqn. (VI) for aldehyde oxidation at 62.5 °C [43]. (a) 300 torr acetaldehyde + 18.8 torr O_2; (b) 100 torr propionaldehyde + 14 torr O_2; (c) 171 torr iso-butyraldehyde + 17.5 torr O_2.

the oxidation. A further embarrassment is the difference between the value for E_{5c} estimated above and the corresponding value (25 kcal . mole^{-1}) derived from the observed temperature dependence of the rate of peracetic acid induced pyrolysis of acetaldehyde [19] for the range 140—190 °C, for which (5c) is considered to be a likely initiation step★. The reason for this divergence is not obvious, and further investigation is needed, although it might be noted that the estimate of E_{5c} is rather sensitive to the values chosen for E_3 and E_a, and that E_a is usually based on pressure measurements.

The period of autocatalysis in the early stages of the reaction and also the studies of the retarded oxidation may also be used to enable comment

★ Combe et al. [19] obtained good agreement between the values for E_{5c} obtained from the oxidation and from the pyrolysis. This agreement is apparent rather than real since they made use of a value for E_3 of 3.5 kcal . mole^{-1} giving E_5 (oxidation) to be 22 kcal . mole^{-1}. Also, they assumed a value for the activation energy for CH_3 + $CH_3CHO \rightarrow CH_4 + CH_3CO$ of 9.0 kcal . mole^{-1} leading to an estimate for E_{5c} (pyrolysis) of 22 kcal . mole^{-1}.

to be passed on the various rate coefficients, particularly k_{1a} and k_{5c}. These approaches are outlined in Sects. 3.1.4 and 3.5.

3.1.3 *Termination*

The only investigations specifically of the termination process appear to be those of McDowell and Sharples [62]. By combining measurements of the rate of initiation in alcohol retarded systems with determinations of the chain length by the rotating sector method, a value of k_{4a} defined by

$$[d(RCO_3)/dt]_{4a} = -2k_{4a}(RCO_3)^2$$

of

$$2 \times k_{4a} = (8.93 \pm 4.2) \times 10^{10} \text{ l.mole}^{-1}\text{.sec}^{-1}$$

was obtained*, the relatively high value showing the process to be very efficient. Further investigations [80] using isotopically enriched oxygen made it possible to show that at room temperature diacetyl peroxide is formed during the oxidation of acetaldehyde, probably via a four-centre transition complex viz.

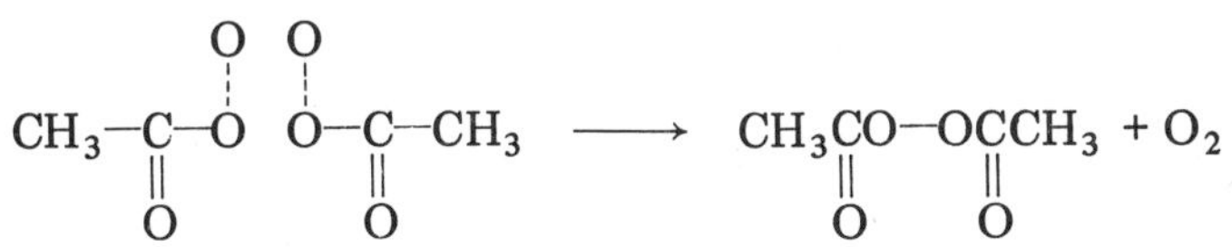

For recent work on the liquid phase bimolecular acetyl peroxy-radical termination see refs. 81 and 83.

Information on the other two possible termination steps is almost entirely speculative. However, in view of (*a*) the high rate coefficient of (2) relative to that of (3) and (*b*) the high collision efficiency of 4(a) it is apparent that (4b) and (4c) are likely to be of significance only when the O_2/fuel ratio is very much less than unity. On the convenient but not unreasonable assumption that $k_{4a} = k_{4b} = k_{4c}$, Griffiths and Skirrow [42] estimated that at 100 °C (4a) will account for less than 80 % of the total termination only when $O_2/\text{fuel} < 3 \times 10^{-1}$. Because of the high temperature coefficient of (3) relative to that of (2) this ratio will increase somewhat with temperature, and the dependence of the oxidation rate on the oxygen pressure noted at higher temperatures may conceivably be accounted for, at least in part, in this way. Reactions (4b) and (4c) may also be of importance in the latter stages of reactions made with the fuel in excess of the oxygen. Thus, if termination occurred exclusively by (4a), the oxidation rate would abruptly fall to zero at the moment of

* McDowell and Sharples definition of k_{4a} appears to be $[d(RCO_3)/dt]_{4a} = -k_{4a}(RCO_3)^2$.

total consumption of the oxygen. In practice, even below 100 °C such an abrupt cut-off is not shown (Fig. 3) and it is possible from an analysis of the shape of reaction—time curves in the final stages of oxidations at low temperatures to draw conclusions on the relative importance of (4a), (4b) and (4c) [43].

3.1.4 *Initiation*

The low rate of the initiation step and the progressive predominance over it of branching as the reaction develops make initiation a difficult subject for study. Information has been sought by (*i*) direct observation of ρ_i [15—19], the initial rate of oxidation, the results being interpreted in terms of reactions (1a), (2), (3) and (4a), (*ii*) observations of the effect on the reaction of retarding or inhibiting additives [63], (*iii*) measurement of the rate of hydrogen peroxide production in reactions proceeding at much higher temperatures where the HO_2 radical is reactive and assuming that the initiation process comprises (1a) followed by (1b) [82], (*iv*) computer matching of reaction—time curves with those obtained experimentally [57] and (*v*) analysis of the early portion of the pressure—time curve of low temperature reactions as the branching progressively takes over from the initiation step [43]. These approaches are discussed below.

(*i*) Despite the technical difficulties of measuring the initial rate when even traces of peracid must be absent for meaningful results to be obtained, Combe et al. [15—19] obtained ρ_i values at 123 °C. They considered that when the initial oxygen was in excess and the peracid concentration effectively zero, the essential steps in the initial oxidation reactions were (1a), (2), (3) and (4a). Combination of these lead to the initial rate expression

$$\rho_i = k_{1a}(\mathrm{RCHO})(\mathrm{O_2}) + k_3 \left[\frac{k_{1a}}{2k_{4a}}\right]^{1/2} (\mathrm{RCHO})^{3/2}(\mathrm{O_2})^{1/2} \qquad \text{(IV)}$$

which was consistent with their observations.

The overall activation energy for the initial rate was 15 kcal . mole^{-1}. This, when combined with the value for E_3 of ca. 7.0 kcal . mole^{-1} and an assumed value of zero for E_{4a} leads to a value for E_{1a} of ca. 16 kcal . mole^{-1}. The corresponding value of k_{1a} at 123 °C can be calculated to be 1.0×10^{-3} l . mole^{-1} . sec^{-1} using the k_3 and k_{4a} values for acetaldehyde given above in Sect. 3.1.1 and 3.1.3★.

(*ii*) A value of 3.57×10^{-3} l . mole^{-1} . sec^{-1} for k_{1a} based on the limiting rate observed in the alcohol retarded oxidation [63] is probably

---

★ Using "averaged" values for k_3 and k_{4a} (p. 396), k_{1a} at 123 °C is calculated to be ca. 7×10^{-5} l . mole^{-1} . sec^{-1}.

open to question since it seems likely that this rate embodies the rate of branching as well as that of initiation (see also Sect. 3.5.1).

(*iii*) Attempts to investigate the initiation process at 320 °C by Sokolova et al. [82] from measurements of the rate of hydrogen peroxide production in a flow system led to a value for E_{1a} of approximately 29 kcal . mole^{-1}. However, this value is based on the assumed mechanism (1a) and (1b) with termination by surface HO_2 destruction. Although this mechanism is consistent with the kinetics observed by these workers, it is not altogether appealing. In support of it, it must be recognized that their estimate of E_{1a} is a good deal closer to the expected 40 kcal . mole^{-1} for a homogeneous process (see Sect. 4.4) than is the value of 16 kcal . mole^{-1} obtained from Combes' work.

(*iv*) Griffiths et al. [72] considered that the best computer matching of experimental pressure—time curves and theoretical reaction—time curves (based on reactions (1a), (2), (3), (4a) and (5c)) was given when k_{1a}/k_{5c} was taken as ca. 1/50 k_{1a} was calculated to be 8.6 x 10^{-5} l . mole^{-1} sec^{-1} when "averaged" values for k_3 and k_{4a} were used (p. 396). Although the curve fitting is somewhat insensitive to the exact value chosen for k_{1a}, it is interesting to note the very large discrepancy between this estimate and that calculated by the methods above.

(*v*) It is possible to obtain information on the initiation step (and also on the branching process) from an analysis of the early stages of the pressure—time curve [43]. Provided that the temperature is sufficiently low to ensure that the stoichiometry approximates to

$$RCHO + O_2 \longrightarrow RCO_3H$$

then a mechanism comprising steps (1a), (2), (3), (4a) and (5c) leads to an instantaneous rate of peracid accumulation given by

$$\frac{d[RCO_3H]}{dt} = \frac{d\Delta P}{dt} = \frac{k_3[RCHO]}{(2k_{4a})^{1/2}} \times \{k_{1a}[RCHO][O_2] + k_{5c}[RCHO][RCHO_3H]\}^{1/2} \quad (V)$$

On replacing [RCHO] and $[O_2]$ by $(A)_0 - (P)_t$ and $(O_2)_0 - (P)_t$, respectively, integration of (V) leads to

$$\left[\frac{\dfrac{k_{1a}(O_2)_0}{(k_{5c}-k_{1a})} + (P)_t}{(A)_0 - (P)_t}\right]^{1/2} = \frac{k_3\,[k_{1a}(O_2) + (k_{5c}-k_{1a})(A)_0]}{2\sqrt{2}k_{4a}^{1/2}k_{5c}^{1/2}} \cdot t + \left[\frac{k_{1a}(O_2)_0}{(k_{5c}-k_{1a})(A)_0}\right]^{1/2} \quad (VI)$$

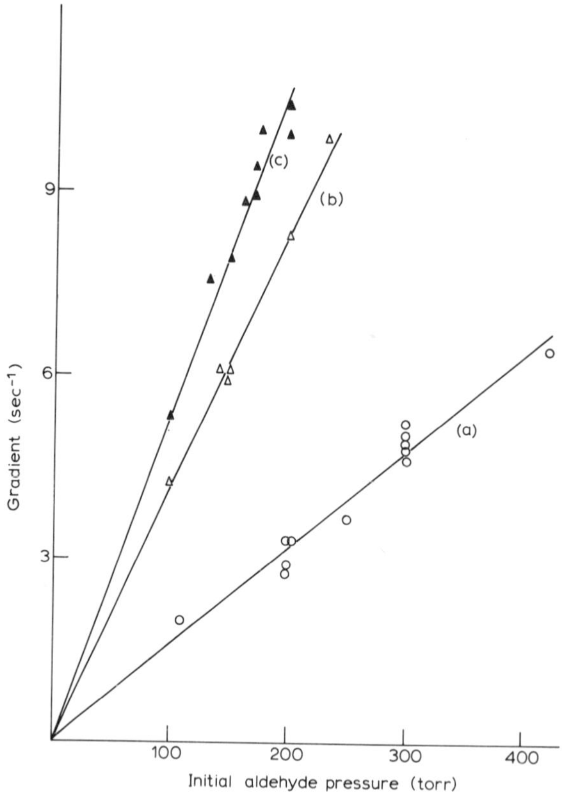

Fig. 4. Variation of the gradient of the eqn. (VI) plots with initial aldehyde pressure at 62.5 °C [43]. (a) Acetaldehyde; (b) propionaldehyde (abs. × 2); (c) iso-butyraldehyde.

where $(A)_0$ and $(O_2)_0$ are the initial aldehyde and oxygen pressures and $(P)_t$ is the instantaneous peracid pressure at time t. Plots of the L.H.S. of

TABLE 5

Values for the rate coefficient of the initiation step in aldehyde low temperature oxidation

Ref.	Temp. (°C)	S/V (cm^{-1})	Rate coefficient (k_{1a}) ($l . mole^{-1} . sec^{-1}$)	Method
15—19	123	ca. 0.6	ca. 7×10^{-5} [a]	Initial rates
63	123	ca. —0.8	3.57×10^{-3}	Alcohol retardation
57	120	ca. 1.2	8.6×10^{-5} [a]	Computer-fitting
43	119	0.6	6.15×10^{-5} [a]	Shape of pressure—time curve

[a] Values of k_3 and k_{4a} averaged over the published values for acetaldehyde and propionaldehyde have been used in the calculation of these values to effect a fair comparison.

(VI) against t give good straight lines (Fig. 3) provided suitable values for $k_{1a}/(k_{5c} - k_{1a})$ are chosen. From the gradients of such plots, values of $(k_{5c} - k_{1a})$ can be obtained if use is made of the k_3 and k_{4a} values discussed earlier and if it is assumed that $k_{1a}(O_2) \ll (k_{5c} - k_{1a})$. Figure 4 shows that the gradients are, in fact, proportional to $(A)_0$.

Results obtained using this procedure confirm that rate determining steps for initiation and branching are, respectively, first-order with respect to both aldehyde and oxygen and first-order with respect to both aldehyde and peracid, and indicate that the initiation is heterogeneous. With an acid-washed pyrex vessel below ca. 120 °C branching is also heterogeneous, marked dependences on S/V ratio being noted for both k_{1a} and k_{5c}. At 120 °C the ratio k_{1a}/k_{5c} is 1/72 reflecting the marked autocatalysis of this system. Taking k_{5c} at this temperature to be 4.23×10^{-3} l . mole^{-1} . sec^{-1}, k_{1a} is calculated* to be ca. 6.15×10^{-5} l . mole^{-1} . sec^{-1}. This is in reasonable agreement with the values derived from initial rate studies and computer matching (Table 5).

Thus there are compelling reasons for regarding the initiation step (1a) as a low activation energy heterogeneous process at low temperatures. In any case from estimates of k_{1a} (homogeneous) for acetaldehyde and propionaldehyde made by Baldwin et al. [21, 22] a value for E_{1a} for 40 kcal . mole^{-1} was calculated. Although this may represent an upper limit (see also Sect. 4.4), it is obvious that the rate of (1a) in the gas phase below 200 °C is far too low to allow the oxidation to get started.

3.1.5 *Further comments on the low temperature oxidation*

Although the scheme comprising (1a), (2), (3), (4a) and (5c) is consistent with the main kinetic features below ca. 150 °C, a number of details remain to be resolved. Thus, although initiation is first-order with respect to both the aldehyde and oxygen, it is unlikely that the overall chemistry of the interaction of aldehyde and oxygen on the surface corresponds to eqn. (1a). The results of a recent examination [43] of the minor products formed in the early stages is shown in Fig. 5. for instance, CO and CH_3OH are formed at relatively high rates in the early stages, and it appears that most of the minor products are formed by heterogeneous processes not involving radicals and irrelevant to the initiation step. However, with present knowledge, detailed comment is speculative.

Reaction (5c) is to be regarded as an over-simplification of the branching process, although there seems little doubt that this is the correct kinetic description. Information was sought by following the yields of products after the consumption of the oxygen for a reaction in which the initial concentration of oxygen was considerably less than that of the aldehyde [43]. Under these circumstances the residual reaction is

---

* Using "averaged" values for k_3 and k_{4a} (p. 396).

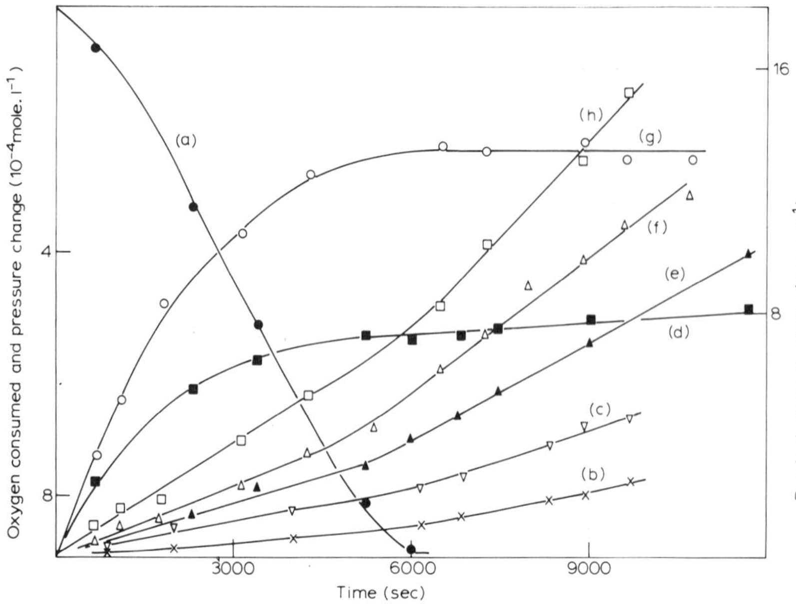

Fig. 5. Oxygen consumed ● and products formed at 71.5 °C [43]. 250 torr aldehyde + 19.5 torr O_2. (a) Pressure decrease; (b) 10^6 C_2H_6; (c) 2 x 10^5 CO_2; (d) 10^6 CO; (e) 10^6 CH_4; (f) 10^5 H_2O; (g) 10^7 CH_3OH; (h) 5 x 10^6 $(CH_3)_2CO$.

essentially the interaction of peracetic acid and acetaldehyde. These results are also illustrated in Fig. 5. The final rate of water generation (R_{H_2O} = 1.5 x 10^{-8} mole . l^{-1}. sec^{-1}) is about 8.1 times the rate of (5c) ($k_{5c} A_\infty P_\infty$) where P_∞ and A_∞ are the final peracid and aldehyde concentrations, respectively, and k_{5c} is calculated from the oxidation kinetics using "averaged" values (p. 396) for k_3 and k_{4a}. Furthermore, experiments using a packed vessel (for which k_{5c} is increased about 4 times) give the constant of proportionality as about 8.2 and hence it seems that water is produced at a rate proportional to (although greatly exceeding) the rate of branching. It was, in fact, found experimentally that R_{H_2O} was proportional to $A_\infty P_\infty$ using the packed vessel. The authors consider that the reactions of peracid and aldehyde are complex and probably include concerted steps [43]. It is outside the scope of this review to discuss these in detail but it can be pointed out that a direct peracid — aldehyde reaction in the gas phase is ruled out.

Other mechanisms can be devised which explain at lease some of the features of the low temperature oxidation. However, generally, these alternatives have serious limitations. For example, one might consider a scheme comprising (1a), (2), (3), (4a) and (5d)

$$RCO_3 + RCO_3H \longrightarrow \text{branching} \qquad (5d)$$

The replacement of (5c) by (5d) has been chosen so as to recognize that as the peracid concentration becomes high attack on it by the radical present in the highest concentration is possible. For the region where the rate of branching is much greater than that of initiation this scheme leads to the instantaneous rate expression

$$\rho = \frac{k_3 k_{5d}}{2k_{4a}} [RCO_3H] [RCHO] \qquad (VII)$$

which, by employing the usual procedure gives a maximum rate expression

$$\rho_{max} = \frac{k_3 k_{5d}}{8k_{4a}} [A]_0^2 \qquad (VIII)$$

where $[A]_0$ is the initial aldehyde concentration. This scheme, although indicating a dependence on the initial aldehyde concentration consistent with that observed, must be rejected since the instantaneous rate expression (VII) to which it leads implies a first-order dependence on both aldehyde and peracid. This is at variance with experimental fact, ρ in the range 50—120 °C being more consistent with branching by (5c) (half-order with respect to peracid and 3/2-order with respect to aldehyde). Furthermore, the scheme including (5d) predicts the maximum rate to be attained at about 50 % reaction instead of the observed 25 % reaction predicted by the more aceptable scheme in which branching is by (5c).

3.2 OTHER SATURATED ALDEHYDES

3.2.1 Propionaldehyde

The determinations by McDowell and Sharples [62] of the rate coefficients for k_3 and k_{4a} for the propionaldehyde—oxygen system can be summarized by

$$k_3(C_2H_5CHO) = 5.2 \times 10^9 \exp(-6{,}800/RT) \text{ l. mole}^{-1}\text{. sec}^{-1}$$

$$2 \times k_{4a}(C_2H_5CHO) = 2.69[\pm 1.35] \times 10^{10} \text{ l. mole}^{-1}\text{. sec}^{-1}$$

These values differ somewhat from the corresponding determinations for acetaldehyde [62]. However, the oxidation rates of the two aldehydes are similar (see Fig. 1), suggesting that either there is a fortuitous combination of k_3, k_{4a} and k_{5c} values for the two systems or, as seems more likely, the difference between the propagation and termination coefficients for the two systems is less than the experimental determinations would suggest.

Figure 22 (p. 412) shows that at low temperatures the oxidation rate for propionaldehyde does not retain its oxygen independence to such low

TABLE 6
The limiting high pressure decomposition of acyl radicals (RCO)

Ref.	R	Temp. range (°C)	$\log_{10} A$ (sec^{-1})	E ($kcal . mole^{-1}$)
85	CH_3	150—240	10.3	15.00
86	C_2H_5	100—175	12.47	11.10
86	n-C_3H_7	80—150	12.50	9.52
86	iso-C_3H_7	80—150	13.04	9.75
87	n-C_4H_9	100—175	13.02	10.25
87	iso-C_4H_9	80—150	12.14	9.69
92	C_6H_5	341—394	14.6	29.4

oxygen pressures as does that of acetaldehyde under comparable conditions. This probably reflects the relative ease of decomposition of the CH_3CO and C_2H_5CO radicals. The limiting high pressure rate coefficients for the decompositions of these and other acyl radicals are given in Table 6.

In other respects, and particularly at lower temperatures, the oxidation of propionaldehyde closely resembles that of acetaldehyde. In particular, Fig. 3 shows that, in the early stages, the course of the reaction is in agreement with eqn. (VI) (p. 383), k_{1a}/k_{5c} being 1/31 at 62.5 °C [43].

3.2.2 *n- and* iso-*Butyraldehydes*

There is little published work on the oxidation of the butyraldehydes, most of what is available being concerned with oxidation in the high temperature region.

It might be expected from the rate coefficients in Table 6 that RCO decomposition would occur more readily with both n- and iso-C_3H_7CO [86] than with either CH_3CO [42, 85] or C_2H_5CO [86]. As a result, the onset of oxygen dependence of the butyraldehyde oxidation rate should occur at lower temperatures and higher oxygen pressures than for the lower aldehydes. Recent unpublished work [84] supports this expectation, and the oxygen dependences of ρ_{max} shown in Fig. 6 can be compared with those of acetaldehyde and propionaldehyde shown in Fig. 22. Nevertheless, at low temperatures eqn. (VI) again represents the early course of oxidation (Fig. 3) [43].

From Table 6 it is seen that valeryl radical decomposition [87] occurs readily. Consequently, the oxidation of valeryl aldehydes at low temperatures would be expected to have a strongly oxygen dependent rate.

3.3 AROMATIC ALDEHYDES

There appears to have been no attention given to the low temperature gas phase uncatalysed oxidation of benzaldehyde, although various studies

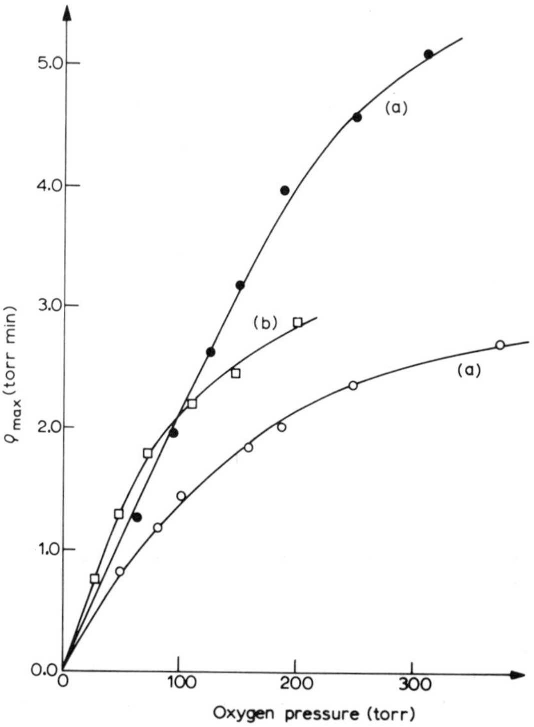

Fig. 6. Variation of the maximum oxidation rate with initial oxygen pressure for *n*- and iso-butyraldehydes [84]. Aldehyde pressure, 100 torr. (a) *n*-Butyraldehyde; ○, 124 °C, ●, 149 °C. (b) iso-butyraldehyde, 124 °C.

of the liquid phase oxidation have been reported [88--91]. This system would be expected to give rise to some experimental difficulty because of the restrictions imposed by the relatively low vapour pressures of the aldehyde and the peracid. The high dissociation energy of the C_6H_5CO radical [92] (Table 6: 29.4 kcal . mole^{-1}) means that oxygen dependence arising from the onset of (6a) or (6b) will be unlikely except at relatively high temperature.

3.4 UNSATURATED ALDEHYDES

Few investigations of the gas phase oxidation of unsaturated aldehydes in the low temperature region have been reported. However, an analytical and kinetic examination of the combustion of crotonaldehyde [48] at 166 °C — probably somewhat above the low temperature region as understood here — suggests that up to this temperature the oxidation is not dissimilar to that of acetaldehyde in the low temperature region. Reaction is accompanied by a pressure decrease and the accumulation of

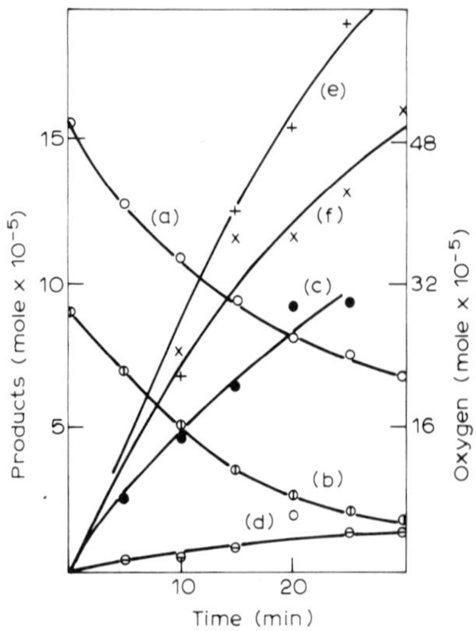

Fig. 7. Analysis throughout the course of the oxidation of crotonaldehyde at 166 °C [48]. (a) O_2; (b) ΔP; (c) CO_2; (d) CH_3CHO; (e) acid; (f) peroxide.

peroxide (percrotonic acid) together with smaller amounts of CO, CO_2 and acetaldehyde (Fig. 7). The rate appears to be oxygen independent and proportional to the square of the aldehyde pressure, although an inert gas effect (accelerating) complicates the kinetic picture. A more detailed examination at temperatures below 166 °C is obviously required, but from what evidence is at present available it seems likely that the low temperature mechanism is not very different from that proposed for the saturated aldehydes, branching proceeding via a peracid—aldehyde reaction.

3.5 EFFECT OF ADDITIVES

Many additives, e.g. N_2, CO_2, H_2O [45], have little or no effect on the low temperature oxidation rate. Others may promote reaction or give rise to retardation or, possibly, inhibition. Promotion or acceleration is usually associated with additives which are themselves directly or indirectly radical sources at the temperature of the system (e.g. ditertiary butyl peroxide [58], peracetic acid [19], HBr [59]), and the effect is understandable in terms of an increased (induced) rate of initiation. The most important additive in this category is peracetic acid. This is a product in the oxidation of acetaldehyde, and the effect of its addition on the oxidation kinetics has been used by Combe et al. [19] to obtain supporting evidence for the now accepted branching step.

Of greater current interest is the effect of those additives which retard

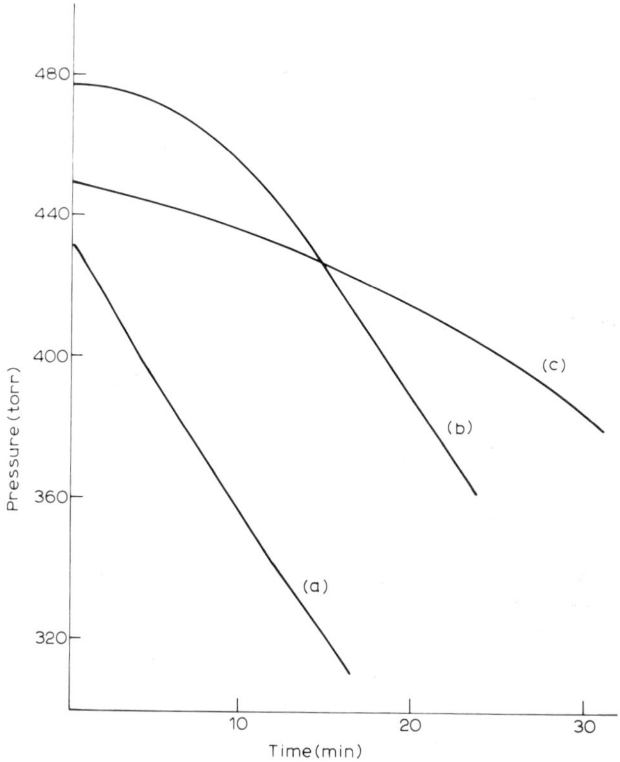

Fig. 8. Pressure—time curves for the retardation of acetaldehyde oxidation by ethanol and iso-propanol at 123 °C [63]. (a) Unretarded; (b) ethanol added; (c) iso-propanol added. No concentrations given.

or inhibit reaction. It must be recognized at the outset that, in general, the mode of action is not fully understood, and it seems likely that no unique mechanism exists. However, the relative simplicity of the low temperature acetaldehyde system means that it is potentially a useful subject for experimental and computer studies of retarder action.

The distinction between retarders and inhibitors is difficult to make with precision. Retarders give no induction period, their presence merely causing a reduction in rate. Inhibitors give rise to an induction period, although the initial period of no detectable reaction is subjective, the limit of detectability depending on the sensitivity of the measuring equipment. The mechanistic distinction usually assumed is that inhibitors interfere with the normal process of initiation; retarders interfere only with the propagation steps.

Published results for acetaldehyde indicate two extreme forms of behaviour seen, for example, by comparing Figs. 8 and 9 with Figs. 10 and 11, and for convenience of discussion the term inhibitor is retained.

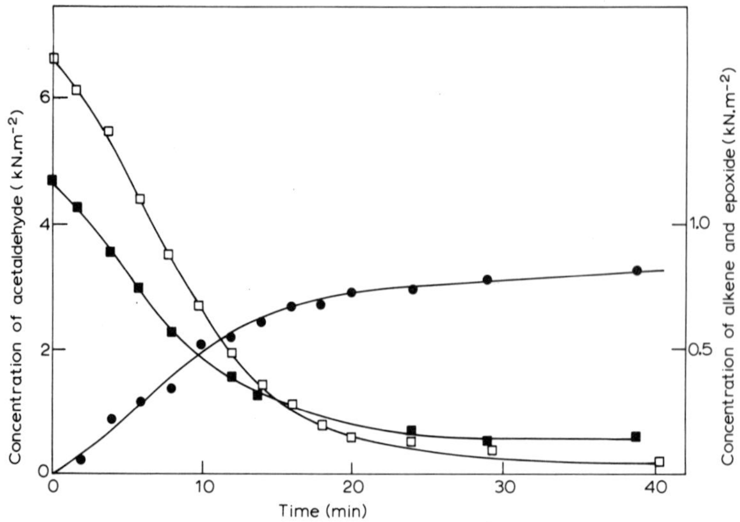

Fig. 9. Retardation of the oxidation of acetaldehyde by *cis*-butene-2 at 184 °C [69, 71]. Oxygen 6.7 KN m^{-2}; acetaldehyde 6.7 KN m^{-2}; *cis*-butene-2 1.17 KN m^{-2}. □, acetaldehyde; ■, *cis*-butene-2; ●, epoxide.

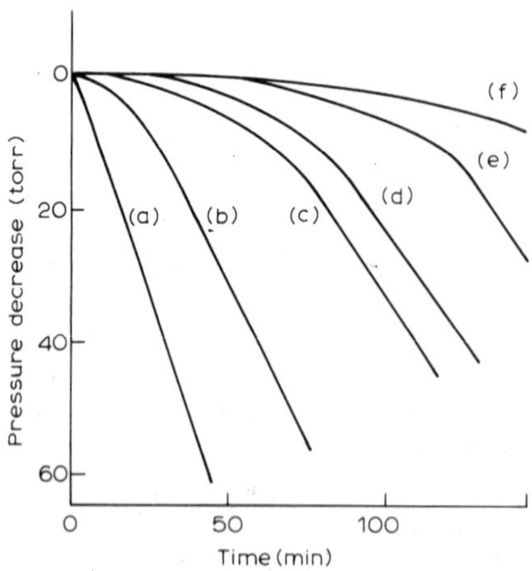

Fig. 10. The influence of primary amines on the oxidation of acetaldehyde at 124 °C [69, 71]. Acetaldehyde pressure 100 torr; oxygen pressure 100 torr. (a) No amine added; (b) 0.95 torr methylamine; (c) 1.79 torr ethylamine; (d) 2.27 torr *n*-butylamine; (e) 2.56 torr *tert*-butylamine; (f) 2.22 torr iso-propylamine.

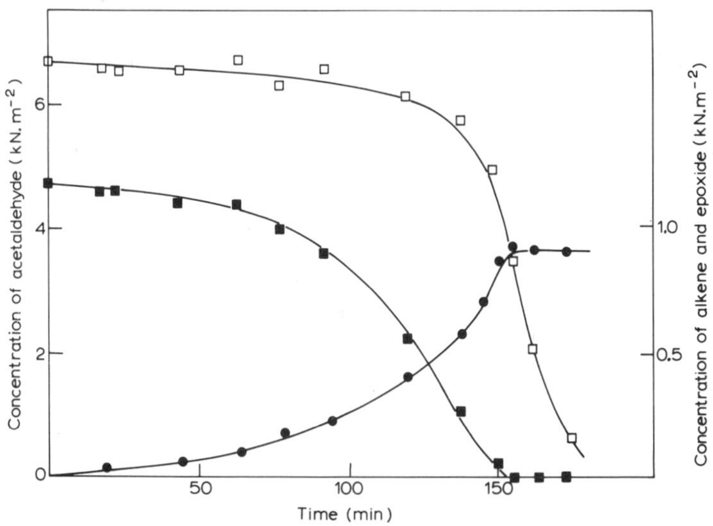

Fig. 11. The effect of 2,3-dimethylbutene-2 addition on the oxidation of acetaldehyde at 184 °C [69, 71]. Oxygen 6.7 KN m^{-2}; acetaldehyde 6.7 KN m^{-2}; 2,3-dimethylbutene-2 1.17 KN m^{-2}. □ acetaldehyde; ■ 2,3-dimethylbutene-2; ● epoxide.

3.5.1 Retarders

For those additives which are purely retarders it is usually found that relationships of the form (XH = retarder)

$$\rho_{max} = K + 1/[XH] \tag{IX}$$

as shown in Fig. 12 (although Farmer and McDowell [63] apparently measured initial rates), or

$$[XH] = K' + 1/\rho_{max} \tag{X}$$

as shown in Fig. 13 apply although sometimes the plot is linearly only over a limited concentration range (Fig. 14). It should be noted that except when K and K' are very small, (IX) and (X) are not strictly mathematically equivalent. It will be seen later that the first of these expressions is the more soundly based theoretically.

For an unretarded reaction, the rate of loss of aldehyde is given by

$$-d[RCHO]/dt = k_{1a}[RCHO][O_2] + k_{5c}[RCHO][RCO_3H] + k_3[RCO_3][RCHO] \tag{XI}$$

and it is evident that retarder action is a consequence of a reduced RCO_3 concentration caused by the presence of the additive. It is reasonable to

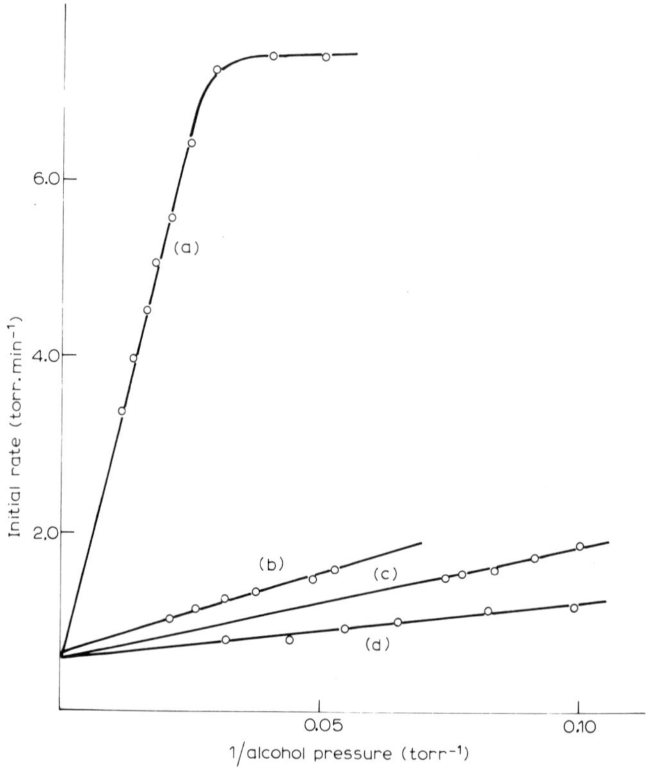

Fig. 12. The relationship between the initial rate of the alcohol-retarded acetaldehyde oxidation and the reciprocal of the alcohol pressure at 123 °C [63]. (a) Methanol; (b) ethanol; (c) *n*-propanol; (d) iso-propanol.

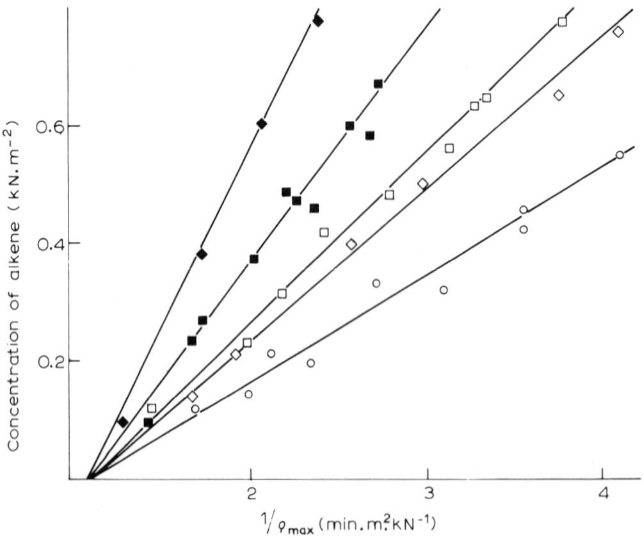

Fig. 13. The effect of the addition of terminal alkenes on the oxidation of acetaldehyde at 184 °C [69, 71]. Acetaldehyde 6.7 KN m^{-2}; oxygen 6.7 KN m^{-2}. ◆, ethylene; ■, propene; ■, butene-1; ○, 2-methylpropene; □, 3,4-epoxybutene-1.

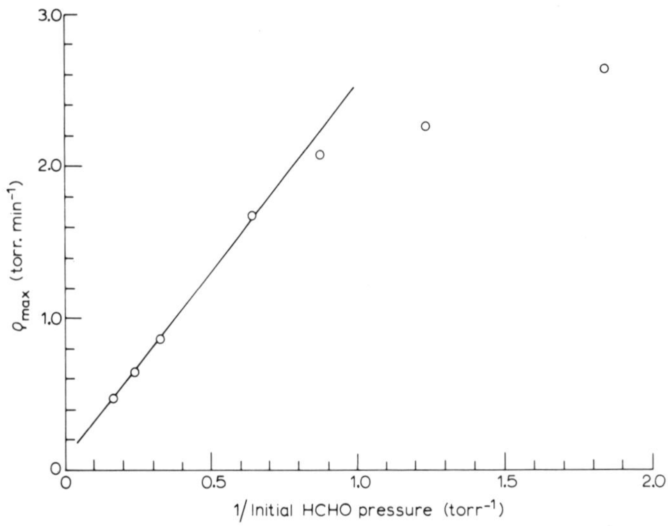

Fig. 14. The relationship between the maximum rate and the reciprocal of the formaldehyde pressure for acetaldehyde oxidation at 188 °C [43]. Initial acetaldehyde pressure = initial oxygen pressure = 40 torr.

suppose that effective retarders should be capable of undergoing reaction with RCO_3 in competition with (3) in such a way as to replace RCO_3 with a species less likely to continue the chain and thereby to lead to termination. Such additives may function by virtue of possessing an easily abstracted hydrogen when the relevant step is

$$RCO_3 + XH \longrightarrow RCO_3H + X \qquad (7)$$

For this class of retarders it might be expected that, for a homologous series, the effectiveness would increase with increased ease of removal of the abstracted hydrogen atom, provided that the fragments X do not differ much in their ability to continue chains. Figure 12 shows that this expectation is realized for the aliphatic alcohols, the effectiveness increasing in the order $CH_3OH < C_2H_5OH < n\text{-}C_3H_7OH < \text{iso-}C_3H_7OH$ which is the same as that for ease of removal of H atoms by abstraction.

An alternative mode of retardation when certain alkenes are used as additives (Fig. 13) has been suggested by Waddington [69, 71]. He considered that they owe their retarding effect (investigated at 183 °C, but presumably also applying at lower temperatures) to the formation of an alkene—acetylperoxy adduct, viz.

$$CH_3CO_3\cdot + \,>C=C<\, \longrightarrow CH_3CO_3\text{—}\overset{|}{\underset{|}{C}}\text{—}\overset{|}{\underset{|}{C}}\cdot \qquad (8)$$

which may either decompose according to the reverse of (8)

$$CH_3CO_3\text{—}\overset{|}{\underset{|}{C}}\text{—}\overset{|}{\underset{|}{C}}\cdot \longrightarrow CH_3CO_3\cdot + \,>C=C<\, \qquad (-8)$$

or react to give epoxides (which were detected and characterized) and a methyl radical which is less reactive than the original peracid radical

$$CH_3CO_3{-}\overset{|}{\underset{|}{C}}{-}\overset{|}{\underset{|}{C}}\cdot \longrightarrow CH_3\cdot + CO_2 + \;>\!\overset{\;O}{C{-}C}\!< \qquad (9)$$

When the rate coefficient for the attack of CH_3CO_3 radicals on the additive is comparable with that for attack on the aldehyde, quite small additions of the retarder may result in a dramatic reduction in the stationary concentration of RCO_3. Thus, in the region of the maximum rate when the contribution of initiation can be ignored, combination of (1a), (2), (3), (4a) and (5c) with (7) leads to

$$[RCO_3] = \frac{[(k_7XH)^2 + 8k_{4a}(k_1\,A\cdot O_2 + k_{5c}P\,.\,A)]^{1/2} - k_7XH}{4k_{4a}} \qquad (XII)$$

where the instantaneous concentrations of RCHO, O_2, RCO_3H, and retarder are designated by A, O_2, P and XH, respectively. In the derivation of (XII) it is assumed that X does not react to generate further chains.

Of course, the effect of a retarder on the rate will depend on the chain length of the unretarded oxidation; if the chains are only short, then the effect will be minimal. By way of illustration, the effect is considered of retarder addition on the stationary $RCO_3\cdot$ concentration for oxidations with different chain lengths but a common value for the unretarded maximum rate (assumed to occur at $P_t = A_i/4$). Three sets of values for k_3 and k_{4a} (and a deduced value for k_{5c} using, for example, expression (III) p. 377) are used. The first two combinations are those using the literature values of k_3 and k_{4a} for the acetaldehyde and propionaldehyde oxidations; the third uses the "averaged" values★ of k_3 and k_{4a} which have been suggested [43] to more fairly describe both the oxidation of acetaldehyde and propionaldehyde. For the oxidation at ca. 120 °C the rate coefficients for the three combinations were calculated [43] for a common maximum rate. Figure 15 shows a plot of $[RCO_3\cdot]$ versus k_7XH using expression (XII) above and ignoring the initiation contribution; clearly, the three curves are very different. The calculated unretarded chain lengths at the maximum rate for the three combinations (using expression (39) of the earlier review) are 5.8 (CH_3CHO values), 415 (C_2H_5CHO values) and 71.4 (averaged values). From Fig. 15 it can be seen, as anticipated, that the higher the chain length the more marked is the effect of the retarder. The authors [9, 43] consider on the basis of

★ Average $k_3 = \{k_3(CH_3CHO,\ T\ ^\circ K) + k_3(C_2H_5CHO,\ T\ ^\circ K)\}/2$; average $k_{4a} = \{k_{4a}(CH_3CHO) + k_{4a}(C_2H_5CHO)\}/2$.

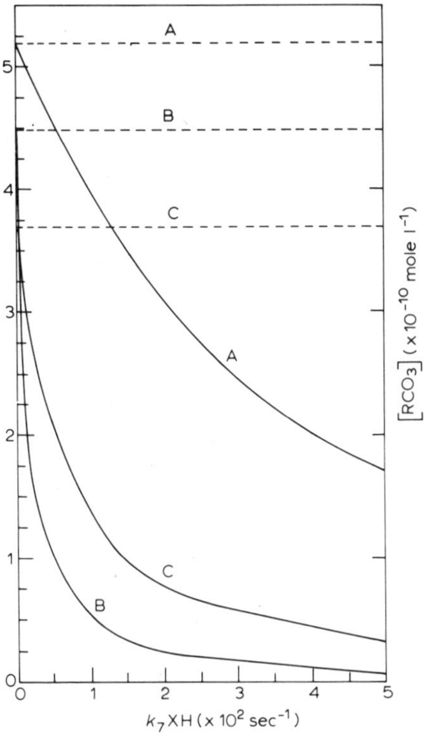

Fig. 15. The effect of retarder addition on the instantaneous $[RCO_3\cdot]$ concentration at 25 % reaction for oxidations with different chain lengths but a common value for the unretarded maximum rate [9]. A, CH_3CHO values for k_3 and k_{4a}; ordinate 4 x scale; $[RCO_3\cdot]$ for XH = 0 is 2.08×10^{-9} mole . l^{-1}. B, C_2H_5CHO values for k_3 and k_{4a}; ordinate = scale $[RCO_3\cdot]$ for XH = 0 is 4.49×10^{-10} mole . l^{-1}. C, "Average" values for k_3 and k_{4a}; ordinate 2 x scale; $[RCO_3\cdot]$ for XH = 0 is 7.39×10^{-10} mole . l^{-1}.

this and other evidence that the averaged values are certainly more satisfactory.

Substitution of (XII) into (XI) gives

$$-\frac{dA}{dt} = k_{1a}A\cdot O_2 + k_{5c}P\cdot A + \frac{k_3 A}{4k_{4a}}\{[(k_7XH)^2 + 8k_{4a}(k_{1a}A\cdot O_2 + k_{5c}P\cdot A]^{1/2} - k_7XH\} \qquad \text{(XIII)}$$

which, for "efficient" retardation [9], i.e. $(k_7XH)^2 \gg 8k_{4a}(k_{1a}A\cdot O_2 + k_{5c}P\cdot A)$, reduces to

$$= k_{1a}A\cdot O_2 + k_{5c}P\cdot A + \frac{k_3A(k_{1a}A\cdot O_2 + k_{5c}P\cdot A)}{k_7XH} \qquad \text{(XIV)}$$

Expression (XIV) can also be derived directly from the above scheme if termination by (4a) is ignored. Further manipulation of this equation in order to obtain an expression for the maximum rate in terms of the initial retarder and reactant concentrations is not possible without making rather severe approximations, and the problem is probably best approached by numerical methods. However, it is possible to show* that in the special case of $k_3 = k_7$, ρ_{max} is attained about half-way through the reaction and that there should be a linear relationship between the maximum rate and the reciprocal of the initial retarder concentration. Plots of ρ_{max} against $1/XH_0$ should intercept the rate axis at a value equal to $k_1\{2(O_2)_0 - A_0\} + k_{5c}(A_0^2/4)$ and not, as is sometimes supposed, at $k_{1a}A{\cdot}O_2$. Experiments designed specifically to test this point do not seem to have been made.

In fact, a recent study [9] of the effect of formaldehyde on the oxidation of acetaldehyde between about 120 and 190 °C has shown that the maximum rate is not a good rate parameter for the retarded reactions and the rate at 25 % reaction ($\rho_{0.25}$) was used. The effect of formaldehyde on $\rho_{0.25}$ is shown in Fig. 16. Small amounts of CH_2O markedly retard the oxidation. At 188 °C, the plots of $\rho_{0.25}$ versus the reciprocal of the initial formaldehyde concentration (F_i) are linear (the gradients being proportional to the cube of the initial acetaldehyde concentration),

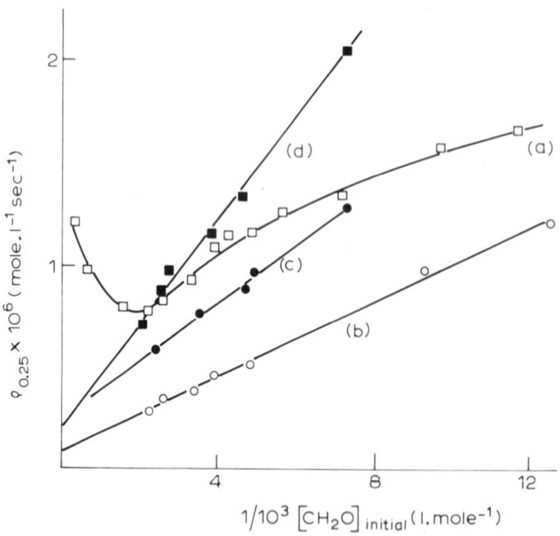

Fig. 16. Effect of formaldehyde on the rate of acetaldehyde oxidation [9]. (a) 200 torr CH_3CHO + 200 torr O_2, 119 °C. (b) (abs. x2) 40 torr CH_3CHO + 40 torr O_2, 188 °C; (c) (ord. x0.5) 50 torr CH_3CHO + 500 torr O_2, 188 °C; (d) 73 torr CH_3CHO + 73 torr O_2, 188 °C.

---

* By assuming that $k_{5c}P{\cdot}A > k_{1a}A{\cdot}O_2$ and substituting $A = A_0 - P$ and $XH = XH_0(A/A_0) = XH(A_{00} - P)/A_0$ in (XIV). The differential of the resultant expression with respect to P is equated to zero, and from this an expression for P in terms of A_0 is obtained which is substituted into (XIV).

TABLE 7

The standard heats of formation for various species [93]

Species	ΔH_f° (kcal . mole^{-1})
O_2	0
CO	26.4
HO_2	5
HCO_3H	−64.8[a]
HCO_3	−26.8[b]
HCO	8.2
CH_3CO_3H	−78.8
CH_3CO_3	−40.8
CH_3CO	−6
CH_3O_2	6

[a] Estimated by using group additivity [94—96].
[b] D(HCO_3—H) taken as 90 kcal . mole^{-1} [97].

whereas at 119 °C the plot is curved and shows a minimum. Analysis for formaldehyde showed that it was removed by the reaction

$$CH_3CO_3 + HCHO \longrightarrow CH_3CO_3H + HCO \tag{10}$$

k_{10} being about $2.4k_3$ at both 119 and 188 °C. The kinetic data at 188 °C are in agreement with a mechanism consisting of reactions (2), (3) and (5c) with termination by (10) instead of (4a). Probably the formyl radicals react with oxygen to give HO_2 radicals, viz.

$$HCO + O_2 \longrightarrow HO_2 + CO \tag{11}$$

which do not propagate the chain. However, at 119 °C reaction (4a) probably contributes to termination and reaction (11) is less important. Most of the formyl radicals add to oxygen at this temperature, viz.

$$HCO + O_2 \longrightarrow HCO_3 \tag{12}$$

and the HCO_3 radicals may abstract hydrogen from CH_3CHO giving performic acid, cf. reaction (3). With relatively large amounts of added formaldehyde it is suggested that the HCO_3H produced can react with acetaldehyde giving radicals, cf. reaction (5c). This extra mode of branching could account for the minimum $\rho_{0.25}$ versus $1/F_i$ plot.

The occurrence of reaction (11) may be, at least partly, the reason why the normal gas phase oxidation of formaldehyde is very slow below about 250 °C, since HO_2 radicals will propagate chains by hydrogen abstraction from HCHO only at temperatures considerably higher than those at which reaction (3) is fast, cf. activation energies (pp. 377 and 407). From the values of the standard heats of formation given in Table 7, the enthalpy changes ($\Delta H°$) for reactions (11) and (12) are calculated to be −29.6 and −35.0 kcal . mole^{-1}, respectively. Both reactions are thus very exothermic as are the corresponding reactions of the acetyl radical, viz.

$$CH_3CO + O_2 \begin{cases} \nearrow CH_3O_2 + CO, & \Delta H° = -26.4 \text{ kcal . mole}^{-1} \\ \searrow CH_3CO_3, & \Delta H° = -34.8 \text{ kcal . mole}^{-1} \end{cases}$$

It might, therefore, be expected that the activation energy of reaction (11) would be small. Presumably CO is not formed from $CH_3CO + O_2$ due to a very unfavourable entropy of activation.

Other retarders may also give fragments capable of reacting further. Possible reactions of the fragments produced during the alcohol retardation of the acetaldehyde/oxygen system have been discussed by Farmer and McDowell [63]. Their suggestion that the methyl, ethyl and *n*-propyl alcohol fragments dimerize whereas the iso-propyl alcohol radicals are lost by a disproportionation is not appealing in view of the relatively high rate of their expected reaction with oxygen. Unpublished results [64] from a mass spectrometric examination of the acetaldehyde oxidation retarded by iso-propyl and *sec*-butyl alcohols showed, respectively, acetone and methyl ethyl ketone to be formed, suggesting that reactions such as

$$(CH_3)_2\dot{C}OH + O_2 \longrightarrow (CH_3)_2CO + HO_2\cdot \tag{13}$$

had occurred.

Alkanes might be expected to give rise to alkyl radicals which would react further, either by peroxidation (possibly leading to some chain continuation), or by the formation of an olefin, e.g.

$$C_3H_7 + O_2 \longrightarrow C_3H_6 + HO_2 \tag{14}$$

The olefin so produced would also be expected to contribute to retardation. Figure 17 illustrates in a simplified form the effect of retarders which act through hydrogen abstraction reactions.

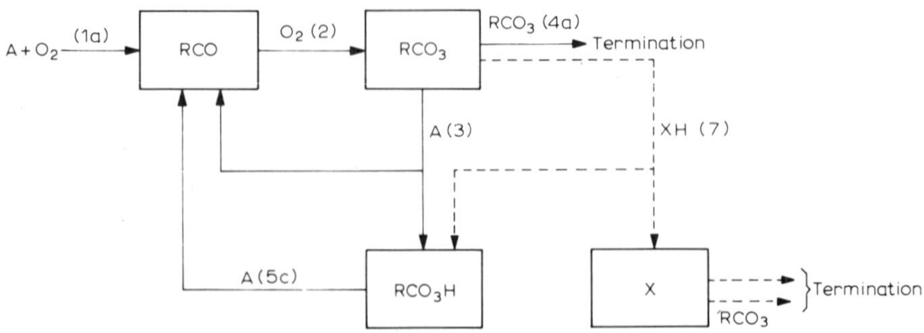

Fig. 17. Block diagram for the retarded gas phase oxidation of acetaldehyde and propionaldehyde at about 120 °C.

Convincing evidence has been given by Waddington for the formation of epoxides during the oxidation of acetaldehyde in the presence of alkenes [69, 71]. In general, epoxide formation was almost quantitative. Figure 9, for example, shows the loss of alkene and formation of epoxide for retardation by the addition of *cis*-butene-2 at 184 °C.

3.5.2 *Inhibitors*

Induction periods or abnormally slow initial stages of reaction characterize the oxidation in the presence of amines [65—69] and some alkenes [69, 71]. Present understanding of inhibitor action is even less clear than is that of retardation, although a number of comments can be made. It is unlikely that these additives function simply by removal of RCO_3 radicals since, unless the rate coefficient for the process were abnormally high compared with that of (3), only retardation would result. Removal of RCO radical by a process competitive with (2) is also unlikely for similar reasons.

The oxidation reaction will fail to develop only if the additive either prevents the generation of peracid, or reacts with and rapidly destroys any peracid that is produced. The initial generation of peracid might be prevented if the initiation process itself is interfered with by, for example, adsorption of the additive on active wall sites which are thereby blocked until the inhibitor is destroyed by an oxidative non-chain-generating process. The reaction will also be abnormally slow to start if the additive is able to react with RCO_3 efficiently in a way which differs from (7) in that no peracid is produced. Alternatively, the additive may react rapidly with any peracid produced thereby destroying it in a non-chain-producing way. At the present time it is not possible to assign described inhibitor results to any specific scheme, although the very marked effect of the addition of small amounts of amines suggests the first alternative. However, Waddington and co-workers [135] have recently reinvestigated the effect of aliphatic amines and have shown that some abstraction of hydrogen atoms from the additives by CH_3CO_3 radicals occurs.

4. Intermediate and high temperature oxidation

For convenience of discussion, this section deals with those aldehyde oxidation phenomena occurring at temperatures above those considered in Sect. 3 apart from cool flame and ignition processes which are examined in Sect. 5. It should be re-emphasized that the oxidation of aldehydes is relatively straightforward only when the temperature is very low (Sect. 3) or very high (>400 °C). In the intermediate region the simultaneous occurrence of reactions characteristic of both these extreme regions complicates the overall mechanism.

An ignition limit diagram for acetaldehyde is shown in Fig. 18. It is seen that the range of reactant pressures which can be used to study the slow oxidation decreases considerably at higher temperatures. Thus, whereas at 150 °C pressures of several hundred torr can be used without problems arising from self-heating or ignition, at 250 °C a pressure of 50/ torr of each reactant leads to cool-flame formation. Above 400 °C the fuel pressure must be no more than a few torr if self-heating and inconveniently high reaction rates are to be avoided.

Considerable modification of the low temperature mechanism is necessary in order to explain observations made at higher temperatures. In competitive pairs of elementary steps, the reaction with the higher activation energy is progressively favoured as the temperature is increased. Decomposition processes become more important, and the high temperature oxidations show enhanced CO yields because of acyl radical decomposition [105]

$$RCO + M \longrightarrow R + CO + M \tag{6a}$$

$$RCO \longrightarrow R + CO \tag{6b}$$

Peracyl radicals and peracid are more unstable and, in any case, cease to have kinetic significance when RCO fails to survive long enough to enable reaction with oxygen to occur. However, there is evidence [133] that peracid is present in the products, and thus may contribute to branching, up to as high as 350 °C. Termination by RCO_3—RCO_3 collisions will decrease in importance, partly because of the reduced concentration of

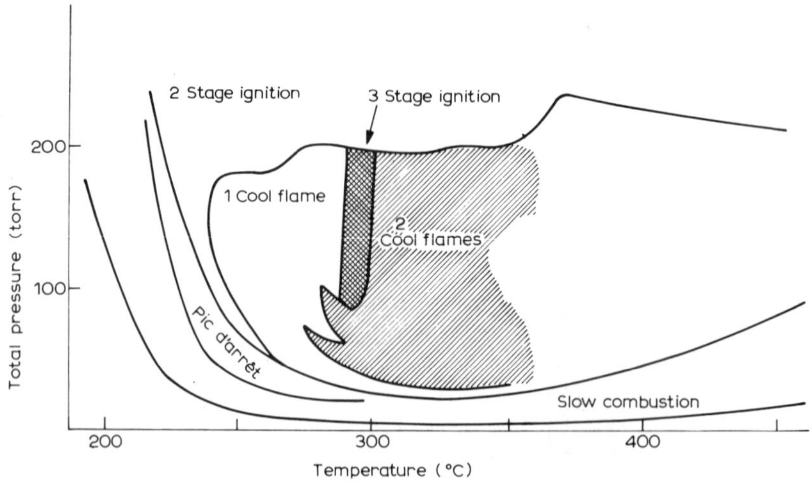

Fig. 18. Combustion diagram for the oxidation of acetaldehyde [51]. Acetaldehyde/ oxygen 1:1.

these radicals, but also because diacyl peroxides (the expected termination products at lower temperatures) are unlikely to be stable at higher temperatures.

Above 400 °C, radicals which play little part in the oxidation below 150 °C may have considerable kinetic importance. Thus, the HO_2 radical will now readily abstract hydrogen atoms

$$HO_2 + RCHO \longrightarrow H_2O_2 + RCO \tag{1b}$$

Furthermore, if the HO_2 concentration is able to become sufficiently high, the reaction

$$HO_2 + HO_2 \longrightarrow H_2O_2 + O_2 \tag{15}$$

must be taken into account. Accumulation of H_2O_2 via (15) and (1b) may result in the branching rate being controlled by hydrogen peroxide decomposition

$$H_2O_2 + M \longrightarrow 2OH + M \tag{16}$$

Further complications will arise from reaction of alkyl radicals, produced by (6), with the reactants and with other radicals.

For intermediate temperature regions some or all of these processes are likely to occur simultaneously with those discussed in Sect. 3, and there will be a progressive transition from a peracid branching mechanism to one involving different mode of branching. When the temperature is sufficiently high, nearly all the acyl radicals will decompose, and it is for this region that Baldwin et al. [21—23, 109] have succeeded in establishing the detailed mechanism of the oxidations of several aliphatic aldehydes.

4.2 FORMALDEHYDE OXIDATION

4.2.1 Characteristic features

A review of earlier observations has been given recently [42], and the following is an outline of the more essential points.

Below about 250 °C the thermal oxidation rate is very low. Above this temperature the reaction proceeds at a convenient rate, but its character depends on the reaction conditions and, in particular, on the type of vessel surface. Thus, the reaction may show a pressure increase throughout its course, have the approximate overall stoichiometry

$$2HCHO + O_2 \longrightarrow 2CO + 2H_2O$$

and have a rate of pressure increase which is a maximum at the start of reaction (Fig. 19). The rate is usually given by

$$\frac{-d[HCHO]}{dt} = k[HCHO]_0^2[O_2]_0^0 \tag{XV}$$

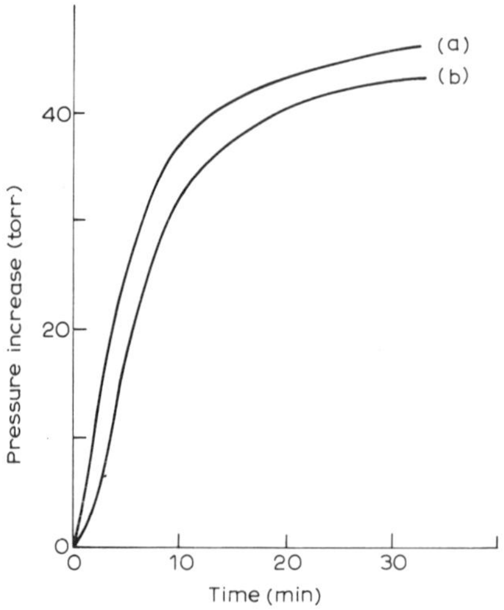

Fig. 19. Oxidation of formaldehyde in nitric acid-washed and in aged silica reaction vessels at 337 °C [30]. (a) Nitric acid-washed vessel; O_2 79.3 torr, HCHO 127.4 torr; (b) aged silica vessel; O_2, 68.6 torr, HCHO 127.6 torr.

where $(HCHO)_0$ is the initial aldehyde concentration, and the second-order aldehyde dependence may be sustained throughout the course of reaction. However, recently it has been shown (see below) that other types of dependence may occur. Major products are CO and H_2O, somewhat smaller amounts of CO_2, H_2 and HCOOH being formed. In addition, relatively small amounts of peroxidic material [stated, or assumed, to be HCO_3H or $(CH_2OH)_2O_2$] are sometimes reported, and recently mass spectrometric evidence [24] has been given for a substance of mass 62 (corresponding to HCO_3H) and of mass 94 (corresponding to $(CH_2OH)_2O_2$).

This type of behaviour characterizes systems described for example by Bone and Gardner [25], Axford and Norrish [27] and Scheer [30], and appears to be typical with newly installed reaction vessels or of reactions carried out in the presence of mercury vapour.

A different type of behaviour, as judged from the pressure—time curve, is indicated by a period of acceleration (Fig. 19) or even an initial period during which little pressure change occurs even though analysis shows the formaldehyde consumption is occurring. During this initial period, peroxidic materials, particularly H_2O_2, but also HCO_3H and $(CH_2OH)_2O_2$, accumulate but decompose during the subsequent period of pressure increase. With this behaviour, noted for example for reactions occurring in aged vessels, the period of low or zero pressure increase was considered by Norrish and Thomas [31] to be a reflection of the overall stoichiometry

$$HCHO + O_2 \longrightarrow CO + H_2O_2$$

Chromic acid-washed molybdenum-glass vessels and boric acid coated surfaces also appear to encourage peroxide formation [34—36], and the hydrogen peroxide so produced must be important in these oxidations since addition of H_2O_2 increases the oxidation rate.

In general, the rate of loss of formaldehyde is greater for those reactions in which appreciable peroxide formation is known to occur. Thus, Scheer [30] showed that

$$-d[HCHO]/dt = k[HCHO]^2 \qquad (XVI)$$

For aged vessels k had a value of 12.8×10^{-4} torr^{-1} min^{-1} and $-d(HCHO)/dt$ was approximately twice the rate of pressure change; for HNO_3 cleaned surfaces or for reactions in the presence of mercury vapour k was 9.1×10^{-4} (torr)$^{-1}$ min^{-1} and the rate of formaldehyde consumption was very much greater than twice the rate of pressure change.

The conclusion that surface effects (and possibly also the presence of mercury vapour) are responsible for many of the differences noted between the results of different groups of workers when experiments are carried out under otherwise the same conditions seems to be inescapable. It also appears that the influence of the surface hinges largely on its attitude towards the destruction or the preservation of peroxides or peroxy radicals. The experiments of Markevitch and Filippova [34, 35] using a flow reactor demonstrate convincingly that surface type can be critical in determining the character of the oxidation, and for certain surfaces (e.g. $K_4B_4O_7$) much of the enthalpy of reaction may be liberated close to the vessel wall. Equally convincing are the experiments of Vardanyan et al. [37, 38] made between about 528 °C and 648 °C, using packed vessels in which the surfaces were (*i*) untreated quarts, (*ii*) boric acid treated, (*iii*) $K_2B_4O_7$ treated and (*iv*) KBr treated. With surfaces (*i*) and (*ii*) the reaction was autocatalytic, appreciable yields of H_2O_2 and smaller quantities of organic peroxides (HCO_3H) being formed. No H_2O_2 was observed with the KBr treated vessel. With both the KBr and the $K_2B_4O_7$ treated vessels there was no evidence of autocatalysis and the rates of HCHO consumption were less than in the untreated or boric acid treated vessels. This latter observation is consistent with that of Scheer [30] noted above.

These observations are also consistent with recent results obtained by Baldwin et al. [23]. They showed that at 440 °C the reaction in unpacked aged boric acid treated vessels is autocatalytic, the autocatalysis being attributed to the H_2O_2 shown to be formed. In unpacked KCl coated vessels the reaction was much slower and non-autocatalytic. This was attributed to efficient surface destruction of H_2O_2, this compound not being detectable amongst the reaction products. It might be noted that with the boric acid coated vessels both initial and maximum rates were of 1.4 order with respect to HCHO; the initial rate depended on the oxygen

TABLE 8
Overall activation energies for formaldehyde oxidation

Ref.	E_a (kcal . mole^{-1})	Comments
129	20.0	
130	17.6	
131	25.0	
27	21.0	Hg vapour in vessel
118, 119	29.4	
30	27.4	Aged vessel
35	26.0	Chromic acid washed molybdemum glass vessel
35	50.0	$K_2B_4O_7$ coated vessel
29	35—45	Range 400—475 °C, Hg vapour in vessel
24	17.8	Range 350—420 °C, flamed washed pyrex vessel
37, 38	30	Range 576—648 °C, untreated quartz vessel
37, 38	30	Range 528—600 °C, boric acid coated quartz vessel
37, 38	49	Range 624—685 °C, $K_2B_4O_7$ coated quartz vessel
37, 38	50	Range 670—720 °C, KBr coated quartz vessel
23		Range 440—540 °C, boric acid coated vessel
23		Range 440—540 °C, KCl coated vessel

pressure but the maximum rate was oxygen independent. In KCl coated vessels the oxidation had an order of 1.8 with respect to HCHO and 0.8 with respect to O_2.

The sensitivity of the relationship between the rates of pressure change and formaldehyde consumption to surface conditions means that it is desirable that the kinetics should be discussed in terms of aldehyde loss rather than of pressure change. This has not always been done, and consequently it is difficult to compare many of the reported activation energies for the oxidation. These (Table 8) cover a large range, the spread being a further indication of the sensitivity of this oxidation to surface and reaction conditions.

Although the rate in the early stages of the oxidation is often oxygen independent (however, see above), high oxygen pressures are sometimes reported to enhance the rate in the latter stages. This may be associated with the occurrence of oxygen induced pyrolysis. It is also possible that for those systems in which peroxidic materials are formed, formaldehyde may continue to be consumed even after consumption of the oxygen. Thus, Hay and Hessam [24] were able to follow mass spectrometrically the disappearance of formaldehyde and hydrogen peroxide which accompanied the formation of formic and performic acids.

4.2.2 Reaction scheme

Despite the apparent inconsistency of many of the reported observations, most workers do agree on the choice of many of the elementary steps, although inevitably speculation is necessary in the setting up of a comprehensive kinetic scheme.

Arguments in favour of accepting the reactions

$$HCO + O_2 \longrightarrow HO_2 + CO \quad (11)$$

$$HCO + O_2 \longrightarrow HCO_3 \quad (12)$$

$$HO_2 + HCHO \longrightarrow H_2O_2 + HCO \quad (17)$$

$$HCO_3 + HCHO \longrightarrow HCO_3H + HCO \quad (18)$$

$$HO_2 + HO_2 \longrightarrow \text{termination} \quad (15)$$

may be outlined as follows. The formyl radical is the expected product of hydrogen abstraction attack on formaldehyde, and although at these temperatures some decomposition by

$$HCO + M \longrightarrow H + CO + M \quad (19)$$

cannot be precluded, this is unlikely to be the main route by which it is lost since oxygen attack by (11) or (12) will be rapid because of the low activation energies for these reactions. The reverse of reaction (12) has not been considered, but probably is of importance.

Electron spin resonance evidence for the occurrence of HO_2 radicals during the oxidation of HCHO has recently been given by Vardanyan et al. [37, 38], and it has been shown that for surfaces which are not particularly destructive towards HO_2 (e.g. B_2O_3), the maximum HO_2 concentration coincides with the maximum H_2O_2 concentration thus giving supporting evidence for the occurrence of (17) for which the rate coefficient

$$k_{17} = 1.9 \times 10^{10} \exp(-10{,}400 \pm 3{,}000/RT) \text{ l . mole}^{-1} \text{. sec}^{-1}$$

for the range 528—600 °C was given [38].

A tentative estimate by Baldwin et al. [23] indicated k_{17} for the range 440—540 °C to be $2.74 \times 10^9 \exp(-12{,}000/RT)$ l . mole^{-1} . sec^{-1} and a numerical analysis of the results obtained from studies [23(a)] of the effect of HCHO on the H_2/O_2 slow reaction at 500 °C indicated that k_{17} (500 °C) = 9.6×10^5 l . mole^{-1} . sec^{-1}. On the assumption of a pre-exponential factor of 10^9 l . mole^{-1} . sec^{-1}, E_{17} was calculated to be 10.7 kcal . mole^{-1}.

These estimates differ mainly in the values for the pre-exponential factors. However, Vardanyan et al. [38] consider their value for the pre-exponential factor to be probably rather high.

The production of H_2O_2 and HCO_3H are most simply explained by the sequences (11) followed by (17) and (12) followed by (18), respectively. The large yield of CO is consistent with (11) being of major importance, and if this is so the HO_2 concentration is likely to be sufficiently high to make (15) important. It has been argued that the relatively high yields of water necessitate the formation and reaction of OH radicals

$$HCHO + OH \longrightarrow H_2O + HCO \tag{20}$$

and Hay and Hessam [24] considered that the OH radicals are generated by

$$HO_2 + HCHO \longrightarrow OH + \text{products } (HCOOH \text{ or } H_2 + CO + H_2O) \tag{21}$$

However, reaction (21) is difficult to reconcile with the observations of high, almost quantitative, hydrogen peroxide yields found by Norrish and Thomas [31], Baldwin et al. [23] and by Russian workers [34, 35, 37, 38]. Clearly under the conditions used by these latter groups of workers (17) takes place readily. If (17) is homogeneous it should also occur readily under the conditions used by Hay and Hessam, and their detection of only small amounts of hydrogen peroxide could be accounted for if their vessel surface (flamed pyrex, distilled water-washed) was destructive towards H_2O_2 and the large water yield was the product of surface hydrogen peroxide decomposition.

The initiation process is usually aassumed to be

$$HCHO + O_2 \longrightarrow HO_2 + CO \tag{22}$$

for which Baldwin et al. [23] estimated the homogeneous rate coefficient at 440 °C to be 10^{-2}—10^{-1} l. mole^{-1}. sec^{-1}. This value considered to be consistent with an activation energy close to the endothermicity and a pre-exponential factor close to 10^{11} l. mole^{-1}. sec^{-1}. It is evident from the variation in detail of the observations of different groups of workers with regard to the products (peroxides or no peroxides), the form of the ΔP—time curves (autocatalysis or no autocatalysis) and the reaction order ($\rho \propto [HCHO]^{1.4} [O_2]^{1/2}$, or $\rho \propto [HCHO]^2 [O_2]^0$) that no single scheme is capable of explaining all of the reported facts.

For those systems in which the rate is independent of oxygen, it seems likely that some branching must occur and possibly predominate over (22) as a chain initiation step since it is otherwise difficult to derive a kinetic equation which does not predict an oxygen dependent rate.

The simplest process is the decomposition of performic acid

$$HCO_3H \longrightarrow OH + \text{products} \tag{23}$$

the rate of branching being controlled by the rate of peracid formation (stationary state concentration with respect to the peracid). In addition, it seems likely that for those reactions in which large yields of H_2O_2 are given, and particularly at higher temperatures, some branching via H_2O_2 decomposition

$$H_2O_2 + M \longrightarrow 2OH + M \tag{16}$$

or via hydrogen peroxide—formaldehyde complexes occurs. (See also ref. 136.)

Hay and Hessam [24] considered that at 350 °C the main reaction sequence comprises (22), (17), (21), (20), (11), (12), (18), (23) and (15). Combination of these equations together with the assumptions that the concentration of the peracid is in a stationary state and that the rate of (11) is greater than that of (12) gives

$$-\mathrm{d}[\mathrm{HCHO}]/\mathrm{d}t = \frac{k_{12}(k_{17} + k_{21})(2k_{21} + k_{17})[\mathrm{HCHO}]^2}{k_{11}k_{15}} \qquad \text{(XVII)}$$

This scheme at least has the merit of explaining the major products and of indicating an overall activation energy not very different from that observed by these workers. However, the reservations indicated above considering the acceptability of (21) should be noted.

For the oxidation in boric acid coated vessels at 440 °C Baldwin et al. [23] proposed a scheme comprising (22), (11), (15), (20) and (16) from which the initial rate expression

$$-\mathrm{d}[\mathrm{HCHO}]/\mathrm{d}t = k_{22}[\mathrm{HCHO}][\mathrm{O_2}] + k_{22}^{1/2}(k_{17}/k_{15}^{1/2})[\mathrm{HCHO}]^{3/2}[\mathrm{O_2}]^{1/2} \qquad \text{(XVIII)}$$

can be obtained. For long chains the expression becomes

$$-\mathrm{d}[\mathrm{HCHO}]/\mathrm{d}t = k_{22}^{1/2}(k_{17}/k_{15}^{1/2})[\mathrm{HCHO}]^{3/2}[\mathrm{O_2}]^{1/2} \qquad \text{(XIX)}$$

As the reaction develops, autocatalysis arising from (16) and (20) becomes important and consequently, (XVIII) and (XIX) no longer apply. For the KCl coated vessels in which the surface destruction of HO_2 radicals is thought to be rapid, the following reactions have to be added to the basic mechanism, viz.

$$\mathrm{H_2O_2}_{\,\mathrm{Surface}} \longrightarrow \mathrm{H_2O} + \tfrac{1}{2}\mathrm{O_2} \qquad (24)$$

$$\mathrm{HO_2}_{\,\mathrm{Surface}} \longrightarrow \tfrac{1}{2}\mathrm{H_2O} + 3/4\,\mathrm{O_2} \qquad (25)$$

Inert gases exert their effect through reactions (16), (24) and (25). Reactions (24) and (25) are diffusion controlled, and consequently the addition of inert gas will lead to an overall accelerating effect which is augmented by the enhanced rate of (16). This accelerating effect of inert gases on the oxidation in the salt coated vessels contrasts with negligible effect which they have on the reaction in boric acid coated vessels in which (24) and (25) are unimportant.

Suggested reaction schemes are given in block diagram form in Fig. 20. In summary, there is good evidence for believing that, under almost all conditions so far studied, the oxidation of HCHO proceeds via HCO and HO_2 radicals. HCO_3 radicals may also be present, even up to 600 °C in

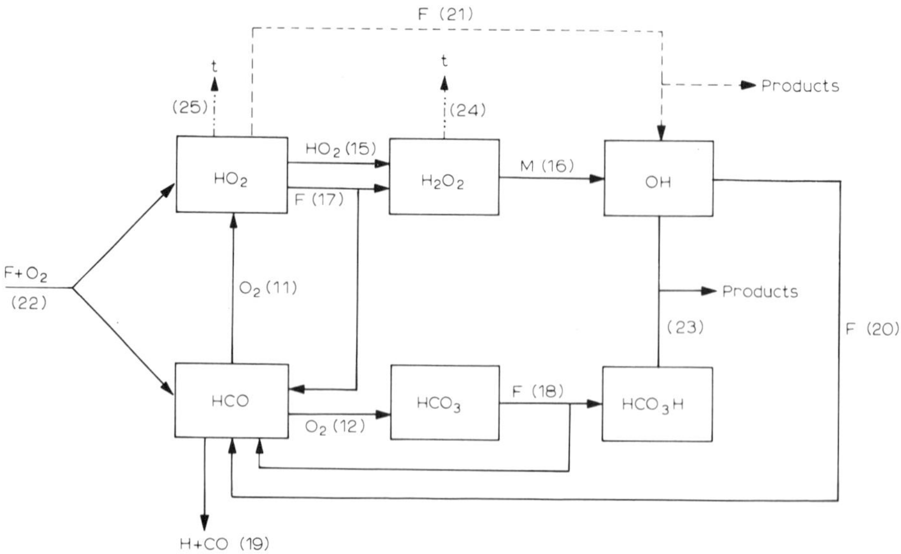

Fig. 20. Block diagram to illustrate the proposed mechanisms of formaldehyde oxidation.

boric acid coated acid vessels [37, 38], and OH radicals are also important in boric acid coated vessels [22, 109]. There is good evidence for the general occurrence of (11), (17) and (22). Reactions (15), (16) and (20) are thought to be important with boric acid coated surfaces. Reactions (12), (18) and (23) probably occur when the temperature is not too high. At very high temperatures (11) would be expected to occur to the exclusion of (12) and branching by (16) will be important. The main effect of the salt-treated surfaces is to cause rapid surface destruction of HO_2 by (25) and of H_2O_2 by (24) thereby reducing the importance of (15), (16) and (20). It might be noted that for packed vessels loss of HO_2 radicals by surface destruction may become important and termination by (25) may predominate over (15). Vardanyan et al. [37, 38] consider that the efficiency of surface destruction decreases in the order $BaBr_2 > KBr > KCl > K_2B_4O_7 >$ untreated quartz $> B_2O_3$.

The role of formaldehyde—peroxide complexes in the oxidation of HCHO is still not resolved [42]. However, in view of the fact that it is possible to explain most of the features of the oxidation in both boric acid and salt coated vessels without including their formation and reactions it seems likely that complexes play only a minor part in the overall oxidation.

4.3 ACETALDEHYDE OXIDATION AT INTERMEDIATE TEMPERATURES

The simplicity of the stoichiometry in the low temperature region means that interpretation of the kinetics as determined by pressure change

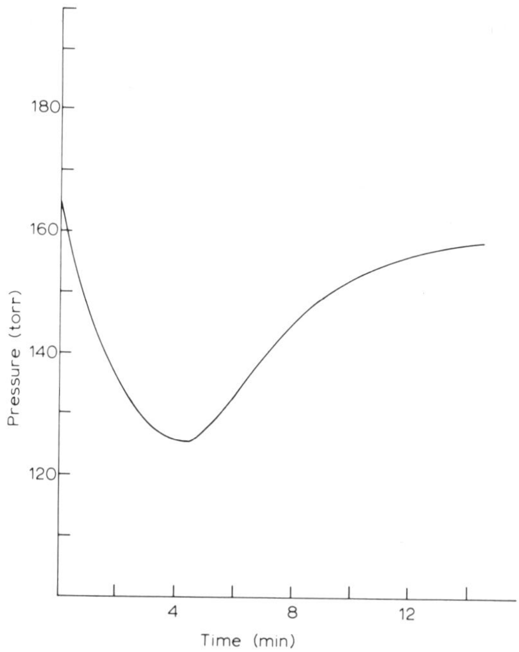

Fig. 21. Pressure—time curve for acetaldehyde oxidation at 182 °C [132]. Initial acetaldehyde pressure = 82.4 torr; initial oxygen pressure = 79.2 torr.

measurements is relatively straightforward. However, the increased rate of decomposition of the peracid [98] and the production of appreciable low molecular weight material as the temperature is increased beyond 150 °C, particularly at the upper end of the intermediate temperature range, means that the simple relationship between pressure change and stoichiometry no longer exists. Reaction at the upper end of this temperature range involves a pressure decrease (Fig. 21) and, strictly, the observed rate should be described in terms of the rate of loss of one of the reactants. This also means that activation energies determined from pressure change measurements over an extended temperature range must be suspect.

The reaction rate at first shows an increase as the temperature is raised. At very high temperatures that is, above the temperature range where the peracid mechanism applies, a region of negative temperature coefficient is noted (see below). A region of negative temperature coefficient of rate is also to be expected at somewhat lower temperatures as the peracid branching mechanism fades out. However, no comprehensive survey of the effect of temperature on rate over the uninterrupted range 100 to above 500 °C appears to have been made.

The enhanced rate of

$$RCO + M \longrightarrow R + CO + M \tag{6a}$$

for which the Arrhenius parameters are given in Table 6 is reflected in an increased CO yield and possibly also in an oxygen dependence of rate

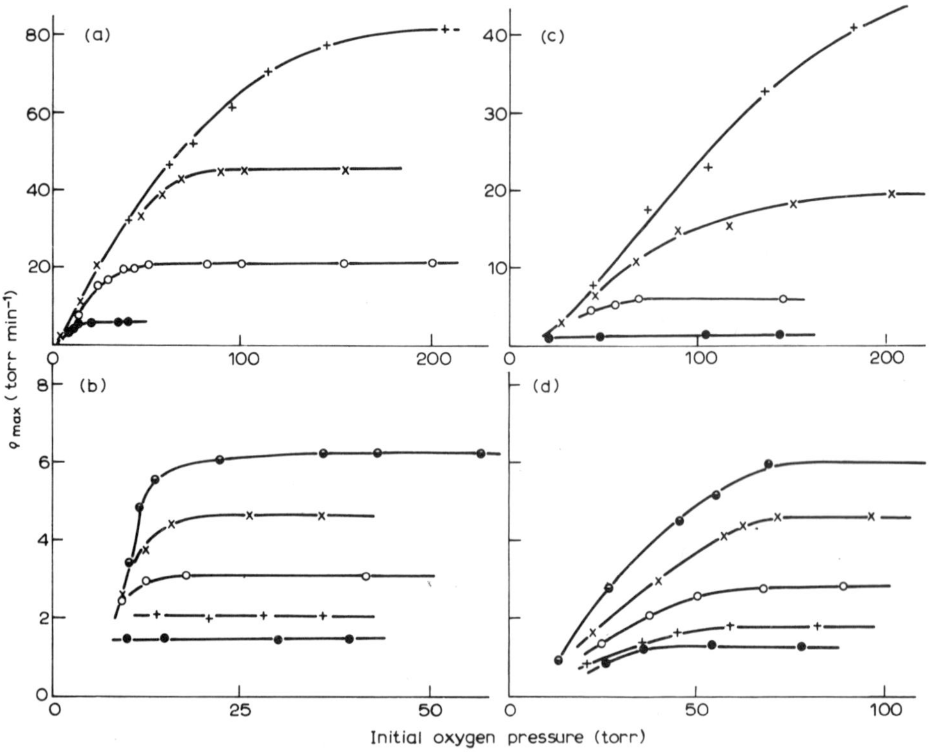

Fig. 22. Variation of the maximum oxidation rate with initial oxygen pressure [44]. (a) Temperature 163 °C; acetaldehyde pressure (torr); +, 400; x, 300; ○, 200; ●, 100. (b) Acetaldehyde pressure 100 torr; temperature (°C); ◒, 163; x, 155; ○, 143; +, 134; ●, 120. (c) Temperature 163 °C; propionaldehyde pressure (torr); +, 300; x, 200; ○, 100; ●, 52. (d) Propionaldehyde pressure, 100 torr; temperature (°C); ◒, 163; x, 155; ○, 143; +, 134; ●, 120.

noted at low O_2/fuel ratios (Fig. 22). For this reason it seems important to include reaction (6a) and/or (6b) together with those discussed in Sect. 3 in order to describe events at the lower end of the intermediate temperature range. A numerical analysis of a scheme comprising (1a) (2), (3), (4a) (5c) and (6) given by Griffiths et al. [57] succeeded in explaining the main kinetic features of the oxidation in this region Competition between (2) and (6) for RCO radicals, gave a predicted oxygen dependence not inconsistent with the observed experimentally (Fig. 23).

Surprisingly, no detailed kinetic examination in the upper end of the intermediate temperature range seems to have been made. It might be expected that as this range is traversed, branching by (5c) would be progressively replaced by the higher activation energy first-order peracid decomposition (5a), and that ultimately the high rate of peracid loss would result in a low, but quasi-stationary RCO_3H concentration. Under

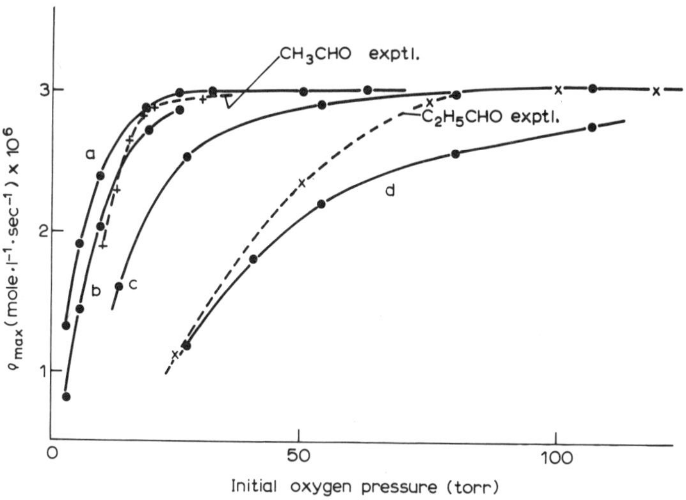

Fig. 23. The influence of termination by (6b) on the computed maximum rate of acetaldehyde oxidation at 155 °C [57]. --------, Experimental curves based on pressure—time data and converted to concentration—time data on the basis of stoichiometry. Other curves are computed on the assumption that a fraction ϕ of the decompositions of RCO radicals lead to termination. (a) $\phi k_{6b} = 0$; (b) $\phi k_{6b} = 2.31 \times 10^1$ sec^{-1}; (c) $\phi k_{6b} = 2.31 \times 10^2$ sec^{-1}; (d) $\phi k_{6b} = 2.31 \times 10^3$ sec^{-1}. Aldehyde pressures, 100 torr.

these conditions an idealized scheme comprising (1a), (2), (3), (4a) and (5a) would have a rate given by

$$-\mathrm{d}[\mathrm{RCHO}]/\mathrm{d}t = (k_3^2/2k_{4a})[\mathrm{RCHO}]^2 \qquad \text{(XX)}$$

although for temperatures above about 200 °C this oxygen-independent rate would be expected only at high oxygen/fuel ratios. Since this rate is also given by $k_{5a}[\mathrm{RCO_3H}]$, an investigation in this region may enable comment on k_3, k_{4a}, and k_{5a} to be made. The values of k_{5a} could then be compared with those obtained directly from the rate of peracetic acid decomposition.

4.4 ACETALDEHYDE OXIDATION AT HIGH TEMPERATURES

Although a few isolated studies of aspects of high temperature acetaldehyde oxidation have been reported, no detailed analytical investigations of the reaction between about 250 and 440 °C appear to have been described. Above 440 °C the useful region of reactant pressures is restricted to only a few torr. In boric acid coated vessels, the reaction is reproducible and occurs with a pressure increase (and possibly with a slight suggestion of an induction period at 440 °C [21]). Effectively, the rate of pressure increase is equal to the rate of aldehyde consumption over

TABLE 9
Orders of reaction for the high temperature acetaldehyde oxidation [21]

Temp. (°C)	Reactant	Pressure range (torr)	Order
540	CH_3CHO	0.25—4.0	1.5
	O_2	0.5—10	0.5
	O_2	10—58	0.8
440	CH_3CHO	0.5—4.0	2.5—3.5
	O_2	1.0—20	1.1
	O_2	30—58	0.3

the first 40 % of reaction, and the order of the rate with respect to the reactants depends on the conditions (Table 9). Over the 440—540 °C interval there is a negative temperature coefficient (Fig. 24). Hydrogen peroxide is formed, presumably by (1a) and (1b)

$$CH_3CHO + O_2 \longrightarrow CH_3CO + HO_2 \qquad (1a)$$

$$CH_3CHO + HO_2 \longrightarrow CH_3CO + H_2O_2 \qquad (1b)$$

CO is a major reaction product, and for a 10 % fuel consumption the CO yield is approximately equal to the aldehyde consumption suggesting that

$$CH_3CO + M \longrightarrow CH_3 + CO + M \qquad (6a)$$

occurs rapidly and to the exclusion of peracetyl formation. CH_4, C_2H_6, HCHO and CH_3OH are expected products of methyl radical reactions in

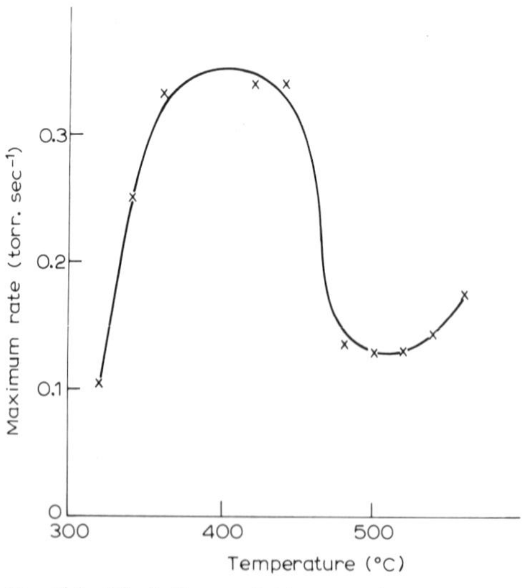

Fig. 24. Variation of the maximum rate of acetaldehyde oxidation in boric acid coated vessels with temperature. Initial pressuress (torr): CH_3CHO 2; O_2 30; N_2 28. (From ref. 21 by permission.)

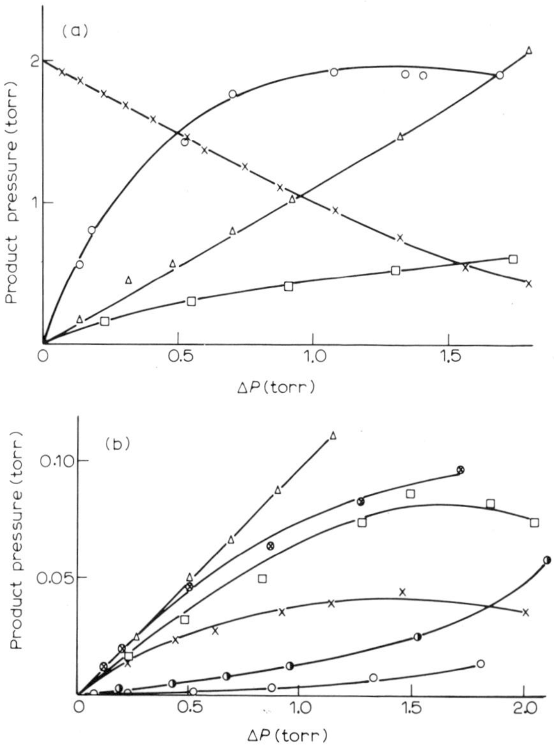

Fig. 25. Variation of oxidation products with pressure change at 540 °C for acetaldehyde oxidation in a boric acid coated vessel. (a) x, CH_3CHO; ○, HCHO 10; △, CO; □, CH_4. (b) △, H_2; ⊗, CH_3OH; □, H_2O_2; x, C_2H_6; ◑, CO_2; ○, C_2H_4. (From ref. 21 by permission.)

systems containing acetaldehyde and oxygen (Fig. 25). The yield of H_2 is high, particularly at 440 °C (Table 10). The rates of methane and ethane

TABLE 10
Product formation during the high temperature acetaldehyde oxidation [21]

		440 °C	540 °C
Initial composition (torr)	CH_3CHO	1.5	2.0
	O_2	30.0	30.0
	H_2	28.5	28.0
Product yield (as % of aldehyde consumed)	CO	100	100
	HCHO	65	32
	CH_4	10	50
	C_2H_6	0.5	6
	CH_3OH	25	7
	H_2O_2	10	5
	H_2	15	7

formation are also high, particularly at 540 °C, and are indicative of a high CH_3 radical concentration. These products can be accounted for by

$$CH_3 + CH_3CHO \longrightarrow CH_4 + CH_3CO \tag{26}$$

and to a lesser extent

$$CH_3 + CH_3CHO \longrightarrow CH_4 + CH_2CHO \tag{27}$$

occurring in competition with

$$CH_3 + CH_3 \longrightarrow C_2H_6 \tag{28}$$

An Arrhenius plot of values of k_{Me} (= $k_{26} + K_{27}$) obtained from

$$k_{Me} = \frac{R_{CH_4} k_{28}^{1/2}}{R_{C_2H_6}^{1/2}[CH_3CHO]} \tag{XXII}$$

where R_{CH_4} and $R_{C_2H_6}$ are the rates of formation of methane and ethane, respectively, was in good agreement with results obtained by previous workers [99, 100] and leads to

$$k_{Me} = (1.6 \pm 0.6) \times 10^9 \exp\{(-8{,}200 \pm 500)/RT\}\ \mathrm{l.mole^{-1}.sec^{-1}}$$

No evidence was found for appreciable attack of the type

$$CH_3 + CH_3CHO \longrightarrow C_2H_6 + HCO \tag{29}$$

A particularly important aspect of this work is the information it gives on the frequently discussed reactions between oxygen and methyl radicals. Much of the HCHO and CH_3OH must arise in this way, possibly indirectly, and the authors considered in depth the possible routes. A key feature in their interpretation was the negative temperature coefficient for the oxidation of methyl radicals. Thus, designating the combined HCHO + CH_3OH yield as "oxidation products", their results show the ratio d(oxidation products)/d[CH_4] to decrease rapidly with temperature increase between 440 and 540 °C with an overall activation energy for

$$CH_3 + O_2 \longrightarrow \text{oxidation products} \tag{30}$$

of about 20 kcal . mole^{-1}. For this reason, the frequently postulated bimolecular step

$$CH_3 + O_2 \longrightarrow HCHO + OH \tag{31}$$

was considered to be unimportant at these temperatures and was rejected in favour of

$$CH_3 + O_2 + M \longrightarrow CH_3OO + M \tag{32}$$

the methyl peroxy radicals being lost by the reverse process

$$CH_3OO + M \longrightarrow CH_3 + O_2 + M \tag{$-$32}$$

which predominates over the competitive step

$$CH_3OO + X \longrightarrow \text{oxidation products} \tag{33}$$

The rate coefficient, k_{ox}, for the methyl radical oxidation is given by

$$k_{ox} = \frac{k_{32}[M]k_{33}[X]}{k_{-32}[M] + k_{37}[X]}, \tag{XXII}$$

and for X = M or CH_3CHO, provided that $k_{-32}[M] \gg k_{33}[X]$, the activation energy for the oxidation, $E_{ox}(= E_{32} + E_{33} - E_{-32})$, will be negative (ca. —25 kcal . mole^{-1}) as E_{32}, E_{-32} and E_{33} will probably be close to 5, —26 and 7 kcal . mole^{-1}, respectively.

This scheme provides a reasonable explanation of the negative temperature coefficient. However, it was not possible to explain quantitatively the relative rates of formation of oxidation products and methane by identifying (33) with any single reaction of type (34)—36)

$$CH_3OO + CH_3 \longrightarrow 2CH_3O \tag{34}$$

$$CH_3OO + CH_3CHO \longrightarrow CH_3OOH + CH_3CO \tag{35}$$

$$CH_3OO + M \longrightarrow HCHO + OH + M \tag{36}$$

Possibly at least two of these reactions contribute.

Although HO_2 radicals are produced by (1a), and possibly also at other stages of the reaction, it seems that, in contrast to the corresponding high temperature oxidation of propionaldehyde (see below), the reaction

$$HO_2 + HO_2 \longrightarrow H_2O_2 + O_2 \tag{15}$$

contributes insignificantly to termination. The most probable termination step at 540 °C is (28) a process consistent with the high stationary concentration of methyl radicals noted above. By assuming negligible branching and equating the rates of initiation and termination, viz.

$$k_{1a}[CH_3CHO][O_2] = k_{28}[CH_3]^2 = R_{C_2H_6} \tag{XXIII}$$

values of k_{1a} were obtained. However there was a trend with aldehyde concentration which suggested that a simple straight-chain mechanism initiated solely by (1a) and termined solely by (28) is probably an over-simplification*. Nevertheless, a value for k_{1a} of 4.0 l . mole^{-1} . sec^{-1} at 540 °C was calculated using the straight-chain approximation. This value probably represents an upper limit, and compares well with the corresponding initiation rate coefficient for the oxidation of propionaldehyde for which the oxidation mechanism is more clearly defined (see

* Recent work by the Hull school indicates the initiation mechanism to be, as suspected, more complex.

below), if the assumption is made that both initiation steps have similar activation energies (ca. 41 kcal . mole^{-1}).

The immediate precursor of methyl alcohol is almost certainly CH_3O. The production of this by (34) or

$$CH_3OO + HO_2 \longrightarrow CH_3OOH + O_2 \quad (37)$$

followed by

$$CH_3OOH + M \longrightarrow CH_3O + OH + M \quad (38)$$

was rejected since the sequence (37) and (38) would give rise to appreciable branching, and the reasonable correspondence found between the rate of initiation (twice that of H_2O_2 formation) and the rate of termination (twice that of ethane formation) indicates branching to be unimportant. Combination (37) and (38) also seems unlikely in view of the low concentration of HO_2 radicals. The alternative possibility

$$CH_3OO + CH_3 \longrightarrow 2CH_3O \quad (39)$$

is attractive in view of the high methyl radical concentration.

There is evidence that

$$CH_3O + CH_3CHO \longrightarrow CH_3CO + CH_3OH \quad (40)$$

is not the only reaction open to the methoxy radical, and that the observed H_2 arises from the abstraction reactions of H atoms produced by

$$CH_3O + M \longrightarrow CH_2O + H + M \quad (41)$$

The efficiency of CH_3CHO in promoting (41) is roughly 14 times that of N_2.

The simplified overall mechanism is represented in block diagram form in Fig. 26.

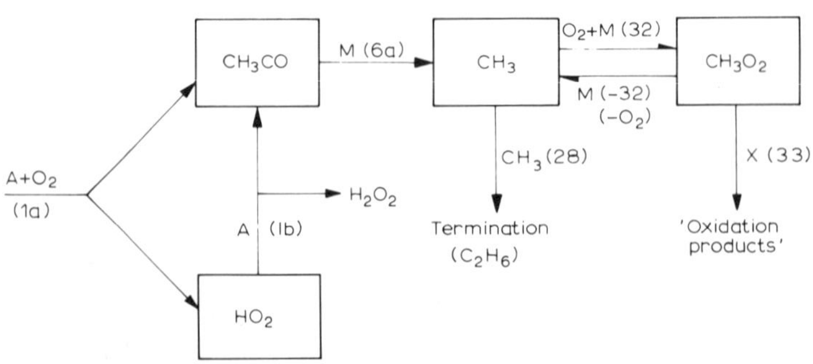

Fig. 26. Block diagram to illustrate the high temperature oxidation of acetaldehyde in boric acid coated vessels [21].

For intermediate temperatures (between about 200 and 400 °C) the oxidation is complicated by the onset of appreciable peracid and peracyl radical decomposition. There is no longer either a simple stoichiometry or a simple relationship between extent of the reaction and the pressure change; appreciable production of low molecular weight material occurs (Fig. 27). The maximum rate of pressure decrease is proportional to the square of the initial aldehyde concentration and is independent of the

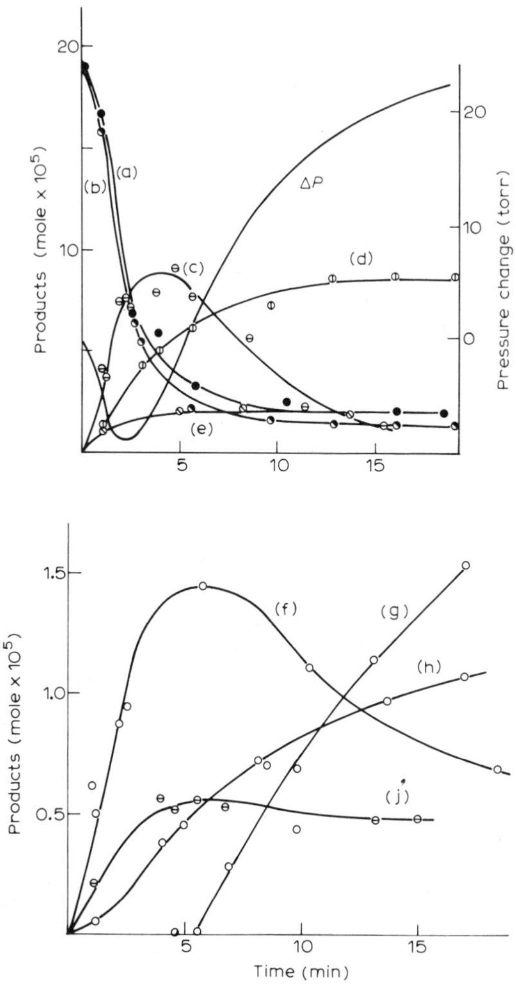

Fig. 27. Analysis throughout the course of the propionaldehyde oxidation at 220 °C [20]. Aldehyde pressure = O_2 pressure = 50 torr. (a) Propionaldehyde; (b) oxygen; (c) peroxide; (d) carbon monoxide; (e) acid; (f) acetaldehyde; (g) ethane; (h) carbon dioxide; (j) ethylene.

initial oxygen concentration only for high oxygen/aldehyde ratios when presumably (2) predominates over (6a) or (6b) and, in general, the rate is oxygen dependent.

Ethylene is produced, presumably by

$$C_2H_5 + O_2 \longrightarrow C_2H_4 + HO_2 \tag{42}$$

Acetaldehyde was suggested by Skirrow and Whim [20] to originate via

$$C_2H_5 + O_2 \longrightarrow CH_3CHO + OH \tag{43}$$

Although Baldwin et al. [22], who also found C_2H_4 and CH_3CHO at 440 °C, have shown that in their system (43) is unlikely to occur and have suggested that at 440 °C CH_3CHO originates via abstraction attack at a secondary hydrogen of the propionaldehyde. However, attack at the secondary hydrogen position at 220 °C is unlikely to be fast enough to explain the observed CH_3CHO/C_2H_4 ratio of ca. 2.0 (see Sect. 4.6). This feature requires further examination.

At 220 °C peracid or peracid/aldehyde branching is still probable, and since for high O_2/C_2H_5CHO ratios the kinetics resemble those obtained below 150 °C, the overall mechanism is almost certainly the same as that at the lower temperatures with additional complications arising from RCO decomposition and the subsequent reactions of C_2H_5 radicals. Since the amount of CO_2 produced after the peracid maximum (Fig. 27) is approximately the same as the amount of peracid lost, it seems reasonable to suppose that, as at the lower temperature, CO_2 is produced in the branching process.

4.6 PROPIONALDEHYDE OXIDATION AT HIGH TEMPERATURES

4.6.1 Main features

The oxidation of propionaldehyde above 440 °C in a boric acid coated vessel is accompanied by a pressure increase, and in addition to the expected CO, the major products are C_2H_4 (80 %) and H_2O_2, although significant amounts of CH_3CHO and CO_2 (10–15 %) and somewhat smaller amounts of C_2H_4O, C_2H_6 and H_2 are obtained (Table 11) [22]. In contrast to the oxidation of acetaldehyde at these temperatures, the process is distinctly autocatalytic (Fig. 28), the branching being attributable to the hydrogen peroxide formed since additions of H_2O_2 at the start of reaction in amounts comparable with those normally produced reduce the time to reach ρ_{max} without having appreciable affect on ρ_{max} itself. Despite difficulties caused by the co-condensation of C_2H_5CHO and H_2O_2 in the sampling traps, Baldwin et al. [22] were able to show that, as with acetaldehyde, ΔP was a valid measure of the extent of reaction (certainly at 440 °C).

TABLE 11
Product formation during the high temperature oxidation of propionaldehyde [22]

Initial pressures (torr): C_2H_5CHO, 4.0; O_2, 30; N_2, 26.
Time of sampling, 0.57 min. C_2H_5CHO reacted, 1.0 torr.

Product	Pressure (torr)	Product	Pressure (torr)
CO	0.95	CO_2	0.095
C_2H_4	0.854	HCHO	0.008
C_2H_4O	0.031	CH_3OH	0.008
CH_3CHO	0.138	H_2O_2	0.84
CH_4	0.014	H_2	0.006
C_2H_6	0.0058		

Between 340 and 400 °C the oxidation shows a negative temperature coefficient (Fig. 29) corresponding to an overall activation energy of about −37 kcal . mole^{-1}. This region of negative temperature coefficient is some 80 °C lower than that for the acetaldehyde system, and may characterize the passage from peracid controlled to H_2O_2 controlled branching. Between 425 and 500 °C the overall activation energy was ca. 30 kcal . mole^{-1}. The order of the maximum rate with respect to

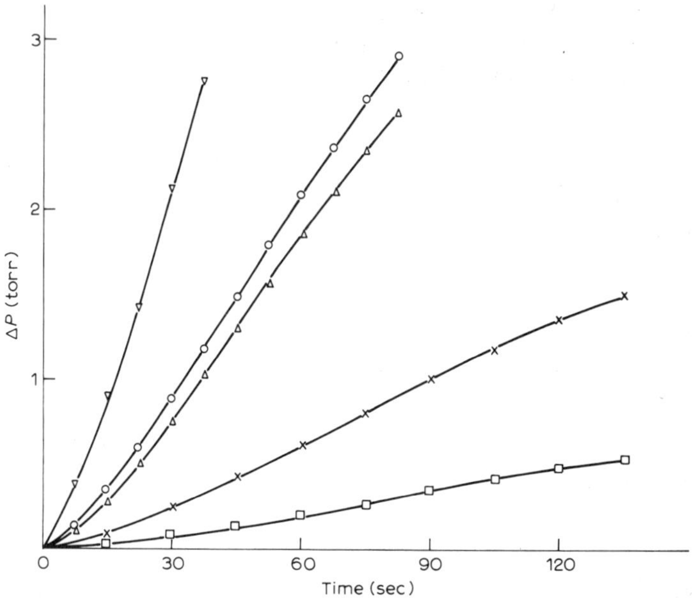

Fig. 28. Typical pressure—time curves for the propionaldehyde oxidation at 440 °C using boric acid coated vessels. O_2, 30 torr; C_2H_5CHO (torr): □, 1; x, 2; ○, 4; ▽, 6. O_2, 8 torr; C_2H_5CHO (torr): △, 4. (From ref. 22 by permission.)

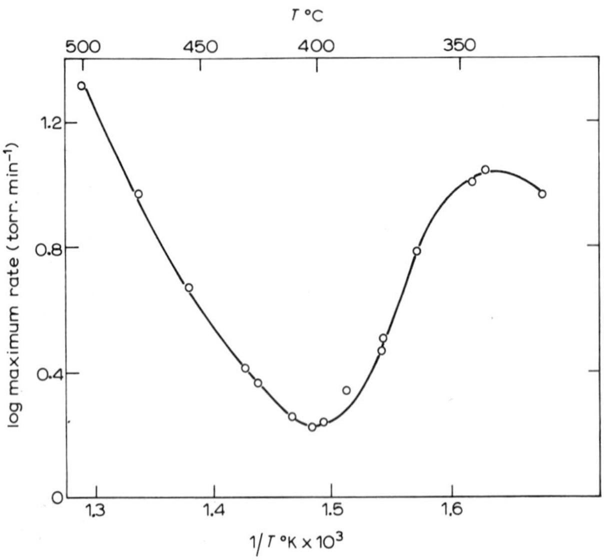

Fig. 29. Variation of the maximum rate of propionaldehyde oxidation with temperature using boric acid coated reaction vessels. (From ref. 22 by permission.)

aldehyde varied from 3/2 (aldehyde pressure < 10 torr) to 5/2 (aldehyde pressures > 10 torr). The order with respect to oxygen was 0.11 except at very low oxygen pressures when it rose to about 0.9. Inert gas had an accelerating effect, the effectiveness increasing in the order $CO_2 > N_2 >$ He. Change of vessel diameter had little influence on ρ_{max}.

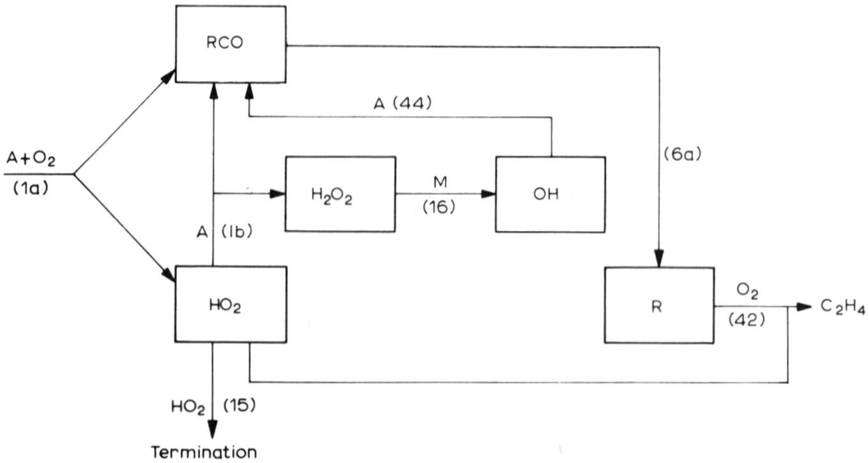

Fig. 30. Block diagram to illustrate the mechanism of propionaldehyde oxidation in boric acid coated reaction vessels.

The results are consistent with the mechanism

$$C_2H_5CHO + O_2 \longrightarrow C_2H_5CO + HO_2 \quad (1a)$$

$$C_2H_5CO + M \longrightarrow C_2H_5 + CO + M \quad (6a)$$

$$C_2H_5 + O_2 \longrightarrow C_2H_4 + HO_2 \quad (42)$$

$$HO_2 + C_2H_5CHO \longrightarrow H_2O_2 + C_2H_5CO \quad (1b)$$

$$HO_2 + HO_2 \longrightarrow H_2O_2 + O_2 \quad (15)$$

$$H_2O_2 + M \longrightarrow 2OH + M \quad (16)$$

$$OH + C_2H_5CHO \longrightarrow C_2H_5CO + H_2O \quad (44)$$

(shown diagrammatically in Fig. 30), and a computer simulation gave predicted pressure—time curves which agreed with those observed experimentally over most of the reaction. The equations

$$\frac{-d[C_2H_5CHO]}{dt} = k_{1a}[C_2H_5CHO][O_2] + \frac{k_{1b}}{k_{15}^{1/2}} G[C_2H_5CHO] + 2k_{16}[H_2O_2](M) \quad (XXIV)$$

$$\frac{d[H_2O_2]}{dt} = k_{1a}[C_2H_5CHO][O_2] + \frac{k_{1b}}{k_{15}^{1/2}} G[C_2H_5CHO] \quad (XXV)$$

where $G^2 = k_{15}[HO_2]^2 = k_{1a}[C_2H_5CHO][O_2] + k_{16}[H_2O_2][M]$, were solved by numerical integration. k_{16} was calculated from

$$k_{16} = 7.17 \times 10^9 \exp(-47{,}000/RT) \text{ l. mole}^{-1}\text{. sec}^{-1}$$

Using this value, the maximum rate of the predicted curve was adjusted by means of the ratio $k_{1b}/k_{15}^{1/2}$, and the general shape by means of k_{1a}. The most satisfactory fit was given when the ratio $k_{1b}/k_{15}^{1/2}$ was 1.167×10^{-3} (l . mole^{-1} . sec^{-1})$^{1/2}$ and k_{1a} was 7.6×10^{-2} l . mole^{-1} . sec^{-1}, k_{1b} being 1.267 l . mole^{-1} . sec^{-1} at 440 °C for M=N_2. A subsequent correction to allow for acetaldeyhde formation reduced the $k_{1b}/k_{15}^{1/2}$ ratio to 9.77×10^{-4} (l . mole^{-1} . sec^{-1})$^{1/2}$.

The values for the rate coefficients which were used in this work seem to be acceptable. Thus, by assuming a pre-exponential factor for (1a) of 10^{10} l . mole^{-1}. sec^{-1} E_{1a} is calculated to be 36 kcal . mole^{-1}, in good agreement with the predicted endothermicity (see Sect. 3.1.4). Evidently, at this temperature initation is homogeneous; this provides support for believing the low temperature initiation to be heterogeneous (Sect. 3.1.4). From the ratio $k_{1b}/k_{15}^{1/2}$ and Foner and Hudson's value [101] for k_{15} (1.8×10^9 l . mole^{-1}. sec^{-1}), k_{1b} at 440 °C was calculated to be 1.85×10^6 l . mole^{-1}. sec^{-1}. This agrees reasonably with values estimated for other peroxy radical attacks on aldehydes. A stationary state treatment of the mechanism (1a), (6a), (42), (1b), (15), (16) and (44) for the initial stages when H_2O_2 accumulation was negligible (and branching by (16) can be ignored) gives the initial rate expression

$$\rho_i = (k_{1b} k_{1a}^{1/2} / k_{15}^{1/2})[\text{RCHO}]^{3/2} [\text{O}_2]^{1/2} \qquad \text{(XXVI)}$$

in satisfactory agreement with the observed dependence on reactant concentrations. Insertion of the value for $k_{1b}/k_{15}^{1/2}$ obtained from the maximum rate studies gave k_{1b} as 1.0×10^{-2} l . mole^{-1} . sec^{-1} in agreement with the value obtained from the maximum rate measurements.

Baldwin et al. [22] rejected

$$C_2H_5 + O_2 \longrightarrow CH_3CHO + OH \qquad (43)$$

as an important acetaldehyde producing step since no pivaldehyde could be detected following additions of neopentane to the $H_2 + O_2$ system at 480 °C. This indicates that the neopentyl radical (which has no conjugate olefin) decomposes rather than reacting to give an aldehyde. Separate experiments showed that the absence of pivaldehyde was not a consequence of its rapid oxidation. The relative yields of C_2H_4 and CH_3CHO found when ethane was added to the $H_2 + O_2$ system [102] at 500 °C also suggest that (43) is not a major source of acetaldehyde. The authors suggested that the secondary C—H bond strength in propionaldehyde is probably only some 4 kcal . mole^{-1} greater than that of the aldehydic C—H bond. The small difference between the activation energies for attack at the secondary and aldehydic positions would favour acetaldehyde formation by

$$HO_2 + CH_3CH_2CHO \longrightarrow H_2O_2 + CH_3CHCHO \qquad (1c)$$

$$CH_3CHCHO + O_2 \longrightarrow \underset{\displaystyle OO}{CH_3\underset{|}{C}HCHO} \longrightarrow CH_3CHO + CO + OH \qquad (45)$$

If equal pre-exponential factors for (42) and (44) are assumed, the predicted relative rates of formation of acetaldehyde and ethylene fall in the range 0.1—0.2, that is, close to the observed ratio of 0.16.

Much of the previous quantitive formation concerning C_2H_5/O_2 reactions is based on observations made close to room temperature, where the rate of

$$C_2H_5 + O_2 \longrightarrow C_2H_5O_2 \qquad (46)$$

is high relative to those of (42) and (43). McMillan and Calvert [106] have pointed out that the activation energy of (42) is greater than that of (46) [103, 104]. This is certainly borne out by observations of the fate of

C_2H_5 produced during the oxidation of C_2H_5CHO both at intermediate and high temperatures. Little information on (43) is available, but rejection of this as the source of acetaldehyde at high temperature raises the question of its origin at the lower temperature of 220 °C (see Sect. 4.5). Assuming tentatively that acetaldehyde produced at all temperatures originates solely by (44) followed by (45), then from the estimated difference between E_{1b} and E_{1c} and the observed CH_3CHO/C_2H_4 ratio at 440 °C, the ratio at 220 °C is calculated to be 0.082. Although this value is probably somewhat low, since there will be some attack of perpropionyl radicals on the aldehyde, the observed ratio at this temperature was ca. 2.0; evidently acetaldehyde is formed by a different mechanism at the lower temperature. It is possible that at the higher temperature ethyl radical reacts with oxygen solely by (42), but that at lower temperatures some ethyl peroxy radical formation by (46) occurs. Decomposition of this to give acetaldehyde and OH is thermodynamically more favourable than its decomposition to give HO_2 and ethylene. However, the mechanism leading to the formation of acetaldehyde at lower temperatures is, at present, uncertain, and it cannot be ruled out that it may be formed via some route which does not involve the ethyl radical directly, for example, by reactions of the perpropionyl radical.

4.6.2 *Minor products*

The finite initial rate of CO_2 formation during the normal oxidation shows that it must, in the early stages at least, be a primary product, although some oxidation of CO in the latter stages is possible. Since at constant total pressure the CO/CO_2 ratio is proportional to the oxygen pressure, but is increased by increase of the total pressure, it seems probable that the ratio is controlled by competition between (6a) and

$$C_2H_5CO + O_2 \longrightarrow C_2H_5CO_3 \qquad (2)$$

followed by either

$$C_2H_5CO_3 \longrightarrow C_2H_4 + CO_2 + OH \qquad (47)$$

or

$$C_2H_5CO_3 \longrightarrow C_2H_5O + CO_2 \qquad (48)$$

The expected instability of the $C_2H_5CO_3$ radical at this temperature makes this origin of CO_2 more attractive, than that proposed for the oxidation at 220 °C (see Sect. 4). The observed CO/CO_2 ratio at 440 °C (0.12) indicates $k_2/k_{6a}[M]$ to be 4×10^{-3} torr^{-1} under the conditions used. From the value of k_{6a} (2.75×10^9 l . mole^{-1} . sec^{-1}, calculated from the data in Table 6) . Baldwin et al. [22] calculated k_2 to be 2.0×10^8 l . mole^{-1} . sec^{-1}. This is probably of the right order (cf. value for $CH_3CO + O_2$, p. 377).

Ethylene oxide is formed autocatalytically, but is also a primary product thought to originate via attack on the primary CH_3 group, viz.

$$CH_3CH_2CHO \xrightarrow{-H} \cdot CH_2CH_2CHO \xrightarrow{O_2}$$

$$\underset{\displaystyle OO\cdot}{\overset{\displaystyle CH_2CH_2CHO}{|}} \longrightarrow \underset{\displaystyle \diagdown O \diagup}{CH_2CH_2} + \cdot OH + CO_2$$

and hydrogen, also formed autocatalytically, is considered to arise from

$$OH + CO \longrightarrow H + CO_2 \tag{49}$$

$$H + C_2H_5CHO \longrightarrow C_2H_5CO + H_2 \tag{50}$$

Minor products (CH_4, HCHO and CH_3OH) presumably arise from reactions of methyl radicals generated during the oxidative degradation of the acetaldehyde.

A mechanism comprising steps (1a), (6a), (42), (1b), (15), (16) and (44) together with

$$C_2H_5 + C_2H_5CHO \longrightarrow C_2H_6 + C_2H_5CO \tag{51}$$

is consistent with the observed relative rates of formation of ethane and ethylene. Thus, the observed linear relationship between $R_{C_2H_4}/R_{C_2H_6}$ and $[O_2]/[C_2H_5CHO]$ gave a value for k_{42}/k_{51} of ca. 41. Brinton and Volman [107] give a value of k_{51} (440 °C) of 2.0×10^6 $l.mole^{-1}.sec^{-1}$, and Sampson's measurements [108] on ethane oxidation gave k_{42} as roughly 10^8 $l.mole^{-1}.sec^{-1}$ at 623 °C. From these values the predicted k_{42}/k_{51} ratio is ca. 50. The agreement with the observed ratio implies that, in contrast to the view expressed by McMillan and Calvert [106] (p. 424), (42) has virtually zero activation energy. If this is so then the formation of ethylene cannot occur via the formation of the ethyl peroxy radical since decomposition of this radical is sufficiently endothermic to make it likely that it would, if formed, react in some way other than giving ethylene.

The propionaldehyde oxidation has also been studied at high temperatures (440 °C) in KCl coated vessels [123]. In contrast to the oxidation in boric acid coated vessels, the reaction shows no autocatalysis and it appears that HO_2 and H_2O_2 are rapidly destroyed heterogeneously. Other heterogeneous processes also occur and the relatively high yield of CH_3CHO (compared with that given in boric acid coated vessels) is probably a consequence of such a reaction.

4.6.3 Comparison of the high temperature oxidations of CH_3CHO and C_5H_5CHO

The striking difference between the behaviour of acetaldehyde and propionaldehyde at higher temperatures (in marked contrast to their similar behaviour at low temperatures) deserves comment. It is clear that this difference arises mainly from the very different ways in which CH_3 and C_2H_5 react with oxygen. Ethyl readily reacts to give C_2H_4 and HO_2 by (42) thereby ensuring the maintenance of low stationary concentration of C_2H_5. The HO_2 which effectively replaces C_2H_5 is in sufficiently high concentration to be responsible for the termination step (15) but also, by virtue of (16), is able to build up a kinetically significant concentration of H_2O_2 responsible by (16) for the branching and autocatalysis. However, CH_3 formed in the acetaldehyde oxidation is not able to undergo any reaction equivalent to (42) and moreover, by virtue of (−32), a high concentration of methyl radicals is maintained. Consequently, methyl is the most abundant radical and termination is by (28). Little HO_2 accumulates and thus branching by (16) is unimportant.

4.7 HIGHER ALDEHYDES

Published work on the oxidation of aldehydes higher than propionaldehyde at intermediate and high temperatures is not yet extensive, although Baldwin et al. [109] have shown that the behaviour of *n*-butyraldehyde is similar to that of propionaldehyde.

It is interesting to speculate on the likely oxidation behaviour at high temperatures of the higher aliphatic aldehydes. Presumably, those aldehydes giving radicals capable of reacting with oxygen to give an olefin and HO_2 will show behaviour resembling that of propionaldehyde; those which give radicals which do not readily give HO_2 (and therefore H_2O_2) will show a different behaviour. Thus, one might expect quite different behaviour to be shown by $CH_3CH_2CH_2CH_2CHO$ and $(CH_3)_2CHCHO$ on the one hand and $(CH_3)_3CHO$ on the other, although it is possible that radical cracking may introduce complications.

4.8 UNSATURATED ALDEHYDES

Apart from some cool-flame studies with acrolein little attention has been given to the oxidation of the lowest unsaturated aldehyde.

Studies of the combustion of crotonaldehyde appear to have been confined mainly to the intermediate temperature region. An analytical and kinetic investigation [48] of this system at 292 °C shows the oxidation to be accompanied by an autocatalytic pressure increase, and the rate, as judged from the maximum rate, to be approximately 3/2-order with respect to the aldehyde and to be zero-order with respect to the oxygen at low aldehyde/oxygen ratios. For higher aldehyde/oxygen ratios the rate is oxygen dependent — a feature reminiscent of the behaviour shown by acetaldehyde and propionaldehyde in the intermediate temperature region. This dependency implies that the RCO radicals decompose as well as react with oxygen [111]. The appreciable yields of low molecular

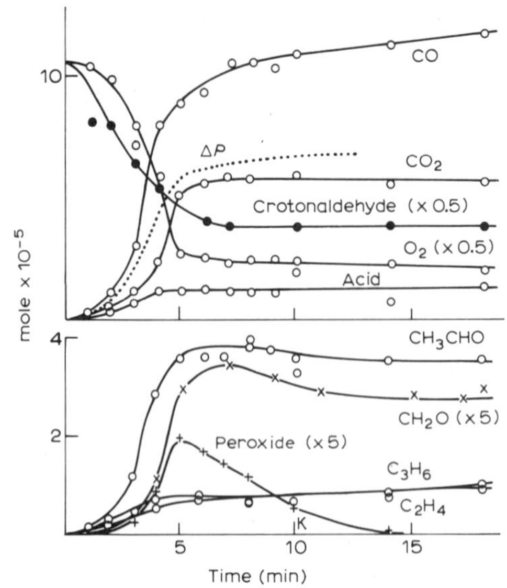

Fig. 31. Analysis throughout the course of the crotonaldehyde oxidation at 292 °C [48].

weight products (Fig. 31) confirm that some decomposition of intermediates occurs. A peroxidic mechanism is probably important at this temperature since the peroxide peak coincides in time with the maximum rate. However, additions of acetaldehyde in amounts comparable with those produced during the reaction enhance the rate, consequently some branching may occur via peracetic acid. The available evidence suggests that the acetaldehyde originates from the terminal CH_3H group of the parent aldehyde, possibly by

$$CH_3CH{=}CHCHO \xrightarrow{-H} CH_3\dot{C}H{\cdot}CH{=}CO \xrightarrow{O_2} CH_3\underset{\underset{\displaystyle CH_3CHO + CHO + CO}{|}}{\underset{OO\cdot}{\underset{|}{C}}}H{\cdot}CH{=}CO$$

although oxygen attack on the $CH_3CH{:}CH$ fragment — a possible product of the RCO decomposition — may also occur. The detailed mechanism is not clear, and a full interpretation requires an understanding of the behaviour of the crotonyl radical in the presence of active radicals. The formation of propene suggests that some decomposition of RCO occurs to give $CH_3CH{:}CH$, although this decomposition would be expected to occur less readily than that of the acetyl radical. Some direct attack on the olefinic bond is likely, although present analytical information is insufficient to allow comment to be made.

5. Cool flames and ignition phenomena

Most organic fuels show a pressure—temperature region in which their oxidation is accompanied by cool-flame formation. In static systems this feature is characterized by the occurrence of one or more pressure pulses, each of which is accompanied by a moderate temperature rise (< 200 °C) and the weak emission of radiation corresponding to the excited HCHO spectrum. The propagation velocity of cool flames is low (ca. 10 cm . sec^{-1}), and often they can be stabilized in flow systems.

Cool flames are difficult subjects for quantitative study since the time scale of events is generally too short to allow the use of conventional sampling. In addition, their non-isothermal character (which implies rate coefficients which change as reaction progresses) makes it difficult to develop theoretical models which satisfactorily describe the more important features (the periodicity and temperature rise). It is outside the scope of this review to discuss the more general theoretical aspects of cool-flame phenomena, and the reader is referred to Vol. 2, Chap. 2 of this Series and also to the work of Yang and Gray [113], Halstead et al. [114, 115] and others [112, 116, 117].

5.1 FORMALDEHYDE

Vanpée [118—121] has observed that between 500 and 600 °C the oxidation of formaldehyde was sometimes accompanied by luminenscence and pressure oscillations. Under certain conditions induction periods of several minutes occurred, particularly when the oxygen was in excess, and during this time the formaldehyde was consumed, the subsequent explosion possibly being that of a $CO—H_2—O_2$ mixture. For mixtures with oxygen containing more than two-thirds HCHO, the induction period was only 5 sec or so, and as ignition occurred after the total consumption of the oxygen it seems likely that peroxides (possibly performic acid or H_2O_2) are involved. The behaviour of formaldehyde—oxygen systems at these temperatures is strongly dependent on the vessel geometry and on the presence of inert gases in a way which suggests that thermal factors — self-heating effects and conductional and convectional heat losses — are important. At the present time information on this region of formaldehyde oxidation is too sparce to allow anything more than general comment.

5.2 ACETALDEHYDE AND HIGHER ALDEHYDES

General features of the ignition regions of acetaldehyde, propionaldehyde, iso- and *n*-butyraldehyde and acrolein have been described by Newitt et al. [110]. These authors were able to show that there was a general similarity in the ignition behaviour of these aldehydes, although

References pp. 435—439

the presence of a side-chain increased the resistance to oxidation (iso-butyraldehyde less reactive than *n*-butyraldehyde). However, detailed comparison of these systems is not possible because of the different conditions used and because of the absence of reliable analytical data.

A more detailed mapping of the ignition diagram for acetaldehyde has been given by Chamboux and Lucquin [51] (Fig. 18). They distinguished several different modes of behaviour depending on the temperature and reactant pressures. Thus, according to Fig. 18, at low pressures and temperatures there is a region described as "reaction d'initiation". The meaning of this appears to be that here the reaction, as judged from the intensity of light emission, shows an induction period. However, from studies of the low temperature oxidation of acetaldehyde (Sect. 3.1), it is clear that branched chain oxidation persists down to temperatures below 100 °C and that nothing equivalent to an induction period is evident when pressure measurements are used to follow the reaction. Thus, this region probably differs only slightly from, and merges into, the region of slow (degenerate) branching described as "reaction en chaines ramifieés".

A further region is recognized by the presence of a "pic d'arrêt", an abrupt but weak emission of light occurring towards the conclusion of reaction. A mechanistic interpretation of this has yet to be given, but this phenomenon indicates a high radical concentration at this stage of reaction.

A region characterized by the occurrence of single cool flames and double cool flames borders on the second stage ignition region which is realized at higher pressures. An unusual feature shown by this diagram is the occurrence of a third stage ignition region existing as a narrow peninsula which extends over a moderate pressure range but only over a small temperature range. It is to be expected that this diagram would be modified somewhat by changes in experimental conditions (surface/volume ratio, vessel diameter, and stirring) because of the importance of

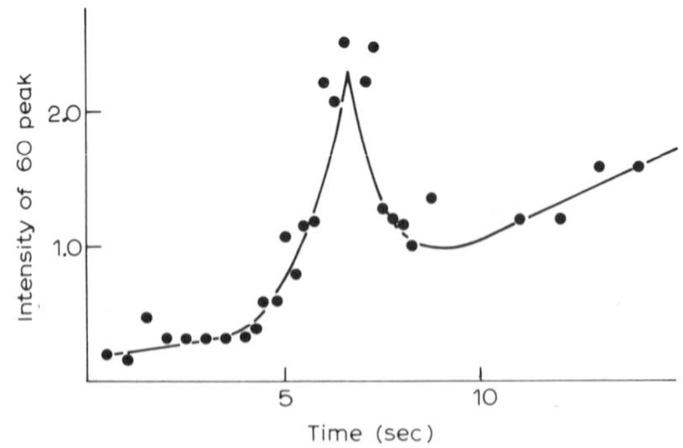

Fig. 32. The variation of the peak at *M/E* 60 (resulting from peracetic acid decomposition in the mass spectrometer) during the cool-flame oxidation of acetaldehyde [47].

self-heating and heat dissipation during the development of the cool flame. For a proper mechanistic interpretation of the chemistry involved in the various regions of Fig. 18 it is essential that the appropriate analytical data be available. In general these data are lacking largely because of the difficulties of sampling from systems in which the time scale is in the order of a few seconds, although sampling has been attempted by Blanchard et al. [47] who monitored events during the development of a cool flame by sampling the contents of a vessel via a pin hole into a fast scanning mass spectrometer. Their results (Fig. 32) show that peracetic acid which accumulates before the flame, decomposes during the passage of the flame. The peracid was not detected by its parent peak, but was presumed to decompose in the spectrometer to give rise to a peak of $m/e = 60$ corresponding to acetic acid. This interpretation was confirmed in separate experiments in which only peracetic acid was present in the vessel.

Flow systems offer the special advantage that by a suitable choice of initiating temperature and flow conditions, a spatial separation along the

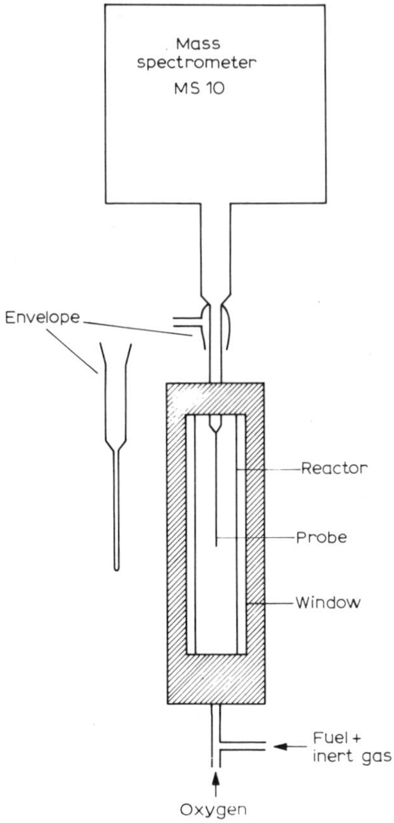

Fig. 33. Vertical flow reactor and mass spectrometer sampling line [57].

reactor tube is obtained of events which, in a static system, are separated in time. This separation greatly facilitates sampling.

Bradley et al. [55, 56] and later Griffiths et al. [57] used sampling via a probe into a mass spectrometer in order to establish the reaction profiles during the cool, second and third stage flames of CH_3CHO and C_2H_5CHO stabilized in a vertical tube reactor (Fig. 33). For CH_3CHO, a tube temperature initially of 200 °C and a low rate of oxygen flow gave a single cool flame in which the temperature was about 425 °C. Increase of the oxygen flow caused a second stage flame (~690 °C) to become stabilized above the cool flame. Further increase in the oxygen flow rate resulted in the formation of a yellow third stage flame. Small amounts of peracetic acid were detected by conventional chemical sampling in the gases before the cool flame. The variations in the yields of the other major products are shown in Fig. 34*, and it was considered that these profiles could be explained in terms of the scheme (1a), (2), (3), followed by

$$CH_3CO_3H \longrightarrow CH_3 + CO_2 + OH \tag{5a}$$

$$CH_3 + CH_3CHO \longrightarrow CH_4 + CH_3CO \tag{26}$$

$$CH_3CO + M \longrightarrow CH_3 + CO + M \tag{6a}$$

$$HCHO + O_2 \longrightarrow HCO + HO_2 \tag{22}$$

$$HCO + O_2 \longrightarrow CO + HO_2 \tag{11}$$

$$HO_2 + CH_3CHO \longrightarrow H_2O_2 + CH_3CO \tag{1b}$$

$$2HO_2 \longrightarrow H_2O_2 + O_2 \tag{15}$$

$$H_2O_2 + M \longrightarrow 2OH + M \tag{16}$$

$$OH + CH_3CHO \longrightarrow H_2O + CH_3CO \tag{52}$$

The temperature of the gases rises as they pass up the heated reactor. Before the cool flame, when the temperature is not very high, it seems reasonable to suppose that the oxidation is sustained by peracid branching, possibly, by the first-order step (5a) which will be rapid above 200 °C, and the increased branching results in a self-heating and consequently accelerating process which culminates in the cool flame. The high radical concentration prevailing in the flame is probably responsible for the high rate of consumption of aldehyde in, and immediately following, the flame. Beyond this point the rate of aldehyde consumption falls somewhat because at the temperature between the flames (6a)

* Recently, Sheinson and co-workers [134] have been able to show the presence of small amount of methyl hydroperoxide in an acetaldehyde cool flame stabilized in a vertical tube reactor.

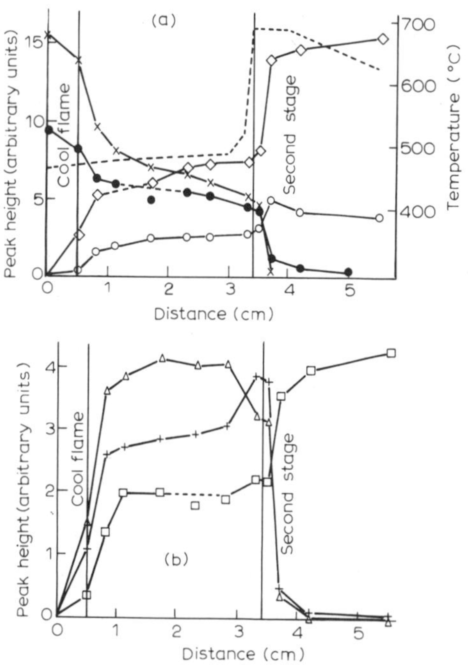

Fig. 34. Composition profile of the cool and second stage flames during acetaldehyde oxidation [56]; (a) ×, Acetaldehyde; ◊, carbon monoxide; ○, methane; ●, oxygen; ----, temperature. (b) □, Carbon dioxide; +, formaldehyde; △, methanol.

predominates over the peroxidation step. Under these conditions formaldehyde and methanol formation occur and reactions (22) and (11) become important. Between the flames branching will be mainly by (22), followed by (11), (1b), (15) and (16). However, it is not until immediately before the second stage that the rate of branching and the overall rate of oxidation and of heat release become sufficiently high for the second stage to develop. At the relatively high temperature of the second stage the H_2O_2 produced will rapidly decompose to give a high radical concentration which results in a rapid consumption of acetaldehyde, formaldehyde and methanol. Acetylene and hydrogen are probably formed by the oxygen catalysed pyrolysis of the methane, and the yellow flame is probably the decomposition flame of the acetylene.

Cool flames were somewhat more difficult to establish with propionaldehyde than with acetaldehyde and it was necessary to use a higher initial tube temperature (270 °C). In its general features, this system resembled that of acetaldehyde, although the second stage flame was less sharply defined. The analytical data were more complex and considerable production of ethylene occurred, presumably via (42), the ethyl radicals being the result of C_2H_5CO dissociation.

References pp. 435–439

Recently Halstead et al. [122, 122a] have proposed a model for acetaldehyde cool flame combustion which is basically similar to that described above. Their treatment accounts for the periodicity and the self-quenching which is attributed to a thermal switch in which the decomposition of CH_3CO

$$CH_3CO + M \longrightarrow CH_3 + CO + M \tag{6a}$$

$E_a \sim 15$ kcal . mole^{-1} competes with the low activation energy sequence (2) followed by (3) responsible for the degenerate branching. During the flame the temperature at first rises. This results in a decreased rate of peracetic acid formation and consequently the temperature falls. This model, which was examined numerically using selected values for the various rate coefficients, accounted for reagent consumption, but gave no account of the spatial distribution. Although, only preliminary results have been described, the approach seems to offer attractive possibilities. This model also shows the sharp transition from slow to cool flame combustion, the negative temperature coefficient in the region of the flame and the dependence of the temperature rise on the initial conditions (thermal capacity, etc.). (See also Chap. 2.)

5.3 EFFECT OF ADDITIVES ON ALDEHYDE COOL FLAMES

Those additives which affect cool flame combustion presumably do so by interfering with critical reaction steps in a manner analogous to the normal retarder action considered in Sect. 3.5.1. The chemistry of this interference is not clear since both retarder action and cool flame phenomena are imperfectly understood. However, additions of C_2H_6 to acetaldehyde—oxygen mixtures [70] in the cool flame region markedly influenced the behaviour as shown in Fig. 35. The transition from cool flame to slow oxidation with increasing amount of additive is probably the result of the replacement of CH_3CO_3 radicals by the less reactive HO_2 through the sequence

$$CH_3CO_3 + C_2H_6 \longrightarrow CH_3CO_3H + C_2H_5 \tag{53}$$

$$C_2H_5 + O_2 \longrightarrow C_2H_4 + HO_2 \tag{42}$$

A similar sharp transition has been noted [43] when formaldehyde is the additive. Furthermore, there is an accompanying shift of the cool flame limits to higher pressures and/or temperatures. Formaldehyde also interferes with the cool flame combustion of acetaldehyde in flow systems [7]. Its addition retards the development of the cool flame but promotes the second stage; this promotion may be caused by the occurrence of

$$X + HCHO \longrightarrow XH + HCO \ (X = \text{active radical})$$

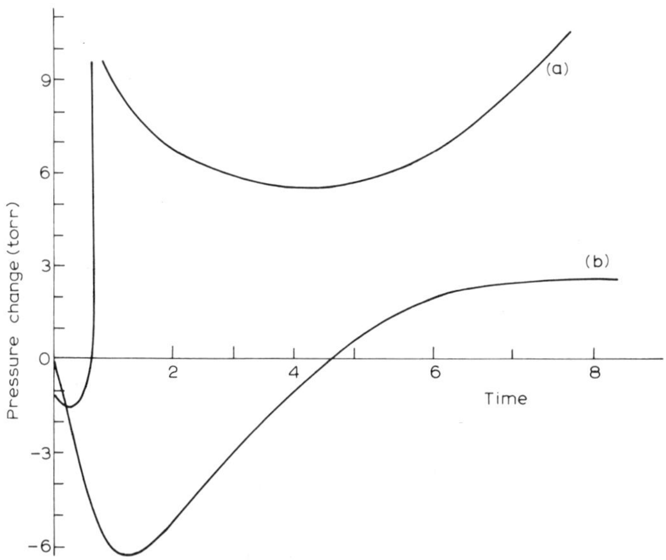

Fig. 35. The effect of ethane addition on the cool-flame oxidation of acetaldehyde in a static system [70]. (a) Pressure—time plot for the cool-flame oxidation of acetaldehyde. Total pressure of a 1:1 acetaldehyde—oxygen mixture = 73 torr. Temperature = 230 °C. (b) As for (a) but with 10.9 torr of ethane present.

followed by (11) and (1b), the H_2O_2 so formed leading to branching.

In contrast to the view that peracetic acid is responsible for the branching in the cool flame temperature region, it has been suggested that methyl hydroperoxide may play an important role [124]. In systems involving CH_3CO radicals and oxygen, CO becomes a major product and CO_2 negligible as the temperature is increased. This may be attributable to the onset of complete acyl radical decomposition, in which case, the rapid methyl radical—oxygen reactions could lead to CH_3OOH formation. In the absence of comprehensive analytical information, further comment cannot be made.

REFERENCES

1 G. J. Minkoff and C. F. H. Tipper, Chemistry of Combustion Reactions, Butterworths, London, 1962.
2 B. Lewis and G. von Elbe, Combustion, Flames and Explosions in Gases, Academic Press, New York, 1961.
3 V. Ya Shtern, The Gas Phase Oxidation of Hydrocarbons, Pergamon, Oxford, 1964.
4 N. N. Semenov, Some Problems of Chemical Kinetics and Reactivity, Pergamon, London, 1958.
5 R. G. W. Norrish and S. G. Foord, Proc. R. Soc. London, Ser. A, 157 (1936) 503.
6 N. S. Enikolopyan, G. V. Korolev and G. P. Savushkina, Zh. Fiz. Khim., 31 (1957) 865.

7 J. F. Griffiths, Thesis, Liverpool, 1967.
8 A. Lemay and C. Ouellet, Can. J. Chem., 35 (1957) 124.
9 D. J. Dixon, Thesis, Liverpool, 1973; D. J. Dixon, G. Skirrow and C. F. H. Tipper, to be published.
10 J. D. Mullen and G. Skirrow, Proc. R. Soc. London, Ser. A, 244 (1958) 312.
11 A. Blundell and G. Skirrow, Proc. R. Soc. London, Ser. A, 244 (1958) 331.
12 G. Skirrow and A. Williams, Proc. R. Soc. London, Ser. A, 268 (1962) 537.
13 K. C. Salooja, Combust. Flame, 9 (1965) 373.
14 J. Brown and C. F. H. Tipper, Proc. R. Soc. London, Ser. A, 312 (1969) 399.
15 M. Niclause, J. Chim. Phys., 49 (1952) 157.
16 A. Boyer, M. Niclause and M. Letort, J. Chim. Phys., 49 (1952) 337.
17 M. Niclause and A. Combe, Rev. Inst. Fr. Pet. Ann. Combust. Liq., 8 (1953) 311.
18 M. Niclause, Rev. Inst. Fr. Pet. Ann. Combust. Liq., 9 (1954) 327, 419.
19 A. Combe, M. Niclause and M. Letort, Rev. Inst. Fr. Pet. Ann. Combust. Liq., 10 (1955) 786, 929.
20 G. Skirrow and B. P. Whim, Combust. Flame, 12 (1968) 269.
21 R. R. Baldwin, M. J. Matchan and R. W. Walker, Combust. Flame, 15 (1970) 109.
22 R. R. Baldwin, R. W. Walker and D. H. Langford, Trans. Faraday Soc., 65 (1969) 792, 806.
23 R. R. Baldwin, personal communication; R. R. Baldwin, A. R. Fuller, D. Longthorn and R. W. Walker, J. Chem. Soc., Faraday Trans. I, 70 (1974) 1257. (a) R. R. Baldwin, A. R. Fuller, D. Longthorn and R. W. Walker, J. Chem. Soc., Faraday Trans. I, 68 (1972) 1362.
24 J. M. Hay and K. Hessam, Combust. Flame, 16 (1971) 237.
25 W. A. Bone and J. B. Gardner, Proc. R. Soc. London, Ser. A, 154 (1936) 297.
26 E. I. Akeroyd and R. G. W. Norrish, J. Chem. Soc. (1936) 890.
27 D. W. E. Axford and R. G. W. Norrish, Proc. R. Soc. London, Ser. A, 192 (1948) 518.
28 R. G. W. Norrish, Nature (london), 162 (1948) 367.
29 A. J. Harding and R. G. W. Norrish, Proc. R. Soc. London, Ser. A. 212 (1952) 291.
30 M. D. Scheer, 5th Symposium on Combustion, Reinhold, New York, 1955, p. 435.
31 R. G. W. Norrish and J. M. Thomas, Nature (London), 210 (1966) 728.
32 P. J. Askey, J. Am. Chem. Soc., 52 (1930) 974.
33 A. M. Markevich and Yu. I. Pecherskaya, Zh. Fiz. Khim., 35 (1961) 697.
34 A. M. Markevich and L. F. Filippova, Zh. Fiz. Khim., 31 (1957) 2649.
35 A. M. Markevich and L. F. Filippova, Zh. Fiz. Khim., 33 (1959) 358.
36 A. A. Anisonyan, S. Ya. Beider, A. M. Markevich and A. B. Nalbandyan, Zh. Fiz. Khim., 33 (1959) 115.
37 I. A. Vardanyan and A. B. Nalbandyan, Kinet. Catal. (U.S.S.R.), 11 (1970) 927.
38 I. A. Vardanyan, G. A. Sachyan and A. B. Nalbandyan, Combust. Flame, 17 (1971) 315.
39 A. D. Jenkins and D. W. G. Style, J. Chem. Soc. (1953) 2387.
40 B. L. Dunicz, D. D. Perrin and D. W. G. Style, Trans. Faraday Soc., 47 (1951) 1210.
41 A. C. Egerton, E. J. Harris and G. H. S. Young, Trans. Faraday Soc., 44 (1948) 745.
42 J. F. Griffiths and G. Skirrow, Oxid. Combust. Rev., 3 (1968) 47.
43 D. J. Dixon, Thesis, Liverpool, 1973; D. J. Dixon, G. Skirrow and C. F. H. Tipper, J. Chem. Soc., Faraday Trans. 1, 70 (1974) 1078, 1090.
44 D. I. Bell, G. Skirrow and C. F. H. Tipper, Combust. Flame, 12 (1968) 557.
45 R. N. Pease, J. Am. Chem. Soc., 55 (1933) 2753.
46 B. J. Poppleton and M. F. R. Mulcahy, Aust. J. Chem., 19 (1966) 65.

47 L. P. Blanchard, J. B. Farmer and C. Ouellet, Can. J. Chem., 35 (1957) 115.
48 A. D. Osborne and G. Skirrow, J. Chem. Soc., (1960) 2750.
49 J. F. Griffiths, B. F. Gray and P. Gray, 13th Symposium on Combustion, The Combustion Institute, Pittsburgh, 1971, p. 239.
50 J. F. Griffiths, P. G. Felton and P. Gray, 14th Symposium on Combustion, The Combustion Institute, Pittsburgh, 1973, p. 453.
51 J. Chamboux and M. Lucquin, J. Chim. Phys. (1962) 797.
52 J. Chamboux and M. Lucquin, J. Chim. Phys. (1963) 521, 527.
53 N. J. H. Small and A. R. Ubbelohde, J. Chem. Soc. (1950) 723.
54 N. J. H. Small and A. R. Ubbelohde, J. Chem. Soc. (1953) 637.
55 J. N. Bradley, G. A. Jones, G. Skirrow and C. F. H. Tipper, 10th Symposium on Combustion, 1965, p. 139.
56 J. N. Bradley, G. A. Jones, G. Skirrow and C. F. H. Tipper, Combust. Flame, 10 (1966) 259.
57 J. F. Griffiths, G. Skirrow and C. F. H. Tipper, Combust. Flame, 13 (1969) 195.
58 R. S. Kelly and G. Skirrow, unpublished results.
59 A. Hardacre, G. Skirrow and C. F. H. Tipper, Combust. Flame, 9 (1965) 53.
60 C. A. McDowell and J. H. Thomas, J. Chem. Soc. (1949) 2208, 2217.
61 C. A. McDowell and J. H. Thomas, J. Chem. Soc. (1950) 1462.
62 C. A. McDowell and L. K. Sharples, Can. J. Chem., 36 (1958) 251, 258, 268.
63 J. B. Farmer and C. A. McDowell, Trans. Faraday Soc., 48 (1952) 624.
64 J. Chapman and G. Skirrow, unpublished results.
65 C. A. Cullis and D. J. Waddington, Trans. Faraday Soc., 53 (1957) 1317.
66 C. A. Cullis and B. A. Khokar, Trans. Faraday Soc., 56 (1960) 1235.
67 C. A. Cullis and D. J. Waddington, Trans. Faraday Soc., 55 (1959) 2069.
68 C. A. Cullis and D. J. Waddington, Proc. R. Soc. London, Ser. A, 244 (1958) 110.
69 D. J. Waddington and D. J. M. Ray, 13th International Symposium on Combustion, The Combustion Institute, Pittsburgh, Penn., 1971, 261.
70 D. Gal, I. Galiba and Z. G. Szabo, Acta Chim. Acad. Sci. Hung., 16 (1958) 39.
71 D. J. Waddington, Oxidation Discussion, Liverpool, 1969.
72 J. F. Griffiths, G. Skirrow and C. F. H. Tipper, Combust. Flame, 12 (1968) 360.
73 L. M. Postnikov, V. Ya. Shlyapintokh and M. N. Shumilina, Kinet. Catal., 6 (1965) 161.
74 L. Bateman, Q. Rev. Chem. Soc., 8 (1954) 147.
75 H. Wieland, Chem. Ber., 54 (1921) 2353.
76 M. K. Kagan and G. D. Lubarsky, J. Chem. Phys., 39 (1935) 837, 847.
77 C. E. H. Bawn and J. B. Williamson, Trans Faraday Soc., 47 (1951) 721.
78 O. P. Yablonskii, M. G. Vinogradov, R. V. Kereselidze and G. I. Kikishin, Izv. Akad. Nauk S.S.S.R., Ser. Khim. (1969) 318.
79 M. G. Vinogradov, R. V. Kereselidze and G. I. Kikishin, Bull. Akad. Sci. U.S.S.R., Div. Chem. Sci. (1970) 998.
80 C. A. McDowell and S. Sifniades, Can. J. Chem., 41 (1963) 300.
81 T. G. Traylor and N. Clinton, 157th Meeting Amr. Chem. Soc., Div. Pet. Chem., Preprint No. 13, 1969.
82 N. A. Sokolova, A. M. Markevich and A. B. Nalbandyan, Russ. J. Phys. Chem., 35 (1961) 415.
83 W. Braun, L. Rajenback and F. R. Eirich, J. Phys. Chem., 66 (1962) 1591.
84 D. R. Cowden, Thesis, Liverpool, 1970.
85 H. E. O'Neale and S. W. Benson, J. Chem. Phys., 36 (1962) 2196.
86 P. Cadman, C. Dodwell, A. F. Trotman-Dickenson and A. J. White, J. Chem. Soc. A (1970) 2371. See also J. A. Kerr and A. C. Lloyd, Trans. Faraday Soc., 63 (1967) 2480.
87 P. Cadman, A. F. Trotman-Dickenson and A. J. White, J. Chem. Soc. A (1970) 3189.

88 T. A. Ingles and H. W. Melville, Proc. R. Soc. London, Ser. A, 218 (1953) 175.
89 E. Raymond, J. Chim. Phys., 28 (1931) 316.
90 E. Raymond, C. R. Acad. Sci., 191 (1930) 616.
91 M. F. R. Mulcahy and I. C. Watt, Proc. R. Soc. London, Ser. A, 216 (1953) 30.
92 R. K. Solly and S. W. Benson, J. Am. Chem. Soc., 93 (1971) 2127.
93 J. A. Kerr, Chem. Rev., 66 (1966) 465.
94 S. W. Benson and R. Shaw, Organic Peroxides, Vol. I, Wiley, New York, 1970, p. 105.
95 S. W. Benson and J. H. Buss, J. Chem. Phys., 29 (1958) 546.
96 S. W. Benson, Chem. Rev., 69 (1969) 279.
97 S. W. Benson, J. Am. Chem. Soc., 87 (1965) 972.
98 C. Schmidt and A. H. Sehon., Can. J. Chem., 41 (1963) 1819.
99 R. E. Dodd, Can. J. Chem., 33 (1955) 699.
100 D. H. Volman and R. K. Brinton, J. Chem. Phys., 20 (1952) 1764.
101 S. N. Foner and R. L. Hudson, in R. F. Gould (Ed.), Advances in Chemistry, Am. Chem. Soc., Washington, 1962, p. 36.
102 R. R. Baldwin, D. E. Hopkins and R. W. Walker, Trans. Faraday Soc., 66 (1970) 189.
103 D. P. Dingledy and J. D. Calvert, J. Am. Chem. Soc., 85 (1963) 856.
104 J. E. Jolley, J. Am. Chem. Soc., 79 (1957) 1537.
105 (a) L. Szirovicza and R. Walsh, J. Chem. Soc., Faraday Trans. 1, 70 (1974) 33. (b) K. W. Watkins and W. W. Ward, Int. J. Chem. Kinet., 6 (1974) 855.
106 G. R. McMillan and J. G. Calvert, Oxid. and Combust. Rev., 1 (1965) 84.
107 R. K. Brinton and D. H. Volman, J. Chem. Phys., 22 (1954) 929.
108 R. J. Sampson, J. Chem. Soc. (1963) 5095.
109 R. R. Baldwin, R. W. Walker and D. A. Yorke, J. Chem. Soc., Faraday Trans. 1, 69 (1973) 826; R. R. Baldwin, C. J. Cleugh and R. W. Walker, J. Chem. Soc., Faraday Trans. 1, 72 (1976) 1715.
110 D. M. Newitt, L. M. Baxt and V. V. Kelkar, J. Chem. Soc. (1939) 1703, 1171.
111 C. A. McDowell and S. Sifniades, J. Am. Chem. Soc., 84 (1962) 4606.
112 J. F. Griffiths, Chem. Commun. (1969) 483.
113 C. H. Yang and B. F. Gray, J. Phys. Chem., 73 (1969) 3395.
114 M. P. Halstead, A. Prothero and C. P. Quinn, Chem. Commun. (1970) 1150.
115 M. P. Halstead, J. Prothero and C. P. Quinn, Proc. R. Soc. London, Ser. A, 322 (1971) 483; Combust. Flame, 20 (1973) 211.
116 D. A. Frank-Kamenetski, Dokl. Akad. Nauk S.S.S.R. (1939) 672.
117 I. E. Salnikoff, Zh. Fiz. Khim., 23 (1949) 258.
118 M. Vanpée, C. R. Acad. Sci., 241 (1955) 951.
119 M. Vanpée, Bull. Soc. Chim. Belg., 62 (1953) 285, 661.
120 M. Vanpée, C. R. Acad. Sic., 242, (1956) 373.
121 M. Vanpée, Thesis, Catholic University of Louvain, 1956.
122 M. P. Halstead, A. Prothero and C. P. Quinn, Chem. Commun. (1970) 1150. (a) M. P. Halstead, A. Prothero and C. P. Quinn, Proc. R. Soc., London, Ser. A, 322 (1971) 483.
123 R. R. Baldwin, M. J. Matcham and R. W. Walker, Trans. Faraday Soc., 67 (1971) 3521.
124 D. E. Hoare and D. E. Lill, J. Chem. Soc., Faraday Trans. 1, 69 (1973) 603.
125 W. H. Hatcher, E. W. R. Steacie and F. Howland, Can. J. Res., 7 (1931) 648.
126 C. A. McDowell and J. H. Thomas, J. Chem. Phys., 17 (1949) 588.
127 W. H. Hatcher, E. W. R. Steacie and F. Howland, Can. J. Res., 7 (1932) 149.
128 M. Bodenstein, Z. Phys. Chem., Abt. B, 12 (1931) 151.
129 R. Fort and C. N. Hinshelwood, Proc. R. Soc. London, Ser. A, 129 (1930) 974.
130 R. Spence, J. Chem. Soc. (1936) 649.
131 F. F. Snowdon and D. G. W. Style, Trans. Faraday Soc., 35 (1939) 426.

132 J. H. Thomas, Thesis, University of Liverpool, 1949.
133 E. O. Ogaressyan, I. A. Vorlanyan and A. B. Nalbandyan, Int. Sym. Gas Kinetics, Brussels, 1973.
134 J. J. DeCorpo, M. V. McDowell, R. S. Sheinson and J. R. Wyatt, J. Chem. Soc., Chem. Commun. (1974) 533.
135 P. W. Jones, K. Selby, M. J. Tidball and D. J. Waddington, Combust. Flame, 22 (1974) 209.
136 I. A. Vardanyan, A. B. Nalbandyan, G. A. Sachyan and A. G. Philiposyan, Combust. Flame, 22 (1974) 153.
137 R. A. Cox, R. G. Derwent, P. M. Holt and J. A. Kerr, J. Chem. Soc., Faraday Trans. 1, 72 (1976) 2061. Study of the photolysis of HONO, NO and NO_2 in CH_3CHO/air.
138 Z. P. Prisyazhnyuk, S. S. Levush and V. V. Shevchuk, Kinet. Catal. (U.S.S.R.), 16 (1975) 1397.

Chapter 4

Gas phase combustion of organic compounds other than hydrocarbons and aldehydes

J. A. BARNARD

1. Alcohols

Early studies of slow combustion and ignition characteristics of organic compounds showed that although methanol, ethanol and propanol are more readily oxidized than methane, ethane and propane, the situation is reversed with butanol and the higher homologues [1—4].

More detailed studies of the slow combustion of the lower aliphatic alcohols have shown that in the early stages the tendency is for reaction to take place more or less exclusively according to the overall equation

$$R_1R_2CHOH + O_2 = R_1R_2CO + H_2O_2$$

the principal products being hydrogen peroxide and the aldehyde or ketone containing the same number of carbon atoms as the original alcohol [5]. The yields of hydrogen peroxide depend on the surface

TABLE 1
The slow combustion of aliphatic alcohols

Fuel	Temp. (°C)	Partial pressure Fuel (torr)	Oxygen (torr)	Induction period (min)	Max. rate of pressure rise (torr . min^{-1})	Ref.
CH_3OH	440	100	100	ca. 2 min. to max. rate	11	7
C_2H_5OH	330	100	50	11.4	24.6	21
n-C_3H_7OH	330	100	50	6.6	39.8	21
i-C_3H_7OH	330	100	50	15.5	1.0	21
n-C_4H_9OH	295	40	60	5.2		
	295	40	200	5.7	21	26
	340	40	60	1.3		
i-C_4H_9OH	295	40	60	31.1		
	295	40	200	17.4	18	26
	400	40	60	0.8		
sec-C_4H_9OH	295	40	60	39.0		
	295	40	200	23.2	10	26
	400	40	60	1.3		
tert-C_4H_9OH	400	40	60	nil	v. small	26
	400	40	200	nil	25	

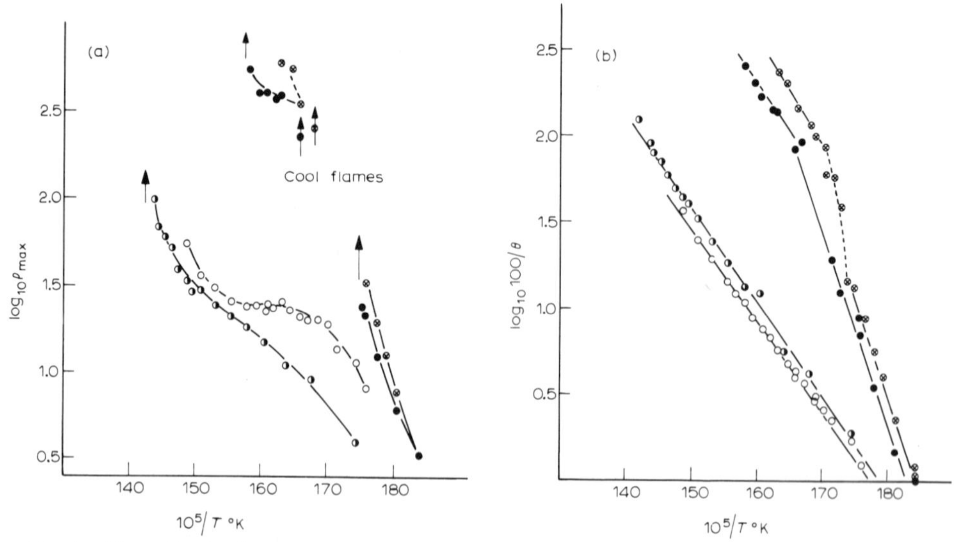

Fig. 1. The effect of temperature on (a) ρ_{max}, the maximum rate of oxidation in torr. min^{-1}; (b) the reciprocal of θ, the induction period in minutes for combustion of alcohols [18]. Reactant concentrations correspond to an alcohol pressure of 40 torr and an oxygen pressure of 200 torr at 323 °C. ◑, Ethanol; ○, propanol; ●, butanol; ⊗, pentanol.

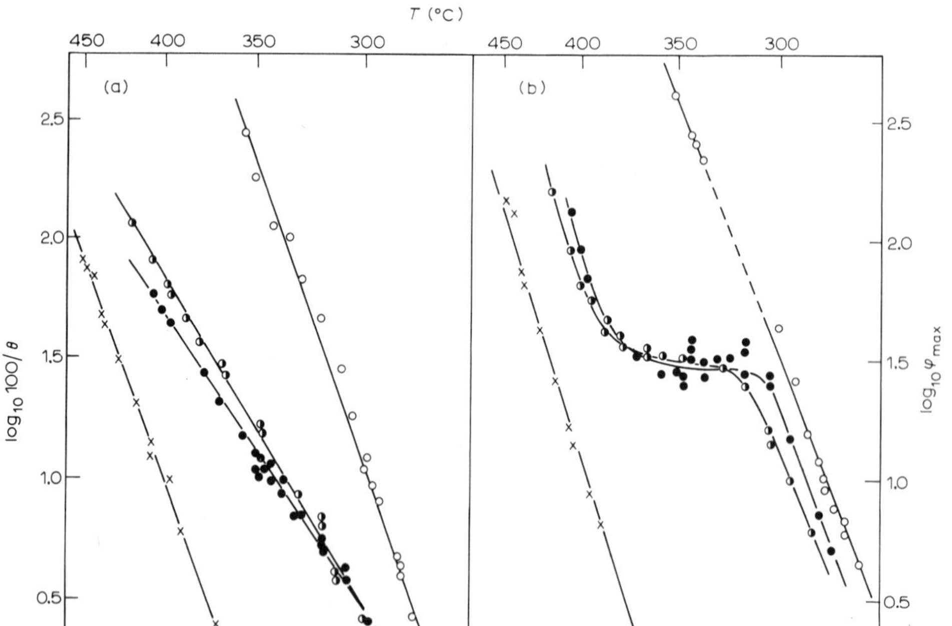

Fig. 2. The influence of temperature on θ, the induction period in minutes, and ρ_{max} maximum rate of pressure change in torr . min^{-1} of combustion of alcohols [26]. Alcohol pressure (at 295 °C), 40 torr; oxygen pressure (at 295 °C) 200 torr. ○, *n*-butanol; ●, iso-butanol; ◑, *sec*-butanol; x, *tert*-butanol.

conditions, being lowest in potassium chloride coated vessels; potassium chloride is, of course, known to be particularly destructive towards peroxides [6].

Once the aldehyde concentration builds up to a critical level (and even in the oxidation of secondary alcohols, small quantities of aldehydes are produced by side-reactions), then autocatalysis is observed. It is suggested that the aldehydes bring about chain-branching by reacting directly with oxygen

$$RCHO + O_2 \longrightarrow R\dot{C}O + HO_2\cdot$$

Table 1 and Figs. 1 and 2 summarize some comparative rate data.

1.1 METHANOL

Methanol is somewhat less reactive than its higher homologues and slow combustion takes place at a conveniently measurable rate only above 390 °C. In uncoated pyrex vessels [7], or vessels coated with boric acid or potassium chloride [8], reaction begins immediately without a true induction period and accelerates to a maximum rate. This maximum is increased by the addition of inert gas and is proportional to the square of the initial methanol concentration, (in boric acid coated vessels this power is about 2.5) but independent of oxygen concentration over a wide range of conditions. The overall activation energy (calculated from the effect of temperature on the maximum rate of pressure change) is about 40 kcal . mole^{-1} in coated vessels and about 53—61 kcal . mole^{-1} in uncoated ones.

The products of reaction in an uncoated vessel included carbon monoxide and water, with smaller amounts of formaldehyde, hydrogen peroxide, carbon dioxide and hydrogen. The formaldehyde concentration rose to a maximum at the maximum rate of pressure change, and this maximum was directly proportional to the initial concentration of methanol but independent of oxygen concentration. The maximum formaldehyde partial pressure also increased with temperature with an activation energy of 12 kcal . mole^{-1}. Similar products, but with some significant variations, were found using vessels coated with boric acid, potassium chloride or sodium hydroxide. One interesting observation with boric acid coated vessels was the sudden small increase in rate near the end of the reaction; this effect was particularly noticeable at relatively low oxygen pressures. Presumably this is the "pic d'arrêt" noticed by Lucquin [9], and connected by Sochet [10] with "oxygen cut-off" observed in liquid phase oxidation.

Bell and Tipper [7] have proposed a mechanism to account for their observations in which a chain reaction is propagated by HO_2 radicals and branching occurs through formaldehyde oxidation. The scheme is

$$CH_3OH + O_2 \longrightarrow \cdot CH_2OH + HO_2\cdot \quad (1)$$

$$CH_3OH + HO_2\cdot \longrightarrow \cdot CH_2OH + H_2O_2 \quad (2)$$

$$\cdot CH_2OH + O_2 \longrightarrow CH_2O + HO_2\cdot \quad (3)$$

$$CH_2O + O_2 \longrightarrow \cdot CHO + HO_2\cdot \quad (4)$$

$$\cdot CHO + O_2 \longrightarrow CO + HO_2\cdot \quad (5)$$

$$CH_2O + HO_2\cdot \longrightarrow \cdot CHO + H_2O_2 \quad (6)$$

$$HO_2 + O_2 \longrightarrow \text{inert products} \quad (7)$$

The effects of surface and of inert gas can be understood if it is assumed that (7), and perhaps (1), are heterogeneous.

By making the justifiable assumption that the rate of (1) is small compared with the maximum rate of reaction, this mechanism leads to a rate equation

$$r_{max} = kp^2_{CH_3OH}$$

in agreement with the experimental findings. The dependence of maximum formaldehyde concentration of methanol concentration is also accounted for.

Methanol is also less reactive than ethanol at very high temperatures; the ignition delay, measured in a shock tube at temperatures between 1570 and 1870 K is about twice that of ethanol. The extent of CH emission is very much smaller than for ethanol and is much less than that expected on the basis of results reported for the spectra of methanol—air flames [14(b)].

The premixed methanol flame [11, 12] does not show the Swan bands of C_2, which are prominent in a methane flame [13]. The base of the flame shows strong emission from excited formaldehyde and further up the flame emission from OH and CH occurs. The burning velocity of a stoichiometric methanol—air flame [12] is about 45 cm . sec^{-1}, and the "global activation energy" and "global order" are 43—47 kcal . mole^{-1} and unity, respectively [14(a)].

1.2 ETHANOL

The combustion of ethanol has been studied between 270 and 370 °C by Cullis and Newitt [15—18]. At first, there is a period of some minutes during which no pressure change is discernible, although acetaldehyde is accumulating in the system [15]. When the acetaldehyde concentration reaches a critical level, the pressure begins to rise autocatalytically and methanol, formaldehyde and carbon monoxide become detectable [16].

The critical concentration of acetaldehyde appears to be independent of reactant ratio or temperature.

The induction period can be eliminated by the addition of acetaldehyde to the reaction mixture, and the minimum quantity required to reduce the induction period to zero is that normally present at its end [16].

During the induction period the only other product detected was hydrogen peroxide; in a potassium chloride coated vessel this was not found, although the other products were unchanged [17].

A simple HO_2 chain reaction was proposed [16]; this gave the main products, viz.

$$CH_3CH_2OH + O_2 \longrightarrow CH_3\dot{C}HOH + HO_2\cdot$$

$$CH_3\dot{C}HOH + O_2 \longrightarrow CH_3CHO + HO_2\cdot$$

$$HO_2\cdot + CH_3CH_2OH \longrightarrow CH_3\dot{C}HOH + H_2O_2$$

Chain-branching (and hence, autocatalysis) occurred via oxidation of acetaldehyde

$$CH_3CHO + O_2 \longrightarrow CH_3\dot{C}O + HO_2\cdot$$

$$CH_3\dot{C}O \longrightarrow CH_3\cdot + CO$$

$$CH_3\cdot + O_2 \longrightarrow \text{products including HCHO, } CH_3OH$$

which also led to the products which appeared after the induction period.

In spite of the fact that there is no evidence of a pronounced negative temperature coefficient of the rate [18], ethanol gives rise to cool flames fairly readily [19]. In a conventional static system, at temperatures between 280 and 330 °C, cool flames occur at pressures above 200 torr for an equimolar ethanol—oxygen mixture. Up to three cool flames can be observed over a narrower range of conditions. However, in all cases there is an induction period of some minutes (e.g. about 24 min for 150 torr ethanol + 150 torr oxygen at 301 °C) during which very little reaction takes place. This is followed by a much shorter period (about 90 sec under the same conditions) during which the rate, as measured by change of either temperature or pressure, accelerates exponentially to the cool flame. The acceleration constant, ϕ, calculated from the slope of the plot of log ΔT against time, was proportional to the total pressure for a given mixture, and the change of ϕ across the cool-flame boundary was continuous, suggesting that the slow combustion and cool flame, at any rate during the acceleration period, were very similar.

The more sensitive analytical techniques available to Brown and Tipper [19] largely confirmed the earlier work of Cullis and Newitt [15, 16] in that during the induction period the small quantities of reactants which were consumed were converted to roughly equal amounts of acetaldehyde and hydrogen peroxide; traces of formaldehyde, ethylene and formic acid were also detected. During the acceleration period, the rate of formation of products rose, and the passage of a cool flame resulted in a large consumption of reactants.

The temperature rise just before a cool flame was remarkably constant (at a constant ambient temperature, T_0) on varying the mixture composition and total pressure and on adding inert gases, but increased slightly with increasing T_0. In slow combustion, the increases were less than that preceding a cool flame.

Although the addition of small amounts of acetaldehyde reduced the induction period markedly, the acceleration period and ϕ were unaffected until the pressure added exceeded that present just before the cool flame. In contrast, apparently, to the results of Cullis and Newitt [16], the addition of quite large amounts of acetaldehyde did not eliminate the induction period completely and only reduced the acceleration period by less than 25 %.

These findings, coupled with these results of experiments in which hydrogen peroxide and acetaldehyde were pre-formed in situ before further reactants were added to give a cool flame, suggest that both acetaldehyde and hydrogen peroxide are involved in chain-branching, perhaps via the equilibrium

$$CH_3CHO + H_2O_2 \rightleftharpoons CH_3CH(OH)OOH$$

followed by decomposition of the hydroxyhydroperoxide

$$CH_3CH(OH)OOH \longrightarrow CH_3CHO + 2OH\cdot$$

This is consistent with the suggestion put forward by Griffiths and Skirrow [20] to explain the kinetics of the oxidation of acetaldehyde alone.

The only condition for cool-flame production with ethanol was that the temperature at the centre of the vessel rose above the ambient by a critical amount, about 20 °C. This suggests that thermal factors are important in cool-flame production, and this was confirmed by the effect of addition of inert gases. The special importance of thermal conductivity is exemplified by the differing effects of helium and xenon, two gases with identical heat capacities but very different thermal conductivities. Thus helium raised the limit, while xenon lowered it.

1.3 *n*-PROPANOL

This compound oxidizes [21] at a rate comparable to that of ethanol; thus the reaction proceeds at a conveniently measurable speed at 264 °C, although there is still an induction period of an hour or so under typical conditions (150 torr oxygen + 150 torr *n*-propanol).

During the induction period propionaldehyde and hydrogen peroxide are the main products, although a little acetaldehyde is also formed. At the end of the induction period the reaction accelerates autocatalytically and formaldehyde, methanol, carbon monoxide, methane and ethane together with acids and peroxides are formed. Hydrogen peroxide is the most abundant peroxidic product, but there are also appreciable amounts of organic peroxides and peroxyacids [17]. These reach concentrations (as does formaldehyde, and to a less marked extent the higher aldehydes) at the time of maximum rate.

As with ethanol, the reaction in the early stages appears to involve a chain propagated by HO_2 radicals in which attack on the fuel occurs principally at the α C—H bond, viz.

$$CH_3CH_2CH_2OH + O_2 \longrightarrow CH_3CH_2\dot{C}HOH + HO_2\cdot$$

$$CH_3CH_2\dot{C}HOH + O_2 \longrightarrow CH_3CH_2CHO + HO_2\cdot$$

$$CH_3CH_2CH_2OH + HO_2\cdot \longrightarrow CH_3CH_2\dot{C}HOH + H_2O_2$$

In the later stages of the reaction, other products may be formed either by attack by more reactive species, e.g. OH, at other positions in the molecule, or by oxidative degradation of propionaldehyde.

1.4 iso-PROPANOL

iso-Propanol oxidizes much less readily than its *n*-isomer [21]. At 330—450 °C the main products in the early stages are acetone and hydrogen peroxide. Later carbon monoxide, methanol, acetaldehyde and formaldehyde make their appearance. The analytical results suggest that methanol results from the oxidation of acetone. Added acetaldehyde reduces the induction period without altering the maximum rate significantly and it appears that this compound, formed in a side-reaction, brings about chain-branching [21].

Experiments with a flow-system confirmed that in the early stages the only process in operation is a linear chain reaction giving acetone and hydrogen peroxide [22—24].

The product analyses, coupled with the fact that the secondary C—H bond is known to be weakened in comparison with the other C—H bonds in the molecule [25], suggest the simple mechanism [21]

$$(CH_3)_2CHOH + O_2 \longrightarrow (CH_3)_2\dot{C}OH + HO_2\cdot$$

$$(CH_3)_2\dot{C}OH + O_2 \longrightarrow (CH_3)_2CO + HO_2\cdot$$

$$HO_2\cdot + (CH_3)_2CHOH \longrightarrow (CH_3)\dot{C}OH + H_2O_2$$

1.5 *n*-BUTANOL

The gas phase oxidation of *n*-butanol proceeds quite rapidly at temperatures around 290 °C. Between 305 and 340 °C Cullis and Warwicker [26] observed single and multiple cool flames, while at still higher temperatures a region of very high maximum rates without ignition was observed. Above and below the cool flame region the overall activation energy was about 39 kcal . mole^{-1}.

At 295 °C, the reaction appeared to take place in three well-defined stages. During the induction period the only products formed are *n*-butyraldehyde and hydrogen peroxide, and these are produced in equivalent amounts. As soon as the pressure starts to rise, however, propionaldehyde, acetaldehyde, formaldehyde and organic peroxides also become detectable. The concentrations of all these compounds eventually pass through maxima and then fall off; roughly at this stage acetone, propene oxide, ethanol and methanol begin to form.

The principal difference between the results at 295 and 340 °C is that, at the higher temperature, formaldehyde is formed in relatively greater yield than the other aldehydes. During the induction period, which was shorter than at the lower temperature, the same products were formed. The initial chain cycle at both temperatures was therefore envisaged by Cullis and Warwicker [26] as being precisely analogous to those already suggested for the lower aliphatic alcohols.

Once the *n*-butyraldehyde has accumulated to a certain level in the system, this compound starts to be oxidized and the total pressure rises. All the lower aldehydes are produced, apparently by non-selective attack on the fuel, perhaps by OH radicals.

The main process taking place in the later stages of the reaction is the formation of products which must arise from the further oxidation or oxygen-catalysed pyrolysis of the intermediate aledhydes.

1.6 iso-BUTANOL (1-HYDROXY-2-METHYLPROPANE)

This alcohol oxidizes somewhat less readily than *n*-butanol [26]. Between 300 and 350 °C the maximum rate of reaction varies only slightly with temperature, while above and below this region the apparent activation energy is about 40 kcal . mole^{-1}. The corresponding variation of induction period with temperature shows no such behaviour, the induction period decreasing smoothly with temperature with an "activation energy" of 21 kcal . mole^{-1}.

In the induction period the now familiar pattern is again observed. At 295 °C the main products are iso-butyraldehyde and hydrogen peroxide with smaller amounts of acetone and formaldehyde. The onset of pressure increase is accompanied by the formation of acetaldehyde and organic peroxides, while towards the end of the reaction propene oxide and methanol appear and the rate of acetone production rises sharply.

At 400 °C, the formation of acetone and formaldehyde (or perhaps formic acid) in the induction period is more important, while in the subsequent stages formaldehyde formation is again increased relative to the higher homologues (cf. *n*-butanol).

The results can again be explained [26] by the propagation of a linear chain by HO_2 radicals which produce hydrogen peroxide and isobutyraldehyde; when sufficient aldehyde has accumulated chain-branching can begin and the reaction accelerates.

1.7 *sec*-BUTANOL (2-HYDROXYBUTANE)

The induction periods and maximum rates of pressure change of iso- and *sec*-butanol are very similar under all conditions examined by Cullis and Warwicker [26]. The products in the induction period at 295 °C are methyl ethyl ketone and hydrogen peroxide, together with a little acetaldehyde, and, at 400 °C, some formaldehyde.

The acceleration of the reaction after the induction period appears to be the result of the build-up of acetaldehyde to a critical level. The amount of acetaldehyde present when the reaction begins to accelerate is about the same as the quantity of butyraldehydes formed at the corresponding stage of reaction of *n*- and iso-butanol. Furthermore, the addition of acetaldehyde to a *sec*-butanol—oxygen mixture results in a considerable decrease in the induction period, whereas the addition of even quite large amounts of methyl ethyl ketone has only a slight effect. This is not unexpected, since, below about 400 °C the combustion of methyl ethyl ketone is preceded by a lengthy induction period (see Sect. 2.2).

1.8 *tert*-BUTANOL

The oxidation of this alcohol takes place only above about 400 °C and is not preceded by an induction period [26]. The products are simpler than those formed during the oxidation of the other butanols. The main product is acetone, and there are smaller amounts of hydrogen peroxide, formaldehyde formed in the early stages. Later in the reaction, acetaldehyde, methanol and organic peroxides also appear.

The sequence of reactions suggested is

$$(CH_3)_3COH + O_2 \longrightarrow (CH_3)_3\dot{C}O + HO_2\cdot$$

$$(CH_3)_3\dot{C}O \longrightarrow (CH_3)_2CO + CH_3\cdot$$

$$CH_3\cdot + O_2 \longrightarrow HCHO \text{ and } CH_3OH, \text{ etc.}$$

$$HO_2\cdot + (CH_3)_3COH \longrightarrow (CH_3)_3\dot{C}O + H_2O_2$$

1.9 DIFFUSION FLAME STUDIES

The diffusion flames of methanol [27], ethanol [27], *n*- and iso-propanol [27], and the four isomeric butanols [28] have been investigated using a quartz probe technique. The alcohol flames were burned on a pyrex wool wick and samples for analysis were taken from various positions in the flame using a quartz microprobe. Thermocouple measurements showed that the temperature varied from 200 °C at the wick to around 1400 °C at the tip and edges of the flame.

The products were extremely complex, but show that pyrolysis of the fuel occurred at the centre of the flame, followed by oxidation of the pyrolysis products in the outer zone.

2. Ketones

The combustion of aliphatic ketones generally resembles that of hydrocarbons, the reactions being autocatalytic and possessing two regimes of slow oxidation, separated by a region of negative temperature coefficient of the rate. Cool flames are also observed under some circumstances.

2.1 ACETONE

This fuel exhibits all the principal features of hydrocarbon oxidation in a particularly clear form and analysis shows that pressure change is an excellent measure of the extent of reactant consumption [29, 30]. There is a well-developed region of negative temperature coefficient of the rate between about 300 and 400 °C. Detailed studies were made above and below this region.

In the high temperature zone [29], the effect of total pressure on the maximum rate corresponded to an order of about 3.6; the order with respect to acetone alone is 1.0, and with respect to oxygen 2.2.

At 498 °C the early stages the reaction may be represented by

$$CH_3COCH_3 + O_2 = 1.6CO + 0.9H_2O + 0.7CH_4 + 0.3HCHO + \\ + 0.07\,(C_2H_4 + CO_2 + H_2 + C_2H_4O)$$

At lower temperatures, the methane yield tended to decline while that of formaldehyde did not alter greatly.

Some methanol, ketene and hydrogen peroxide (as well as a few other minor products) were also detected, while at 400 °C Hoare and Ting-Man Li [31] found the first of these to be relatively a much more important product. These workers also found acetic acid, but this could have been formed from ketene.

The formaldehyde concentration built up to a maximum which was reached at the time of maximum rate, and thereafter declined. The addition of formaldehyde reduced the induction period, and when an amount was added approximately equal to that present at the maximum rate, oxidation commenced immediately at the maximum rate. Consequently, it was believed that formaldehyde was the intermediate responsible for degenerate branching [29].

The simplest scheme which accounts for the main products is

$$CH_3COCH_3 + O_2 \longrightarrow CH_3COCH_2\cdot + HO_2\cdot$$

$$CH_3COCH_2\cdot \longrightarrow CH_3\cdot + CH_2CO$$

$$CH_2CO + O_2 = CO, CO_2, H_2O, HCHO$$

$$CH_3COCH_2\cdot + O_2 \longrightarrow \underset{\displaystyle |\atop \displaystyle \cdot O{-}O}{CH_3COCH_2} \begin{cases} \nearrow\ CH_3O\cdot + CO + CH_2O \\ \searrow\ C_2H_4 + CO + \cdot OH \end{cases}$$

$CH_3COCH_3 + X\cdot(HO_2\cdot, CH_3\cdot, \cdot OH, \text{etc.}) \rightarrow CH_3COCH_2\cdot + HX(H_2O_2, CH_4, H_2O, \text{etc.})$ with branching by

$$HCHO + O_2 \longrightarrow HO_2\cdot + \cdot CHO$$

$$\cdot CHO + O_2 \longrightarrow CO + HO_2\cdot$$

In the low temperature region [30], the kinetics and mechanism were quite different. The order of reaction was 2.0 with respect to total pressure; with respect to acetone it was 1.6 and with respect to oxygen 0.2. The products in the early stages of reaction at 284 °C corresponded to the stoichiometric equation

$$CH_3COCH_3 + 2.5O_2 = 1.75H_2O + 1.25CO + 0.65CO_2 + 0.6CH_3OOH + 0.5HCHO + 0.1H_2O_2 + 0.1H_2$$

At 330 °C using an HF-washed pyrex vessel the products found by Hoare and Ting-Man Li [32] were similar, but methanol was an early product (in the other study it appeared only at the time of maximum rate) and no methyl hydroperoxide was detected. Possible causes of this discrepancy include the different reaction vessel surfaces employed (although Barnard and Honeyman [30] found the low temperature reaction to be remarkably insensitive to reactor surface) and the rather higher temperature employed by Hoare and Ting-Man Li. Increasing temperature displaces the equilibrium

$$CH_3\cdot + O_2 \rightleftharpoons CH_3O_2\cdot$$

to the left [33, 34]. Consequently, the yield of methyl hydroperoxide formed by hydrogen abstraction

$$CH_3O_2\cdot + RH \longrightarrow CH_3OOH + R\cdot$$

will be diminished.

The methyl hydroperoxide concentration increased to a maximum at the time of maximum rate of reaction and it was concluded that this compound was responsible for chain-branching. This was subsequently confirmed [35] using the theoretical treatment developed by Knox [36]. Values of the rate coefficient of the branching reaction at several temperatures were obtained from the intercepts of the plots of the acceleration constant (ϕ) plotted against acetone concentration. The variation of rate coefficient with temperature was expressed by the equation [35]

$$k = 4.3 \times 10^{12} \exp(-38{,}500/RT) \text{ sec}^{-1}$$

which is in reasonable agreement with the thermochemistry of the reaction

$$CH_3OOH \longrightarrow CH_3O\cdot + \cdot OH \qquad (1)$$

The main findings can be explained by the mechanism

$$CH_3COCH_3 + O_2 \longrightarrow CH_3COCH_2\cdot + HO_2\cdot$$

$$CH_3COCH_2\cdot + O_2 \longrightarrow CH_3COCH_2O_2\cdot$$

$$CH_3COCH_2O_2\cdot \longrightarrow CH_3\cdot + CO_2 + HCHO$$

$$CH_3COCH_2O_2\cdot \longrightarrow CH_3O\cdot + CO + HCHO$$

$$CH_3\cdot + O_2(+M) \longrightarrow CH_3O_2\cdot(+M)$$

$$R\cdot(CH_3O_2\cdot, CH_3O\cdot, OH\cdot) + CH_3COCH_3 \longrightarrow RH + CH_3COCH_2\cdot$$

together with branching by (1).

The relative importance of the two modes of acetonylperoxy radical decomposition are discussed below (see Sect. 2.7).

Under suitable conditions up to three cool flames can be observed [37(a)] in acetone—oxygen mixtures and the cool flame limits are shown in Fig. 3. An ignition diagram for acetone–air mixtures at pressures above atmospheric is given by Maccormac and Townend [38]. Acetone cool flames are rather slow, and when there are multiple flames they are of decreasing amplitude. They are accompanied by appreciable temperature rises (sometimes as much as 100 °C) and by emission of the characteristic blue light which is also visible during slow combustion. The relationship between the temperature rise and the negative temperature coefficient has been thoroughly explored [37(a), 37(b)] and there seems little doubt that

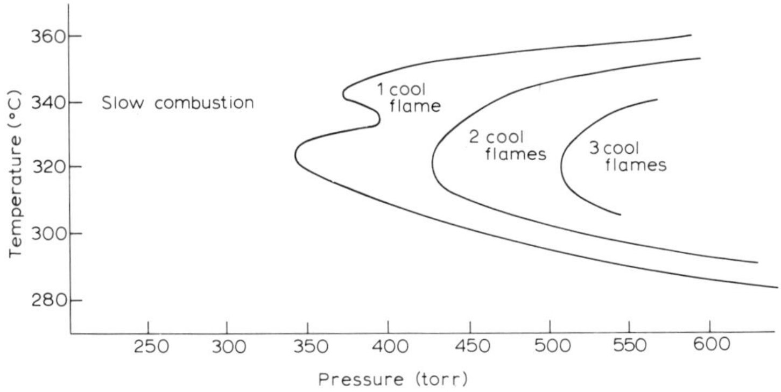

Fig. 3. Cool flame limits for an equimolar acetone—oxygen mixture. (Silica reaction vessel, diameter 9 cm.)

the periodicity of acetone cool flames is connected with the exothermic nature of the low temperature combustion reaction accelerating the reaction and carrying it into the region of negative temperature coefficient of the rate. The importance of methyl hydroperoxide in the branching process and in the processes leading up to cool flames was also confirmed, its concentration waxing and waning with the reaction rate [37(a)].

The importance of methyl hydroperoxide as a branching agent in the combustion of acetone and other simple ketones is also emphasized by Hoare and Lill [37(c)].

2.2 METHYL ETHYL KETONE

This was the first ketone to be subjected to a detailed examination [39–42] and it has recently been studied again [43(a)–(c)].

Bardwell and Hinshelwood [39] established the existence of a region in which the rate of reaction decreased with rising temperature; the regions of temperature and pressure in which slow reaction, single and multiple cool flames and explosion occurred were also mapped out [40] (Fig. 4).

The rate of reaction between 300 and 400 °C exhibits a complex dependence on oxygen concentration increasing rapidly and then declining to a constant value as the oxygen partial pressure is raised, while the dependence on ketone concentration is very high corresponding at 328 °C and a constant oxygen pressure of 400 torr, to an apparent order in the neighbourhood of six [39].

Later work [37(a)] has shown that under these conditions there are appreciable temperature changes, even in slow combustion, and therefore the detailed interpretation originally offered [40] cannot be wholly correct; nevertheless there is good evidence that peroxide branching is important in the low temperature reaction and Bardwell and Hinshelwood suggested [40, 42].

The most striking feature of the slow combustion of methyl ethyl

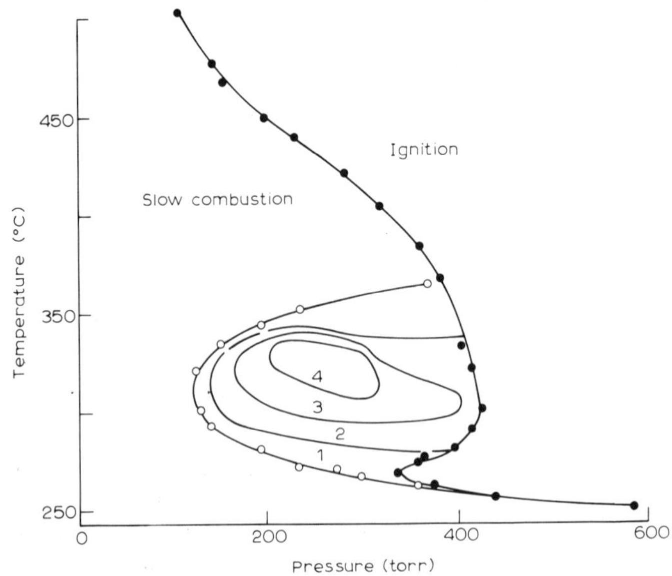

Fig. 4. Conditions of temperature and pressure for inflammation of an equimolar mixture of methyl ethyl ketone and oxygen [40]. The numbers shown indicate the number of cool flames observed.

ketone is the extremely long and somewhat irreproducible induction periods which precede reaction in the low temperature regime. Thus, Akbar and Barnard [43(a)] found that in certain cases at about 250 °C approximately 7 h elapsed before appreciable pressure change occurred, reaction then being complete within 10 min. Hoare and Ting-Man Li [31] also mention induction periods of 5 h. However, at 250 °C with 100 torr of an equimolar ketone—oxygen mixture the induction period was never less than 1 h. During this time, consumption of reactants is exceedingly small and only traces of products are detectable. These include hydrogen peroxide, formaldehyde, acetaldehyde, ethylene oxide, methanol and propene oxide [43].

The length of the induction period is greatly reduced by the addition of acetaldehyde, but this does not affect either the maximum rate or the maximum peroxide concentration, which is reached at the time at which the rate of reaction is also greatest [41].

Detailed chemical analysis [43(a)] showed that pressure change was a valid measure of reactant consumption, and further that at 250 °C the reaction in the early stages following the induction period could be represented by

$$\begin{aligned} CH_3CH_2COCH_3 + 2.5O_2 = {} & 1.3CO_2 + 0.6CO + 0.65HCHO + \\ & + 0.75H_2O + 1.0CH_3OH + \\ & + 0.04CH_3COOOH + 0.04CH_3OOH + \\ & + 0.03H_2O_2 + 0.05CH_3CHO + \\ & + \text{minor products} \end{aligned}$$

A plausible mechanism involves the initiation and propagation steps

$$CH_3CH_2COCH_3 + O_2 \longrightarrow CH_3\dot{C}HCOCH_3 + HO_2\cdot \qquad (1)$$

$$CH_3\dot{C}HCOCH_3 + O_2 \longrightarrow CH_3CH(OO\cdot)COCH_3 \qquad (2)$$

$$CH_3CH(OO\cdot)COCH_3 \longrightarrow CH_3CHO + CO + CH_3O\cdot \qquad (3)$$

$$CH_3CH(OO\cdot)COCH_3 \longrightarrow CH_3CHO + CO_2 + CH_3\cdot \qquad (4)$$

$$CH_3CH(OO\cdot)COCH_3 \longrightarrow CH_2CHCOCH_3 + HO_2\cdot \qquad (5)$$

$$CH_3\cdot + O_2 + M \rightleftharpoons CH_3O_2\cdot + M \qquad (6)$$

$$R\cdot(CH_3\cdot, CH_3O_2\cdot, CH_3O\cdot, OH\cdot) + CH_3CH_2COCH_3 \longrightarrow CH_3COCH_2CH_2\cdot \text{ (or } \cdot CH_2COCH_2CH_3) + RH \qquad (7)$$

$$CH_3COCH_2CH_2\cdot + O_2 \longrightarrow CH_3COCH_2CH_2(O{-}O\cdot) \qquad (8)$$

$$CH_3COCH_2CH_2(O{-}O\cdot) \longrightarrow CH_3\dot{C}O + 2CH_2O \qquad (9)$$

$$CH_3COCH_2CH_2(O{-}O\cdot) \longrightarrow CH_3CO_3\cdot + C_2H_4 \qquad (10)$$

$$CH_3CO_3\cdot + RH \longrightarrow CH_3CO_3H + R\cdot \qquad (11)$$

$$\cdot CH_2COCH_2CH_3 + O_2 \longrightarrow (O{-}O\cdot)CH_2COCH_2CH_3 \qquad (12)$$

$$(O{-}O\cdot)CH_2COCH_2CH_3 \longrightarrow CH_2O + CO_2 + \cdot CH_2CH_3 \qquad (13)$$

$$(O{-}O\cdot)CH_2COCH_2CH_3 \longrightarrow CH_2O + CO + \cdot OCH_2CH_3 \qquad (14)$$

together with branching by peroxide decomposition.

At higher temperatures the induction periods are much less and the product distribution is quite different [43(a)]. In the early stages at 450 °C, the stoichiometry is approximately

$$CH_3COCH_2CH_3 + 1.3O_2 = 1.5CO + 0.5H_2O + 0.7HCHO + 1.1CH_3OH + 0.3C_2H_4 + 0.15CH_4 + \text{minor products}$$

These products can be accounted for by the increasing importance of (3), (9) and

$$CH_3CO_3\cdot \longrightarrow CH_3O\cdot + CO_2$$

together with pyrolysis of the butanonyl radicals

$$\cdot CH_2CH_2COCH_3 \longrightarrow C_2H_4 + CH_3\cdot + CO$$

One somewhat surprising minor product is 1:2-epoxypropane. Its formation parallels the formation of ethylene oxide and 1:1-epoxybutane in the slow combustion of acetone and diethyl ketone, respectively. It may perhaps be formed via

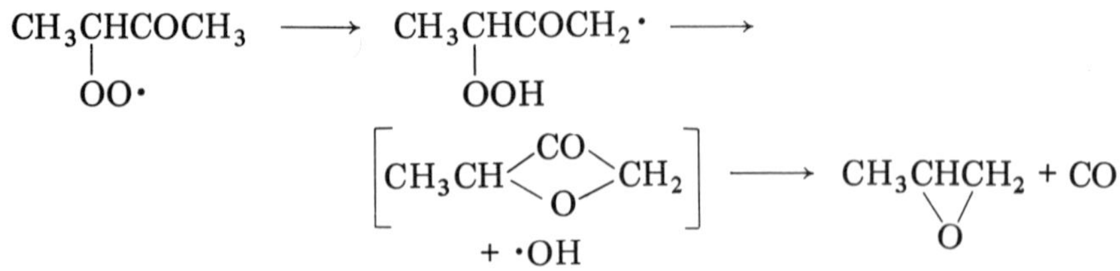

There is good evidence that chain branching at 400 °C involves formaldehyde because (*i*) its concentration reaches a peak at the time of maximum rate (*ii*) the addition of formaldehyde reduces the induction period without affecting the maximum rate, and (*iii*) the amount of formaldehyde needed to reduce the induction period to its minimum value is almost exactly that normally present at the maximum rate.

2.3 DIETHYL KETONE

The slow combustion of diethyl ketone is easily studied using a nitric acid-washed pyrex reaction vessel, the rate being reproducible and conveniently measurable at temperatures between 250 and 450 °C [44]. There is the usual well-developed region of negative temperature coefficient of the rate, but the induction periods are relatively short and in both temperature regimes the order of reaction for an equimolar mixture of fuel and oxygen is 1.4 with respect to fuel, 1.0 with respect to oxygen and 2.3 with respect to total pressure. The addition of inert gases has

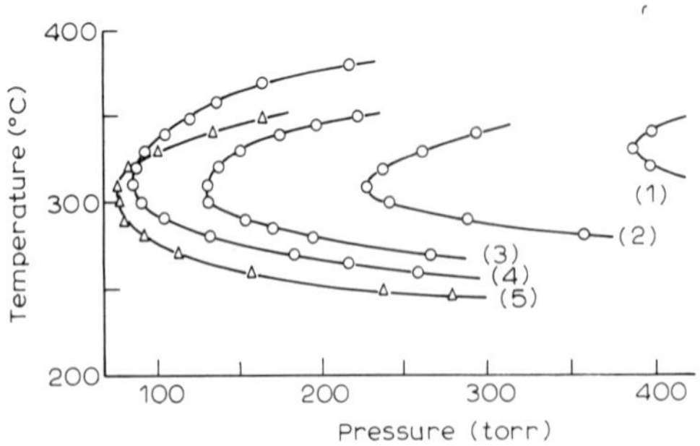

Fig. 5. The cool flame limit curves of equimolar ketone—oxygen mixtures in an HF treated spherical pyrex vessel, diameter = 10.1 cm [45]. (1) Acetone; (2) methyl iso-propyl ketone; (3) methyl ethyl ketone; (4) methyl *n*-propyl ketone; (5) diethyl ketone.

little effect on the pressure–time curves and there is little evidence of any pronounced surface effects [44].

Diethyl ketone gives rise to a single and multiple cool flames under relatively mild conditions, and the cool flame limit diagram [45] is shown in Fig. 5. The cool flames are accompanied by considerable temperature rises [37(a)]

The products of reaction in the high temperature region (400 °C) are rather complex and include carbon monoxide, water, ethylene and hydrogen peroxide, together with numerous minor products. It appears that formaldehyde is once again responsible for degenerate branching [44]. A scheme of reactions analogous to those already discussed for acetone and methyl ethyl ketone and involving pyrolysis and oxidation of ketyl radicals can be written to account for all the observed products [31, 44].

At low temperatures (250 °C) the products are different, there being a large amount of carbon dioxide and far less ethylene. Peroxides are also present, Akbar and Barnard [44] reporting moderate yields of ethyl hydroperoxide while Hoare and Ting-Man Li [32] found only traces of methyl hydroperoxide and perpropionic acid. Since the yield of ethyl hydroperoxide passed through a maximum at the time of maximum rate of reaction Akbar and Barnard [44] believe that it is involved in degenerate chain branching. Otherwise, both sets of workers are in substantial agreement over the mechanism involved which is analogous to that proposed for the lower ketones.

2.4 METHYL iso-PROPYL KETONE

The slow combustion of this fuel has been studied at 310 and 400 °C, these temperatures being representative of the low and high temperature regimes [46].

At 310 °C, the pressure–time curves were of an unusual shape; initially the rate of pressure rise accelerated smoothly and exponentially to a maximum which was sustained for some time before reaction ceased abruptly. Individual product–time curves showed similar behaviour. The primary products included hydrogen peroxide, formaldehyde, carbon dioxide, acetone, acetaldehyde and 1:2-epoxypropane.

At 400 °C, the pressure–time curves were S-shaped and propene was a much more important product; hydrogen peroxide and formaldehyde production was also increased.

Reactions of the ketonyl and ketonylperoxy radicals, e.g. $CH_3\dot{C}OC(CH_3)_2$ and $CH_3CO\underset{\cdot OO}{\underset{|}{C}}(CH_3)_2$, can be written to account for all the observed products.

The limits for cool flame propagation in an equimolar ketone–oxygen mixture are shown in Fig. 5.

References pp. 496–500

2.5 METHYL *tert*-BUTYL KETONE

This ketone is unique amongst those studied in that it apparently exhibits no region of negative temperature coefficient of the rate and no cool flames have been observed [45]. The pressure—time curves were similar to those of methyl iso-propyl ketone at 310 °C, the reaction accelerating smoothly and then stopping suddenly [46]. Analyses of the combustion products have been made at various stages of the reaction at 270, 310 and 350 °C. Carbon monoxide, carbon dioxide, hydrogen peroxide, iso-butene-1:2-oxide, methanol, methyl ethyl ketone and acetaldehyde were all detected, and towards the end of the reaction iso-butene and methane were also formed.

At 400 °C, the latter two hydrocarbons were major products, while the formaldehyde and hydrogen peroxide concentrations passed through sharp maxima which coincided with the maximum rate of pressure rise.

The reaction scheme at low temperatures may be

$$CH_3COC(CH_3)_3 + O_2 \longrightarrow \cdot CH_2COC(CH_3)_3 + HO_2\cdot \qquad (1)$$

$$\cdot CH_2COC(CH_3)_3 + O_2 \longrightarrow \underset{\displaystyle OO\cdot}{\underset{|}{C}}H_2COC(CH_3)_3 \qquad (2)$$

$$\underset{\displaystyle OO\cdot}{\underset{|}{C}}H_2COC(CH_3)_3 \longrightarrow (CH_3)_3C\cdot + CO_2 + CH_2O \qquad (3)$$

$$\underset{\displaystyle OO\cdot}{\underset{|}{C}}H_2COC(CH_3)_3 \longrightarrow (CH_3)_3CO\cdot + CO + CH_2O \qquad (4)$$

$$(CH_3)_3CO\cdot \longrightarrow CH_3\cdot + (CH_3)_2CO \qquad (5)$$

$$(CH_3)_3C\cdot + O_2 \longrightarrow (CH_3)_3CO_2\cdot \qquad (6)$$

$$(CH_3)_3C\text{–}O\text{–}O\cdot \longrightarrow (CH_3)_2C\underset{O}{\diagdown\!\diagup}CH_2 + \cdot OH \qquad (7)$$

$$(CH_3)_3C\text{–}O\text{–}O\cdot \longrightarrow CH_3COCH_3 + CH_3O\cdot \qquad (8)$$

At 400 °C the iso-butene yield is enhanced, but the carbon dioxide yield is not; the iso-butene does not therefore appear to be formed from the *tert*-butyl radical produced in (3), and Anderson and Hoare [46] suggest another reaction

$$CH_3CO\overset{\displaystyle CH_3}{\underset{\displaystyle CH_3}{C}}\text{–}CH_2\cdot \longrightarrow CH_3\cdot + CO + (CH_3)_2C{=}CH_2 \qquad (9)$$

2.6 OTHER KETONES

Biacetyl and acetylacetone have been studied by Salooja [47(a)] in a flow system. Biacetyl was rather reactive, appreciable reaction beginning at 350 °C and ignition occurring at about 530 °C under the experimental conditions employed; acetylacetone began to react above 400 °C but ignited at 480 °C. Biacetyl was anomalous in that it did not appear to exhibit a zone of negative temperature coefficient of the rate of combustion, probably because no stable olefinic intermediates are formed in the oxidation process. Some measurements on the rate of slow combustion of methyl vinyl ketone have also been reported [47(b)].

2.7 SUMMARY

The reactions of the simple ketones can be understood if it is assumed that there are several possible routes by which ketonyl radicals can react: the competition between these routes is determined by the structure of the radical and by the temperature.

Possible reaction for α-ketonyl radicals include

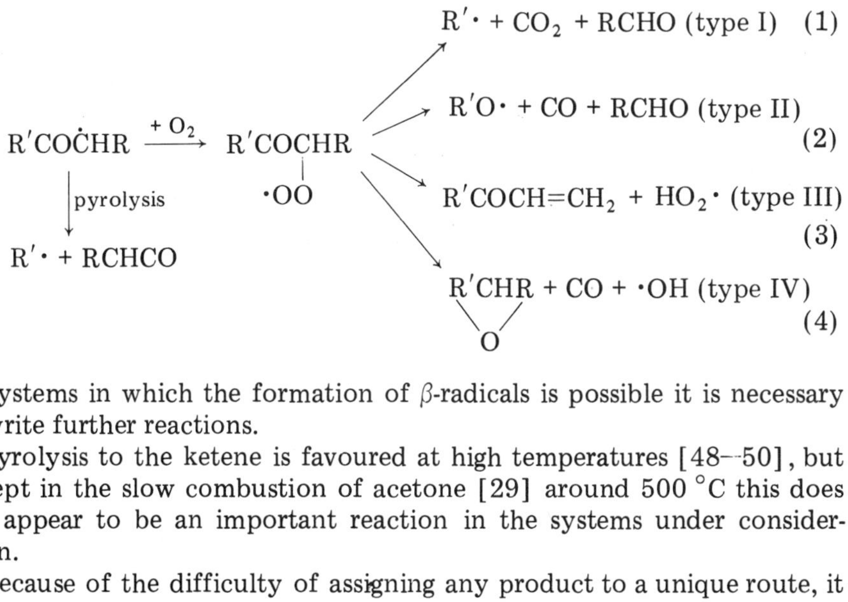

In systems in which the formation of β-radicals is possible it is necessary to write further reactions.

Pyrolysis to the ketene is favoured at high temperatures [48—50], but except in the slow combustion of acetone [29] around 500 °C this does not appear to be an important reaction in the systems under consideration.

Because of the difficulty of assigning any product to a unique route, it is not easy to establish even relative rate coefficients for the competing reactions of the ketonyl peroxy radicals. However, carbon dioxide yields are greatest in the low temperature regime, and therefore type I reactions are favoured under these conditions.

The photo-oxidation of acetone [51], methyl ethyl ketone [52] and diethyl ketone [53, 54] at temperatures between 100 and 250 °C has yielded some further information on these reactions and Hoare and

TABLE 2
The slow combustion of aliphatic ketones
Reactor: nitric acid-washed pyrex. Induction period defined as intercept on the time axis of the tangent to the pressure—time curve at the maximum rate.

Fuel	Temp. (°C)	Partial pressure		Induction period (min)	Max. rate of pressure change (torr . min^{-1})	Activation energy (of max. rate) (kcal . $mole^{-1}$)		Ref.
		Fuel (torr)	Oxygen (torr)			High temp. reaction	Low temp. reaction	
Acetone	284	100	100	0	0.25		approx. 37	29
	498	100	100	0	29.5	37		
Methyl ethyl ketone	250	50	50	30	1.5		24	43
	450	50	50	1.25	62	27		
Diethyl ketone	256	40	40	0.25	10		24	44
	400	40	40	1.5	19.5	33		

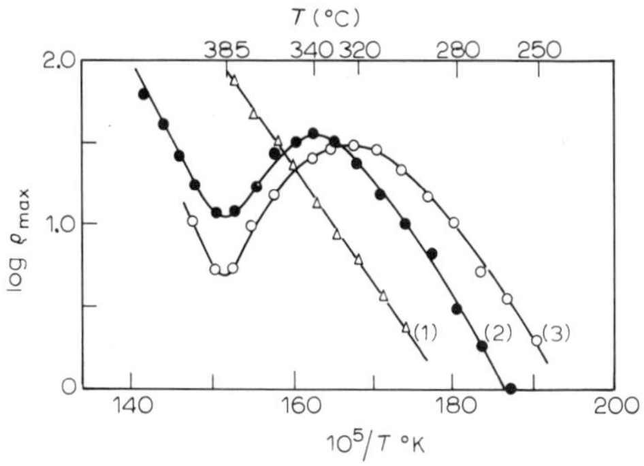

Fig. 6. The log ρ_{max} vs. $1/T$ plots for equimolar ketone—oxygen mixtures [45]. ρ_{max} is the maximum rate of reaction in torr . min^{-1}. (1) Methyl *t*-butyl ketone, *P*(total) = 300 torr; (2) acetone, *P*(total) = 300 torr; (3) diethyl ketone, *P*(total) = 65 torr.

Whytock [51] have calculated values of relative rate coefficients and activation energy differences based on carbon monoxide and dioxide yields. Thus for acetone (R′ = CH_3, R = H) k_1/k_2 = 1.0 at 250 °C and $E_2 - E_1$ may be about 14 kcal . $mole^{-1}$. However, these photo-oxidations are also chain reactions and the mechanisms are complex making it difficult to obtain quantitative results. Qualitatively the indications from all these systems are the same; namely that at low temperatures reactions of type I are favoured relative to type II.

Some data on the rates of oxidation of simple ketones are collected in Table 2. Further information on the effect of temperature of the maximum rate of reaction is given in Fig. 6, while Figs. 5 and 7 give the cool flame and hot ignition limits.

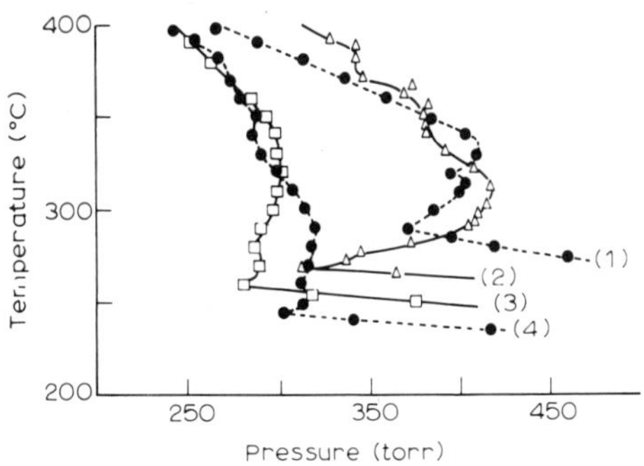

Fig. 7. The hot flame limit curves of equimolar ketone—oxygen mixtures in an HF treated spherical pyrex vessel, diameter = 10.1 cm [45]. (1) Methyl iso-propyl ketone; (2) methyl ethyl ketone; (3) methyl *n*-propyl ketone; (4) diethyl ketone.

3. Ketene

Although ketene is a likely intermediate in the combustion of some hydrocarbon derivatives, and is perhaps a precursor of the acids sometimes detected in combustion products, its combustion has only recently been investigated [55—59].

The ignition diagram (Fig. 8) between 300 and 450 °C for a CH_2CO: 2 O_2 mixture shows regions of 1, 2 and 3 cool flames, as well as explosion, between 125 and 225 torr [56]. In the early stages of reaction, formaldehyde was observed [56], confirming the previous work of Barnard and Kirschner [57]. These workers [57] studied the slow combustion of ketene between 300 and 500 °C and found the order of reaction with respect to oxygen was 1.2 and with respect to ketene 1.7. The reaction was extremely rapid, although not strongly temperature dependent (see Fig. 9), the apparent activation energy above about 350 °C being only 6 kcal . mole^{-1}. The main products were formaldehyde, carbon monoxide, carbon dioxide and water. At 475 °C, there were also small amounts of methane, hydrogen, acetic acid, acetaldehyde, methanol and hydrogen peroxide, while at 307 °C there were, in addition, traces of acetone, 1:2-epoxypropane and methyl ethyl ketone, although methane was absent. Significantly, ethylene was always absent.

Michaud and Ouellet [58, 59] in a more extensive investigation [58] confirmed the previous work, finding in addition [59] at 280 °C—350 °C a low temperature regime in which methyl hydroperoxide was formed.

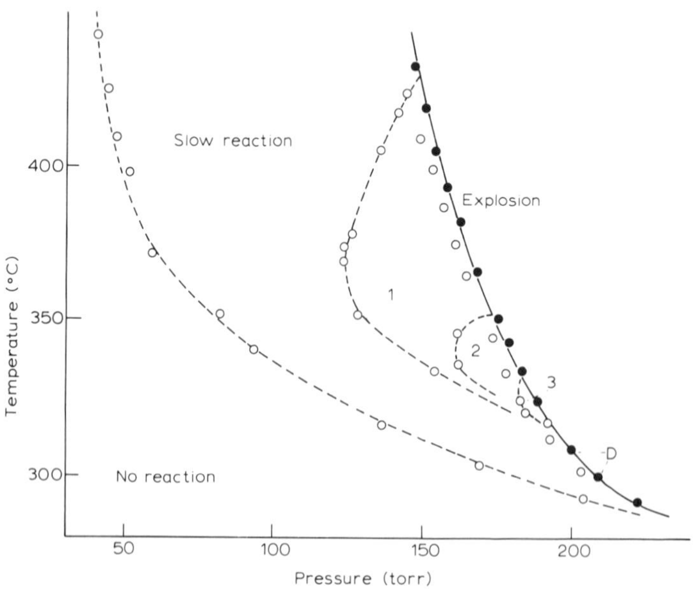

Fig. 8. Regions of no reaction, slow reaction, 1, 2 and 3 cool flames, explosions and detonations (D) for CH_2CO:$2O_2$ mixtures [56].

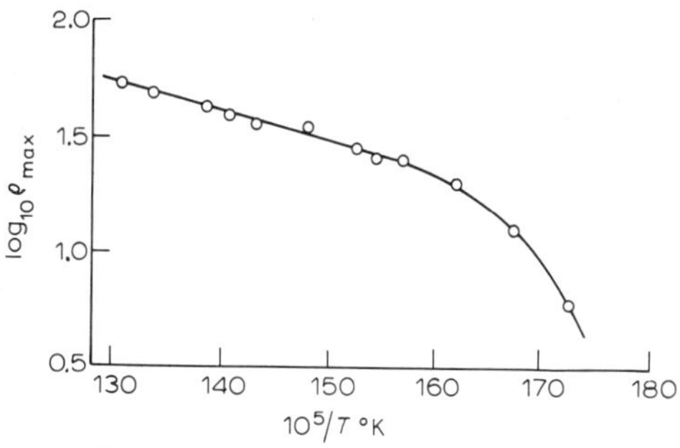

Fig. 9. The variation of ρ_{max}, the maximum rate of reaction in torr . min^{-1}, with temperature for an equimolar ketene—oxygen mixture [57]. Initial pressure 55 torr.

At high temperatures, the formaldehyde concentration passed through a maximum at the time of maximum rate, suggesting that formaldehyde is involved in the branching process.

The results are consistent with simultaneous molecular and free radical chain mechanisms, viz.

$$CH_2CO + O_2 \longrightarrow CH_2(O{-}O\cdot){-}\dot{C}{=}O \longrightarrow CH_2O + CO_2 \quad \text{(Molecular reaction)}$$

$$CH_2(O{-}O\cdot){-}\dot{C}{=}O \longrightarrow \cdot CHO + \cdot OH + CO \quad \text{Initiation}$$

Linear Propagation:

$$\cdot OH + CH_2CO \rightarrow H_2C(OH){-}\dot{C}{=}O \longrightarrow HCHO + \cdot CHO$$

$$\cdot CHO + O_2 \longrightarrow CO + HO_2\cdot$$

$$HO_2\cdot + CH_2CO \longrightarrow H_2C(OOH){-}\dot{C}{=}O \longrightarrow HCHO + CO + \cdot OH$$

$$\cdot OH + HCHO \longrightarrow H_2O + \cdot CHO$$

$$HCHO + O_2 \longrightarrow HO_2\cdot + \cdot CHO \quad \text{Branching}$$

$$HO_2\cdot \longrightarrow \text{inert at wall} \quad \text{Termination}$$

$$2HO_2\cdot \longrightarrow H_2O_2 + O_2$$

Michaud and Ouellet [58] believe that branching may involve other reactions of formaldehyde, perhaps

$$HCHO + O_2 \longrightarrow HCOOH + O$$

$$O + HCHO \longrightarrow \cdot CHO + OH$$

They also propose reactions to account for minor products.

At lower temperatures [59] there is good evidence for chain branching by pyrolysis of methyl hydroperoxide and it is suggested that methyl radicals formed by

$$\cdot OH + CH_2CO \longrightarrow CH_3\cdot + CO_2$$

and which give rise to methane at higher temperature, react instead with oxygen (cf. acetone slow combustion, Sect. 2.1)

$$CH_3\cdot + O_2 + M \rightleftharpoons CH_3O_2\cdot$$

followed by

$$CH_3O_2\cdot + CH_2CO \longrightarrow CH_3OOH + \cdot CHCO$$

$$CH_3OOH \longrightarrow CH_3O\cdot + \cdot OH \qquad (1)$$

and

$$\cdot CHCO + O_2 \longrightarrow 2CO + \cdot OH$$

A stationary state analysis of the system [59] leads to a value for the rate coefficient for pyrolysis of methyl hydroperoxide k_1, of 2×10^{-2} sec^{-1} at 330 °C in very satisfactory agreement with earlier work [35].

4. Oxirans

4.1 ETHYLENE OXIDE

4.1.1 Slow combustion and cool flames

In a quartz vessel an equimòlar ethylene oxide–oxygen mixture gives rise to cool flames quite readily between about 260 and 380 °C [60]. The slow combustion has been studied in detail above and below the optimum conditions for cool flame formation, and the kinetics in the two regions are quite different [61]. At 420 °C the rate obeyed the law

$$-d[O_2]/dt = k[C_2H_4O]^2[O_2]^{1/2}$$

and neither added nitrogen nor variation of the surface : volume ratio of the reaction vessel had much influence on the reaction rate. Simultaneously with combustion, an independent decomposition reaction appeared to take place.

At 298 °C the rate equation is

$$-d[O_2]/dt = k[C_2H_4O]$$

while added nitrogen retards and increased surface accelerates the reaction slightly.

The reaction rate does not show any marked negative-temperature coefficient, and the apparent activation energy calculated from the slope of the log $(d[O_2]/dt)$ versus $1/T$ graph was about 14 kcal . mole^{-1}.

In addition to carbon monoxide, carbon dioxide and water, the products included methanol and a hydrocarbon assumed to be methane. Acids and peracids were detected, while at 298 °C hydrogen peroxide was present.

4.1.2 Decomposition flame

Burden and Burgoyne [60] measured not only the cool flame and hot ignition limits for mixtures of ethylene oxide with air and oxygen, but also observed blue flames and self-decomposition flames.

The velocity of the decomposition flame was first measured by Gerstein et al. [62] using upward flame propagation in a tube. The experimental value of 12.5 cm . sec^{-1}, corrected to 1 atm and room temperature, was in reasonable agreement with values calculated from theories of flame propagation using the Arrhenius parameters obtained by Mueller and Walters [63] for the first-order pyrolysis of ethylene oxide at much lower temperatures.

Subsequent work by Friedman and Burk [64] using an Egerton--Powling burner is in sharp disagreement with these results. The measured burning velocity at 70 °C was 5 cm . sec^{-1} decreasing slightly with increasing pressure; the value corrected to 25 °C and 1 atm was 2.7 cm . sec^{-1}. Neither the pressure nor temperature dependence of the burning velocity was consistent with first-order kinetics of the decomposition reaction.

The products of decomposition were analysed (44.9 % CO, 24.2 % CH_4, 20.6 % H_2, 10.3 % C_2H_4) and the flame temperature was found to be 1182 °K, compared with a value calculated from the observed product distribution and thermodynamic data of 1217 °K.

These results can be understood if the burning rate is governed by the same first-order reaction which governs the low temperature decomposition, but subsequent flame reactions governing the product distribution depend on the pressure and initial temperature.

4.1.3 Ethylene oxide—oxygen flame

The composition and temperature profiles in low-pressure fuel-rich flames of ethylene oxide have been studied by Bradley et al. [65]. The major products were carbon monoxide, hydrogen, ethylene, methane, acetylene, butadiene and vinylacetylene, with traces of propene and propane. The unsaturated products were formed marginally later than the others, and ethane showed a maximum which coincided with the almost complete removal of fuel and oxygen. Acetylene and vinylacetylene continued to increase above the flame, although other products remained constant.

The adiabatic flame temperature calculated from the stoichiometry was 1000 °K, while the measured value was 923 °K the difference being almost certainly due to heat losses from the 0.05 cm diameter thermocouple used. The residence time in the reaction zone under the conditions employed was approximately 3×10^{-2} sec which was long enough for appreciable self-decomposition to have taken place. For a brief discussion of the likely mechanism, see Sect. 4.2

4.2 PROPENE OXIDE

No work has been reported on the slow combustion or ignition of this compound. The propene oxide—oxygen flame has been studied under low-pressure fuel-rich conditions [66]. The products in the post-flame gases are remarkably similar to those from the corresponding ethylene oxide flame, while the final flame temperature is only slightly higher. Just as the ethane concentration showed a maximum in the ethylene oxide flame, so did the ethane, propane and propene profiles show maxima corresponding to the almost complete removal of oxygen.

Methane and ethane production in the ethylene oxide flame suggest the participation of methyl radicals

$$C_2H_4O + X\cdot \longrightarrow \cdot C_2H_3O + XH$$

$$\cdot C_2H_3O \longrightarrow CH_3\cdot + CO$$

$$C_2H_4O \longrightarrow CH_3\cdot + \cdot CHO$$

followed by

$$CH_3\cdot + C_2H_4O \longrightarrow CH_4 + \cdot C_2H_3O, \text{ etc.}$$

In the propene oxide flame analogous processes

$$C_3H_6O + X\cdot \longrightarrow \cdot C_3H_5O + XH$$

$$\cdot C_3H_5O \longrightarrow C_2H_5O\cdot + CO$$

$$C_3H_6O \longrightarrow CH_3\cdot + \cdot C_2H_3O$$

lead to methyl and ethyl radicals which form methane and ethane by H-abstraction from the fuel. Absence of butane demonstrates that radical—radical reactions are unimportant, and the authors suggest that propane may arise by addition of ethyl radicals to the epoxide ring followed by H-transfer and decomposition, viz.

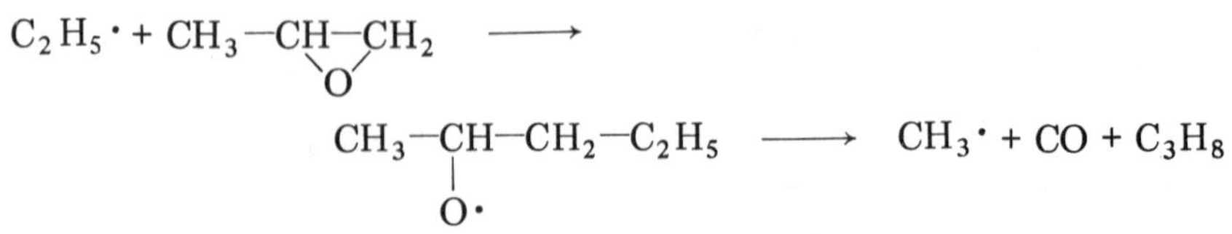

Ethane formation in the ethylene oxide flame might then arise in a similar sequence, viz.

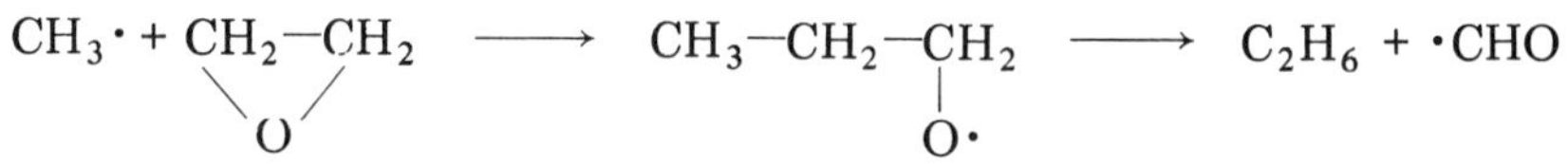

although the possibility that it arises from methyl combination cannot be entirely ruled out.

With both flames, the formation of butadiene is strong evidence for the existence of C_2H_3 radicals. As no methanol was found in the propene oxide flames it was concluded that pyrolysis of the fuel

$$C_2H_4O \longrightarrow C_2H_3\cdot + \cdot OH$$

and

$$C_3H_6O \longrightarrow C_2H_3\cdot + CH_3O\cdot$$

was not the source of vinyl radicals, and these were formed instead by reaction of hydrogen atoms with the C_2H_3O radical or with ethylene oxide itself, viz.

$$\cdot C_2H_3O + H\cdot \longrightarrow C_2H_3\cdot + \cdot OH$$

$$C_2H_4O + H\cdot \longrightarrow C_2H_3\cdot + H_2O$$

5. Ethers

5.1 DIMETHYL ETHER

In a study of the comparative case of oxidation of six aliphatic ethers, Eastwood and Hinshelwood [67] found the dimethyl ether was the most resistant to oxidation, only giving rise to cool flames at about 250 °C, whereas the others all gave cool flames at 150—200 °C (Fig. 10).

The variation of rate of oxidation with changing oxygen and ether pressures and the temperature dependence of the rate were all studied. Finally, it was shown that peroxides were formed in slow combustion, but not completely destroyed in the cool flame.

5.2 DIETHYL ETHER

The ignition diagram for diethyl ether—oxygen mixtures has been determined [68] and the slow combustion has been studied in some detail. The low temperature oxidation becomes appreciable at 120 °C and Lemay and Ouellet [69], working at 160—175 °C which was below the lower limit of cool flame for in their pyrex reaction vessel, observed an initial pressure drop during which oxygen was consumed by a zero-order process with an activation energy of about 50 kcal . mole^{-1}. Peroxides

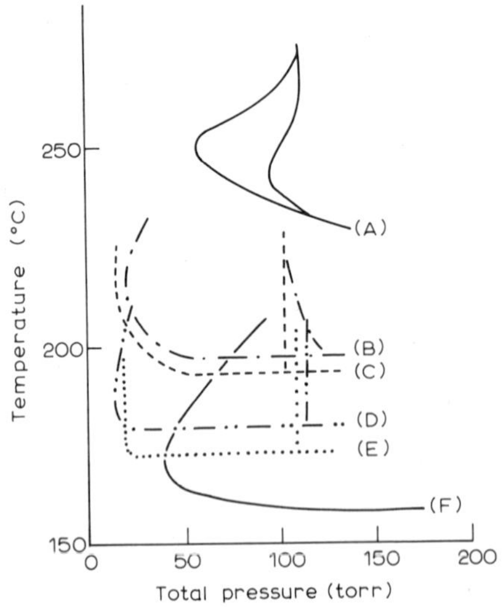

Fig. 10. The pressure—temperature explosion diagrams for equimolar ether—oxygen mixtures [67]. (A) Dimethyl ether. (B) Methyl *n*-propyl ether. (C) Ethyl methyl ether. (D) Di-*n*-propyl ether. (E) Diethyl ether. (F) Di-iso-propyl ether.

and acids were detected in the products. The pressure rise which follows this reaction is secondary to the main oxidation and does not require the presence of oxygen.

Waddington [70] found that the pressure—time curves at 153 °C were of the familiar sigmoidal shape, an appreciable time elapsing between the admission of reactants and the occurrence of a measurable pressure change. During this induction period considerable amounts of acetaldehyde, ethanol and peracetic acid together with smaller amounts of some other products are formed. The acetaldehyde and peracetic acid concentrations reach maxima at the end of the induction period, defined as the time for a pressure change of 2 torr to occur. After this, the concentrations of acetaldehyde, peracetic acid, organic hydroperoxides and hydrogen peroxide all fall, while large amounts of acetic acid and ethanol are formed together with smaller amounts of formaldehyde and methanol.

The apparent activation energy between 150 and 190 °C, the temperature at which cool flames appear, was about 23 kcal . mole^{-1} for a mixture of 100 torr diethyl ether and 50 torr oxygen. Packing the reaction vessel with lengths of pyrex tubing slightly increased the induction period and lowered the rate to about 2/3 of its value in an empty vessel.

While the order of reaction with respect to ether was about unity, excess oxygen tended to decrease the rate and to lengthen the induction period. This behaviour is in marked contrast to that commonly observed in hydrocarbon combustion.

A mechanism was proposed by Waddington [70] which involved formation of peracetic acid from acetaldehyde, itself a breakdown product of the radical formed by attack of oxygen on the fuel, viz.

$$CH_3CH_2OCH_2CH_3 + O_2 \longrightarrow CH_3\dot{C}HOCH_2CH_3 + HO_2\cdot \qquad (1)$$

$$CH_3\dot{C}HOCH_2CH_3 \longrightarrow CH_3CHO + CH_3CH_2\cdot \qquad (2)$$

$$CH_3CHO + O_2 = CH_3COOOH$$

Chain branching was due to breakdown of the peracid

$$CH_3COOOH \longrightarrow CH_3CO_2\cdot + RH$$

$$CH_3CO_2\cdot + RH \longrightarrow CH_3COOH + R$$

Ethanol arose from oxidation of the ethyl radical formed in (2), viz.

$$CH_3CH_2\cdot + O_2 \longrightarrow CH_3CH_2O_2\cdot$$

$$CH_3CH_2O_2\cdot \longrightarrow CH_3CHO + \cdot OH$$

$$CH_3CH_2O_2\cdot + RH \longrightarrow CH_3CH_2OOH + R$$

$$CH_3CH_2OOH \longrightarrow CH_3CH_2O\cdot + \cdot OH$$

$$CH_3CH_2O\cdot \longrightarrow CH_3\cdot + HCHO$$

$$\searrow CH_3CHO + H\cdot$$

$$CH_3CH_2O\cdot + RH \longrightarrow CH_3CH_2OH + R\cdot$$

Subsequent work [71, 72] has suggested that the ether radical formed in (1) is quite stable, and will normally react with oxygen rather than pyrolyse. The following mechanism seems more likely, viz.

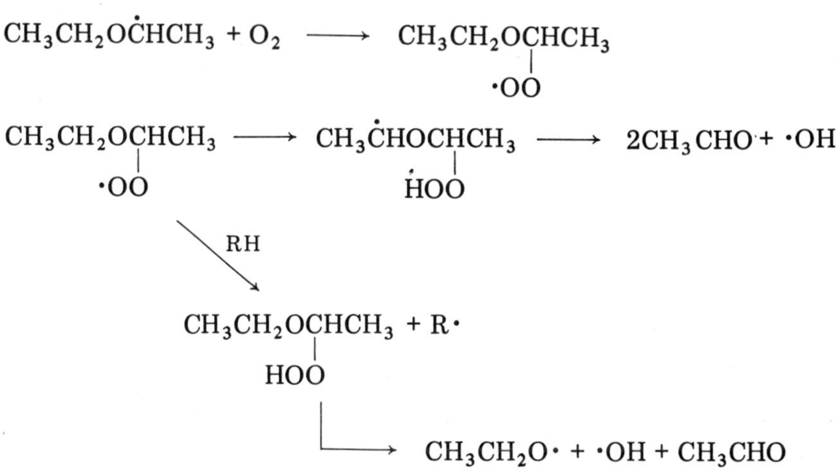

The light emission from ether cool flames has been studied by Ouellet and Ouellet [73]. The total emission increases with reactant concentration, and the duration increases with diameter of the reaction vessel. Raising the ambient temperature modifies the way the emission varies with time.

A diethyl ether cool flame, followed by a second-stage flame can be stabilized in a tube [74] or above a burner [75--77], and Agnew and Agnew [78] have used a quartz probe to remove samples from various positions in these flames. Numerous products were identified including not only carbon monoxide, carbon dioxide, water, various saturated and unsaturated hydrocarbons, acetaldehyde, formaldehyde, methanol, ethanol and acetic acid, but also ethyl formate, ethyl acetate, acetone, propionaldehyde and 2-methyl-1:3-dioxacyclopentane. The main features of the analytical results were

(*a*) about 30 % of the reactants were converted by the time the gases entered the first-stage flame and 75 % by the time they entered the second;

(*b*) the concentration of acetaldehyde, methanol, formaldehyde, acetic acid, ethyl acetate, ethyl formate and 2-methyl-1:3-dioxacyclopentane reach maxima between the two stages of the flame.

It appeared that the transfer of heat back from the highly exothermic final stages to the earlier parts was crucial in establishing the stable flame. Most of the products could be accounted for by reactions of types frequently postulated in the combustion of hydrocarbons and their derivatives. Thus, for example, 2-methyl-1:3-dioxacyclopentane could arise by

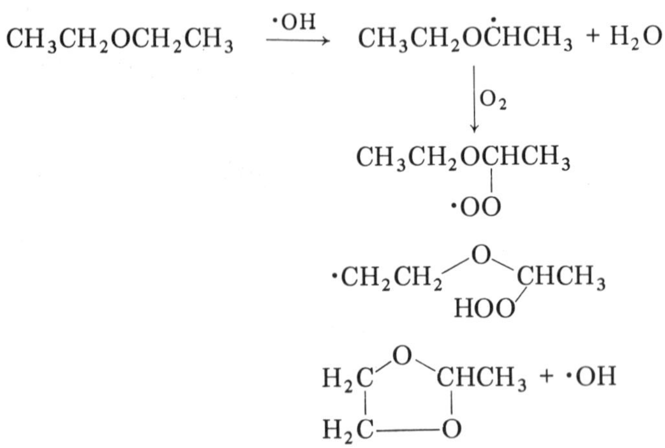

The emission spectrum of the first-stage flame is truly that of excited formaldehyde [78, 79] whereas that of the second may be due to either the same emitter or to the formyl radical, depending primarily on the initial mixture ratio.

Spectroscopic evidence points to the presence of carbon suboxide in these two-stage flames [80]. It suggested that this is formed via acetone which pyrolyses to ketene which in turn gives carbon suboxide, viz.

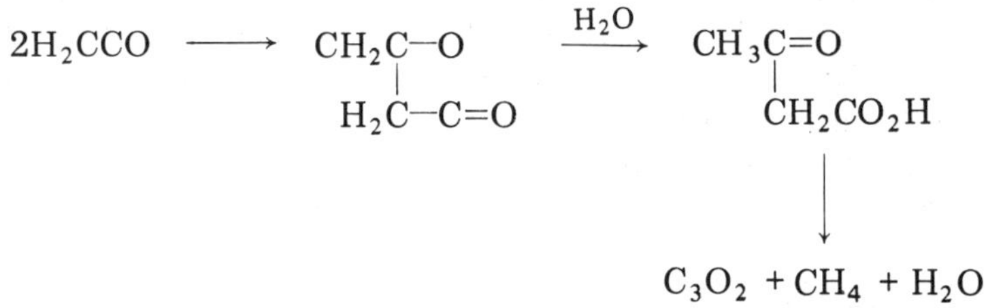

The diethyl ether diffusion flame has also been studied [81], and in common with other diffusion flames [27, 28], it was found that pyrolysis was the primary process occurring in the inner regions of the flame. In the central zone where most of the ether disappears and in which the measured temperature is 400—800 °C, the products are acetaldehyde, ethane, ethanol and ethylene, suggesting that pyrolysis of the fuel was taking place according to the two alternative overall reactions

$$CH_3CH_2OCH_2CH_3 = CH_3CHO + C_2H_6 \quad (3)$$

$$CH_3CH_2OCH_2CH_3 = C_2H_5OH + C_2H_4 \quad (4)$$

which are known [82] to contribute to comparable extents to diethyl ether decomposition at about 500 °C. In the hotter parts of the central region, reaction (4) which is known to have a higher overall activation energy, was predominant, no acetaldehyde or ethane being detected in those parts of the flame where the temperature was appreciably above 600 °C.

The formation of moderate amounts of methane throughout the flame probably indicates the presence of methyl radicals, but these do not appear to be the precursor of methanol (which is present in highest concentration along the boundary between the blue and smoky parts of the flame) since it is not a product in the diffusion flames of acetaldehyde or acetone, both of which should give rise to high concentrations of methyl radicals. One suggested source of methanol is through decomposition of the peroxidized ether radical, viz.

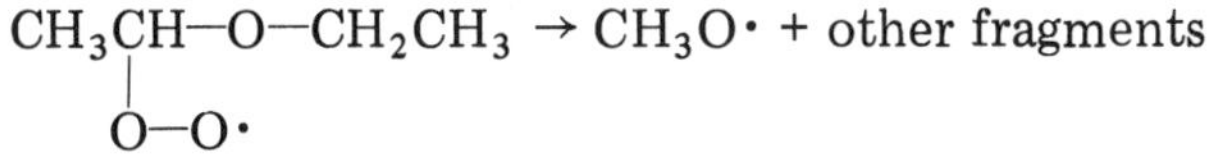

Vovelle et al. [83] showed that the cool-flame region for diethyl ether was considerably extended by the addition of di-*tert*-butyl peroxide. The products of the normal combustion, cool flames and two-stage ignition were all examined [84] and found to include carbon monoxide and dioxide, water, acetaldehyde, methanol and methane, ethylene and acetylene.

References pp. 496—500

Aliphatic amines inhibit the combustion of diethyl ether [85], the secondary and tertiary compounds being more powerful inhibitors than the primary amines [86]. The effectiveness of $CH_3CD_2NH_2$ as an inhibitor has been compared with that of $CH_3CH_2NH_2$ [87].

Aromatic compounds also affect the cool flame and hot ignition characteristics of diethyl ether and the relationship between the antiknock behaviour of aromatic additives and the influence of the side-chain on the electronic properties of the aromatic ring has been discussed by Malherbe and Walsh [88].

5.3 DI-iso-PROPYL ETHER

Chamberlain and Walsh [89, 90] have shown that this ether oxidizes by two mechanisms. In the high-temperature regime between 360 and 460 °C the pressure—time curves are sigmoidal, whereas at 220 °C in the low temperature zone they are not. At 360 °C the rate of reaction varied according to the equation

$$(\mathrm{d}P/\mathrm{d}t)_{max} = kP^2_{ether}P_{O_2}P_{total}$$

and the overall activation energy was 22 kcal . mole^{-1}. The addition of aromatic compounds retarded the high temperature reaction, but had less effect on the processes occurring at low temperatures.

5.4 DIETHYL ACETAL

This compound oxidizes by two mechanisms. At low temperatures cool flames and slow oxidation (accompanied by luminescence) are observed, while at higher temperatures normal ignition occurs [91a)].

5.5 1,3-DIOXALANE

The kinetics of the slow reaction and the cool-flame and explosion limits for the gas phase oxidation of this compound between 240 and 340 °C have been investigated, and the proposed mechanism involves hydroperoxides as the intermediates responsible for degenerate branching [91(b)];

6. Esters

Although some aspects of the slow combustion of a number of esters have been studied, there is, as yet, no clear understanding of the elementary reactions involved. Most of the work has been directed towards product analyses and to determining the ignition characteristics.

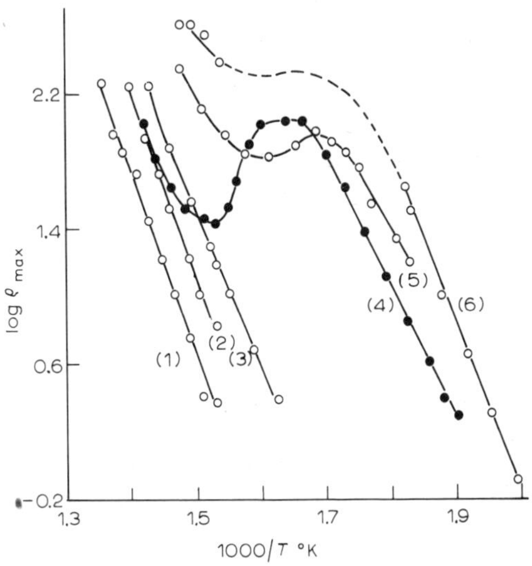

Fig. 11. The effect of temperature on the maximum rate of oxidation (in torr . min^{-1}) of esters [93]. (1) Methyl acetate. (2) Methyl formate. (3) Methyl propionate. (4) Ethyl formate. (5) Methyl butyrate. (6) Propyl formate.

The first experiments by Parsons et al. [92—94] revealed that the oxidations were autocatalytic and that the rates and induction periods were complex functions of the reactant concentration. The effect of temperature on the oxidation of a number of esters is shown in Figs. 11 and 12.

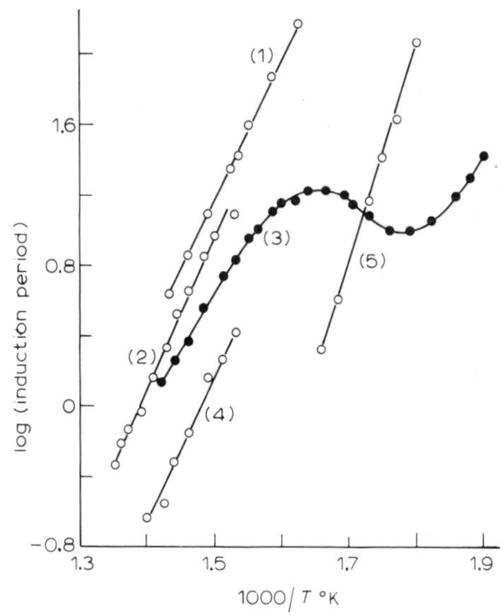

Fig. 12. The effect of temperature on the induction period (min) for the oxidation of esters [93]. (1) Methyl propionate. (2) Methyl acetate. (3) Ethyl formate. (4) Methyl formate. (5) Methyl butyrate.

6.1 METHYL FORMATE

The slow combustion [93] is measurable at 380 °C, but there is no low temperature mechanism, nor have cool flames been observed [45]. At 560 °C, in a flow system, mixtures of air and methyl formate ignite with explosive violence [47(a)]. The preflame reaction produces methane and methanol.

6.2 ETHYL FORMATE

This compound oxidizes in a flow system at temperatures as low as 120 °C and possesses a region of negative temperature coefficient between about 330 and 370 °C. Products include acetaldehyde, formaldehyde, formic acid and peroxides. Below 250 °C, acetaldehyde is the sole non-peroxidic organic product [95]. This is formed by a 1:5 intramolecular H-transfer followed by β-scission.

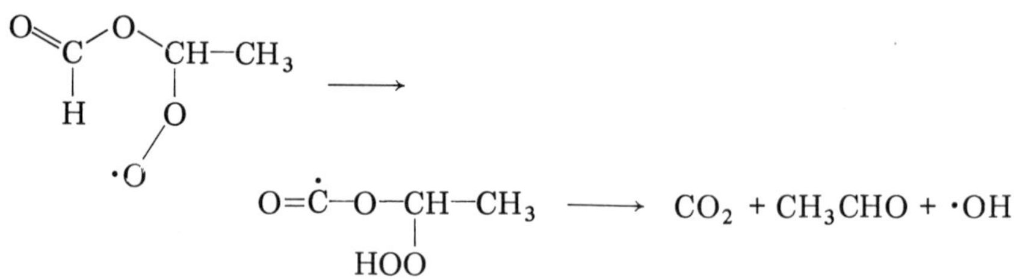

At much higher temperatures, Salooja [47(a)] found ethylene in the combustion products.

6.3 *n*-PROPYL FORMATE

Propyl formate is fairly readily oxidized, and there are apparently both high temperature and low temperature mechanisms [92, 93].

6.4 METHYL ACETATE

There is little evidence for a low temperature mechanism in the combustion of this fuel, although Fish and Waris [95] have demonstrated that below 350 °C acetaldehyde and organic peroxides are formed, while above 400 °C the main products are acetic and formic acids. Cool flames have not been observed [45].

6.5 ETHYL ACETATE

The combustion of ethyl acetate in a flow system was studied by Fish and Waris [96]. Below 250 °C there was no oxidation under the conditions employed, (equimolar fuel and oxygen flows, residence time

134 sec), while above 450 °C pyrolysis predominated. Between 320 and 360 °C the rate of reaction decreased with increasing temperature. Below 320 °C acetic acid and formaldehyde were the major products with smaller amounts of organic peroxides, peroxyacids and formic acid. Acetaldehyde was initially absent, but above 350 °C it was a major product. Anhydrides and very small amounts of hydrogen peroxide were also formed in the high temperature region, together with products which were also formed at low temperatures. No alcohols were detected under any conditions.

The combustion of ethyl acetate has also been studied in a flow system by Salooja [47(a)] and in a static system by Hoare et al [45, 97]. Under the conditions employed, Salooja found that ethyl acetate was slightly more resistant to oxidation than methyl acetate, but that it ignited at a considerably lower temperature with explosive violence.

The results using a static system [97] confirmed the existence of a region of negative temperature coefficient of the rate between 320 and 360 °C, and the limits for cool-flame formation for an equimolar mixture were determined. In contrast to the earlier work [96], methanol was found to be a major product at both 300 and 360 °C. Some iso-propanol and methane were detected, the yield of the latter being greater at the higher temperature. No carbon dioxide was observed, but carbon monoxide was a major product.

6.6 *n*-PROPYL ACETATE

The slow combustion of this compound [95] also possesses a region of negative temperature coefficient between 320 and 360 °C and it very readily gives rise to cool flames [45]. The major products are aldehydes and acids, propionic acid and propionaldehyde predominating at high temperatures. At low temperatures considerable amounts of hydrogen peroxide are formed.

6.7 iso-PROPYL ACETATE

This ester oxidizes very easily [95], reaction being perceptible even below 140 °C. Above about 300 °C, pyrolysis to propene and acetic acid also takes place. It too, gives cool flames [47]. Fish and Waris [95] detected only acetone, organic peroxides and peroxyacids in the products. Between 280 and 360 °C, Hoare and Kamil [97] found a wider range of products including hydrogen peroxide, formaldehyde, methanol, iso-propanol, acetic acid and at 320 °C and below, acetaldehyde. Propene and acetone were found at 360 °C but organic peroxides and peroxy-acids were always absent.

References pp. 496—500

6.8 *tert*-BUTYL ACETATE

Neither cool flames nor a region of negative temperature coefficient have been observed, and it appears that at 220 °C this ester pyrolyses rather than combusts [97].

6.9 METHYL PROPIONATE

The combustion of this ester is unusual in that it yields considerable amounts of hydrogen peroxide [97]. Other products include methyl acrylate, carbon monoxide and dioxide, methanol and formaldehyde. There is no low temperature mechanism, and no cool flames have been observed.

6.10 ETHYL PROPIONATE

Below about 300 °C the main products of combustion are acetic acid and propionaldehyde together with some propionic and formic acids and acetaldehyde. At about 400 °C, there is much more acetaldehyde than propionaldehyde and the yield of propionic acid exceeds that of acetic acid [95].

6.11 METHYL *n*-BUTYRATE

The Arrhenius graph for the combustion of this compound possesses a region of negative temperature coefficient, and the cool flame limits have been determined [45].

6.12 SUMMARY

The fact that ethyl formate, ethyl acetate, ethyl propionate, *n*-propyl formate, *n*-propyl acetate, iso-propyl acetate and methyl *n*-butyrate all

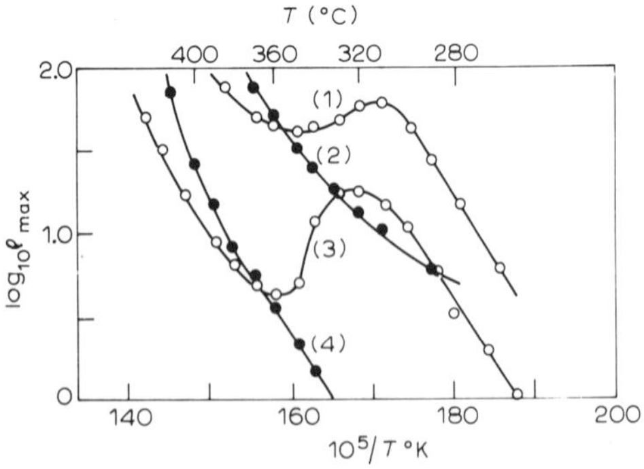

Fig. 13. The log ρ_{max} versus $1/T$ plots for equimolar ester—oxygen mixtures [45]. ρ_{max} is the maximum rate of reaction in torr . min^{-1}. (1) Methyl *n*-butyrate, *P*(total) = 240 torr. (2) Methyl propionate, *P*(total) = 500 torr. (3) Ethyl acetate, *P*(total) = 160 torr. (4) Methyl acetate, *P*(total) = 500 torr.

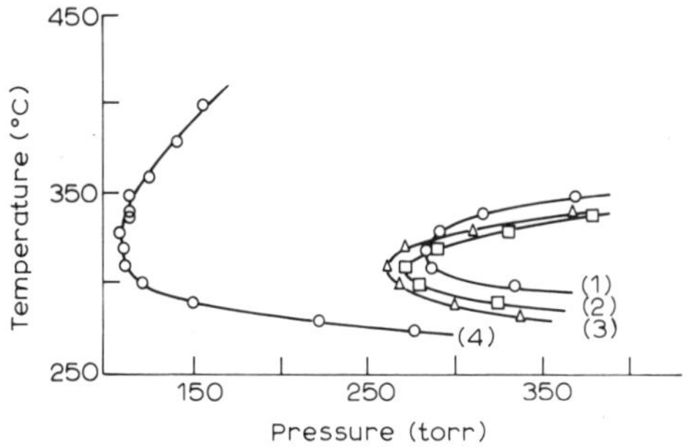

Fig. 14. The cool flame limit curves of equimolar ester—oxygen mixtures in an HF-treated spherical pyrex vessel, diameter = 10.1 cm [45]. (1) Ethyl acetate. (2) iso-Propyl acetate. (3) Methyl *n*-butyrate. (4) *n*-propyl acetate.

possess a region of negative temperature coefficient of the rate (and hence presumably are capable of giving rise to cool flames) while methyl formate, methyl acetate, methyl propionate and *tert*-butyl acetate do not (Figs. 13 and 14), suggests that 1:6 and 1:7 transfer of primary hydrogen in the peroxy radicals occurs too slowly at these temperatures to give a cool flame, but that 1:6 transfer of secondary hydrogen or 1:5 or 1:4 transfers of any hydrogen are sufficiently rapid to do so [98].

7. Peroxides

7.1 DIETHYL PEROXIDE

Nearly all peroxides decompose readily, and many of the lower members are explosive. The decomposition of diethyl peroxide has been studied under both non-explosive and explosive conditions [99—102(b)]. The reaction is first-order and the variation of rate coefficient with temperature (uncorrected for any self-heating) is represented by $k = 1.6 \times 10^{14} \exp(-34{,}000/RT)$ sec^{-1}. At around 200 °C the course of the slow reaction may be represented by [102(a)]

$$\begin{aligned} C_2H_5OOC_2H_5 = {} & 0.84C_2H_5OH + 0.04HCHO + 0.385CO \\ & + 0.375CH_3CHO + 0.273C_2H_6 \\ & + 0.24CH_4 + 0.33H_2 \end{aligned}$$

This reaction is exothermic ($\Delta H = -47$ kcal . mole^{-1}) and at 180 °C and above is accompanied by self-heating.

Above a critical pressure (about 8 torr at 180 °C falling to about 2 torr at 200 °C) diethyl peroxide explodes spontaneously, emitting a flash of blue light. This explosion, which is preceded by an induction period during which the reactant self-heats, results in decomposition more extensive than occurs in the slow reaction, the stoichiometry being

$$C_2H_5OOC_2H_5 = 0.96HCHO + 0.49C_2H_6 + 0.42CO + 0.40CH_4$$
$$+ 0.32CH_3CHO + 0.30C_2H_5OH + 0.23H_2$$
$$(\Delta H = -38 \text{ kcal. mole}^{-1})$$

All the major products of both the slow decomposition and explosive reaction can be accounted for qualitatively by reactions of the ethoxy radicals formed in the initial step which involves fission of the O—O bond. The reaction scheme is

$$C_2H_5OOC_2H_5 \longrightarrow 2C_2H_5O\cdot$$
$$C_2H_5O\cdot \longrightarrow CH_3\cdot + HCHO$$
$$C_2H_5O\cdot + RH \longrightarrow C_2H_5OH + R\cdot$$
$$2C_2H_5O\cdot \longrightarrow C_2H_5OH + CH_3CHO$$

followed by

$$CH_3\cdot + CH_3\cdot\ (+M) \longrightarrow C_2H_6\ (+M)$$
$$H\cdot, CH_3\cdot + RH \longrightarrow H_2, CH_4 + R\cdot$$
$$CH_3CHO \longrightarrow CH_3\cdot + H\cdot + CO$$

At about 500 °K disproportionation is favoured as the products are mainly ethanol and acetaldehyde. Explosion leads to higher temperatures, and more ethoxy radicals decompose yielding more ethane and formaldehyde.

The thermal aspects of the reaction have been thoroughly investigated and both temperature—time histories and temperature profiles across the vessel have been measured using very fine thermocouples [102(a), 103].

The results have been used to test the theory of thermal explosions in several ways. The agreement is excellent; for example, Frank—Kamenetski's theory predicts that the maximum temperature rise in a system which just fails to explode should be $1.61RT_0^2/E$ (where T_0 is the initial temperature, and E the activation energy of the reaction). At 204 °C, this corresponds to 20.5 °C, while the measured values was 20 °C.

7.2 *tert*-BUTYL HYDROPEROXIDE

This compound is capable of supporting both oxidation and decomposition flames, [104] but there have been no detailed investigations of either the flames or explosive reaction.

7.3 DI-*tert*-BUTYL PEROXIDE

The inflammability limits for flame propagation in a vertical tube of mixtures of this peroxide with air have been measured [105].

8. Sulphur compounds

The kinetics of combustion of gaseous sulphur compounds has recently been reviewed by Cullis and Mulcahey [106(b)].

8.1 THIOLS

Methane and ethane thiols are more readily oxidized than their parent alkanes [106(a), 107].

Methane thiol oxidizes at a convenient rate at 200—275 °C, the reaction being mildly autocatalytic and accompanied by a pressure decrease [107]. The rate is enhanced by increased oxygen concentration, but retarded by excess thiol. The main products include sulphur dioxide, carbon monoxide, formaldehyde, acetaldehyde and methane but no hydrogen sulphide, carbonyl sulphide or free sulphur. Unless excess oxygen is present, these products do not account for all the sulphur consumed, and Cullis and Roselaar [107] attributed this to formation of dimethyl disulphide although later work [108] has shown that this explanation was unlikely.

Ethane thiol reacts more readily than the methyl compound, but the main features of the reaction (e.g. the effect of reactant concentration) are similar. The products include some acid (assumed to be peracetic or acetic) as well as sulphur dioxide and acetaldehyde; the large sulphur deficit is again ascribed to disulphide formation. In oxygen-rich mixtures all the sulphur is converted to sulphur dioxide.

A mechanism has been proposed in which initial attack occurs at the —SH group, viz.

$$RSH + O_2 \longrightarrow RS\cdot + HO_2\cdot$$

The RS radicals react with oxygen to yield sulphur dioxide and an alkyl radical

$$RS\cdot + O_2 \longrightarrow R\cdot + SO_2$$

The alkyl radicals react rapidly with oxygen to yield aldehydes and alcohols (or in very oxygen-deficient systems) the hydrocarbon. Chain termination occurs by recombination

$$RS\cdot + RS\cdot \longrightarrow RS\,SR$$

8.2 DIALKYL SULPHIDES

Mixtures of dimethyl sulphide and oxygen explode with a blue flash at 210 °C [109]. Sulphur is deposited, and other products include sulphur dioxide, carbon monoxide, carbon dioxide, methane and methane thiol but methanol and dimethyl disulphide were not found.

Below this temperature, the pressure—time curves are S-shaped and the stoichiometry of the reaction is represented by [110]

$$CH_3SCH_3 + 2\tfrac{1}{2}O_2 = CH_3OH + CO + SO_2 + H_2O$$

The rate of reaction (after the induction period) is linearly dependent on the initial oxygen pressure, but substantially independent of the initial sulphide pressure. The reaction is inhibited by sulphur dioxide, and ceases before all the reactants are consumed.

Harkness and Murray [109], while agreeing that in slow combustion all the sulphur is converted to sulphur dioxide, failed to find methanol. Carbon dioxide and methane were also absent, but a little formaldehyde was present in the products.

The combustion of diethyl sulphide proceeds explosively at 170 °C whereas at 155 °C the reaction is very quickly self-inhibited. Analytical studies using a packed vessel at 195 °C showed that the products were sulphur dioxide, acetaldehyde, acetic acid and water [110].

8.3 DIMETHYL DISULPHIDE

The combustion of dimethyl disulphide was studied at 240 °C. The reaction is autocatalytic, the principal products being sulphur dioxide, methanol and carbon monoxide with smaller amounts of formaldehyde, methane thiol and an acid [111].

9. Nitrogen compounds

9.1 AMINES

The slow combustion of the lower aliphatic amines occurs at temperatures and pressures comparable with those under which the corresponding hydrocarbons also react.

9.1.1 Primary and secondary amines

A comparative study of the oxidation of a number of primary and secondary aliphatic amines under standard conditions was made by Cullis et al. [112—114]. The results are summarized in Figs. 15 and 16.

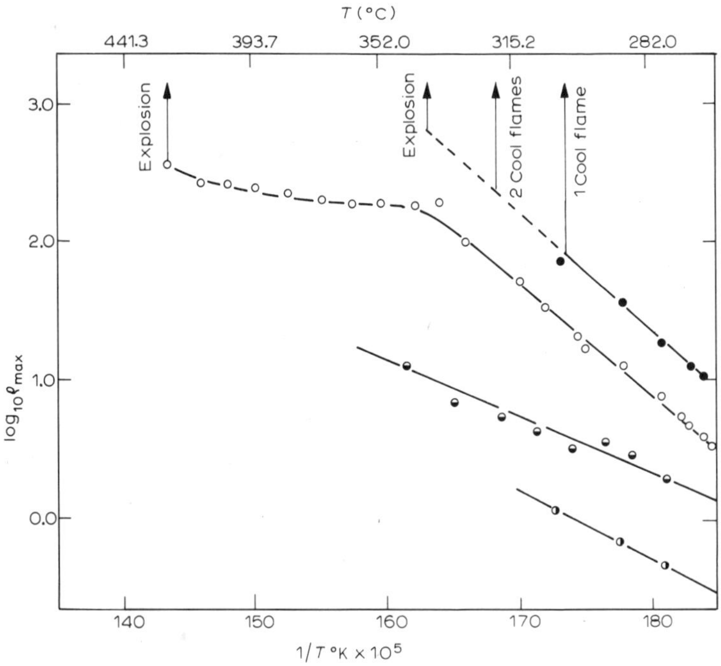

Fig. 15. The variation of ρ_{max}, the maximum oxidation rate in torr . min^{-1} of some primary aliphatic amines with temperature [112]. Amine pressure 50 torr; oxygen pressure 200 torr. ◑, Methylamine; ◒, ethylamine; ○, propylamine; ●, butylamine.

Reaction began without appreciable induction period (except in the case of iso-butyl amine) and the rate accelerated smoothly to a maximum. In the series Me-, Et-, *n*-Pr-, *n*-Bu-amine the oxidation rate increased with lengthening hydrocarbon chain, while secondary amines were more readily oxidized than the corresponding primary compounds.

At temperatures around 300 °C, the maximum rate of oxidation of methylamine [113] and dimethylamine depends on both the oxygen and amine pressures to a power of less than one, whereas for ethylamine and *n*-propylamine the order with respect to amine is greater than unity.

Analytical studies revealed considerable peroxide formation, the concentration of peroxide building up to a maximum at the time of maximum rate (see Fig. 17). Formaldehyde was also a product together with ammonia, nitrogen oxides and, in some cases, hydrogen cyanide.

A chain mechanism involving peroxy radicals was put forward to account for these results. In the case of ethylamine, the steps suggested are

$$CH_3CH_2NH_2 + O_2 \longrightarrow CH_3\dot{C}HNH_2 + HO_2\cdot$$

$$CH_3\dot{C}HNH_2 + O_2 \longrightarrow \underset{\displaystyle OO\cdot}{CH_3CHNH_2}$$

References pp. 496–500

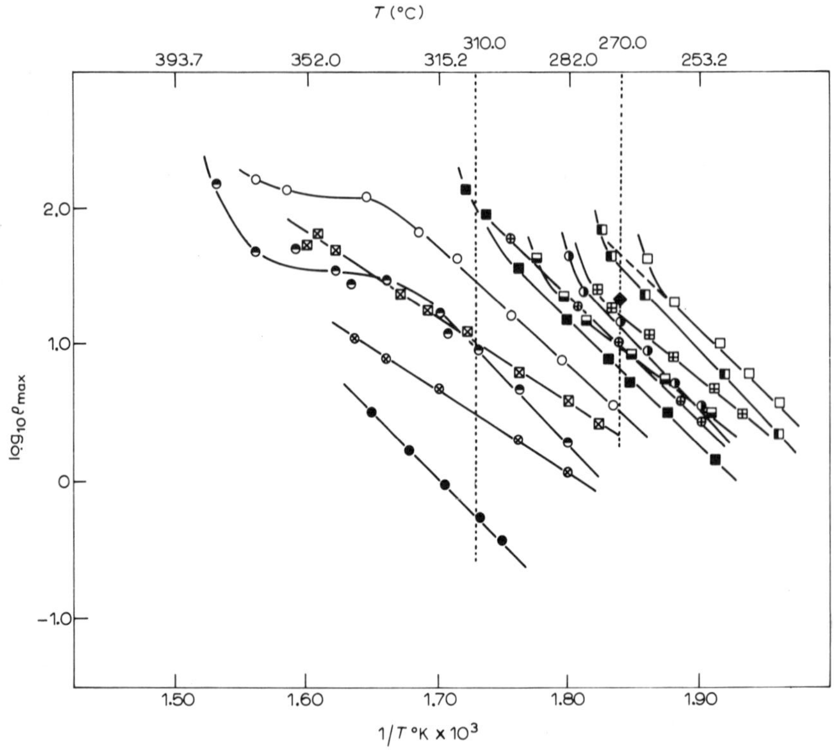

Fig. 16. The variation of ρ_{max}, the maximum rate of oxidation in torr . min^{-1} of some primary and secondary aliphatic amine with temperature [114]. Amine pressure 500 torr; oxygen pressure 200 torr. ●, Methylamine; ⊗, ethylamine; ◓, iso-propylamine; ⊠, dimethylamine; ○, *n*-propylamine; ■, di-iso-propylamine; ⬓, ethylmethylamine; ⊕, iso-butylamine; ◑, *n*-butylamine; ⊞, diethylamine; ◧, methyl-*n*-propylamine; □, di-*n*-propylamine; ◆, *n*-butylchloride.

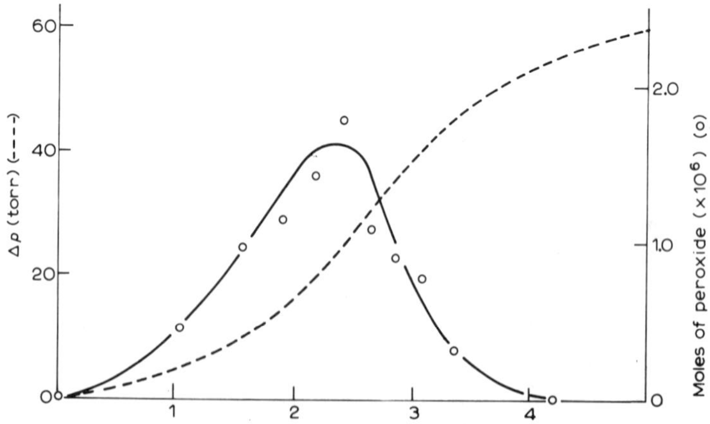

Fig. 17. The variation of peroxide concentration during the oxidation of methylamine at 350 °C [113]. Methylamine pressure, 100 torr; oxygen pressure, 300 torr.

$$CH_3CH(OO\cdot)NH_2 + CH_3CH_2NH_2 \longrightarrow CH_3CH(OOH)NH_2 + CH_3\dot{C}HNH_2$$

$$CH_3CH(OOH)NH_2 \longrightarrow CH_3CHO + NH_2\cdot + \cdot OH$$

$$X\cdot(NH_2\cdot, \cdot OH) + CH_3CH_2NH_2 \longrightarrow XH(NH_3, H_2O) + CH_3\dot{C}HNH_2$$

$$NH_2\cdot + O_2 \longrightarrow \text{inert products}$$

The combustion limits and burning velocity of methylamine in air have been determined [115] using a horizontal tube technique [116]. The burning velocity was greatest (25 cm . sec^{-1}) in a mixture containing 9 % amine in air. The reaction products include hydrogen cyanide, but ammonia, methane and hydrogen were absent. In the very high temperature region before the combustion zone the amine pyrolysed, viz.

$$CH_3NH_2 = HCN + 2H_2$$

and the predominant combustion reaction was that between hydrogen and oxygen.

9.1.2 Tertiary amines

The tertiary amines present a rather more complex picture [117]. Although trimethylamine begins to react at about 165 °C, the reaction is rapidly inhibited as products accumulate and ceases when the major part of the reactants are still unconsumed [118]. Both the rate and extent of oxidation are reduced by increased surface, but additions of inert gas have no influence on the reaction. The products include formaldehyde and dimethylamine with smaller amounts of nitrogen and carbon monoxide. No methylamine, ammonia or nitrogen oxides were found.

The main products are accounted for by a hydroperoxide mechanism

$$N(CH_3)_3 \xrightarrow{X\cdot} N(CH_2\cdot)(CH_3)_2 + HX$$

$$N(CH_2\cdot)(CH_3)_2 + O_2 \longrightarrow N(CH_2OO\cdot)(CH_3)_2$$

$$N(CH_2OO\cdot)(CH_3)_2 + RH \longrightarrow N(CH_2OOH)(CH_3)_2 + R\cdot$$

$$N(CH_2OOH)(CH_3)_2 \longrightarrow CH_2O + \cdot OH + \cdot N(CH_3)_2$$

$$\cdot N(CH_3)_2 + RH \longrightarrow (CH_3)_2NH + R\cdot$$

Although the product dimethylamine is an inhibitor for the oxidation of trimethylamine, its effect is not powerful enough to account for the results, and it was suggested that the much more powerful inhibitor *N,N*-dimethylhydroxylamine was responsible.

Triethylamine begins to oxidize around 200 °C and this reaction is also self-inhibited [119]. The slow reaction persists to about 280 °C above which is a region of explosive reaction. Between 360 and 400 °C slow combustion takes place, while above 400 °C, explosion finally intervenes. Although the reaction is sensitive to the surface of the reaction vessel, addition of nitrogen was without effect on the rate.

At low temperatures, the initial rate of reaction is directly proportional to oxygen concentration and dependent on a power of the amine concentration greater than one. Later in the reaction, the rate becomes far less dependent on amine pressure. The main products are the primary and secondary amine and acetaldehyde, and there is evidence for two simultaneous but independent reactions taking place

$$(C_2H_5)_3N + O_2 = C_2H_5NH_2 + 2CH_3CHO$$

$$(C_2H_5)_3N + \tfrac{1}{2}O_2 = (C_2H_5)_2NH + CH_3CHO$$

The first reaction involves formation of peroxy radicals, which undergo internal hydrogen abstraction followed by decomposition, viz.

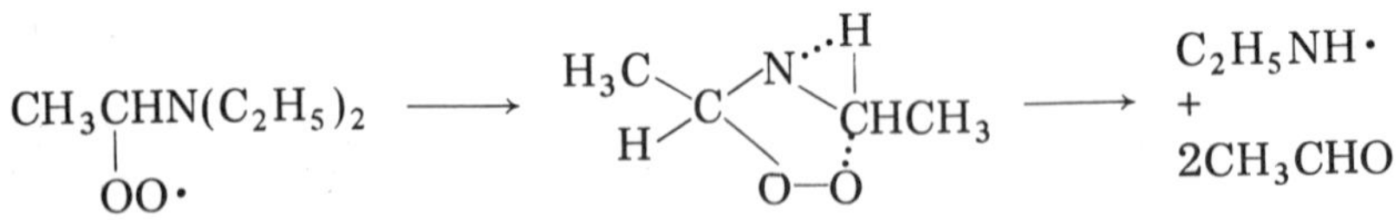

The resulting C_2H_5NH radical stabilizing itself by H-abstraction from a further triethylamine molecule. The other reaction proceeds via hydroperoxide formation

$$CH_3CH(OO\cdot)N(C_2H_5)_2 \xrightarrow{RH} CH_3CH(OOH)N(C_2H_5)_2$$

the hydroperoxide then breaking down ,

$$CH_3CH(OOH)N(C_2H_5)_2 \longrightarrow CH_3CHO + \cdot OH + (C_2H_5)_2N\cdot$$

$$(C_2H_5)_2N\cdot + RH \longrightarrow (C_2H_5)_2NH + R\cdot$$

The combustion of *N*-methyldiethylamine resembles triethylamine, whilst that of *N*-ethyldimethylamine shows a striking similarity to trimethylamine [120]. In both cases, oxidation leads to a secondary amine, but is rapidly self-inhibited.

At 211 °C, the initial rate of oxidation of *N*-methyldiethylamine is linearly dependent on oxygen concentration, but varies with a higher power of the fuel concentration. In the early stages the products include both primary and secondary amines and also acetaldehyde and formaldehyde, though later in the reaction the products are almost exclusively acetaldehyde and the primary amine.

The two concurrent overall reactions suggested are

$$(C_2H_5)_2NCH_3 + \tfrac{1}{2}O_2 = \begin{cases} (C_2H_5)NHCH_3 + CH_3CHO \\ \text{or} \\ (C_2H_5)_2NH + HCHO \end{cases}$$

and

$$(C_2H_5)_2NCH_3 + O_2 = CH_3NH_2 + 2CH_3CHO$$

The second of these proceeds by a peroxy radical mechanism involving internal hydrogen abstraction followed by decomposition analogous to that suggested above. The radical produced in this process is CH_3NH which in turn yields methylamine.

As with trimethylamine [118], only a single mechanism appears to be operative with *N*-ethyldimethylamine, partly perhaps for steric reasons. Intramolecular attack in the peroxy radicals

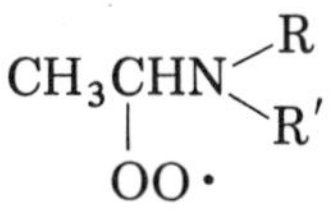
and
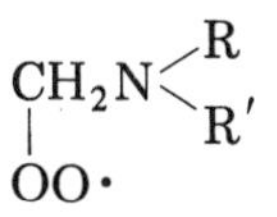

is apparently difficult when either or both the groups R and R′ is a methyl [121]. Hence, only a reaction producing aldehyde and a secondary amine, which rapidly becomes self-inhibited, takes place.

9.2 NITRITE ESTERS

9.2.1 Methyl nitrite

Methyl nitrite burns vigorously with oxygen at atmospheric pressure giving a very fast flame. It will also support a very feeble, slow self-decomposition flame [122–124]. The latter is orange-red and the burning velocity [125], S_u relative to the unburnt gas at 18 °C, is about 3.5 cm . sec^{-1}, while the "global" activation energy [126] is 37.5 kcal . mole^{-1}. The final products of the flame are represented by the stoichiometric equation

$$CH_3ONO = 0.72CO + 0.56H_2 + 0.54NO + 0.42H_2O + 0.14N_2 +$$
$$+ 0.13CH_3OH + 0.10CH_2O + 0.06NH_3 + 0.05N_2O +$$
$$+ 0.02\ \text{"}CH_2NOH\text{"} + 0.02CH_4 + 0.01CO_2$$

The flame temperature calculated on the basis of this stoichiometry is about 1100 °C, in excellent agreement with the observed value [127].

These results suggest that a considerable degree of self-heating occurs when the thermal decomposition of methyl nitrite is studied in a static system. At about 250 °C, the products of thermal decomposition are largely formaldehyde, methanol and nitric oxide together with a little nitrous oxide and water. The decomposition is accompanied by chemiluminescence, and the glow is ascribed to excited formaldehyde [128]. Added argon lowers the pressure limit for glow by hindering diffusion of excited species to the wall where they are deactivated. The decomposition almost certainly involves methoxy radicals

$$CH_3ONO \longrightarrow CH_3O\cdot + NO$$

which can then propagate a chain reaction [129, 130].

9.2.2 Ethyl nitrite

In its chemiluminescent decomposition and decomposition flame, ethyl nitrite is very similar to methyl nitrite [128, 131].

9.3 NITRATE ESTERS

9.3.1 Methyl nitrate

This is the most explosive of the nitrate esters. Not only will it burn in an atmosphere of oxygen, nitric oxide or nitrogen dioxide, but also it can support a stationary decomposition flame which can be stabilized on a burner [122]. At low pressures the various zones of the decomposition flame are clearly separated and the early stages show strong formaldehyde bands in emission.

The fuel molecule breaks down in the pre-heat zone to give methoxy radicals

$$CH_3ONO_2 \longrightarrow CH_3O\cdot + NO_2$$

and it is interesting to observe [128] that the glow from stoichiometric mixtures of methyl nitrite + nitrogen dioxide and methyl nitrate + nitric oxide is very similar, lending support to the belief that both systems yield the same products

$$CH_3ONO + NO_2 \longrightarrow CH_3O\cdot + NO + NO_2 \longleftarrow CH_3ONO_2 + NO$$

In a static system, the decomposition can proceed in three ways, explosion, chemiluminescent reaction and decomposition without glow [131]. The glow which is probably due to excited formaldehyde is detectable even when the nitrate is diluted between 10^3 and 10^6 times with inert gas.

A careful study of self-heating during the spontaneous ignition of methyl nitrate vapour has given results in very satisfactory agreement with the predictions of thermal explosion theory [132(a), (b)].

9.3.2 *Ethyl nitrate*

This ester resembles its methyl homologue in possessing three modes of decomposition [131]. It also supports a self-decomposition flame, the multiple reaction zones of which are clearly separated at low pressures [122, 123, 125]. Temperature and composition profiles in the low-pressure decomposition flame have been measured [133]. The products include formaldehyde, acetaldehyde and ethanol with smaller amounts of methane and nitromethane. The activation energy derived from the variation of flame speed with final flame temperature was 38 kcal . mole^{-1}, close to the dissociation energy of the $RO{-}NO_2$ bond. The controlling reaction is believed to be unimolecular in its low pressure regime, and the rate coefficient calculated from the heat-release profile is

$$k = 5 \times 10^{17} \exp(-38{,}000/RT) \, \mathrm{l.\,mole^{-1}\,.\,sec^{-1}}$$

Most of the ethyl nitrate decomposes in the high temperature region (740—800 °K) of the flame where it is believed that unimolecular thermal decomposition [129] is the rate controlling process. The rate equation given above corresponds to a unimolecular reaction in its low-pressure region and the pre-exponential term suggests that the activation energy is distributed over about 10 square terms.

In the low temperature region of the flame the activation energy is lower and, up to 740 °K, radical attack is significant, viz.

$$C_2H_5ONO_2(+M) \longrightarrow C_2H_5O\cdot + NO_2(+M)$$

$$C_2H_5O\cdot + X\cdot \longrightarrow C_2 \text{ products}$$

$$C_2H_5O\cdot \longrightarrow CH_3\cdot + CH_2O$$

$$CH_3\cdot + NO_2 \longrightarrow CH_3O\cdot + NO$$

$$CH_3\cdot, CH_3O\cdot + X\cdot \longrightarrow C_1 \text{ products}$$

The stationary flame, stabilized above a liquid surface has also been examined [134]. Immediately above the liquid there is a dark space, followed by an orange glow which in turn, gives way to a faint greenish glow. The maximum temperature 1175 °K was reached about 10 mm above the liquid surface. Numerous products were identified and most can be accounted for by the following scheme

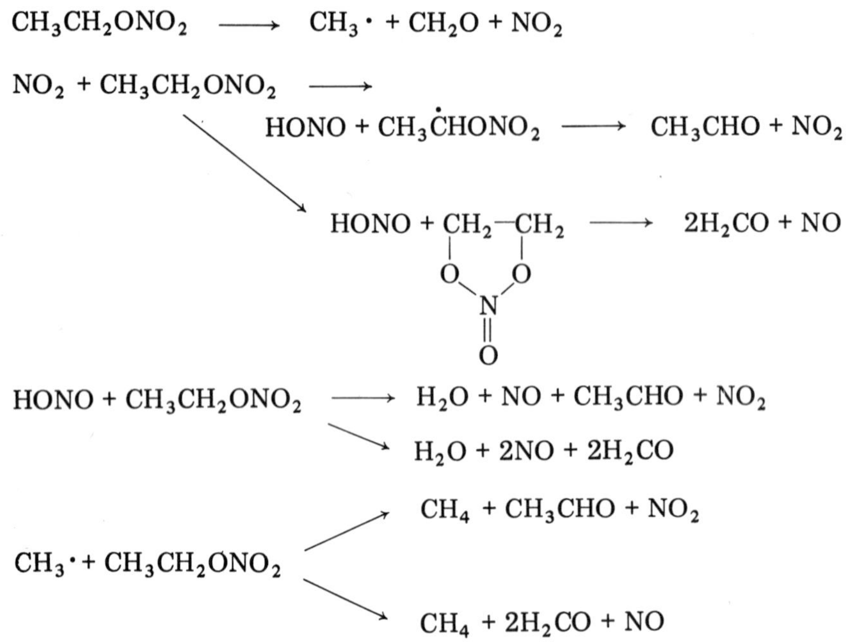

9.3.3 Substituted ethyl nitrates

The flame decompositions of 2-hydroxyethyl nitrate, 2-methoxyethyl nitrate and 2-ethoxyethyl nitrate have been studied using a flat flame burner [135]. The major products of very rapid reaction in the flame front are nitric oxide, carbon monoxide, water, formaldehyde, methyl formate, methanol and a large amount of unidentified material. The absence of 2-methoxyethanol and of nitrogen dioxide, and presence of only minor amounts of dimethyl ether is of some importance.

The initiating step is considered to be rupture of the weakest band in the molecule, followed by breakdown of the resulting alkoxy radical, viz.

$$CH_3OCH_2CH_2ONO_2 \longrightarrow CH_3OCH_2CH_2O\cdot + NO_2$$

$$\downarrow$$

$$CH_3OCH_2\cdot + H_2CO$$

$$\downarrow$$

$$CH_3\cdot + H_2CO$$

The methyl formate could arise via reactions analogous to those proposed for ethyl nitrate, viz.

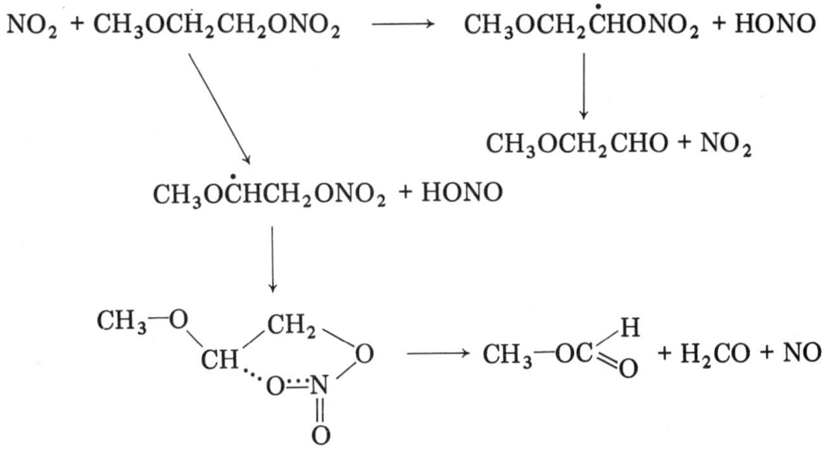

9.3.4 Propyl nitrates

The mechanisms of *n*- and iso-propyl nitrate decomposition flames appear to be the same as that of the ethyl nitrate flame, the main attack on the nitrate ester being by nitrogen dioxide and nitrous acid [136].

For iso-propyl nitrate the main products acetaldehyde and CH_3NO_2 are formed in the following sequence

$$(CH_3)_2CHONO_2 \longrightarrow (CH_3)_2CHO\cdot + NO_2$$

$$(CH_3)_2CHO\cdot \longrightarrow CH_3\cdot + CH_3CHO$$

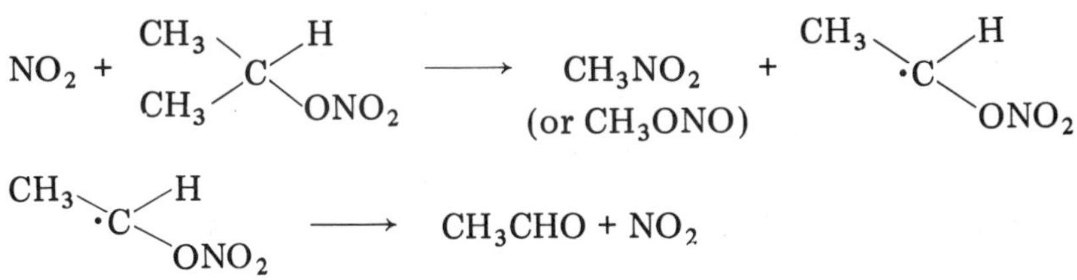

The less important product, acetone, could be formed by

$$NO_2 + (CH_3)_2CHONO_2 \longrightarrow (CH_3)_2\dot{C}ONO_2 + HONO$$

$$(CH_3)_2\dot{C}ONO_2 \longrightarrow (CH_3)_2CO + NO_2$$

or by

$$HONO + (CH_3)_2CHONO_2 = (CH_3)_2CO + H_2O + NO + NO_2$$

More recent work has shown that, below about 200 °C, the pyrolysis of iso-propyl nitrate follows first-order kinetics. The main products are iso-propyl nitrite, methyl nitrite, nitromethane, acetaldehyde, acetone, nitrogen oxides and carbondioxide [136(b)];

9.3.5 Dinitrates

The burning rates of butane-2,3 and 1,4-diol dinitrates are much less than those of mononitrates, and their combustion takes place by a different mechanism. Powling and Smith [137] suggest unimolecular decompositions

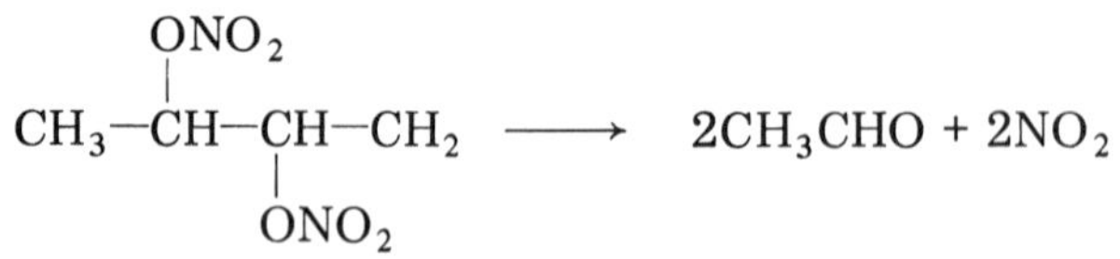

and

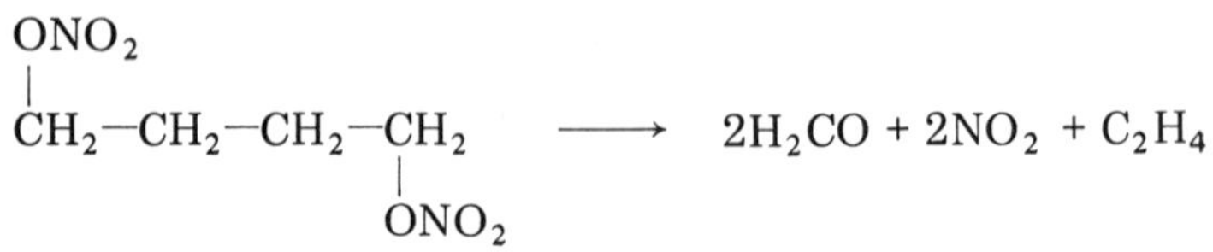

9.4 NITROMETHANE

Flame velocities of and temperatures in nitromethane flames supported by oxygen have been measured. The global activation energy is surprisingly low, viz. 10—16 kcal . mole^{-1}, depending on the fuel/oxygen ratio [126].

9.5 AZOMETHANE

The pyrolysis of azomethane which yields mainly nitrogen and ethane, has been extensively studied [138, 139]. At high temperatures the decomposition is explosive [140]. The evidence suggests that the explosion is of thermal origin above 636 °K, while below this temperature there are indications that chain-branching plays a significant part.

This reaction has been studied over an unusually wide range of temperature [141] and the activation energy is known to be about 53 kcal . mole^{-1}, from about 500 to 1300 °K. The results from the study of explosive reaction [140], however, lead to a value of 32 kcal . mole^{-1}.

10. Halogen compounds

10.1 FLUOROCARBONS

The combustion behaviour of freons and fluorocarbons has been reviewed by Fletcher [142(b)].

Fluorocarbons are far less reactive towards oxygen than hydrocarbons. Thus mixtures of tetrafluoromethane, hexafluoroethane, perfluoropropane and perfluorobutane and oxygen cannot be ignited in a burner at atmospheric pressure and laboratory temperature. However, some fluorocarbons will support combustion and two significant differences between the flame propagation characteristics of fluorocarbon—oxygen and hydrocarbon—oxygen mixtures, have been noted. Firstly, the maximum burning velocity of the former is considerably lower than that of the corresponding hydrocarbon—oxygen mixture and secondly the maximum burning velocity of fluorocarbon—oxygen mixtures, unlike that of hydrocarbon—oxygen mixtures, cannot be correlated with maximum adiabatic flame temperature [142(a)]. It was suggested that flame propagation in these systems is dominated by diffusion of F atoms from the hot products into the unburnt reactants, rather than by thermal considerations.

The burning velocities of mixtures of tetrafluoroethylene [143], perfluoropropane [142(a)], perfluorocyclobutene [142(a)] and perfluorocyclobutane [142(a)] with oxygen at atmospheric pressure have been measured using a burner technique. The results for perfluorocyclobutane are in reasonable agreement with measurements using a flat flame deflagration tube. In a tube, lean perfluorocyclobutane mixtures with oxygen ignited giving a flame with a blue leading edge and a pinkish-blue body. The body became orange and finally yellow as the mixtures were made richer [144]. The fundamental flame speed of these flames has been measured for various mixtures and it was noted that above 40 mole % C_4F_8, pronounced soot formation occurred [145].

In a bomb [146], hexafluoroethane will not react with either hydrogen or oxygen separately. With a hydrogen—oxygen mixture, however, a rapid reaction occurs according to the equation

$$C_2F_6 + 3H_2 + 2O_2 = 6HF + 2CO_2$$

At 100 °C, tetrafluoroethylene reacts slowly with oxygen [147]. Thus, in an equimolar mixture after 14 h, about 90 % of the oxygen and 85 % of the tetrafluoroethylene have disappeared, the main products being approximately equal amounts of carbonyl fluoride and hexafluorocyclopropane together with a little tetrafluoroethylene oxide.

Flame speeds and quenching distances in the perfluoroethane—perfluoropropane and perfluorocyclobutane—chlorine trifluoride systems have been measured. Perfluoroethane burns much more weakly than perfluoropropane, while perfluorocyclobutane burns very vigorously [148].

Perfluorocyclobutane—fluorine mixtures detonated readily and detonation velocities and limits have been measured [149(a)].

The flammability limits of several H_2—O_2— fluorocarbon mixtures have been reported by McHale et al. [149(b)] and for some fluorobenzenes by Pollard [149(c)].

Other studies have been concerned with the burning velocities of trifluorochloromethane and trifluorobromomethane flames in fluorine [150]. Addition of hydrogen to flames of difluorodichloromethane and trifluorochloromethane in fluorine raised the flame temperature considerably [151(a)]. With no hydrogen, the products include only compounds with one carbon atom; when hydrogen is added then C_2 and even C_3 compounds appear.

It appears therefore that in the absence of hydrogen the concentration of carbon containing radicals, e.g. CF_3, which can recombine or add to fuel molecules is small and the main reactions occurring are direct substitution e.g.

$$CCl_2F_2 + F\cdot \longrightarrow CClF_3 + Cl\cdot$$

When hydrogen is present, then H atoms are formed which abstract rather than substitute

$$CCl_2F_2 + H\cdot \longrightarrow \cdot CClF_2 + HCl$$

$$\searrow \quad \cdot CCl_2F + HF$$

and the carbon containing radicals can then undergo reaction leading to higher halocarbon products.

From studies of the concentration profiles through dichlorodifluoromethane/fluorine flames at low pressures, Homann and MacLean have proposed a chain mechanism involving fluorine and chlorine atoms as chain carriers [151(b)].

10.2 METHYL CHLORIDE

Methyl chloride will burn in air or in oxygen with a rather slow flame in which the Swan C_2 bands are prominent [12]. The burning velocity of a stoichiometric methyl chloride—air mixture is about 10 cm . sec^{-1} and the effect of pressure on the burning velocity has been studied [152].

10.3 METHYLENE CHLORIDE

The slow combustion of methylene chloride is a degenerately branched chain reaction; it proceeds by a mechanism similar to that involved in the pyrolysis of the same compound which takes place at a slightly higher temperature [153]. The primary chains are the same and several of the chlorinated hydrocarbon minor products are identical. Oxygen is only involved in the conversion of the intermediate dichloroethylene to the final products hydrogen chloride and carbon monoxide.

At 533 °C in a vessel whose internal diameter was 34 mm, the maximum rate of reaction was given by

$$-(d[CH_2Cl_2]/dt)_{max} = k[CH_2Cl_2]^{2.0}[O_2]^{0.9}$$

The initial stages of the reaction showed an exponential acceleration, according to the usual equation for degenerately branched chain reaction

$$\Delta p = N \exp(\phi t)$$

and the apparent activation energy of the acceleration constant ϕ, was 26 kcal . mole^{-1}. Above 600 °C an explosion limit was observed, the variation of this limit in a 25 mm diameter vessel with temperature being expressed by

$$\log_{10}(p/\text{torr}) = 5100/T + 2.8$$

for an equimolar mixture.

The overall stoichiometry of the reaction tended to

$$CH_2Cl_2 + \tfrac{1}{2}O_2 = CO + 2HCl$$

but the products also included carbon dioxide, $CHCl_2CHCl_2$, $CHCl{=}CCl_2$, $CHCl_3$, CCl_4, *cis*- and *trans*-$C_2H_2Cl_2$, $CH_2{=}CCl_2$, C_2Cl_4, H_2O, HCHO and a trace of chlorine.

The primary chain is

$$CH_2Cl_2 + Cl\cdot \longrightarrow \cdot CHCl_2 + HCl$$

$$\cdot CHCl_2 + CH_2Cl_2 \longrightarrow CHCl{=}CHCl + HCl + Cl\cdot$$

The secondary oxidation chain involves radical attack and oxidation of the dichloroethylene, viz.

$$CHCl{=}CHCl + Cl\cdot \longrightarrow \cdot C_2HCl_2 + HCl$$

$$\cdot C_2HCl_2 + O_2 \longrightarrow 2CO + HCl + Cl\cdot$$

with branching by

$$CHCl{=}CHCl + O_2 \longrightarrow H_2O + CO + \cdot CCl_2$$

$$\cdot CCl_2 + CH_2Cl_2 \longrightarrow 2\dot{C}HCl_2$$

Termination steps may include

$$\cdot CHCl_2 \xrightarrow{\text{wall}} \text{Inert}$$

$$\cdot CHCl_2 + Cl\cdot \longrightarrow CHCl_3$$

$$2Cl\cdot \xrightarrow{\text{wall}} Cl_2$$

10.4 TRICHLOROMETHANE (CHLOROFORM)

The flames of methyl chloride, methylene chloride and chloroform have been briefly studied. The latter is pale blue-green, suggesting that Cl_2O may be involved [154, 155].

References pp. 496–500

The flame of the former compound may be stabilized, and exhibits two zones, the first of which is orange-brown and the second, pale blue-green. The temperature in the first region is around 1100—1300 °C and in the second about 200 °C higher. The flame colours suggest Cl_2O is involved in the flame reaction [154, 155].

The chlorine photosensitized oxidation of trichloroethylene has been thoroughly investigated [156], the main reaction being

$$CHCl{=}CCl_2 + \tfrac{1}{2}O_2 = CHCl_2{\cdot}COCl$$

Minor products include trichloroethylene epoxide, phosgene, chloroform and carbon tetrachloride.

The rate of reaction is represented by

$$d[CHCl_2{\cdot}COCl]/dt = I_a[O_2]/(k + k'[O_2])$$

where I_a is the intensity of the absorbed light. It was found that

$$\log\{k(\text{mole}\,.\,\text{l}^{-1})\} = 3.24 - 14.520/4.576T$$

and

$$\log k' = -4.33 + 3400/4.576T$$

The mechanism proposed is

$$Cl_2 \xrightarrow{h\nu} 2Cl{\cdot}$$

$$Cl{\cdot} + CHCl_2CCl_2 \longrightarrow {\cdot}C_2HCl_4$$

$${\cdot}C_2HCl_4 + Cl_2 \longrightarrow C_2HCl_5 + Cl{\cdot}$$

$$2C_2HCl_4{\cdot} \longrightarrow \text{Termination}$$

$${\cdot}C_2HCl_4 + O_2 \longrightarrow C_2HCl_4O_2{\cdot}$$

$$C_2HCl_4O_2{\cdot} + {\cdot}C_2HCl_4 \longrightarrow C_2HCl_4O_2C_2HCl_4$$

$$C_2HCl_4O_2{\cdot} + C_2HCl_4O_2{\cdot} \longrightarrow C_2HCl_4O_2C_2HCl_4 + O_2$$

$$\searrow 2C_2HCl_4O{\cdot} + O_2$$

$$C_2HCl_4O{\cdot} \longrightarrow C_2HCl_3O + Cl{\cdot}$$

Tetrachloroethylene burns feebly with a very slow flame in pure oxygen, but not at all in air.

10.6 CHLOROMETHANES AND NITROGEN DIOXIDE

Chloromethanes react with nitrogen dioxide in the gas phase. Carbon tetrachloride reacts slowly at 385 °C whereas the other chloromethanes are more reactive. The kinetics of the reactions involving the three lower

TABLE 3
The reaction of some chloromethanes with nitrogen dioxide

Chloromethane	Temp. range (°C)	Order[a] with respect to		log A[a] (mole, l, sec units)	E_a[a] (cal . mole^{-1})	Ref.
		NO_2	Chloromethane			
CH_3Cl	250—300	1.22	1.30	10.94	24,500	157
		0.90	1.01	11.82	30,860 (packed vessel)	
CH_2Cl_2	289—380	1.32	0.75	9.55	23,800	158
$CHCl_3$	272—322	0.81	0.83	9.37	28,300	158

[a] The orders of reaction, and the Arrhenius parameters all refer to the $(-d[NO_2]/dt)_{max}$.

compounds have been studied in some detail by Thomas et al. [157, 158]. In each case the reaction is autocatalytic, and the principal kinetic features are summarized in Table 3.

The products include

CH_3Cl: NO, NOCl, HCl, H_2O and CO

CH_2Cl_2: NO, NOCl, HCl, H_2O and CO

together with CCl_4 and $COCl_2$

$CHCl_3$: NO, NOCl, HCl, H_2O and CO

together with CCl_4, $COCl_2$ and CO_2

The observations suggest a chain reaction initiated by H-atom abstraction

$$CH_3Cl + NO_2 \longrightarrow \cdot CH_2Cl + HNO_2$$

This is supported by the appearance of CH_2ClNO_2 but no CH_3NO_2 in the reaction products from methyl chloride, and by the fact that carbon tetrachloride is relatively unreactive compared with the other compounds.

The initiation step is followed by

$$\cdot CH_2Cl + NO_2 \longrightarrow CH_2NO_2Cl \longrightarrow CH_2O + NOCl$$

$$CH_2O + NO_2 = CO + H_2O + NO$$

Autocatalysis is due to chlorine atoms resulting from

$$Cl_2 + NO \longrightarrow NOCl + Cl\cdot$$

$$HCl + NO \longrightarrow HNO_2 + Cl\cdot$$

$$NOCl \longrightarrow NO + Cl\cdot$$

The chlorine atom then reacts with, for example, methyl chloride

$$Cl + CH_3Cl \longrightarrow \cdot CH_2Cl + HCl$$

The minor products can all be satisfactorily accounted for. Very similar mechamsims are proposed for the two other chloromethanes.

10.7 METHYL BROMIDE

Although methyl bromide is a flame inhibitor [159—161], it undergoes slow combustion between 280 and 327 °C. The pressure—time curves are sigmoidal and at high oxygen partial pressures there is an initial pressure decrease which is followed by an increase [162]. The main products of the reaction are hydrogen bromide, bromine and vinyl bromide and the maximum rate of pressure change is proportional to the square of the methyl bromide pressure and the first power of the oxygen pressure.

The mechanism probably includes the steps

$$CH_3Br + O_2 \longrightarrow \cdot CH_2Br + HO_2\cdot$$

$$\cdot CH_2Br + CH_3Br \longrightarrow \cdot CH_2CH_2Br + HBr$$

$$\cdot CH_2CH_2Br \longrightarrow CH_2CHBr + H\cdot$$

$$H\cdot + CH_3Br \longrightarrow HBr + CH_3\cdot$$

$$CH_3\cdot + O_2 \longrightarrow HCHO + \cdot OH$$

$$\cdot OH + CH_3Br \longrightarrow H_2O + \cdot CH_2Br$$

10.8 METHYL IODIDE

The photolysis of this compound gives methyl radicals, and the reaction of these with oxygen has been studied at low temperatures by several workers [163—165].

The combustion of methyl iodide induced by flash photolysis has also been studied up to 400 °C. The spectrum of OH radicals was observed immediately after the flash, and this led McKellar and Norrish [166] to infer that the methyl peroxy radical, if formed, must be highly excited and have an exceedingly short lifetime. At somewhat higher pressures, the spectrum of formaldehyde was detected; although the OH spectrum could no longer be seen due to the extension of the methyl iodide spectrum to longer wavelengths and perhaps also because of scavenging of OH radicals by formaldehyde.

REFERENCES

1 R. Fort and C. N. Hinshelwood, Proc. R. Soc. London, Ser. A, 129 (1930) 284.
2 W. A. Bone and J. B. Gardner, Proc. R. Soc. London, Ser. A, 154 (1936) 297.

3 G. P. Kane, E. A. C. Chamberlain and D. T. A. Townend, J. Chem. Soc. (1937) 436.
4 T. E. Layng and M. A. Youker, Ind. Eng. Chem., 20 (1928) 1048.
5 C. F. Cullis, Petroleum (London), 27 (1964) 34.
6 D. E. Cheaney, D. A. Davies, A. Davis, D. E. Hoare, J. Protheroe and A. D. Walsh, 7th Symp. Combust. (1959) 183.
7 K. M. Bell and C. F. H. Tipper, Proc. R. Soc. London, Ser. A, 238 (1956) 256.
8 K. M. Bell and C. F. H. Tipper, Trans. Faraday Soc., 53 (1957) 982.
9 M. Lucquin, J. Chem. Phys., 55 (1958) 827.
10 L. R. Sochet and M. Lucquin, Combust. Flame, 13 (1969) 319.
11 W. H. Wiser and G. R. Hill, 5th Symp. Combust. (1955) 553.
12 H. T. Henderson and G. R. Hill, J. Phys. Chem., 60 (1956) 874.
13 A. G. Gaydon and H. G. Wolfhard, 3rd Symp. Combust. (1949) 504.
14 (a) E. de Wilde and A. van Tiggelen, Bull Soc. Chim. Belg., 77 (1968) 67.
(b) D. F. Cooke, M. G. Dodson and A. Williams, Combust. Flame, 16 (1971) 233.
15 C. F. Cullis and E. J. Newitt, Proc. R. Soc. London, Ser. A, 237 (1956) 530.
16 C. F. Cullis and E. J. Newitt, Proc. R. Soc. London, Ser. A, 242 (1957) 516.
17 C. F. Cullis and E. J. Newitt, 6th Symp. Combust. (1957) 827.
18 E. J. Newitt, 5th Symp. Combust. (1955) 558.
19 J. Brown and C. F. H. Tipper, Proc. R. Soc. London, Ser. A, 312 (1969) 399.
20 J. Griffiths and G. Skirrow, Oxid. Combust. Rev., 3 (1968) 64.
21 C. F. Cullis and E. J. Newitt, Proc. R. Soc. London, Ser. A, 256 (1960) 402.
22 A. R. Burgess, C. F. Cullis and E. J. Newitt, J. Chem. Soc. (1961) 1884.
23 A. R. Burgess and C. F. Cullis, J. Chem. Soc. (1961) 3041.
24 A. R. Burgess, J. Appl. Chem., 11 (1961) 235.
25 R. Walsh and S. W. Benson, J. Am. Chem. Soc., 88 (1966) 3480.
26 C. F. Cullis and E. A. Warwicker, Proc. R. Soc. London, Ser. 264 (1961) 392.
27 S. R. Smith and A. S. Gordon, J. Phys. Chem., 60 (1956) 1059.
28 S. R. Smith, A. S. Gordon and M. H. Hunt, J. Phys. Chem., 61 (1957) 553.
29 J. A. Barnard and T. W. Honeyman, Proc. R. Soc. London, Ser. A, 279 (1964) 236.
30 J. A. Barnard and T. W. Honeyman, Proc. R. Soc. London, Ser. A, 279 (1964) 248.
31 D. E. Hoare and Ting-Man Li, Combust. Flame, 12 (1968) 145.
32 D. E. Hoare and Ting-Man Li, Combust. Flame, 12 (1968) 136.
33 J. H. Knox, Combust. Flame, 9 (1965) 297.
34 S. W. Benson, J. Am. Chem. Soc., 87 (1965) 972.
35 E. Kirschner, Thesis, London, 1965.
36 J. H. Knox, Trans. Faraday Soc., 55 (1959) 1362.
37 (a) J. A. Barnard and A. Watts, 12th Symp. Combust. (1969) 365.
(b) J. A. Barnard and A. Watts, Combust. Sci. Technol., 6 (1972) 125.
(c) D. E. Hoare and D. E. Lill, J. Chem. Soc., Faraday Trans. I, 69 (1973) 603.
38 M. Maccormac and D. T. A. Townend, J. Chem. Soc. (1938) 238.
39 J. Bardwell and Sir C. N. Hinshelwood, Proc. R. Soc. London, Ser. A, 203 (1950) 26.
40 J. Bardwell and Sir C. N. Hinshelwood, Proc. R. Soc. London, Ser. A, 205 (1951) 375.
41 J. Bardwell and Sir C. N. Hinshelwood, Proc. R. Soc. London, Ser. A, 207 (1951) 461.
42 J. Bardwell, Proc. R. Soc. London, Ser. A, 207 (1951) 470.
43 (a) M. Akbar and J. A. Barnard, Trans. Faraday Soc., 64 (1968) 3035.
(b) J. A. Barnard and M. A. Sheikh, Pak. J. Sci. Ind. Res., 16 (1973) 93.
(c) J. A. Barnard and M. A. Sheikh, Pak. J. Sci. Ind. Res., 17 (1974) 55.

44 M. Akbar and J. A. Barnard, Trans. Faraday Soc., 64 (1968) 3049.
45 D. E. Hoare, A. D. Walsh and Ting-Man Li, 11th Symp. Combust. (1967) 879.
46 D. Anderson and D. E. Hoare, Combust. Flame, 13 (1969) 511.
47 (a) K. Salooja, Combust. Flame, 10 (1966) 11.
(b) J. A. Barnard and M. A. Sheikh, Pak. J. Sci. Ind. Res., 15 (1972) 140.
48 R. C. Ferris and W. S. Haynes, J. Am. Chem. Soc., 72 (1950) 893.
49 W. S. Haynes and P. D. Gardner, J. Am. Chem. Soc., 73 (1951) 3065.
50 R. K. Brinton, J. Am. Chem. Soc., 83 (1961) 1541.
51 D. E. Hoare and D. A. Whytock, Can. J. Chem., 45 (1967) 865.
52 D. E. Hoare and D. A. Whytock, Can. J. Chem., 45 (1967) 2741.
53 D. E. Hoare and D. A. Whytock, Can. J. Chem., 45 (1967) 2841.
54 A. S. Kallend and J. N. Pitts, Jr., J. Am. Chem. Soc., 91 (1969) 1269.
55 K. Salooja, Combust. Flame, 10 (1966) 17.
56 P. Michaud, J. Lebel and C. Ouellet, Combust. Flame 12 (1968) 395.
57 J. A. Barnard and E. Kirschner, Combust. Flame, 11 (1967) 496.
58 P. Michaud and C. Ouellet, Can. J. Chem., 49 (1971) 294.
59 P. Michaud and C. Ouellet, Can. J. Chem., 49 (1971) 303.
60 F. A. Burdon and J. H. Burgoyne, Proc. R. Soc. London, Ser. A, 199 (1949) 328.
61 J. H. Burgoyne and P. K. Kapur, Trans. Faraday Soc., 48 (1952) 234.
62 M. Gerstein, G. E. McDonald and R. L. Schalla, 4th Symp. Combust. (1953) 375.
63 K. H. Mueller and W. D. Walters, J. Am. Chem. Soc., 73 (1952) 1458.
64 R. Friedman and E. Burke, 5th Symp. Combust. (1955) 596.
65 J. N. Bradley, R. Gilbert and G. A. Jones, Combust. Flame, 13 (1969) 33.
66 J. N. Bradley, R. Gilbert and G. A. Jones, Combust. Flame, 13 (1969) 668.
67 T. A. Eastwood and Sir C. N. Hinshelwood, J. Chem. Soc. (1952) 733.
68 G. H. N. Chamberlain and A. D. Walsh, 3rd Symp. Combust. (1949) 368, 375.
69 A. Lemay and C. Ouellet, Can. J. Chem., 33 (1955) 1316.
70 D. J. Waddington, Proc. R. Soc. London, Ser. A, 252 (1959) 260.
71 J. Long and G. Skirrow, Trans. Faraday Soc., 58 (1962) 1403.
72 K. Salooja, Combust. Flame, 9 (1965) 33.
73 L. Ouellet and C. Ouellet, Can. J. Chem., 29 (1951) 76.
74 J. E. C. Topps and D. T. A. Townend, Trans. Faraday Soc., 42 (1946) 345.
75 R. J. Foresti, 5th Symp. Combust. (1955) 582.
76 R. E. Donovan and W. G. Agnew, J. Chem. Phys., 23 (1955) 1592.
77 W. G. Agnew and J. T. Agnew, Ind. Eng. Chem., 48 (1956) 2224.
78 W. G. Agnew and J. T. Agnew, 10th Symp. Combust. (1965) 123.
79 W. G. Agnew, J. T. Agnew and K. Wark, Jr., 5th Symp. Combust. (1955) 766.
80 L. H. S. Roblee, Jr., J. T. Agnew and K. Wark, Jr., Combust. Flame, 5 (1961) 65.
81 J. A. Barnard and C. F. Cullis, 8th Symp. Combust. (1962) 481.
82 G. R. Freeman, Proc. R. Soc. London, Ser. A, 245 (1958) 751.
83 C. Vovelle, R. Delbourgo and P. Laffitte, C. R. Acad. Sci., Ser. C, 264 (1967) 2003.
84 C. Vovelle, R. Delbourgo and P. Laffitte, C. R. Acad. Sci., Ser. C, 265 (1967) 899.
85 D. J. Waddington, 7th Symp. Combust. (1959) 165.
86 D. J. Waddington, Proc. R. Soc. London, Ser. A, 265 (1962) 436.
87 P. W. Jones and D. J. Waddington, Chem. Ind. (London), 1969, 492.
88 F. E. Malherbe and A. D. Walsh, Trans. Faraday Soc., 46 (1950) 835.
89 G. H. N. Chamberlain and A. D. Walsh, Rev. Inst. Fr. Pet. Ann. Combust. Liq., 4 (1949) 301.
90 G. H. N. Chamberlain and A. D. Walsh, Trans. Faraday Soc., 45 (1949) 1032.
91 (a) P. Dugleux, Rev. Inst. Fr. Pet. Ann. Combust. Liq., 5 (1950) 178.
(b) M. J. Molera, J. A. Garcia Dominguez and A. U. Acuna, J. Chem. Soc. B (1971) 1916.

92 B. I. Parsons and C. J. Danby, J. Chem. Soc. (1956) 1795.
93 B. I. Parsons and Sir. C. N. Hinshelwood, J. Chem. Soc. (1956) 1799.
94 B. I. Parsons, J. Chem. Soc. (1956) 1804.
95 A. Fish and A. Waris, J. Chem. Soc. (1963) 820.
96 A. Fish and A. Waris, J. Chem. Soc. (1962) 4513.
97 D. E. Hoare and M. Kamil, Combust. Flame, 15 (1970) 61.
98 A. Fish, in D. Swern (Ed.), Peroxide Mechanisms, Vol. I, Wiley, New York, 1970, pp. 141–198.
99 E. J. Harris and A. C. Egerton, Proc. R. Soc. London, Ser. A, 168 (1938) 1.
100 K. Moriya, Rev. Phys. Chem. Jpn., Horiba volume (1946) 143.
101 P. L. Hanst and J. G. Calvert, J. Phys. Chem., 63 (1959) 104.
102 (a) D. H. Fine, P. Gray and R. Mackinven, Proc. R. Soc. London, Ser. A, 316 (1970) 241, 255.
(b) P. Gray, D. T. Jones and R. Mackinven, Proc. R. Soc. London, Ser. A, 325 (1971) 175.
103 D. H. Fine, P. Gray and R. Mackinven, 12th Symp. Combust. (1969) 545.
104 P. Rutledge, Combust. Flame, 8 (1964) 77.
105 M. F. Giqueaux-Duval, R. Delourgo and P. Laffitte, C. R. Acad. Sci., Ser. C, 260 (1965) 4757.
106 (a) C. F. Cullis and L. C. Roselaar, Rev. Inst. Fr. Pet. Ann. Combust. Liq., 13 (1958) 466.
(b) C. F. Cullis and M. F. R. Mulcahy, Combust. Flame, 18 (1972) 225.
107 C. F. Cullis and L. C. Roselaar, Trans. Faraday Soc., 55 (1959) 272.
108 A. C. Harkness and F. E. Murray, Air Water Pollut., 10 (1966) 245.
109 A. C. Harkness and F. E. Murray, Atmos. Environ., 1 (1967) 491.
110 C. F. Cullis and L. C. Roselaar, Trans. Faraday Soc., 55 (1959) 1554.
111 C. F. Cullis and L. C. Roselaar, Trans. Faraday Soc., 55 (1959) 1562.
112 C. F. Cullis and L. S. A. Smith, Trans. Faraday Soc., 46 (1950) 42.
113 C. F. Cullis and J. P. Willsher, Proc. R. Soc. London, Ser. A, 209 (1951) 218.
114 C. F. Cullis and I. Isaac, Trans. Faraday Soc., 48 (1952) 1023.
115 G. R. Hill and D. G. Wilson, Pyrodynamics, 3 (1966) 235.
116 M. Gerstein, O. Levine and E. L. Wong, J. Am. Chem. Soc., 73 (1951) 418.
117 C. F. Cullis and D. J. Waddington, 5th Symp. Combust. (1955) 545.
118 C. F. Cullis and D. J. Waddington, Proc. R. Soc. London, Ser. A, 246 (1958) 91.
119 C. F. Cullis and D. J. Waddington, Proc. R. Soc. London, Ser. A, 244 (1958) 110.
120 C. F. Cullis and D. J. Waddington, Proc. R. Soc. London, Ser. A. 248 (1958) 136.
121 A. Fish, in D. Swern (Ed.), Organic Peroxides, Vol. I, Wiley-Interscience, New York. 1970, p. 182.
122 P. Gray, A. R. Hall and H. G. Wolfhard, Proc. R. Soc. London, Ser. A, 232 (1955) 389.
123 A. R. Hall and H. G. Wolfhard, 6th Symp. Combust. (1957) 190.
124 E. A. Arden and H. G. Wolfhard, 6th Symp. Combust. (1957) 177.
125 P. Gray and M. W. T. Pratt, 6th Symp. Combust. (1957) 183.
126 S. de Jaegere and A. van Tiggelen, Combust. Flame, 3 (1959) 187.
127 P. Gray and A. Williams, 8th Symp. Combust. (1962) 496.
128 P. Gray, Proc. R. Soc. London, Ser. A, 221 (1954) 462.
129 J. B. Levy, Ind. Eng. Chem., 48 (1956) 762.
130 P. Gray, R. Shaw and J. C. J. Thynne, Prog. React. Kinet., 4 (1967) 63.
131 P. Gray and A. D. Yoffe, Proc. R. Soc. London, Ser. A, 200 (1949) 114.
132 (a) P. Gray and P. R. Lee, 11th Symp. Combust., (1967) 1123.
(b) H. Goodman, P. Gray and D. T. Jones, Combust. Flame, 19 (1972) 157.
133 J. A. Hicks, 8th Symp. Combust. (1962) 487.
134 D. P. Needham and J. Powling, Proc. R. Soc. London, Ser. A, 232 (1955) 337.
135 J. Powling, W. A. W. Smith and J. C. J. Thynne, Combust. Flame, 4 (1960) 201.

136 (a) J. Powling and W. A. W. Smith, Combust. Flame, 1 (1957) 308.
(b) J. F. Griffiths, M. F. Gilligan and P. Gray, Combust. Flame, 24 (1975) 11.
137 J. Powling and W. A. W. Smith, Combust. Flame, 2 (1958) 157.
138 C. Steele and A. F. Trotman-Dickenson, J. Chem. Soc. (1959) 975.
139 W. Forst and O. K. Rice, Can. J. Chem., 41 (1963) 562.
140 N. J. Gerri and F. Kaufman, 10th Symp. Combust. (1965) 227.
141 G. Chiltz, C. F. Aten, Jr. and S. H. Bauer, J. Phys. Chem., 66 (1952) 1426.
142 (a) R. A. Matula, D. I. Orloff and J. T. Agnew, Combust. Flame, 14 (1970) 97.
(b) E. A. Fletcher, Nat. Bur. Stand. (U.S.), Spec. Publ. 1972, No. 357, 153.
143 R. A. Matula and J. T. Agnew, Combust. Flame, 13 (1969) 101.
144 E. A. Fletcher and D. B. Kittelson, Combust. Flame, 12 (1968) 164.
145 L. E. Fuller and E. A. Fletcher, Combust. Flame, 13 (1969) 434.
146 E. F. Croomes, Combust. Flame, 10 (1961) 71.
147 V. Caglioti, A. Delle Site, M. Lenzi and A. Mele, J. Chem. Soc. (1964) 5430.
148 E. A. Fletcher and L. L. Ambs, Combust. Flame, 8 (1964) 275.
149 (a) E. A. Fletcher and D. B. Kittelson, Combust. Flame, 12 (1968) 119.
(b) A. T. McHale, R. W. Geary, G. von Elbe and C. Huggett, Combust. Flame, 16 (1971) 167.
(c) R. T. Pollard, Combust. Flame, 17 (1971) 337.
150 D. J. Parks and E. A. Fletcher, Combust. Flame, 13 (1969) 487.
151 (a) K. H. Homann and D. I. Maclean, Combust. Flame, 14 (1970) 410.
(b) K. H. Homann and D. I. Maclean, J. Phys. Chem., 75 (1971) 3645.
152 S. Novikoff and J. Combourieu, Bull. Soc. Chim. Fr. (1962) 1159.
153 M. R. Hoare, R. G. W. Norrish and G. Whittingham, Proc. R. Soc. London, Ser. A, 250 (1959) 197.
154 B. Kaesche-Krischer, Combust. Flame, 6 (1962) 183.
155 B. Kaesche-Krischer, Chem. Ing. Tech., 35 (1963) 856.
156 G. Huybrechts and L. Meyers, Trans. Faraday Soc., 62 (1966) 2191.
157 D. R. Thomas and J. H. Thomas, Trans. Faraday Soc., 57 (1961) 266.
158 D. V. E. George and J. H. Thomas, Trans. Faraday Soc., 58 (1962) 262.
159 R. F. Simmons and H. G. Wolfhard, Trans. Faraday Soc., 52 (1956) 53.
160 C. F. Cullis, A. Fish and R. B. Ward, Proc. R. Soc. London, Ser. A, 276 (1963) 527.
161 A. Fish, Combust. Flame, 8 (1964) 84.
162 C. F. Cullis, A. Fish and R. B. Ward, Combust. Flame, 7 (1963) 303.
163 J. R. Bates and R. Spence, J. Am. Chem. Soc., 53 (1931) 1689.
164 M. I. Christie, Proc. R. Soc. London, Ser. A, 244 (1958) 411.
165 G. R. McMillan and J. G. Calvert, Oxid. Combust. Rev., 1 (1965) 83.
166 J. R. McKellar and R. G. W. Norrish, Proc. R. Soc. London, Ser. A, 263 (1961) 51.

Index

B

C

F

G

H

I

N

O

P

Q

R

S

T

U

V

W

X